BASIC MECHANISMS OF SOLAR ACTIVITY

INTERNATIONAL ASTRONOMICAL UNION
UNION ASTRONOMIQUE INTERNATIONALE

SYMPOSIUM No. 71
HELD IN PRAGUE, CZECHOSLOVAKIA,
25–29 AUGUST 1975

BASIC MECHANISMS OF SOLAR ACTIVITY

EDITED BY

V. BUMBA AND J. KLECZEK

Astronomical Institute of the Czechoslovak Academy of Sciences, Ondřejov

D. REIDEL PUBLISHING COMPANY

DORDRECHT-HOLLAND/BOSTON-U.S.A.

1976

Library of Congress Cataloging in Publication Data

Main entry under title:

Basic mechanisms of solar activity.

    (Symposium – International Astronomical Union;
    no. 71)
    Includes bibliographies and index.
    1. Solar activity – Congresses. I. Bumba, Václav.
II. Kleczek, Josip. III. Series: International Astronomical
Union. Symposium; no. 71.
QB524.B37   523.7   76-21689
ISBN-13:978-90-207-0681-2   e-ISBN-13:978-94-010-1481-6
DOI: 10.1007/978-94-010-1481-6

Published on behalf of
the International Astronomical Union
by
D. Reidel Publishing Company, P. O. Box 17, Dordrecht, Holland

Sold and distributed in the U.S.A., Canada, and Mexico
by D. Reidel Publishing Company, Inc.
Lincoln Building, 160 Old Derby Street,
Hingham, Mass. 02043, U.S.A.

TABLE OF CONTENTS

PREFACE

Our first attempt to organize a Symposium on solar activity was made at the IAU General Assembly in Brighton 1970. There, at the session of Commission 10, we proposed to organize a Symposium which would stress the observational aspects of solar activity. It was our hope that such a Symposium might stimulate studies of those important problems in solar physics which for a long time had been neglected in overall scientific discussion. Although a provisional date for the Symposium was then decided, it did not take place to avoid overlapping with other IAU activities.

At the session of Commission 10 in Sydney – on the occasion of the XVth IAU General Assembly in 1973 – we repeated our proposal and forwarded the invitation of the Czechoslovak Academy of Sciences to organize the Symposium in Prague. Both were accepted. During the discussions about the programme of the Symposium – enthusiastically promoted by the late president of Commission 10, Prof. K. O. Kiepenheuer – it was decided to change slightly its subject. The theoretical problems were stressed and the majority of the Scientific Organizing Committee agreed not to deal with short-lived phenomena of the solar activity or with individual active regions.

Symposium No. 71 was held in Prague from August 25 to August 29, 1975. Its Organizing Committee consisted of V. Bumba (Chairman), W. Deinzer, R. G. Giovanelli, R. Howard, K. O. Kiepenheuer, M. Kopecký, T. Krause, M. Kuperus, G. Newkirk and J. Vitinsky. The Local Organizing Committee of the Symposium was represented by M. Kopecký (Chairman), V. Rajský (Secretary), V. Bumba, J. Kleczek and J. Sýkora.

We wish to express our gratitude to all members of the Scientific Organizing Committee for their advice and assistance. It is our pleasure to thank J. O. Stenflo, V. E. Stepanov, M. Kopecký, L. Mestel, W. Deinzer, N. O. Weiss and B. R. Durney, who helped to organize the individual sessions and kindly served as their chairmen. Our special thanks are due to Dr Gordon Newkirk for organizing and presiding over the Concluding Summarizing Discussion. We are also obliged to P. Kotrč and his assistants, who kindly recorded all the discussions and persistently chased the discussion participants to get their questions and remarks in a definitive form. In most cases they succeeded.

V. BUMBA
J. KLECZEK

LIST OF PARTICIPANTS

Akinian, S. T., Izmiran, Moscow, U.S.S.R.

Ambrož, P., Astronomical Institute, 251 65 Ondřejov, C.S.S.R.

Antalová, A., Chair of Technical Physics, Technical University for Transport, 010 88 Žilina, C.S.S.R.

Antonucci, E., Physical Institute, I-10125, Torino, Italy.

Anzer, U., Max-Planck-Institut, 8000 München, Föhringer Ring 6, F.R.G.

Artus, H., VEB Carl Zeiss, Jena, D.D.R.

Aurass, H., Observatorium für solare Radioastronomie, DDR-1501, Tremsdorf, D.D.R.

Balklavs, A. E., Radioastronomical Institute, Riga, Latvian S.S.R., U.S.S.R.

Beck, H. G., VEB Carl Zeiss, Jena, D.D.R.

Bednářová, B., Geophysical Institute, 141 31 Praha, C.S.S.R.

Belvedere, G., Astrophysical Observatory, I-95125 Catania, Italy.

Blanco, C., Astrophysical Observatory, I-95125 Catania, Italy.

Böhme, A., Sonnenobservatorium Einsteinturm, 15 Potsdam, Telegrafenberg, D.D.R.

Bonov, A. D., University of Sofia, Sofia, Bulgaria.

Bruzek, A., Fraunhofer Institut, D 78 Freiburg im Breisgau, Schöneckstrasse 6, F.R.G.

Bumba, V., Astronomical Institute, 251 65 Ondřejov, C.S.S.R.

Chistjakov, V. F., Service of Sun, Ussurijsk 692533, U.S.S.R.

Chvojková, E., Astronomical Institute, 120 23 Praha, C.S.S.R.

Cimakhovich, N. P., Radioastronomical Institute, Riga, Latvian S.S.R., U.S.S.R.

Csada, I. K., Konkoly Observatory, Konkoly-Thege U. 13-17, 1525 Budapest, XII., Hungary.

Dara-Papamargariti, H., Research Centre for Astronomy and Applied Mathematics, 14 Anagnostopoulou Street, Athens (136), Greece.

Deinzer, W., Universitäts-Sternwarte, D-34 Göttingen, Geismarlandstrasse 11, F.R.G.

Demkina, L. B., Izmiran, Moscow, U.S.S.R.

Deubner, F. L., Fraunhofer Institut, Schöneckstrasse 6, 7800 Freiburg im Breisgau, F.R.G.

Dezsö, L., Heliophysical Observatory, H-4010 Debrecen, P.O. Box 30, Hungary.

Dravins, D., Lund Observatory, Svanegatan 9, S-222 24, Lund, Sweden.

Durney, B. R., National Center for Atmospheric Research, P.O. Box 1470, Boulder, Colo. 80302, U.S.A.

Durrant, Ch.-J., Fraunhofer Institut, Schöneckstrasse 6, D-78 Freiburg im Breisgau, F.R.G.

Elwert, G., Lehrstuhl für Teoretische Astrophysik, 7400 Tübingen, Auf der Morgenstelle 10, F.R.G.

Fárník, F., Astronomical Institute, 251 65 Ondřejov, C.S.S.R.

Fischer, S., Astronomical Institute, 120 23 Praha, C.S.S.R.

Fortini, T., Astronomical Observatory, 00136 Roma, Italy.

Fossat, E., Astrophysical Institute, 98bis Boulevard Arago, Paris 14e, France.

Gelfreikh, G. B., Main Astronomical Observatory, Pulkovo, 196140 Leningrad, U.S.S.R.

Ghabrus, R. A., Helwan Observatory, Helwan, near Cairo, U.A.R.

Gigolashvili, M. S., Abastumani Astrophysical Observatory, Abastumani, Georgia, U.S.S.R.

Gilman, P. A., Advanced Study Program, NCAR, P.O. Box 3000, Boulder, Colo. 80302, U.S.A.

Giovanelli, R. G., CSIRO, National Standards Laboratory, Division of Physics, University Grounds, City Road, Chippendale, N.S.W., 2008, Sydney, Australia.

Gleissberg, W. K. H., Senckenberganlage 23, 6000 Frankfurt, F.R.G.

Gnevyshev, M. N., Astronomical Observatory, Pulkovo, 196 140 Leningrad, U.S.S.R.

Gnevysheva, R. S., Astronomical Observatory, Pulkovo, 196 140 Leningrad, U.S.S.R.

Godoli, G., Astrophysical Observatory, I-95125 Catania, Italy.

Golub, L., American Science and Engineering, Inc., 955 Massachusetts Avenue, Cambridge, Mass. 02139, U.S.A.

Grigoriev, V. M., Sibizmiran, Irkutsk 664033, U.S.S.R.

Gurtovenko, Main Astronomical Observatory, Kiev, U.S.S.R.

Gutcke, H., VEB Carl Zeiss, Jena, D.D.R.

Halenka, J., Geophysical Institute, 141 31 Praha, C.S.S.R.

Hamatschek, R., VEB Carl Zeiss, Jena, D.D.R.

Hartmann, R., Astronomical Institute, Senckenberg-Angle 23, 600 Frankfurt/Main 1, F.R.G.

Hedeman, E. R., McMath-Hulbert Observatory, 895 N. Lake Angelus Road, Pontiac, Michigan 48055, U.S.A.

Hejna, L., Astronomical Institute, 251 65 Ondřejov, C.S.S.R.

Howard, R., Hale Observatories, 813 Santa Barbara Street, Pasadena, Calif. 91101, U.S.A.

Jäger, F. W., Zentralinstitut für Solar-Terrestrische Physik, DDR-15 Potsdam, Telegrafenberg, D.D.R.

Jakimiec, J., Astronomical Institute, Kopernika 11, Wroclaw, 51-622, Poland.

Karabin, M., Astrophysical Department, Faculty of Sciences, Studentski trg 16, P.O. Box 550, 11001 Beograd, Yugoslavia.

Kasinskij, V. V., Sibizmiran, Irkutsk 664033, U.S.S.R.

Kato, S., University of Kyoto, Kyoto, Japan.

Khetsuriani, Abastumani Astrophysical Observatory, Abastumani, Georgia, U.S.S.R.

Kim, I. S., Izmiran, Moscow, U.S.S.R.

Kleczek, J., Astronomical Institute, 251 65 Ondřejov, C.S.S.R.

Klvaňa, M., Astronomical Institute, 251 65 Ondřejov, C.S.S.R.

Knoška, Š., Astronomical Institute, 059 60 Tatranská, Lomnica, C.S.S.R.

Kopecký, M., Astronomical Institute, 251 65 Ondřejov, C.S.S.R.

Kotrč, P., Astronomical Institute, 251 65 Ondřejov, C.S.S.R.

Kovács, A., Heliophysical Observatory, P.O. Box 30, H-4010, Debrecen, Hungary.

Krause, F., Zentralinstitut fur Astrophysik, 15 Potsdam, Telegrafenberg, D.D.R.

Křivský, L., Astronomical Institute, 251 65 Ondřejov, C.S.S.R.

Krüger, A., Zentralinstitut für Solar-Terrestrische Physik, 1199 Berlin-Adlershof, Rudower Chaussee 5, D.D.R.

Kubičela, A., Astronomical Observatory, Volgina 7, 11050 Beograd, Yugoslavia.

Kuklin, G. V., Sibizmiran, Irkutsk 664033, U.S.S.R.

Kulidzhanishvili, V. I., Abastumani Astrophysical Observatory, Abastumani, Georgia, U.S.S.R.

Künzel, H., Sonnenobservatorium Einsteinturm, 15 Potsdam, Telegrafenberg, D.D.R.

Lê-Bach-Yén, Astronomical Institute, 251 65 Ondřejov, C.S.S.R.

Leftus, V., Astronomical Institute, 251 65 Ondřejov, C.S.S.R.

Lielausis, O. A., Astronomical Observatory, Riga, Latvian S.S.R., U.S.S.R.

Macák, P., Astronomical Institute, 251 65 Ondřejov, C.S.S.R.

Macháček, M., Astronomical Institute, 251 65 Ondřejov, C.S.S.R.

Macris, C. J., Research Center for Astronomy and Applied Mathematics, 14 Anagnostopoulou Street, Athens (136), Greece.

Marilli, E., Astrophysical Observatory, I-95125 Catania, Italy.

Mestel, L., Astronomy Centre, Physics Building, University of Sussex, Falmer, Brighton BN1 9QH, England.

Mogilevskij, E. I., Izmiran, Moscow, U.S.S.R.

Motta, S., Astrophysical Observatory, I-95125 Catania, Italy.

Newkirk, G. A., Jr., N.C.A.R., High Altitude Observatory, P.O. Box 1470, Boulder, Colo. 80302, U.S.A.

Niță, I., Observatoire de Bucarest, 5 rue Cutitul de Argint, Bucarest 28, Roumania.

Noci, G., Astrophysical Observatory Arcetri, Largo E. Fermi 5, 50125 Firenze, Italy.

Obridko, V. N., Izmiran, Moscow, U.S.S.R.

Oetken, L., Zentralinstitut für Astrophysik, 15 Potsdam, Telegrafenberg, D.D.R.

Paciorek, J. M., Astronomical Observatory, Kopernika 11, 51-622 Wroclaw, Poland.

Pallavicini, R., Astrophysical Observatory Arcetri, Largo E. Fermi 5, 50125 Firenze, Italy.

Paluš, P., Faculty of Sciences, University Bratislava, Trnavská Str. 1, 800 00 Bratislava, C.S.S.R.

Parker, E. N., Laboratory for Astrophysics and Space Research, 933 East 56th Street, Chicago, Ill. 60637, U.S.A.

Paterno, L., Astrophysical Observatory, I-95125 Catania, Italy.

Pflug, K., Zentralinstitut für Solar-Terrestrische Physik, Sonnenobservatorium Einsteinturm, 15 Potsdam, Telegrafenberg, D.D.R.

Pirronello, V., Astrophysical Observatory, I-95125 Catania, Italy.

Pospíšil, M., Astronomical Institute, 251 65 Ondřejov, C.S.S.R.

Priest, E. R., Department of Applied Mathematics, North Haugh, St. Andrews, KY16 9SS, U.K.

Rädler, K.-H., Zentralinstitut für Astrophysik, 15 Potsdam, Telegrafenberg, D.D.R.

Rajský, V., Astronomical Institute, 120 23 Praha, C.S.S.R.

Rodonò, M., Astrophysical Observatory, I-95125 Catania, Italy.

Rompolt, B., Astronomical Observatory, Kopernika 11, 51-622 Wroclaw, Poland.

Roxburgh, I. W., Queen Mary College, University of London, Department of Applied Mathematics, Mile End Road, London E1 4NS, England.

Rüdiger, G., Zentralinstitut für Astrophysik, 15 Potsdam, Telegrafenberg, D.D.R.

Rušín, V., Astronomical Institute, 059 60 Tatranská Lomnica, C.S.S.R.

Ruždjak, V., Institute of Physics of the University, P.O. Box 304, 41001 Zagreb, Yugoslavia.

Ružičková-Topolová, B., Astronomical Institute, 251 65 Ondřejov, C.S.S.R.

Rybanský, M., Astronomical Institute, 059 60 Tatranská Lomnica, C.S.S.R.

Schröter, E. H., Universitäts–Sternwarte, D-34 Göttingen, Geismarlandstrasse 11, F.R.G.

Shmeleva, O. P., Izmiran, Moscow, U.S.S.R.

Šidlichovský, M., Astronomical Institute, 120 23 Praha, C.S.S.R.

Slottje, C., Radio Observatory, Dwingeloo, The Netherlands.

Smith, D. F., N.C.A.R., High Altitude Observatory, P.O. Box 3000, Boulder, Colo. 80302, U.S.A.

Staude, J., Zentralinstitut für Solar-Terrestrische Physik, Sonnenobservatorium Einsteinturm, 15 Potsdam, Telegrafenberg, D.D.R.

Stenflo, J. O., Lund Observatory, Svanegatan 9, S-222 24, Lund, Sweden.

Stepanian, N. N., Crimean Astrophysical Observatory, Nauchny, Crimea 334413, U.S.S.R.

Stepanov, V. E., Sibizmiran, Irkutsk 664033, U.S.S.R.

Stix, M., Universitäts-Sternwarte, 3400 Göttingen, Geismarlandstrasse 11, F.R.G.

Suda, J., Astronomical Institute, 251 65 Ondřejov, C.S.S.R.

Sýkora, J., Astronomical Institute, 059 60 Tatranská Lomnica, C.S.S.R.

Tlamicha, A., Astronomical Institute, 251 65 Ondřejov, C.S.S.R.

Tritakis, B., Research Center for Astronomy and Applied Mathematics, 14 Anagnostopoulou Street, Athens (136), Greece.

Touminen, J. V., Observatory and Astrophysics Laboratory, Tähtitorninmäki, SF-00130, Helsinki 13, Finland.

Vainshtein, S. I., Sibizmiran, Irkutsk 664033, U.S.S.R.

Valníček, B., Astronomical Institute, 251 65 Ondřejov, C.S.S.R.

Vandakurov, Yu. V., Physical and Technical Institute A. F. Ioffe, Leningrad, 194021, U.S.S.R.

Vertlib, A. B., Sibizmiran, Irkutsk 664033, U.S.S.R.

Vitinskij, Yu. I., Main Astronomical Observatory, Pulkovo, Leningrad, 196140, U.S.S.R.

Vujnović, V., Institut of Physics of the University, P.O. Box 304, 41001 Zagreb, Yugoslavia.

Wagner, W. J., Sacramento Peak Observatory, Air Force Cambridge Research Laboratories, Sunspot, N.M. 88349, U.S.A.

Weiss, N. O., Department of Applied Mathematics and Theoretical Physics, Silver Street, Cambridge, CB3 9EW, England.

Wilcox, J. M., Institute for Plasma Research, Via Crespi, Stanford, Calif. 94305, U.S.A.

Yoshimura, H., Hale Observatories, California Institute of Technology, 813 Santa Barbara Street, Pasadena, Calif. 91101, U.S.A.

Zappala, R., Astrophysical Observatory, I-95125 Catania, Italy.

Zwaan, C., Astronomical Observatory 'Sonnenborgh', Zonnenburg 2, Utrecht, The Netherlands.

INAUGURAL ADDRESS*

JAROSLAV KOŽEŠNÍK
President of the Czechoslovak Academy of Sciences

Ladies and gentlemen, allow me to welcome you most sincerely on behalf of the Czechoslovak Academy of Sciences at the 71st Symposium of the International Astronomical Union whose subject matter is the basic mechanisms of solar activity.

The Czechoslovak Academy of Sciences esteems it an honour that the International Astronomical Union has accepted its invitation to organize this symposium in the Czechoslovak Socialist Republic. The presidium of the Czechoslovak Academy of Sciences has commissioned the Astronomical Institute to organize such an important international event. The Astronomical Institute in deciding to organize this symposium in Prague – a city with a rich astronomical tradition – certainly made a correct decision.

After all, astronomy took root in the Czech lands more than 600 years ago. Astronomy helped to establish a university in Prague as early as the turn of the 13th and the 14th centuries. In the Middle Ages, the Prague Astronomical School was a source of astronomical knowledge for the whole of Central Europe. The Polish research workers have recently proved that the Prague Astronomical School gave birth to the Cracovian School which, at the end of the 15th century, gave the world its greatest pupil Nicolas Copernicus. It was Prague again that played an important role in disseminating the teaching of Copernicus. The Copernicus teaching found its supporters especially in a Czech family of Tadeáš Hájek of Hájek, known under the name of Hagecius. Four hundred years ago, in 1574, the Hagecius book *Dialexis de novae et prius incognitae stellae apparitione* was published. This book sharply criticised the very base of the medieval and Aristotelian interpretation of the Universe. It became the most famous of all 16th-century writings. Thanks to Hagecius, born 450 years ago, Tycho Brahé and Johannes Kepler, the best astronomers in the world, came to Prague at the turn of the 16th and the 17th centuries and Prague became the most important centre of astronomical research all over the world.

The institutes in the Czechoslovak Socialist Republic follow in their work this glorious tradition. The Astronomical Institute of the Czechoslovak Academy of Sciences, the oldest and the largest of all our institutes, celebrated last year its 250 years of existence. It is one of the oldest scientific astronomical institutions in our country.

Therefore, we are glad that astronomers from the whole world meet again in Prague, after the International Astronomical Union General Assembly in 1967 and the COSPAR congress in 1969. A close international cooperation has a long tradition in the field of astronomy. As you probably know, a meeting of solar astronomers became the predecessor of the IAU General Assemblies. This close

---

* The Inaugural Address was presented by corresponding member of the Czechoslovak Academy of Sciences Vl. Guth.

Bumba and Kleczek (eds.), Basic Mechanisms of Solar Activity, xv–xvi. All Rights Reserved.
Copyright © 1976 by the IAU.

cooperation between scientists from the whole world is important not only for scientific progress itself but it also is an important element in the present day detente, in the struggle to strengthen peaceful cooperation in the spirit of the recent Helsinki Conference results.

We appreciate also the fact that the subject of our symposium is the basic mechanisms of solar activity in the first place. The study of these questions concerns not only solar activity itself but it also is of great importance for other scientific branches as well. The Sun is nothing but the nearest star and the knowledge of the Sun is of great importance for stellar astrophysics and cosmogony taken as a whole. Knowledge of basic mechanisms of solar activity is closely connected with plasma physics, with problems of plasma and magnetic-field interaction, with problems of nuclear reactions and energy release in general. For this reason solar physics can bring a lot of new stimuli in this field and thus can contribute to the solution of important technical and economic problems of mankind. Knowledge of basic mechanisms of solar activity plays an important role in the prognosis of solar activity, physically justified and thus reliable. These prognoses are ever more necessary since the importance of the solar activity influence upon the Earth is ever more evident in the sphere of geophysics, technology and of the biosphere. The actual trend shows that the importance of solar activity research for everyday life is steadily increasing.

Ladies and gentlemen, I am convinced that this symposium will represent another step in our efforts to know better the laws of nature and at the same time will enhance further cooperation between scientists all over the world. I wish you a lot of success in your work and a pleasant stay in Prague – the capital of the Czechoslovak Socialist Republic.

INTRODUCTION

THE ENIGMA OF SOLAR ACTIVITY

E. N. PARKER

Dept. of Physics, Dept. of Physics and Astronomy,
University of Chicago, Chicago, Ill. 60637, U.S.A.

1. Introduction

The review lectures that make up the basic program of this symposium will cover the most recent observational results, and the present state of theoretical knowledge, of solar activity. It seems, therefore, that the most useful role for the introductory lecture would be to review the outstanding puzzles presented to us by the activity of the Sun so that we may have those numerous dilemmas clearly in mind as the speakers review the accumulated facts and theories.

It has become clear in the last ten years that the cause of all the many different forms of solar activity can be traced to the convection and circulation within the Sun. The convective zone of the Sun is a giant heat engine which converts a small fraction of the outward flowing heat into convective motions, and from there into magnetic fields and hydrodynamic and hydromagnetic waves. From these basic ingredients (of low entropy) there then arises the sunspot, the prominence, the flare, the corona and solar wind, etc.

The most obvious circulation within the Sun is the differential rotation of the visible surface, in which the equator rotates nearly 50% faster than the poles. This nonuniform rotation cannot be an artifact of the formation of the Sun, some 5×10^9 yr ago, for the eddy viscosity of the convective zone would long since have destroyed any initial nonuniform rotation. The present nonuniform rotation is an integral part of the present convection and circulation within the Sun, maintained today by the contemporary thermal gradients and heat fluxes.

The theory of convection, circulation, and nonuniform rotation is fundamental to the understanding of solar activity. Unfortunately, the enormous density variation across the convective zone, from 2×10^{-1} gm cm^{-3} at the bottom (at a depth of 2×10^5 km) to 5×10^{-7} gm cm^{-3} at the top, makes the theoretical treatment of the problem exceedingly difficult. What is difficult but possible in the Boussinesq approximation (uniform density) becomes a formidable task in the real stratified convective zone of the Sun. Some of the review speakers in this symposium will go into the problem in detail. I want to emphasize that the convection and circulation problem is fundamental to our understanding of any, and all, solar activity.

Let me begin, then, with the statement that we now know so much about the Sun that nearly every aspect of the Sun presents a dilemma. There is no other star about which we know enough to be so puzzled.

The most fundamental dilemma with the Sun is the failure to detect the expected neutrinos from the core (Davis and Evans, 1973). That problem, although not obviously central in questions of solar activity, is nonetheless so fundamental that we cannot ignore it. The neutrino dilemma involves the theory of weak interactions, opacity, radiative transfer, circulation and convection and, indeed, the whole physical basis for the theory of stellar structure (Bahcall *et al.*, 1973; Ulrich, 1974). We must not forget that our understanding of the convective zone – particularly its depth – is based in large measure on models of the solar interior. What would be the implications for solar activity if the Sun were convective all the way to its center? The explanation of the dilemma may, or may not, prove to be superficial, so far as the Sun is concerned. For instance, the luminosity of the Sun may vary by 5% over 10^4–10^6 yr (Fowler, 1972, 1973). Or it may be only that neutrinos are unstable (i.e., have nonvanishing rest mass) decaying before reaching Earth. This would have tremendous impact on the physics of elementary particles, but might well affect the theory of the solar interior very little. Or there may be some exotic effect that reduces opacity slightly, such as an absence of metals in the core of the Sun, or a convective core. But until the neutrino question is resolved, we cannot be sure of our knowledge of the interior structure of the Sun, and hence cannot be sure that we understand the convective origin of solar activity.

There are some curious questions of climatology that suggest that our knowledge of the solar interior, and the general evolution of a star on the main sequence, is less than complete. For instance, the conventional theory of evolution of the solar interior predicts that 10^9 yr ago the Sun was some 10% less luminous than we find it today; 4×10^9 yr ago it was 30% less luminous. Now the most sophisticated numerical atmospheric models of Earth predict that if the Sun were 6% less luminous, the surface of Earth would freeze over completely, increasing the albedo and further reducing the heating effect of the Sun, etc. But paleoclimatological studies are emphatic in the conclusion that Earth was not cooler 10^9 yr ago. Indeed, the indications are that it was, if anything, a few degrees warmer. Clearly we must keep an open mind when confronted with this problem. We know so little of the Sun and terrestrial climatology that the resolution could lie anywhere, and perhaps everywhere. But clearly something is out of line.

The historical sunspot record shows another gap in our understanding of the convective zone. Sunspots were first discovered and studied in the western world in 1610 with Galileo's application of the telescope to astronomy. Sunspots were considered at the time to be of no intrinsic interest in themselves (after the first trauma of their appearing as a blemish on the face of the 'perfect' sphere of the Sun) and so were not studied systematically. But there were enough records kept to show that the number of sunspots went through two distinct maxima after 1611, and then fell to a minimum at about 1645. The records go on to show that the Sun remained in a state of extreme minimum activity for about 70 yr thereafter, until approximately 1715, after which time activity resumed in the form of the familiar 11-yr cycle that we know so well today (Maunder, 1894, 1922). During the 70 yr of inactivity there was occasionally a sunspot or two, but long years with none at all; there was no white light corona visible during total eclipse by the Moon, whereas the corona is usually so conspicuous then; there were only a few significant auroral events, which are normally so common in clear skies over Scandinavia and Northern England. In view of the absence of a white light corona, we may conjecture whether the Sun was entirely shrouded in a coronal hole, yielding a fast, steady solar wind, or whether there was simply no solar wind at all. I would guess the former, but I know of no way to prove the answer.

The occurrence of the 70 yr minimum (sometimes called the Maunder minimum), indicates that there is available to the Sun a convective mode of circulation different from its present state. The other mode – let us call it the Maunder mode – is such as to be less effective in the generation of magnetic field. Evidently the Sun can flip-flop back and forth between the Maunder mode and the present mode. Since 1610, some 365 yr ago, the Sun has spent 70 yr in the Maunder mode and about 300 yr in the present mode. On this basis I am tempted to call the present mode the 'normal' mode, but we must be careful not to allow so preliminary an appellation to color our thinking about the physics of the convection. The point is that future theoretical studies of the convection and circulation in the Sun must look not merely for one mode, but for two or more modes, perhaps of distinctly different form.

2. Solar Convection, Circulation and the Dynamo

There are a number of questions that arise concerning the convective zone of the Sun. Present theoretical models of the Sun place the bottom of the convective zone at a depth of about 2×10^5 km (Spruit, 1974), where $\rho \cong 0.2$ gm cm^{-3} and $T \cong 2 \times 10^6$ K. The more rapid rotation of the equatorial surface of the Sun has been explained as a consequence of meridional circulation within the convective zone (Kippenhahn, 1963; Weiss, 1965; Cocke, 1967; Durney, 1968, 1970, 1971, 1972; Osaki, 1970; Busse, 1970; Kohler, 1970; Durney and Roxburgh, 1970; Yoshimura and Kato, 1971; Yoshimura, 1972; Gilman, 1972, 1974; Gierasch, 1974). Unfortunately, the calculations indicate that the meridional circulation must be so strong that a pole-equator difference $\delta\omega/\omega$ in angular velocity is accompanied by a pole-equator energy flux difference $\delta F/F \sim \delta\omega/\omega$ i.e., at least 10%. No such flux difference is observed. Indeed, the recent measurements of Dicke and Goldenberg (1974; Dicke, 1974) indicate that the brightness of the solar disk is circular to within a fraction $\delta R/R \le 4.5 \times 10^{-5}$. Thus if the brightness W varies with radius r as $W(r)$ out the limb at $r = R$, we have

$$\frac{\delta W}{W} = \frac{1}{W} \frac{\mathrm{d}W}{\mathrm{d}r} \delta R$$

$$< 4.5 \times 10^{-5} \frac{R}{W} \frac{\mathrm{d}W}{\mathrm{d}r}.$$

But (Minnaert, 1953)

$$\frac{R}{W} \frac{\mathrm{d}W}{\mathrm{d}r} \sim 20$$

near the limb of the Sun, so that, very roughly, $\delta F/F \cong \delta W/W < 10^{-3}$.

The brightness at the pole and equator is the same to within one part in 10^3! What then of the meridional circulation and equatorial acceleration? We should note that the calculations to date are based on the Boussinesq approximation, ignoring the enormous density variation across the convective zone. Convection in a stratified layer is *very difficult* to treat mathematically, and progress is only just beginning to be made. But we must have *qualitative* differences to extricate us from the dilemma.

Mere quantitative corrections can hardly be expected to make up the factor of 10^2 in the discrepancy.

There are other difficulties. The hydrodynamic models for the circulation and differential rotation predict that the angular velocity ω at low latitudes declines inward from the surface ($d\omega/dr > 0$) toward the reduced values observed at the surface at high latitudes. There is no direct observational objection to this result. Indeed, it gives the simplest internal variation of ω consistent with the motion of the surface. But there is a severe problem in understanding the generation of the magnetic fields of the Sun. The azimuthal fields beneath the surface of the Sun, whose rise to the surface produces the bipolar magnetic regions, are estimated to be of the order of 10^2 G and are believed to be generated by the combined dynamo effects of the nonuniform rotation and the cyclonic rotation of the rising and falling convective cells (Parker, 1955b, 1970a, b, 1971, 1972, 1975; Steenbeck, Krause and Radler, 1966; Steenbeck and Krause, 1969; Leighton, 1969, Gilman, 1969; Deinzer and Stix, 1971; Deinzer, Kusserow and Stix, 1974; Stix, 1974; Yoshimura, 1972b, 1973).

It is an observed fact that the fields appear first at middle latitudes and then migrate toward the equator. According to the dynamo equations, this requires that the product of $d\omega/dr$ and the helicity $\langle \mathbf{v} \cdot \text{curl } \mathbf{v} \rangle$ of the convective motions be positive. But in the northern hemisphere of the Sun a rising convective cell is expanding laterally. We would expect the coriolis forces to cause it to rotate more slowly than its surroundings (Steenbeck, Krause, and Radler, 1966) so that $\langle \mathbf{v} \cdot \text{curl } \mathbf{v} \rangle$ is negative. If, then, $d\omega/dr$ is positive, the product has the wrong sign and we would expect the fields to migrate away, rather than toward, the equator (Parker, 1972). We are plagued again with a *qualitative* difficulty.

But there are still more problems. One is the magnetic buoyancy of the fields. A magnetic field is buoyant because the magnetic field exerts pressure and expands, reducing the density of the gas within it (Parker, 1955a). A magnetic flux tube of field density B has a pressure $B^2/8\pi$, which causes a pressure reduction $\Delta p = B^2/8\pi$ in the gas inside the tube. The density reduction is $\Delta\rho/\rho = \Delta p/p = B^2/8\pi p$. There is, then, a buoyancy force $g\Delta\rho = B^2/8\pi\Lambda$ dyne cm^{-3} where $\Lambda \equiv kT/Mg$ is the pressure scale-height of the atmosphere. If the flux tube has a circular cross section of radius a, then the force per unit length is $B^2 a^2/8\Lambda$, causing the tube to rise rapidly through the convective zone. The velocity v of rise of a horizontal flux tube is restrained by the aerodynamic drag $C\rho v^2 a$, where the coefficient C is of the order of unity. Hence the terminal speed of rise is $V_A(\pi/2C)^{1/2}(R/\Lambda)^{1/2}$, i.e., of the order of the Alfvén speed. The Alfvén speed computed in 10^2 G at the base of the convective zone is about 60 cm s^{-1}, rising 10^5 km in 5 yr. Higher in the convective zone the rate of rise is faster and the field is lost in periods much less than 5 yr. The characteristic time in which the azimuthal field is generated from the poloidal (meridional) field by $d\omega/dr$ is typically 5 yr. Hence the only place that the magnetic field can possibly remain long enough to be regenerated is near the bottom of the convective zone. The solar dynamo does not extend below the convection, and its possible overshoot, if for no other reason than the absence of turbulent diffusion.

Altogether, then, it appears that, if the solar dynamo is to function at all, it must be in the lowest level of the convective zone (Parker, 1975b). According to Spruit's (1974) model of the convective zone, this would be at a depth of 1.5–2×10^5 km.

Higher up in the convective zone the magnetic flux tubes are merely rising to the surface with little time for regeneration. We see the activity at the surface caused by the continual arrival of tubes of magnetic flux from below.

What then, is the resolution of these many problems and contradictions? How must our ideas be modified to make sense of the rotation, convection, and dynamo effects in the Sun? Is it possible, for instance, that the circulation was in another mode, perhaps with a 5–10% pole-equator brightness difference from 1645 to 1715? Could it have been noticed by the sharp eyes of the 17th-century astronomer looking at an image of the Sun projected from the eyepiece of his telescope onto a screen? Perhaps in a few more years when adequate codes are available to explore the properties of convection in a rotating, deeply stratified, convecting atmosphere it will be possible to explore this question (see review by Gilman, these Proceedings). The general point that I want to make here is that we must search over a wide range of possibilities if we are ever to develop an understanding of what was happening at the Sun during the Maunder minimum

Turning to more concrete ideas, Durney (1975) (see review, these Proceedings) suggests that there may perhaps be a simple resolution of the whole puzzle. His first point is that there appear to be solutions of the hydrodynamic equations in a stratified rotating sphere in which the equatorial surface rotates more rapidly than the rest, but which exhibits no pole-equator difference in convected energy flux. The solutions are of the nature of rotation in cylinders, with ω a function only of the distance $\tilde{\omega}$ from the axis of rotation. If correct, this resolves the question of $\delta\omega/\omega$ without $\delta F/F$. Durney goes on to point out that the work of Yoshimura (1975) suggests that the direction of rotation of the cyclonic motions in the *lowest* level of the convective zone is *reversed* from the conventional considerations on local Coriolis force. If this is correct, then the product of $d\omega/dr$ and the helicity is positive in the lower convective zone where we now think the dynamo functions, and the migration of sunspots toward the equator follows from the dynamo equations. Thus, the annoying restriction of dynamo activity to the lowest levels of the convective zone, together with the resolution of the equatorial acceleration problem, appears to resolve the dynamo dilemma with the migration of solar fields toward, rather than away from, the equator. Durney's synthesis points the way for the development of a complete, deductive, self consistent theory.

3. Sunspots and Intense Flux Tubes

The activity that we see at the Sun is caused by the continual emergence of magnetic fields through the surface of the Sun. The fields come up through the surface in complicated forms, contorted by the fluid motions in the convective zone from which they spring. It can be shown that the topology of most field configurations admits of no hydrostatic equilibrium, there being instead rapid reconnection and dissipation (Parker, 1972). It is the dissipation of these nonequilibrium fields, sometimes by explosive reconnection, that produces the boisterous activity where the fields are freed at the photosphere. One of the most remarkable properties of the magnetic field at the photosphere is the tendency to compress itself into extremely dense flux tubes, with little or no field in the regions between tubes.

When we remember that the stresses in a magnetic field consist of a tension $B^2/4\pi$ along the lines of force, and an isotropic pressure $B^2/8\pi$, we expect the magnetic field to behave much like a gas, expanding to fill all the available space.

What is it on the Sun, then, that causes the fields to contract into isolated bundles? The effect is remarkable and merits serious attention.

The sunspot is the most conspicuous example of the self-confinement of the magnetic field, producing fields of 1500 G in pores, to 3000–4500 G in the fully developed spot. One of the most startling developments in solar physics has been the growing realization over the past decade that even the general 1–2 G magnetic field of the Sun in quiet regions is almost entirely composed of flux tubes of 2000 G, or more, compressed into tubes of 400 km diameter. This conclusion is not a direct observation, of course. It is a sophisticated inference drawn from the theoretical interpretation of a number of independent observational studies of the Zeeman broadening of various spectral lines in both weak and strong fields (Sheeley, 1967; Livingston and Harvey, 1969, 1971; Sawyer, 1971; Simon and Noyes, 1971; Howard and Stenflo, 1972; Frazier and Stenflo, 1972; Chapman, 1973; Stenflo, 1975). But the conclusion now appears to be inescapable. I am sure that we will hear more about it in this meeting.

Why, or how, can a magnetic field gather itself into a dense bundle, in opposition to its own enormous pressure? The pressure of a 3000 G field is a little more than the gas pressure at the surface of the Sun, where the number density N is 2×10^{17} cm^{-3} and $T = 6 \times 10^3$ K.

The sunspot provides what appears to be the basic clue. The sunspot is cool, some 3900 K at the surface within the magnetic field. The reduced temperature means a reduced scale height, presumably over a depth of several scale heights, so that the gas within the field drops down out of the magnetic field, and the field is compressed into its dense form by the surrounding gas (Parker, 1955a; Schlüter and Temesvary, 1958). In this way we can understand an equilibrium configuration in which the total pressure, composed of the magnetic pressure of the vertical magnetic field $B(x, y)$ and the gas pressure $p(x, y)$ is uniform across the photosphere $p(x, y) + B^2(x, y)/8\pi = constant$ (ignoring the tension and the curvature of the lines of force).

But what makes the gas cool? It has been suggested (Biermann, 1941) that the magnetic field inhibits the convective transport of heat in the sunspot. That is to say, the region of intense field is a thermal insulator. The result is clearly a reduction of temperature at the surface. Unfortunately it is not always appreciated that it also means an enhanced temperature *under* the insulator. When we put on a coat, we become warm underneath, rather than cool. I have examined a number of models of reduced heat transport and have been unable to construct one in which the cooling is of such form as to cause the field to concentrate. The enhanced temperature beneath the region of concentrated field *increases* the gas pressure extending up along the field and *disperses* the field. A deep inverted cone of reduced heat transport seems to come closest to solving the problem, because the heat flow is easily diverted around to the sides. Perhaps someone with deeper insight can construct a situation where reduced heat transport is able to concentrate the field. This has not been done so far, and it appears to me that either it must be done soon or we are forced to the view that some other effect is largely responsible for the cooling. The only available idea is overstability, with vigorous production of Alfvén waves in the upper $1-2 \times 10^3$ km of the convective zone immediately below the photosphere. The thermal convective forces act as a heat engine converting a major function of the heat flow into Alfvén waves and actively refrigerating the gas in the field. The overstability was pursued originally by Danielson (1965), Musman (1967), and Savage (1969) as an explanation for the missing flux from a sunspot produced by the inhibition of convective heat transport. We have used their calculations (Parker, 1974, 1975a) as a basis for arguing that the production of Alfvén waves is the *principal cause* of the sunspot. The spot is cool because the energy flux is converted from heat into mobile waves, so that the heat transport is *enhanced* rather than inhibited. The Alfvén waves carry the energy (some 5×10^{10} erg cm^{-2} s^{-1}) out of the region both upward and downward along the magnetic lines of force. The waves are dissipated elsewhere around, and in, the Sun. One fundamental question in the theory of the sunspot, then, is the reality of the Alfvén waves presumed to be responsible for the cooling. If they exist, then the cooling and intense magnetic field can be understood (see, for instance, the recent observations of Phillis, 1975; Beckers, 1975). If they do not exist, then we must find another explanation and *demonstrate* it.

There is another fundamental question, however, to which we must address ourselves. Until it is answered, understanding of the sunspot cannot be complete. It is not enough to establish the existence of a theoretical equilibrium configuration to understand the sunspot. We must also show how the configuration can be assembled from the gas and field in the first place, and why it is stable once assembled. These two points are probably closely related; if we knew the answer to one, we would probably be able to construct the answer to the other. The difficulty is that a magnetic field constricted by gas pressure to a small throat (the umbra of the sunspot) is unstable to the hydromagnetic exchange instability (fluting instability). The magnetic lines of force are concave toward the gas, so that the tension along the lines of force tends to pull individual flux tubes out of the larger tube as sketched in Figure 1. The characteristic time for the instability is the Alfvén transit time across the tube, of the order of an hour. The effect is well known in the plasma laboratory where one tries to confine a gas by wrapping a field around it. We must not confuse the equilibrium and the instability. The magnetic field is compressed by the reduced pressure of the cool sunken gas level within the field. The reduced gravitational potential energy of the depressed area of cool gas within the field compensates for the increased energy of the magnetic field, and the gas confines a magnetic flux tube of *circular* cross section in a state of *hydrostatic equilibrium*. However, if the circular cross section is perturbed, the balance is upset by the tension along the magnetic lines of force. The field is free to break up into a number of individual tubes which separate from each other and shorten, thereby reducing their energy. We would expect that a sunspot in the course of an hour or so should split across and break into many small spots, each carrying its cool depressed umbra with it, and each breaking into smaller tubes, quickly obliterating the intense fields. When a sunspot is young, the opposite happens. The individual magnetic knots stream into the spot and add to its field (Beckers and Shröter, 1969). When the spot is old, it breaks up into small pieces, but

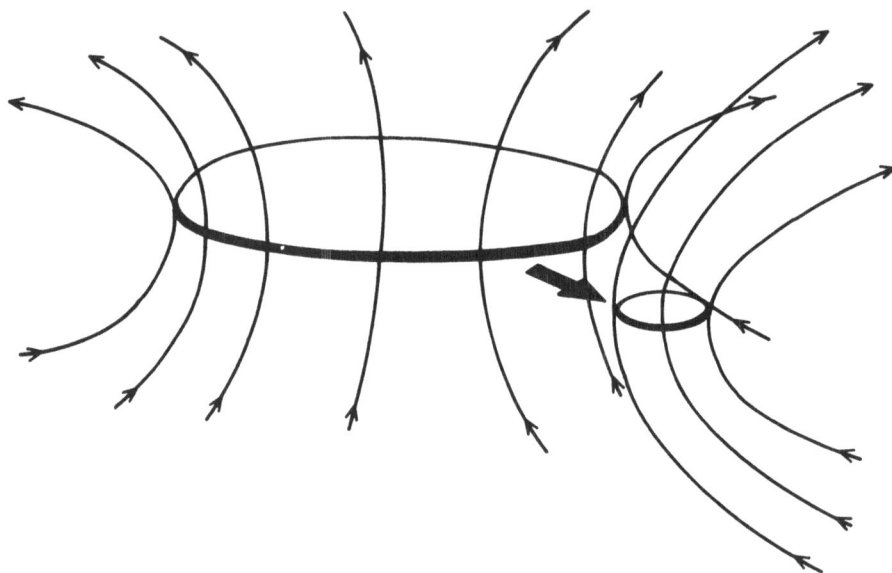

Fig. 1.   A sketch of the magnetic field through a sunspot, illustrating the tendency for the tension along
the magnetic lines of force to pull tubes of flux out the side of the sunspot.

even this final disintegration takes days instead of hours. The mechanism for
stabilizing the spots observed on the Sun has not been demonstrated, so the sunspot
is an enigma.

Recently Meyer *et al.* (1974) have proposed that a swift (0.5 km s^{-1}) converging,
horizontal flow, and associated downdraft, at depths of $1-2 \times 10^4$ km where $\rho \cong$
2×10^{-4} gm cm^{-3}) play a fundamental role in assembling the magnetic field into the
concentrate sunspot. They suggest that the otherwise inexplicable 'attraction' of
magnetic knots toward each other and toward the sunspot is one of the direct
consequences of the converging flow. Certainly some such drastic assumption is
needed to account for the behavior of the spot. They also suggest that in some way
the geometry of the depressed umbra is such as to aid in stabilization of the sunspot
once it is formed.

It is difficult to think of alternatives to their suggestions (Parker, 1975a). Unfortunately any such hypothesis of subsurface dynamical effects can be established only by
theoretical calculation, and that is a difficult task in the unstable stratified convective
zone of the Sun.

Now consider the isolated flux tubes of 2000 G that make up the general field of
the Sun. They appear in the supergranule boundaries. Their magnetic pressure is
comparable to the gas pressure in the photosphere, and very much in excess of the
dynamical pressure of any of the fluid motions observed at the surface. For instance
the 0.5 km s^{-1} of the supergranule in the surface density of $\rho = 3 \times 10^{-7}$ gm cm^{-3};
corresponding to a dynamical pressure of $p = 0.7 \times 10^3$ dyne cm^{-2}, is equivalent to
the pressure $B^2/8\pi$ of a field of about 10^2 G. Granule motions of 3 km/sec
correspond only to 700 G. How, then, are we to understand such intense concentration of magnetic field?

I have spent a long period of time exploring the many 'beautiful' ideas for producing intense magnetic fields (Parker, 1976). Some have suggested that the flux tubes are self-confined force-free magnetic fields. Unfortunately, there is no such thing. Magnetic fields expand. They do not confine themselves. In each case I have been obliged to abandon the 'beautiful' idea as inadequate, for one reason or another, except the possibility that the small flux tubes (of 400 km diameter and 3×10^{13} Mx) are cooled and concentrated in much the same manner as sunspots. Roberts (1976) has shown that the overstability is just as strong in a slender flux tube as in a broad one, so the same basic idea as the sunspot appears to be tenable. But, of course, the same questions of initial assembly and long term stability arise too.

Presumably no cool spot in the small flux tubes is visible at the surface because the cooling need not be as severe as in the sunspot, and because the active cooling does not extend above the top of the active convective zone, terminating several hundred km beneath the visible surface of the photosphere. The surrounding photosphere closes in over the small 200 km radius of the flux tube, partially obliterating the coolness. A temperature reduction of 500–800 K over a diameter of 400 km at the surface would not be conspicuous in the general granule pattern of 500 K temperature variation. It would be interesting to see what a careful search might turn up, because we need an explanation for the general occurrence of intense isolated flux tubes in the solar photosphere. If active cooling by the overstable production of Alfvén waves is *not* the correct explanation, then we need to discover what is. And even if cooling is the answer for their compressed equilibrium, what provides the necessary stability of that equilibrium?

4. Flares, X-rays, and Eruptions

The solar flare has for decades occupied a prominent position in the thoughts of both observers and theoreticians. It is generally agreed – through the absence of alternatives – that the flare is caused by the annihilation of magnetic field, presumably the Dungey-Sweet-Petschek mechanism of rapid reconnection of opposite fields. The theory of rapid reconnection has been carried forward in the past few years (Dungey, 1955; Sweet, 1958; Parker, 1963, 1973a; Petschek, 1964; Green and Sweet, 1967; Petschek and Thorne, 1967; Sonnerup, 1970, 1971; Yeh and Axford, 1970; Priest, 1972a, b, 1973; Fukao and Tsuda, 1974a, b; Vasyliunas, 1975) to the point that it has now been demonstrated that there are circumstances under which opposite fields can reconnect and merge at a significant fraction of the Alfvén speed (computed in the opposite fields). It is not entirely clear just what external boundary conditions yield the highest reconnection rates, but speeds of the order of $0.1 V_A$ are to be expected, and $0.5 V_A$ can be accomplished under special circumstances. Altogether, then, the idea of magnetic merging as the cause of the solar flare is not without a substantial theoretical foundation. However, it is not clear to me that there is anything that can be called observational proof. Perhaps we will hear more on that question in this Symposium.

It is abundantly clear that flares have complicated personalities, providing endless combinations of temporal and spatial form, radio emission, X-ray emission, fast particle emission, and interplanetary blast waves. For instance, flares occurring at the prow of moving sunspots appear to produce more type III radio bursts (Zirin and Lazareff, 1975; Takakura and Yousef, 1975). The outward shock waves (observed in Hα) are closely associated with type II bursts (Harvey *et al.* 1974). Particles appear to be accelerated in two or three stages, producing X-rays after the first stage and 'cosmic rays' after the second (Kane, 1974; Rust and Hegwer, 1975; Tanaka and Enome, 1975). At least half of the energy release of the flare is in the form of fast particles. Theory does not provide any ready explanation for such complete conversion of magnetic energy into accelerated particles. There are a variety of ideas on particle acceleration in solar flares, beginning with the Fermi mechanism (Fermi, 1949, 1954; Parker, 1958) over twenty-five years ago, acceleration in plasma turbulence (Kadomtsev and Tsytovich, 1969; Tsytovich, 1973) and acceleration in neutral sheets (Speiser, 1965, 1967; Coppi and Friedland, 1971; Low, 1975). All of these ideas have merit and may contribute, but in no case, of which we are aware, has it been possible to demonstrate that more than a tiny fraction of the energy release goes into fast – often relativistic – electrons and protons. It is clear that there is much work yet to be done in the theory of particle acceleration that is so central to the outbursts of solar activity.

The discovery by Skylab of repeated eruptions from the Sun into interplanetary space, in the absence of visible flaring on the surface, adds a new dimension to the quandary. The X-ray bright spots, associated with the tiny bipolar magnetic regions – the pepper and salt effect in magnetograms – are another curiosity (Krieger *et al.*, 1971; Vaiana, Krieger, and Timothy, 1973; Harvey and Martin, 1973; Golub *et al.*, 1974; Harvey *et al.*, 1975; Parker, 1975c) behaving much like little bipolar sunspot groups, including miniature flares.

5. General Comments

Altogether, I am awed and challenged by the tricks displayed by the Sun. The Solar magician is clever indeed. Sunspots have been studied, and thought about, for the better part of a century, and I cannot tell you that I understand much of their behavior in terms of physics. The flare is a turbulent phenomenon, which, therefore, may never be reduced to quantitative theoretical understanding. But I think we can understand more about the magnetic activity of the Sun than at present if we put serious effort into recognizing and distinguishing the fundamental problems. We must develop the habit of recognizing what is not understood, and what promises, therefore, to teach us new physics. Unfortunately, the literature is full of folklore on the various aspects of the activity of the Sun. There is no phenomenon that does not have an 'explanation', and a retinue of followers of that explanation.

Perhaps the most illusive aspect of all in the great riddle of solar activity is the seeming relation between activity on the Sun and unusual environmental conditions at the surface of Earth (Willett, 1965; Bray, 1968; Wilcox, 1968; Woodbridge, 1971; Shapiro, 1972; Hines, 1973; Roberts and Olsen, 1973a, b; Bandeen and Maran, 1974). It is noteworthy that the annual tree rings were unusually uniform during the Maunder minimum (Douglass, 1919; Maunder, 1922) while the Baltic sea froze over in the winter, something that has not happened at any other time within historical memory. No physical connection has yet been established, leaving one with the uneasy feeling that much of it may be coincidence. But to explain all as coincidence appears to be even more difficult than hunting for missing physical connections. It is a complicated subject to approach, but its fundamental importance to our lives, as well as our scientific knowledge, requires that the subject be pursued until it can be discarded with confidence, or established and understood. Success will involve the cooperation of individuals working in the physics of the Sun, interplanetary space, and the terrestrial magnetosphere, ionosphere, and atmosphere, as well as the geophysicist and biologist studying conditions at ground level.

Altogether the fundamental problems that confront us are (a) the convection and circulation in the ionization zone of the Sun, (b) the generation of magnetic fields, (c) the properties of the merging fields, forming active regions, (d) the frequent solar eruptions, (e) the coronal hole and the high speed wind, and the suppression of the coronal hole by magnetic fields, and finally (f) the complicated climatological effects of the solar luminosity and solar activity. Together these problems make solar physics the most exciting and challenging field of astrophysics. Unlike the other subjects in astrophysics we have learned enough to get our teeth into the physics of the Sun. The basic theoretical problems are difficult but not impossible. We have a good idea of the crucial observations that have yet to be carried out, from coordinated high resolution X-ray, XUV, visible, Doppler, and magnetograph studies of flares and active regions, to global studies of the circulation at the visible surface, global studies of the corona and solar wind from the equator to the poles, to accurate (1 part in 10^3) absolute synoptic measurement of the solar luminosity. The necessary instrumentation and technology is presently tractable. Spacecraft observations and very high resolution ground based observations are essential, with careful coordination between them on many occasions. If there ever was an urgent program that needs worldwide cooperation for its solution it is the present problem of understanding the Sun. It is in recognition of this state of urgency and complexity that we are all here in Prague. I look forward very much to the next few days to hear how the various aspects of the problem are being developed in so many places around the world.

References

Bahcall, J. N., Heubner, W. F., Magee, N. H., Merts, A. L., and Ulrich, R. K.: 1973, *Astrophys. J.* **184**, 1.
Bandeen, W. R. and Maran, S. P.: 1974, *Possible Relationships between Solar Activity and Meteorological Phenomena, NASA Symposium*, Goddard Space Flight Center (X-901-74-156).
Beckers, J. M.: 1975, *Astrophys. J.* (in press).
Beckers, J. M. and Schröter, E. H.: 1969, *Solar Phys.* **4**, 192, 303.
Biermann, L.: 1941, *Vierteljahrsschr. Astron. Gessellsch.* **76**, 194.
Bray, J. R.: 1968, *Nature* **220**, 672.
Busse, F. H.: 1970, *Astrophys. J.* **159**, 629.
Chapman, G. A.: 1973, *Astrophys. J.* **191**, 255.
Cocke, W. J.: 1967, *Astrophys. J.* **150**, 1041.
Coppi, B. and Friedland, A. B.: 1971, *Astrophys. J.* **169**, 379.
Danielson, R. E.: 1965, *IAU Symp.* **22**, 315; *Handbuch der Physik*, **52**, 151.
Davis, R. and Evans, J. M.: 1973, *Proc. 13th Int. Cosmic Ray Conf.*, Denver, Colorado.
Deinzer, W., Kusserow, H. U. V. and Stix, M.: 1974, *Astron. Astrophys.* **36**, 69.

Deinzer, W. and Stix, M.: 1971, *Astron. Astrophys.* **12**, 111.
Dicke, R. H.: 1974, *Astrophys. J.* **190**, 187.
Dicke, R. H. and Goldenberg, H. M.: 1974, *Astrophys. J. Suppl.* No. 241.
Douglass, A. E.: 1919, 'Climatic Cycles and Tree Growth', Carnegie Institution.
Dungey, J. W.: 1953, *Phil. Mag.* **44**, 725.
Durney, B.: 1968, *J. Atmos. Sci.* **25**, 372.
Durney, B.: 1970, *Astrophys. J.* **161**, 1115.
Durney, B.: 1971, *Astrophys. J.* **163**, 353.
Durney, B.: 1972, *Solar Phys.* **26**, 3.
Durney, B.: 1975, *Astrophys. J.* (in press).
Durney, B. R. and Roxburgh, I. W.: 1971, *Solar Phys.* **16**, 3.
Fermi, E.: 1949, *Phys. Rev.* **75**, 1169.
Fermi, E.: 1954, *Astrophys. J.* **119**, 1.
Fowler, W. A.: 1972, *Nature* **238**, 24; 1973, *Nature* **242**, 424.
Frazier, E. N. and Stenflo, J. O.: 1972, *Solar Phys.* **27**, 330.
Fukao, S. and Tsuda, T.: 1973a, *Planetary Space Sci.* **21**, 1151.
Fukao, S. and Tsuda, T.: 1973b, *J. Plasma Phys.* **9**, 409.
Gierasch, P. J.: 1974, *Astrophys. J.* **190**, 199.
Gilman, P. A.: 1969, *Solar Phys.* **8**, 316; **9**, 3.
Gilman, P. A.: 1972, *Solar Phys.* **27**, 3.
Gilman, P. A.: 1974, *Ann. Rev. Astron. Astrophys.* **12**, 47.
Golub, L., Krieger, A. S., Silk, J. K., Timothy, A. F., and Vaiana, G. S.: 1974, *Astrophys. J. Letters* **189**, L93.
Green, R. M. and Sweet, P. A.: 1967, *Astrophys. J.* **147**, 1153.
Harvey, K. L., Harvey, J. W., and Martin, S. F.: 1975, *Solar Phys.* **40**, 87.
Harvey, K. L. and Martin, S. F.: 1973, *Solar Phys.* **32**, 389.
Harvey, K. L., Martin, S. F., and Riddle, A. C.: 1974, *Solar Phys.* **36**, 151.
Hines, C. O.: 1973, *J. Atmospheric Sci.* **30**, 739.
Howard, R. and Stenflo, J. O.: 1972, *Solar Phys.* **22**, 402.
Kadomtsev. B. B. and Tsytovich, V. N.: 1969, *IAU Symp.* **39**, 108.
Kane, S. R.: 1974, 'Coronal Disturbances', *IAU Symp.* **57** (Dordrecht: D. Reidel), ed. by G. Newkirk, p. 105.
Kippenhahn, R.: 1963, *Astrophys. J.* **137**, 664.
Kohler, H.: 1970, *Solar Phys.* **13**, 3.
Krieger, A. S., Vaiana, G. S., and van Speybroeck, L. P.: 1971, *IAU Symp.* **43**, 397.
Leighton, R. B.: 1969, *Astrophys. J.* **156**, 1.
Livingston, W. and Harvey, J.: 1969, *Solar Phys.* **10**, 294.
Livingston, W. and Harvey, J.: 1971, *IAU Symp.* **43**, 51.
Low, B. C.: 1974, *Astrophys. J.* **189**, 353.
Maunder, E. W.: 1894, *J. British Astron. Assoc.* **5**, 47.
Maunder, E. W.: 1922, *J. British Astron. Assoc.* **32**, 140, 223.
Meyer, F., Schmidt, H. U., Weiss, N. O., and Wilson, P. R.: 1974, *Monthly Notices Roy. Astron. Soc.* **169**, 35.
Minnaert, M.: 1953, in G. P. Kuiper (ed.), *The Sun*, University of Chicago Press, Chicago, Chap. 3, p. 100.
Musman, S.: 1967, *Astrophys. J.* **149**, 201.
Osaki, Y.: 1970, *Monthly Notices Roy. Astron. Soc.* **131**, 407.
Parker, E. N.: 1955a, *Astrophys. J.* **121**, 491.
Parker, E. N.: 1955b, *Astrophys. J.* **122**, 293.
Parker, E. N.: 1958, *Phys. Rev.* **109**, 1328.
Parker, E. N.: 1963, *Astrophys. J. Suppl.* **8**, 177.
Parker, E. N.: 1970a, *Astrophys. J. Suppl.* **160**, 383; **162**, 665.
Parker, E. N., 1970b, *Ann. Rev. Astron. Astrophys.* **8**, 1.
Parker, E. N.: 1971, *Astrophys. J.* **164**, 491.
Parker, E. N.: 1972, *Astrophys, J.* **176**, 213.
Parker, E. N.: 1973a, *Astrophys. J.* **180**, 247.
Parker, E. N.: 1973b, *J. Plasma Phys.* **9**, 49.
Parker, E. N.: 1974, *Solar Phys.* **36**, 249; **37**, 127.
Parker, E. N.: 1975a, *Solar Phys.* **40**, 275; 291.
Parker, E. N.: 1975b, *Astrophys. J.* **198**, 205.
Parker, E. N.: 1975c, *Astrophys. J.* **201**, 494.
Parker, E. N.: 1976, *Astrophys. J.* **204**, 259.

Petschek, H. E.: 1964, *Physics of Solar Flares*, AAS-NASA Symposium (NASA, SP-50), p. 425.
Petschek, H. E. and Thorne, R. M.: 1967, *Astrophys. J.* **147**, 1157.
Phillis, G. L.: 1975, *Solar Phys.* **41**, 71.
Priest, E. R.: 1971a, *Quart. J. Mech. Appl. Math.* **25**, 319.
Priest, E. R.: 1972b, *Monthly Notices Roy. Astron. Soc.* **159**, 389.
Priest, E. R.: 1973, *Astrophys. J.* **181**, 227.
Roberts, B.: 1976, *Astrophys. J.* **204**, 268.
Roberts, W. O. and Olson, R. H.: 1973a, *Rev. Geophys. Space Phys.* **11**, 731.
Roberts, W. O. and Olson, R. H.: 1973b, *J. Atmospheric Sci.* **30**, 135.
Rust, D. M. and Hegwer, F.: 1975, *Solar Phys.* **40**, 141.
Savage, B. D.: 1969, *Astrophys. J.* **156**, 707.
Sawyer, C.: 1971, *IAU Symp.* **43**, 316.
Schluter, A. and Temesvary, St.: 1958, *IAU Symp.* **6**, 263.
Shapiro, R.: 1972, *J. Atmospheric Sci.* **29**, 1213.
Sheeley, N. R.: 1967, *Solar Phys.* **1**, 171.
Simon, G. W. and Noyers, R. W.: 1971, *IAU Symp.* **43**, 663.
Sonnerup, B. U. O.: 1970, *J. Plasma Phys.* **4**, 161.
Sonnerup, B. U. O.: 1971, *J. Geophys. Res.* **76**, 8211.
Speiser, T. W.: 1965, *J. Geophys. Res.* **70**, 1717, 4219.
Speiser, T. W.: 1967, *J. Geophys. Res.* **72**, 3919.
Spruit, H. C.: 1974, *Solar Phys.* **34**, 277.
Steenbeck, M. and Krause, F.: 1969, *Astron. Nachr.* **291**, 49, 271.
Steenbeck, M., Krause, F., and Radler, K. H.: 1966, *Z. Naturforsch.* **21a**, 369.
Stenflo, J. O.: 1973, *Solar Phys.* **32**, 41.
Stix, M.: 1974, *Astron. Astrophys*, **37**, 121.
Sweet, P. A.: 1958, *Nuovo Cimento* **8**, 188.
Takakura, T. and Yousef, S.: 1975, *Solar Phys.* **40**, 421.
Tanaka, H. and Enome, S.: 1975, *Solar Phys.* **40**, 123.
Tsytovich, V. N.: 1973, *Annual Rev. Astron. Astrophys.* **11**, 363.
Ulrich, R. K.: 1974, *Astrophys. J.* **188**, 369.
Vaiana, G. S., Krieger, A. S., and Timothy, A. F.: 1973, *Solar Phys.* **32**, 81.
Vasyliunas, V. M.: 1975, *Rev. Geophys. Space Physics* **13**, 303.
Weiss, N. O.: 1965, *Observatory* **85**, 37.
Wilcox, J. M.: 1968, *Space Sci. Rev.* **8**, 258.
Willett, H. C.: 1965, *J. Atmospheric Sci.* **22**, 120.
Woodbridge, D. D.: 1971, *Planetary Space Sci.* **19**, 821.
Yeh, T. and Axford, W. I.: 1970, *J. Plasma Phys.* **4**, 207.
Yoshimura, H.: 1972a, *Solar Phys.* **22**, 20.
Yoshimura, H.: 1972b, *Astrophys. J.* **178**, 863.
Yoshimura, H.: 1973, *Solar Phys.* **33**, 131.
Yoshimura, H.: 1975, *Astrophys. J. Suppl.* **29**, 467.
Yoshimura, H. and Kato, S.: 1971, *Publ. Astron. Soc. Japan* **23**, 57.
Zirin, H. and Lazareff, B.: 1975, *Solar Phys.* **41**, 425.

DISCUSSION

Gilman: Let me comment on your second dilemma, that of the long sunspot minimum in the 17th century. Jack Eddy, Dorothy Trotter and I have been looking at the sunspot rotation rate just prior to the beginning of that period. From data collected by J. Hevelius in 1643–45, we find no strong differences in rotation rate from modern values.

Stix: You said that the long activity minimum during the 17th century might have been caused by a different mode of *convection*. Could the cause also be a different mode of the *magnetic field*, with the motion field remaining the same?

Parker: That is certainly another possibility. A slight change in the level of convection might well shift the dynamo to another mode.

Mestel: Your remark about a magnetic field's finding it hard to reach hydrostatic equilibrium: doesn't this apply primarily to convective zones with a nearly adiabatic pressure-density relation? In a sub-adiabatic, relative zone, arbitrary fields of moderate strength may be balanced by pressure and gravity, provided temperature and density vary independently. This may be the situation in the envelopes of early-type stars, and indeed in the core of the Sun. The question then arises of the stability of such fields. For dynamical stability (against adiabatic motions), one almost certainly needs a complex topology, with linked flux. Stability against the much slower modes which depend on heat exchange – of which magnetic buoyancy in a radiative zone is an example – may require an inward gradient of mean molecular weight. The question is not irrelevant to a Symposium on 'Solar Activity': magnetic flux from the solar core could be significant for surface phenomena.

Parker: Generally speaking, the magnetic fields in nature lack the perfect symmetry that is necessary for equilibrium. Any deviations from perfect symmetry permit their buoyancy to carry them upward to the surface of the Sun, or other astrophysical body. I should add that any field topology that varies along the lines of force, produces a nonequilibrium in the form of neutral point reconnection of the lines of force. It is on this general basis that I made the statement that magnetic fields have no equilibrium and so necessarily produce activity.

PART 1

BASIC OBSERVED PARAMETERS
OF THE SOLAR CYCLE

DIFFERENTIAL ROTATION AND
GLOBAL-SCALE VELOCITY FIELDS

ROBERT HOWARD and HIROKAZU YOSHIMURA

Hale Observatories, Carnegie Institute of Washington,
California Institute of Technology, Pasadena, Calif., U.S.A.

Abstract. A review is given of the observational and theoretical background of global-scale velocity fields on the solar surface. A newly-developed method of reduction of the Mount Wilson velocity data is described, and the results from this new method are compared with the results of the old method. A preliminary analysis is made of the new results over a short time interval. Small-scale latitude irregularities in the differential rotation are shown to exist. Variations in time which occur in the rotation rate are broadly distributed in latitude and longitude. Although a non-solar (instrumental) cause cannot be found for these variations, such a cause cannot be ruled out at this time. Global-scale non-axisymmetric velocity field patterns intermediate between solar diameter and super-granular scale are shown to exist on the solar surface as predicted by theory.

1. Introduction
Review of the Problems, Both Observational and Theoretical

In 1961, Babcock showed, after the suggestion of Cowling (1953), that the stretching of the general magnetic field in longitude inside the Sun by differential rotation (equatorial acceleration) can explain the basic phenomenon of the solar cycle, i.e., Hale's polarity rule. This idea was formulated by Leighton (1969) into a mathematical form, numerical solutions of which demonstrated the basic behavior of the solar cycle. Thus the study of the differential rotation has come to be of great importance for understanding the solar cycle, which is the basic underlying process of solar activity. However, Cowling (1933) has shown that a steady axisymmetric velocity field, such as the differential rotation, cannot maintain the general axisymmetric magnetic field by itself. Among other things, the dynamo action of fluid motions of a non-axisymmetric nature has come to be considered a necessary supplementary mechanism for generating and maintaining the general magnetic field (Parker, 1955; Steenbeck and Krause, 1969). However, the nature of the non-axisymmetric motions has been unknown. Earlier, small-scale turbulent convection, such as granulation and supergranulation, were considered possible candidates. However, the necessary supplementary mechanism, called the regeneration action in general, or α-effect, or Γ-effect, requires the influence of rotation on the fluid motions, and since this influence from the smaller space and time scales is small, there is doubt whether the granulation and supergranulation can provide the necessary regeneration action.

Another kind of large-scale velocity field has been suspected to exist inside the convection zone – so large that we call it global-scale. Two independent suggestions of such a global-scale velocity field were made in 1965. One is by Bumba and Howard (1965), who suspected its existence from the organized behavior of the global-scale magnetic field. The other is by Ward (1965), who studied the proper motions of sunspots and found a correlation in their movements: spots rotating faster than the average tend to move toward the equator, and those rotating slower move

Bumba and Kleczek (eds.), Basic Mechanisms of Solar Activity, 19–35. All Rights Reserved.
Copyright © 1976 by the IAU.

toward the poles. There still remains some question about whether these motions reflect motions of the fluid; however, if we regard this as a valid correlation in the fluid motions, it is evident that angular momentum is being transported toward lower latitudes, and this can act as a mechanism to create the equatorial acceleration (Starr and Gilman, 1965).

The nature of the global-scale velocity fields has been a subject of controversy because there are three possible modes of fluid motions in general in a rotating spherical shell, *i.e.*, acoustic or sound waves, Rossby waves, and the convection mode (Yoshimura, 1974). Since the space and time scales of this velocity field are large, the possibility of a sound wave mode is ruled out. The early investigators regarded it as a Rossby wave mode (Ward, 1964, 1965; Kato, 1969; Kato and Nakagawa, 1969; Gilman, 1969a, b). However, these motions of the Rossby wave mode, flowing mainly horizontally, are difficult to excite in the solar situation because there is no major force to move them horizontally. Meanwhile the study of the third possibility – the convection mode – was started by Durney and Skumanich (1968), and Durney (1968a, b). Then Busse (1970), Durney (1970), and Yoshimura and Kato (1971) found in succession that the convective modes that are most easily excited by the superadiabatic gradient in the convection zone, in other words which have the greatest growth rate and lowest critical Rayleigh number, can actually excite and maintain the equatorial acceleration. The convective modes with smaller growth rates can accelerate the higher latitudes and not necessarily the equator (Yoshimura and Kato, 1971; Yoshimura, 1972a). Moreover, this convective mode, especially the mode with the greater growth rate, which has a sector pattern in longitude, was shown by Yoshimura (1971) to be able to explain the organized behavior of the global-scale magnetic-field distribution, such as the existence and characteristics of unipolar magnetic regions and of active longitudes, especially their rigid body-like rotation (Babcock and Babcock, 1955; Bumba and Howard, 1965; Howard *et al.*, 1967; Wolfer, 1897; Losh, 1939; Brunner-Hagger, 1944; Eigenson *et al.*, 1948; Becker, 1955; Warwick, 1965; Dodson and Hedeman, 1968; Svestka, 1968a, b). Since the sector structure of the magnetic field in interplanetary space is also due to the organized solar-surface magnetic field (Wilcox, 1968), the dynamics of this interplanetary sector structure can also be regarded as subject to the dynamics of the global-scale velocity field of the convection mode. Besides this global-scale ordering of solar activity, many local characteristics of the structure or dynamics of active regions have been found to be organized, i.e., the faster rotation of sunspots than the fluids (Howard and Harvey, 1970), the preponderance of preceding spots of bipolar sunspot groups (Grotrian and Künzel, 1950), the tilt of the bipolar axes of the groups (Brunner, 1930), the forward inclinations of the normal axes of the magnetic fields of sunspots (Maunder, 1907; Minnaert, 1946), the association of the characteristics of active regions with the presence of older active regions in the vicinity (Martres, 1970), and the correlations among the characteristics of active regions stated above (Weart, 1970, 1972; Sawyer and Haurwitz, 1972). All these effects are shown to be explainable by the velocity and magnetic fields associated with the global-scale convection mode (Yoshimura, 1973). Thus, the existence of a global-scale velocity field inside the Sun may be regarded as fairly certain.

If we admit the existence of the global-scale convection, then the necessary supplementary process to drive the solar cycle can be provided by convection. This has been shown by Yoshimura (1972b), and the numerical simulation of the solar cycle due to the dynamo action of the global convection has succeeded in explaining many basic characteristics of the solar cycle (Yoshimura, 1975a); and moreover has predicted a characteristic evolution of the general magnetic field at the surface, which was confirmed by analyzing the Mount Wilson magnetic synoptic chart data (Yoshimura, 1975b).

In recent years the techniques of the observational study of differential rotation have been improved. Earlier studies depended mainly on the movement of tracers, such as sunspots. Thus the detection of short time-scale variations in the rotation was impossible. Modern electronic technology has made it possible to detect the velocity field of the solar surface to a sensitivity approaching 5 m s^{-1}. A systematic study of the differential rotation using such sensitive measurements was first made by Howard and Harvey (1970). One of the results of this study is that the differential rotation varies day by day and week by week.

Temporal variations on the rotation rate of the Sun have been noted almost since the first days of spectroscopic rotation determinations. Halm (1904) observed differences between the rotation rates he determined in 1901–02 and 1903. H. H. Plaskett (1916) observed variations in the rotation rate of the Sun over intervals of a few days with an amplitude of about 0.15 km s^{-1}. He also proposed a secular variation of the rotational velocity. Evershed (1931) also found evidence for variation of the rotational velocity with time.

Some investigators have proposed that the observed variations are instrumental in origin (i.e., J. S. Plaskett, 1912). In particular De Lury (1939) suggested that the variations observed could be explained by the varying presence of scattered light near the solar limb – the early spectroscopic observations were invariably made near the limb of the Sun. Such a sweeping object is difficult to answer after the fact, and the effect of De Lury's paper seems to have been to discourage further work in the field for many years. Apparently no attempt was made to verify experimentally that scattered light was an important factor in such measurements.

Hart (1954) showed that scattered light need not be an important factor in such measurements, and discovered variations in line-of-sight velocity in the solar surface which we now know to correspond to the supergranulation pattern. She attributed the previously reported variations in rotation rate to the existence of this pattern. It seems unlikely, however, that a velocity pattern with a scale as small as that of the supergranulation could be responsible for rotation rate changes such as those reported by Halm (1904), Plaskett (1916), and Evershed (1931), whose results show a broad latitude dependence.

Howard and Harvey's (1970) and Howard's (1971) results are derived from sensitive photoelectric measurements, and hence are even accurate near the central meridian, thus greatly reducing the risk of influence by scattered light. Also, these measurements generally cover the solar surface, so that the influence by small-scale velocity fields is greatly reduced. The same sort of temporal variations seen earlier are also seen in these photoelectric measurements. Variations in the rotation rate may occur in a period of a few days, or over a month or two. In addition, there is some evidence for a secular change in the rotation rate of the Sun. The nature of the variations seen by Howard and Harvey (1970) is not clear since, in their method, data over the full disk are used and fitted by a least-squares technique to a functional form of the differential rotation, including an approximation to the radial dependence on the disk of the limb redshift. Any global-scale non-axisymmetric velocity field will affect the resulting rotation and limb redshift results in an unpredictable fashion.

This paper presents in a preliminary way a new method to analyze the Mount Wilson Doppler data taken at the 150-ft Solar Tower Telescope. This new method separates the effects of the differential rotation, limb redshift, and global-scale velocity field in order to derive more accurately the characteristics of each of these phenomena. The analysis is still in progress, and the results presented here are preliminary ones.

2. A New Method to Analyze Global-Scale Surface Velocity Fields: the Differential Rotation

The Mount Wilson magnetograph observations have been described elsewhere (Howard and Harvey, 1970), and the reduction of the velocity data to obtain rotational velocity as a function of latitude was described in the same paper in great detail.

In the present study we shall describe a newly-developed technique of reduction for the same kind of data. This new method provides us with a check on many aspects of the earlier reduction method, and provides us with a comprehensive method of examining the global-scale velocity patterns on the solar surface.

The corrections for the background velocity effects – the orbit and axial rotation of the Earth – are essentially identical to the earlier reduction technique. We start, then, with a grid of about 10 000 points covering the entire solar disk. (At the time the observations for this study were made – summer 1974 – the scanning aperture size was 17.5″ square. Currently the aperture size is 12.5″ square.) For each point we have a velocity signal which represents the velocity of the photosphere observed in the wings of the Fe I λ 5250.2 line. The 'zero' level of the velocity signal is unknown. The scan of the Sun is an east-west raster, starting at the north or south pole of the Sun and requiring a total of about 90 min to complete. Thus any slow drift of the background velocity, due to the instrument or a change in the barometric pressure or to some other cause, can result in a change in the background velocity, which then will vary from pole to pole on the Sun. However, in the time required for a single east-west scan – approximately one minute – there is no chance of an instrumental drift that can significantly affect the velocity signal. Thus the derived rotation rate will be unaffected.

At any latitude, if the intrinsic rotational velocity, which is assumed not to depend on central meridian distance, L, is denoted by V, then the line-of-sight component of this velocity, which is the quantity measured by the magnetograph, is given by

$$V_m = V \sin L. \tag{1}$$

Thus, if we plot V_m against $\sin L$, we expect to see a straight line with slope of V.

A complicating factor in this analysis is the limb redshift, which is very difficult to account for precisely because its magnitude and position dependence are not well known. A remarkable feature of the new method of reduction that is described here is that the effect of this shift is largely eliminated in the rotation determination.

The data are divided into 5 latitude zones, starting from the equator in each hemisphere. In each of these latitude zones a background 'zero' velocity level is calculated by averaging all the velocity signals (converted to km s^{-1}) within 50°, or in some cases 15°, of the central meridian. This reference zero signal is subtracted from all the velocity signals for that 5° zone. Thus the velocity field considered after this step is the velocity field referenced to the central meridian region. The next step is to convert negative values of sin L to positive sin L, and for those negative sin L points the signs of the velocity signals are reversed. This has the effect of a 180° rotation about the origin in the plane of the $V - \sin L$ plot of the points with negative sin L. (Refer to Figure 1.) From this array of points, a slope is determined, assuming that

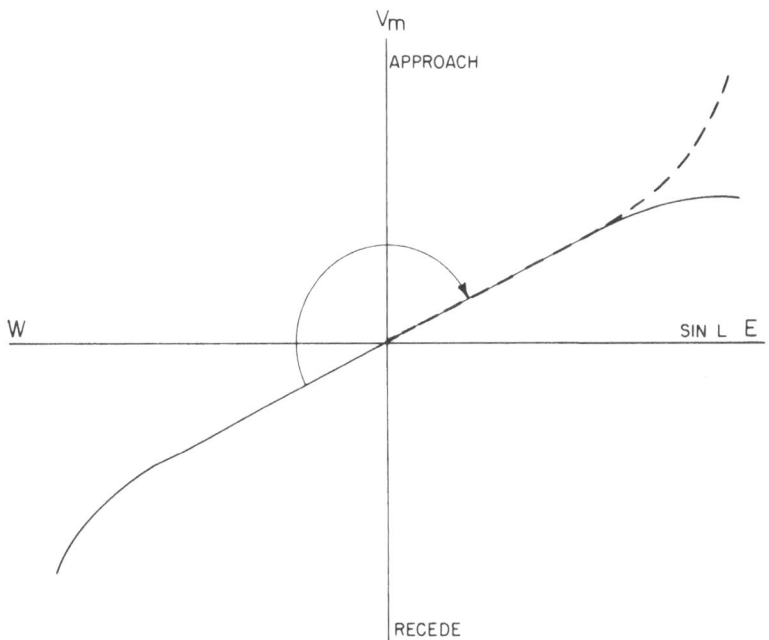

Fig. 1.   A schematic representation of the velocity-sin L curve showing the limb redshift and the rotation at one latitude zone. The dashed curve represents the effect of rotation of the left half of the curve about the axis in the plane of the plot in order to minimize the effect of the limb redshift in the process of the least squares fitting of the rotation curve to the observed data. The observed points are scattered around the solid curve and have an arbitrary zero point. In this figure the zero has been corrected.

the straight-line solution passes through the origin. Points with $|\sin L| > 0.8$, i.e., $|L| > 53°$, were not included in this slope determination. The advantage of this folding technique is that the effect on the slope of the line of the limb redshift is essentially cancelled if the redshift is symmetric about the central meridian. Limiting the slope solution to points within 53° of the central meridian further reduces the influence of the limb redshift because the magnitude of this shift is greatly reduced at this distance from the limb – to about 50 m s^{-1} at the equatorial latitudes.

In the solutions using the method derived above, the probable errors in the slopes for the latitude zones equatorward of 50° were rarely as much as 10 m s^{-1} and generally about half that figure. This is less than 0.5% of the rotational velocity. Poleward of 50° the errors were a little larger, but rarely greater than 25 m s^{-1}. The number of points involved in each independent slope determination was generally greater than 600 for the equatorial latitudes, around 400 at 30°, and down to about 30 at 70°.

The slope of each 5° latitude zone gives directly the average rotational velocity of that zone in km s^{-1}. The solution in each zone is independent of the others; no points are included in more than one zone. Table I gives the results for a typical day for the various latitude zones.

TABLE I

Slope results for 8/14/75

Lat. zone	No. of pts.	V km s^{-1}	p.e. km s^{-1}	ω μ rad s^{-1}
+75–+80	43	0.288	0.013	1.91
+75–+75	83	0.439	0.014	2.10
+65–+70	116	0.632	0.012	2.37
+60–+65	180	0.756	0.012	2.35
+55–+60	226	0.934	0.010	2.50
+50–+55	272	1.078	0.009	2.54
+45–+50	350	1.196	0.015	2.54
+40–+45	395	1.357	0.008	2.64
+35–+40	443	1.527	0.008	2.77
+30–+35	483	1.629	0.007	2.78
+25–+30	540	1.751	0.007	2.84
+20–+25	555	1.836	0.007	2.86
+15–+20	607	1.915	0.006	2.88
+10–+15	613	1.986	0.006	2.92
+ 5–+10	595	2.040	0.006	2.96
0–+ 5	618	2.024	0.006	2.91
− 5– 0	620	2.051	0.006	2.95
− 10–− 5	593	2.028	0.007	2.94
− 15–− 10	553	2.000	0.006	2.94
− 20–− 15	526	1.922	0.006	2.90
− 25–− 20	489	1.858	0.007	2.89
− 30–− 25	432	1.759	0.008	2.85
− 35–− 30	374	1.643	0.008	2.80
− 40–− 35	347	1.520	0.009	2.75
− 45–− 40	274	1.352	0.010	2.64
− 50–− 45	221	1.218	0.010	2.59
− 55–− 50	163	1.048	0.011	2.47
− 60–− 55	115	0.899	0.014	2.40
− 65–− 60	77	0.765	0.014	2.38

3. Comparison with the Results from the Old Reduction Technique

In the original reduction technique described by Howard and Harvey (1970), all the points on the solar disk are analyzed in one large least-squares solution to obtain the coefficients of an expansion in latitude:

$$\omega = a + b \sin^2 B + c \sin^4 B. \tag{2}$$

An additional term, $e(1 - \cos \rho)^2$, was included to account for the limb redshift.

Such an equation has some disadvantages because any global-scale velocity field will be mixed into the solution in a complicated way, and the real cause of variation in any of the coefficients cannot be determined without addition analysis. However, such a formula has been used for the representation of the differential rotation since the earliest days of Doppler measurements, and it turns out to provide a rather accurate description of the average large-scale latitude dependence of the rotation that is determined by the new method.

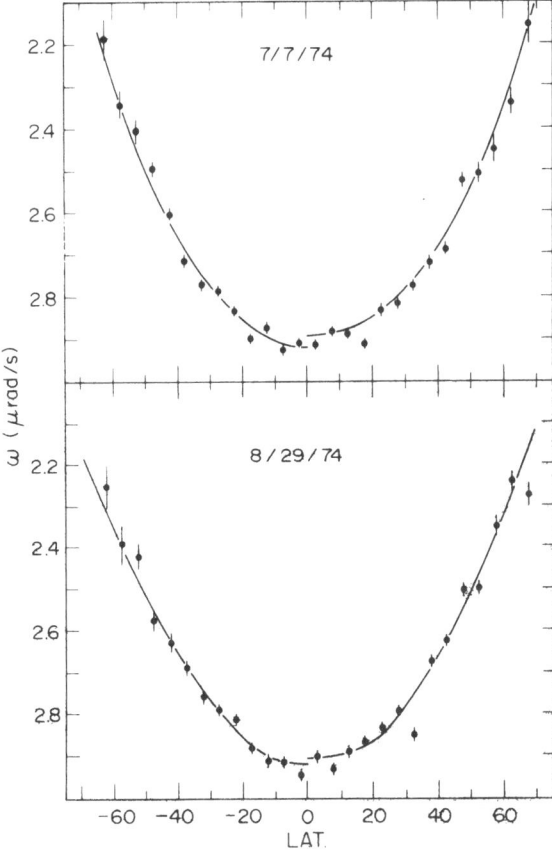

Fig. 2. Rotation results for the two dates indicated. The solid curves represent the old solution for solar rotation for each hemisphere separately, *i.e.*, the solution for ω using the derived coefficients a, b, and c in Equation (2). The filled circles are the slope solutions for each 5° latitude zone. The vertical lines represent probable errors in the slopes.

It is reasonable to suppose, however, that such a formulation cannot account for possible small-scale structure of the rotation. Figure 2 shows two examples of the results from the old analysis using Equation (2) and the new method using the slope reductions. There are evidently small-scale latitude features that cannot be represented by Equation (2). The large-scale shape of the latitude dependence, however, is well represented by the old analysis.

The smallest scale features seen in Figure 2 are clearly significant because they are at least several times greater than the errors. It is also evident that they are not long-lived because they are not similar on the two examples shown. A further investigation into this small-scale latitude dependence of the rotation is planned.

Such a fine structure in the latitude dependence has been noted by Deubner *et al.* (1975). It appears from Figure 2 that averaging the data north and south of the equator, as these authors have done, is not justified. Furthermore, their conclusion that there are humps in the rotation rate at active latitudes seems to be not verified in the small sample of data shown in Figure 2 and other samples analyzed so far.

4. Preliminary Analysis of the Rotation Slope Results

Slope reductions as described above were run for the regular Mount Wilson full-disk magnetogram data for the interval 1974, July 15 through September 14. Figure 3 shows a plot of the equatorial velocities – the latitude zone $-5°$ to $+5°$ – in this interval. On the same plot is the equatorial velocity derived from the old analysis – the parameter a. Clearly the agreement between these two reduction methods is good.

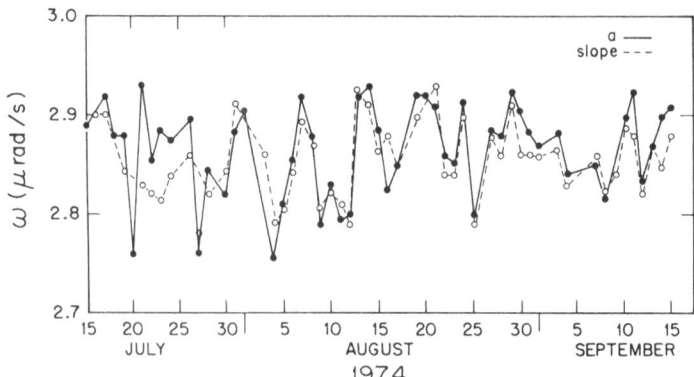

Fig. 3.   The variation with time of the equatorial rotational velocity of the Sun. The filled circles and solid line represent the coefficient a from Equation (2), and the open circles and dashed line represent the slope solutions for the latitude zone $+5°$ to $-5°$. The same raw data are used for both solutions.

The most striking feature of this plot is the regular variation seen in the equatorial velocity. The angular velocity seems to oscillate, especially in the interval July 25 to August 20, with a period of about 6 or 7 days. The amplitude of this variation is about 4%, which corresponds to about 80 m s^{-1}.

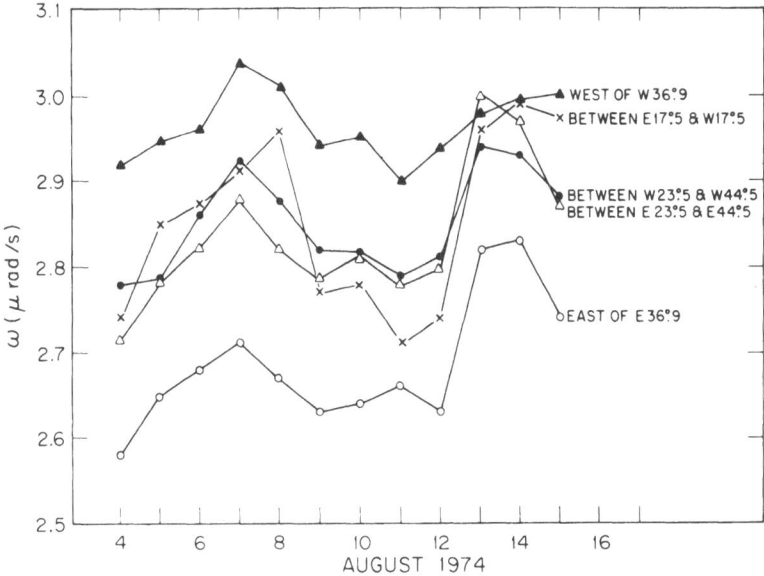

Fig. 4.   The variation with time of the slopes of the velocity-sin L curves determined at the various zones of different central meridian distances indicated. These data are from the latitude zone $+15°$ to $-15°$.

In order to investigate this effect in more detail, we determined the slopes from segments of the $V_m - \sin L$ curves for the equatorial region. Figure 4 shows for the interval August 4 to August 15 the angular velocities for the various zones of central meridian distance indicated. Although there are some differences between the shapes of the curves, the basic location of the two peaks seen in this interval is the same.

If the variations seen in Figure 3 were due to a stationary or quasi-stationary global-scale pattern of east-west motions rotating with the Sun, then one should expect that, as this pattern rotated, the slopes in different parts of the disk would peak at different times. In other words, in Figure 4 we would expect to see phase differences in the location of the peaks as we look at different central meridian zones. The fact that there are no such phase differences indicates that the whole zone from ± 0.8 in $\sin L$ changed its slope together. Thus the whole visible equatorial region accelerated and decelerated roughly in unison.

Note that the differences in average ω of the various curves in Figure 4 are caused by the fact that no correction is made for the limb redshift in this analysis – there is no folding of the data about the central meridian here – and the redshift affects the slopes of the curves differently in the east and west hemispheres. One may presume that the redshift does not vary with time and thus does not affect the shapes of the curves in Figure 4, only their constant vertical displacements.

Figure 4 refers to the latitude range $\pm 15°$. We examined the latitude dependence of the velocity variations by examining the latitude range 35° to 65° in both the north and south hemispheres. Figure 4 shows the rotation velocities determined from the slopes in various central meridian zones in these latitude ranges. Although there are some minor differences between these curves and those in Figure 4, the overall

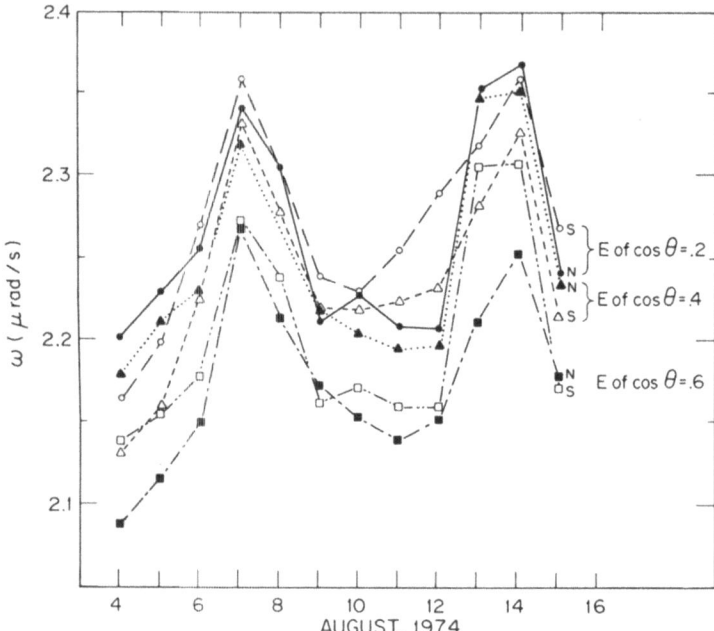

Fig. 5.   The variation with time of the slopes of the velocity-sin curves determined at the various zones of different central meridian distances and in the latitude zone 35° to 65°, north or south as indicated.

appearance of the two diagrams is quite similar; the peaks appear at the same times.

The tentative conclusions to be drawn from Figures 4 and 5 are, first, that the short-period variations in rotational velocity seen during this period resulted mainly from axisymmetric accelerations and decelerations of the photospheric gas, and effects from a global-scale velocity pattern are small compared to that of the axisymmetric variations, if present at all; and, second, that these accelerations and decelerations extended over a wide range of latitude and longitude. Further study of this phenomenon is planned.

An alternative explanation, of course, is that the axisymmetric variations are due to some instrumental effect. That is, that there may be some problem with calibration, instrument alignment, or some other factor not originating on the Sun, which causes the variations seen in Figures 4 and 5. At the moment a nonsolar cause for the velocity variations cannot be ruled out. Some tests have been made in an attempt to isolate an instrumental origin for the variations, but no means of reproducing the magnitude of the effect has been found. Altering the balance on the photomultiplier tubes at the exit slits so as to offset the line profile can change the derived rotation velocity, but to achieve the magnitude of the observed effect requires an unreasonably large offset – so large that at the extrema it is difficult to hold the line with the line-centering servo. Similarly, decollimating the telescope will alter the resulting rotation velocity, but to achieve the observed effect requires an unreasonably large decentering. Variations in the spectrograph dispersion could account for the effect, but the dispersion is measured daily as a part of the calibration procedure, and the day-to-day variations in this quantity are rarely more than a fraction of a percent. The measured dispersion is used in velocity determinations for each day's observations. The dispersion enters directly into the calculation for the velocity in km s^{-1}, so an error of a fraction of a percent in the dispersion will result in an error of the same percentage in the relative velocities, and hence of the same percentage in the slope of the $V_m - \sin L$ curve. For any of these instrumental effects, even if the magnitude of the velocity change could reasonably be accounted for, it is difficult to conceive how such a regular pattern of variations could result. Also, the fact that the earlier observers noted variations quite similar to those seen in Figures 4 and 5 is an argument in favor of their solar origin. Nevertheless, until more analysis has been done we cannot completely rule out some nonsolar explanation for these velocity variations.

5. A New Method to Analyze Global-Scale Surface Velocity Fields: the Non-axisymmetric Velocity Field

The main purpose of the slope analysis is to derive the differential rotation from one day's observation. However, the differential rotation thus obtained is not necessarily the true differential rotation, which is by definition the axisymmetric part of the longitudinal component of the velocity field of the Sun. We observe on one day less than one half the circumference of the Sun, and the true differential rotation should be derived from data around the whole circumference at one instant. Thus the effects of a nonaxisymmetric velocity field of a scale of the order of the solar diameter, the largest-scale non-axisymmetric velocity field, can alter the appearance of the differential rotation derived from one day's observation.

If the differential rotation does not vary on a time scale of a rotation or less, then we can separate the differential rotation from the largest-scale non-axisymmetric field. This could be done by constructing a velocity field synoptic chart by subtracting the average rotational velocity at each latitude from each day's results. Some examples of such synoptic charts will be presented below; however, if the differential rotation does vary with a time scale smaller than one month, then one cannot derive the true rotation without observing the Sun simultaneously from several directions in interplanetary space.

If the lifetime of the largest-scale non-axisymmetric field is also longer than one rotation, then it is possible to separate in the analysis the differential rotation from this large-scale non-axisymmetric field. In the pattern of the non-axisymmetric field there must, by definition, be boundaries in the velocity distribution, which at some times will lie near the central meridian. At these times, the slopes derived from the eastern and western sections of the solar disk should be different because in these two areas the direction of the field is opposite. Thus, by analyzing the differences of the slopes of the eastern and western data, we can examine the largest-scale non-axisymmetric velocity field, even though the axisymmetric differential rotation may vary. This approach is complicated somewhat by the effects of the limb redshift, which will increase the derived slope on one side of the central meridian and decrease it on the other side. In order to analyze the data in this manner, the instrumental effects that may affect the velocity signal must be minimized. This problem will be discussed again below.

The analysis presented in this paper has not proved decisive in sorting out the axisymmetric and the largest-scale non-axisymmetric components of the global-scale velocity field. This is due partly to the remaining doubt about the possible influence of unknown instrumental effects and partly to the small data sample analyzed here. A more comprehensive analysis is planned for the near future, which should enable us to isolate the variation of the differential rotation from the effect of the largest-scale non-axisymmetric velocity field. For the moment we can say only that the effect of this non-axisymmetric field, if it is present at all, is small in amplitude compared with the axisymmetric variations in the results of the slope analysis.

Such difficulties affect only our analysis of the differential rotation and the largest-scale non-axisymmetric velocity field. We are able to analyze the non-axisymmetric fields with scale smaller than the solar diameter, using data from one day's observation. That is, by subtracting from the velocity data the differential rotation derived from the slope analysis and the limb redshift, it is possible to obtain a non-axisymmetric velocity-field pattern with a scale smaller than the solar diameter but larger than supergranules, which manifests itself as 'residual velocities' in this procedure. The limb redshift correction presents us with some difficulties, however. If it is constant in time, which seems a reasonable assumption, then we can obtain a value of the limb redshift as a function of central meridian distance by averaging the data of many days. However, if it is not constant in time, then we cannot determine whether the non-axisymmetric velocity field is due to a real velocity field or to whatever effect causes the limb redshift. If these two effects vary with different time scales, it still might be possible to separate them.

At any rate, assuming the limb redshift to be a constant effect, we can analyze the non-axisymmetric velocity field of scale smaller than a solar diameter. There are two ways to display this field from the data we have. One is to plot the residual velocities (after subtracting the differential rotation and the limb redshift) for one day's observation. This method has been used by Howard (1971) and Hendl (1974). The fact that the large-scale fields are horizontal somewhat limits the usefulness of this approach, because one views a different component of the large-scale motions in different parts of the solar disk. The second method is to construct a synoptic chart of the non-axisymmetric velocities at some central meridian distance zone – of the east or west limb of the central meridian – for each day's observation. Such a synoptic chart from the limb data shows mainly the longitudinal component of the non-axisymmetric field, and a chart obtained from central meridian data shows the latitudinal component of the field.

A plot showing synoptic charts from limb data is shown in Figure 6. With data as complicated as these velocity measurements, one must be very careful that instrumental effects are not present; however, in the case of the analysis of these non-axisymmetric velocity fields with scales smaller than that of the diameter of the Sun, we feel confident that they represent the true velocity field on the Sun, since the subtraction is done from data of one day and at least the questions of variation of calibration day by day does not affect the results very much.

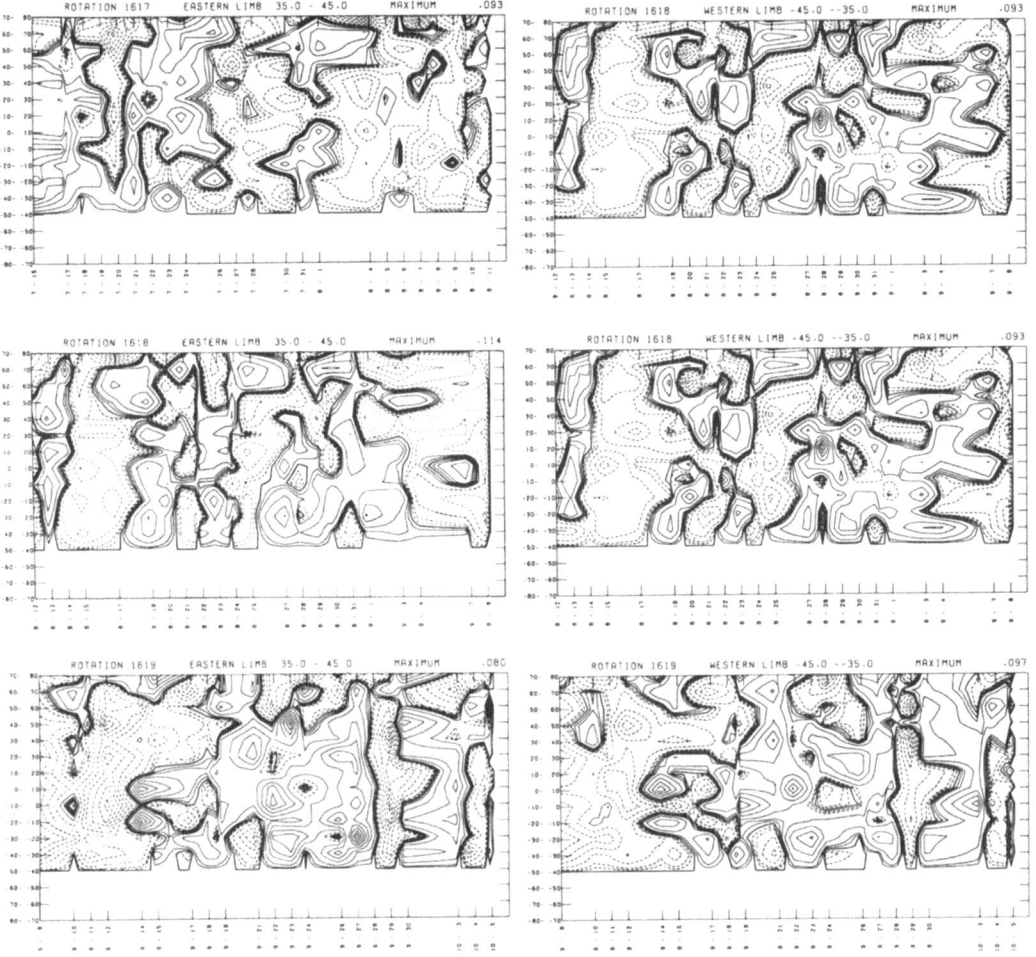

Fig. 6. Synoptic chart representations of the non-axisymmetric velocity fields with scales smaller than the solar diameter but larger than the scale of supergranules. The synoptic charts are for rotation numbers 1617, 1618, and 1619 and for eastern and western longitude zones of $\pm 35°$ to $\pm 45°$. The dates of observation are designated at the bottom of each Figure while the latitudes are shown on the left-hand side. In this case $10° \times 10°$ square averaging in longitude and latitude was done to cancel the small scale velocity noise due to such effects as the 5 min oscillation, supergranulation, and granulation. The zero-reference point was adopted as that of the central meridian zone from $-15°$ to $15°$ in longitude. The data southwards of $-50°$ in latitude were omitted because in this season, the Sun tilts northwards and south polar regions were not observable. The effect of the limb redshift was corrected after subtracting the averaged velocity in longitude. Note that there are coherent features elongated in latitude and that these features remained for 3 rotations although the shape of the features was affected. Note also that the data points were obtained after subtracting the effect of rotation from the daily observation and hence the variation of the rotational results shown in Figure 7 did not affect the results of the feature. The positive feature in the eastern (western) part shows the region rotation faster (slower) than the averaged rotation derived from the one day observation, and vice versa. These features are presumed to be real features of the global-scale non-axisymmetric velocity field which has been predicted by theories. Especially its elongated feature confirms that the mode expressed by the sectorial associated legendre function P_n^n dominates, although several other modes seem to exist.

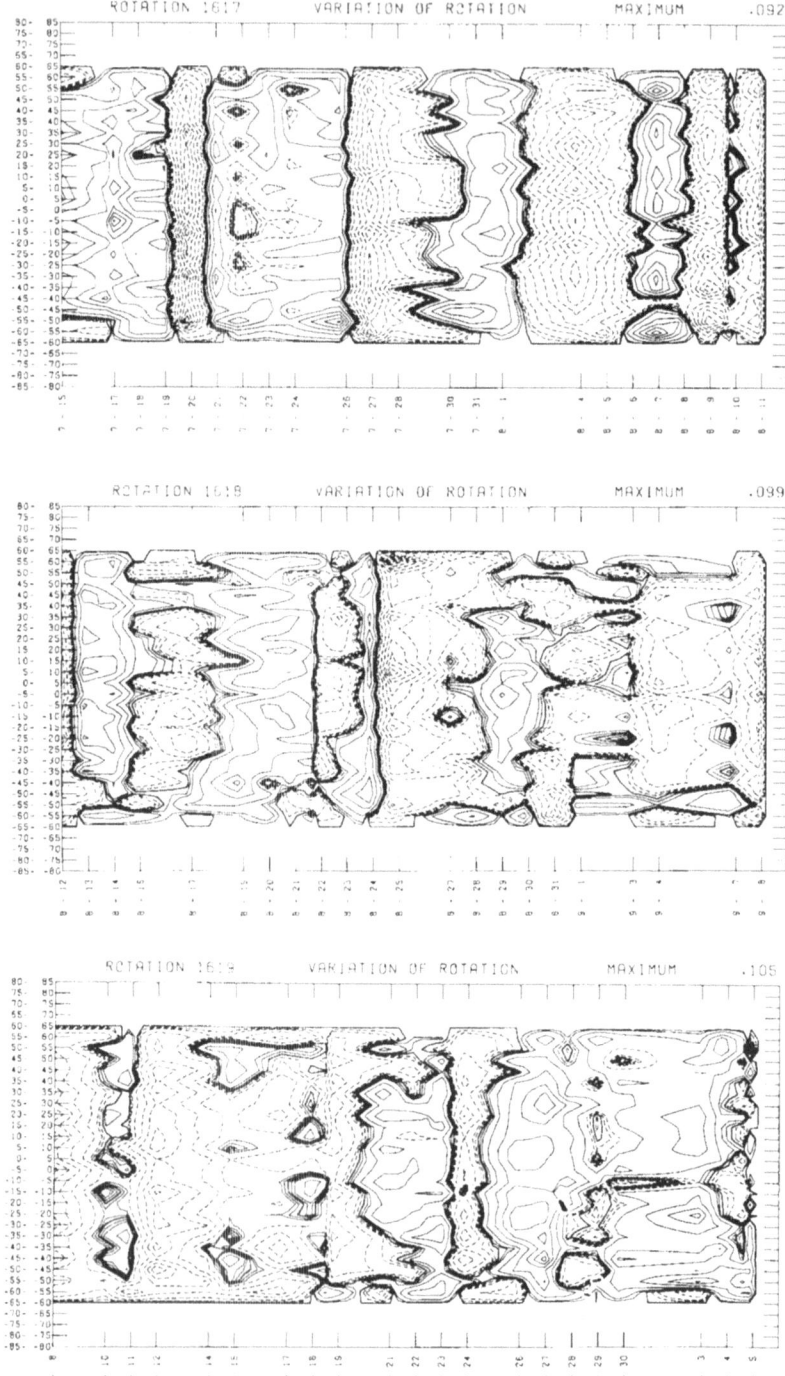

Fig. 7. Synoptic charts of the variations of differential rotation for the same period as Figure 6. However, the rotation is derived for each 5° latitude zone. The charts show the deviation from the average rotation of each rotation number. The data poleward of 60° were omitted because the results there are less well determined. Note that the variations occur in unison over the whole latitude range and that the feature does not last for 3 rotations. This means that such variations do not represent a true feature of the non-axisymmetric velocity field and cast some doubts on the reality of this variation of the rotation.

We see in Figure 6 that there are regions where the horizontal velocity differs significantly from the rotational velocity. If there were no systematic non-axisymmetric velocity field larger than the supergranules, then these synoptic charts would consist of only weak random patterns. However, it is evident from Figure 6 that there are velocity features that are coherent in space (latitude) and time (longitude), and that some of the features persist over at least three rotations. The dimensions of these features are about 30°, or 200 000 km. If these features represent more than one mode of the velocity field that has been studied theoretically, then their dynamics could be quite complicated, since different modes have different propagation velocities.

Figure 7 shows synoptic charts of the variation of solar rotation as derived by the slope analysis. These charts show the deviation from the average rotation rate for each Carrington rotation. It is clear that variations occur often in unison over a wide range of latitude, and that such 'features' do not persist for even 3 rotations. Such behavior raises serious doubts about the reality of these velocity changes, as mentioned above. Such changes are not due to a global-scale non-axisymmetric velocity field, and may not originate on the Sun. A further search for instrumental effects will be pursued. Note that these features should be detected as a correlation of a, b, and c, in the old analysis of the northern and southern hemispheres. However, as explained above, these correlations do not necessarily reflect the global-scale non-axisymmetric velocity field.

We may conclude from Figure 6 that a large-scale surface velocity field exists on a scale larger than the supergranules. Such a field has been predicted to exist deep inside the convection zone from the work of various investigators as mentioned above. The preliminary results described here show that this non-axisymmetric field also appears (overshoots) on the surface. This is encouraging because we have a means of studying what is going on inside the convection zone by examining the surface dynamics of the Sun. The observational study of the large-scale dynamics of the solar photosphere will be continued with the Mount Wilson data. It will be necessary to study the data of many years because the time scale of the phenomenon is long.

This research was supported in part by a contract (N00014-66-C-0239) from the U.S. Office of Naval Research and a grant (NGR 09-140-015) from the U.S. National Aeronautics and Space Administration.

References

Babcock, H. W.: 1961, *Astrophys. J.* **133**, 572.
Babcock, H. W. and Babcock, H. D.: 1955, *Astrophys. J.* **121**, 349.
Becker, U.: 1955, *Z. Astrophys.* **37**, 47.
Brunner, W.: 1930, *Astron. Mitt. Zürich* **124**, 67.
Brunner-Hagger, W.: 1944, *Publ. Eidgen. Sternwarte Zürich* **8**, 31.
Bumba, V. and Howard, R.: 1965, *Astrophys. J.* **141**, 1502.
Busse, F. H.: 1970, *Astrophys. J.* **159**, 629.
Cowling, T. G.: 1933, *Monthly Notices Roy. Astron. Soc.* **94**, 39.
Cowling, T. G.: 1953, *The Sun* (ed. by G. P. Kuiper), University of Chicago Press, p. 575.
De Lury, R. E.: 1939, *J. Roy. Astron. Soc. Can.* **33**, 345.

Deubner, F.-L., Vasquez, M., and Schröter, E. H.: 1975, *Solar Phys.* (in press).
Dodson-Prince, H. W. and Hedeman, E. R.: 1968, in K. O. Kiepenheuer (ed.), *IAU Symp.* **35**, 56.
Durney, B. and Skumanich, A.: 1968, *Astrophys. J.* **152**, 255.
Durney, B.: 1968a, *J. Atmospheric Sci.* **25**, 372.
Durney, B.: 1968b, *J. Atmospheric Sci.* **25**, 771.
Durney, B.: 1970, *Astrophys. J.* **161**, 1115.
Eigenson, M. S., Gnevyshev, M. N., Ohl, A. I., and Rubashev, B. M.: 1948, *Solnechnaya Aktivnost i Yeyo Zemnyye Proyavleniya* (Moscow).
Evershed, J.: 1931, *Monthly Notices Roy. Astron. Soc.* **92**, 105.
Gilman, P. A.: 1969a, *Solar Phys.* **8**, 316.
Gilman, P. A.: 1969b, *Solar Phys.* **9**, 3.
Grotrian, W. and Künzel, H.: 1950, *Z. Astrophys.* **28**, 28.
Halm, J.: 1904, *Trans. Roy. Soc. Edinb.* **41**, Part 1, No. 5.
Hart, A. B.: 1954, *Monthly Notices Roy. Astron. Soc.* **114**, 17.
Hendl, R. G.: 1974, *Thesis*, Massachussetts Inst. of Tech.
Howard, R.: 1971, *Solar Phys.* **16**, 21.
Howard, R., Bumba, V., and Smith, S. F.: 1967, Carnegie Inst. of Washington, Publ. No. 626, Washington, D.C.
Howard, R. and Harvey, J.: 1970, *Solar Phys.* **12**, 23.
Kato, S.: 1969, *Astrophys. J.* **157**, 827.
Kato, S. and Nakagawa, Y.: 1969, *Solar Phys.* **10**, 476.
Leighton, R. B.: 1969, *Astrophys. J.* **156**, 1.
Losh, H. M.: 1939, *Publ. Observ. Univ. Michigan* **7**, 127.
Martres, M. T.: 1970, *Solar Phys.* **11**, 258.
Maunder, A. S. D.: 1907, *Monthly Notices Roy. Astron. Soc.* **67**, 451.
Minnaert, M.: 1946, *Monthly Notices Roy. Astron. Soc.* **106**, 98.
Parker, 1955
Plaskett, H. H.: 1916, *Astrophys. J.* **43**, 145.
Plaskett, J. S.: 1912, *Astrophys. J.* **42**, 373.
Sawyer, C. and Haurwitz, M. W.: 1972, *Solar Phys.* **23**, 429.
Steenbeck and Krause, 1969.
Svestka, Z.: 1968a, *Solar Phys.* **4**, 18.
Svestka, Z.: 1968b, *IAU Symp.* **35**, 287.
Ward, F.: 1964, *Pure Appl. Geophys.* **58**, 157.
Ward, F.: 1965, *Astrophys. J.* **141**, 534.
Warwick, C. S.: 1965, *Astrophys. J.* **141**, 500.
Weart, S. R.: 1970, *Astrophys. J.* **162**, 987.
Weart, S. R.: 1972, *Astrophys. J.* **177**, 271.
Wilcox, J. M.: 1968, *Space Sci. Rev.* **8**, 258.
Wolfer, A.: 1897, *Publ. Sternw. Eidgen. Polytechnikums Zürich* **1**, XII.
Yoshimura, H.: 1971, *Solar Phys.* **18**, 417.
Yoshimura, H.: 1972a, *Solar Phys.* **22**, 20.
Yoshimura, H.: 1972b, *Astrophys. J.* **178**, 863.
Yoshimura, H.: 1973, *Solar Phys.* **33**, 131.
Yoshimura, H.: 1974, *Publ. Astron. Soc. Japan* **26**, 9.
Yoshimura, H.: 1975a, *Astrophys. J. Suppl. Series* **29**, 467.
Yoshimura, H.: 1975b *Solar Physis.* (in press).
Yoshimura, H. and Kato, S.: 1971, *Publ. Astron. Soc. Japan* **23**, 57.

DISCUSSION

Gilman: (1) What is the magnitude of the residual velocities? (2) Can you see any correlations between north and south hemispheres in either the variations of differential rotation with time, or in residual velocities?

Howard: (1) The maximum residual velocity in each diagram is about 0.1 km s^{-1}, and the contour lines are some tens of meters per second. (2) The rotation variations are symmetric about the equator, the residual velocity variations appear not to show this symmetry.

Deubner: Did you also compare the position in latitude of irregularities of tne differential rotation with zones or position of centers of activity? It appears from recent results of Vazques and myself, that the latitude of both these features is strongly correlated. It is of interest in this context to note that in measurements of differential rotation obtained during the period of June this year when no spots were present at all, the irregularities mentioned before were much less conspicuous.

Howard: We found no such correlation in this interval.

Kuklin: (1) I had at my disposal an old series of your observed velocities published in *Solar Physics* in 1970. It seems to me that the increase of rotation velocity is connected with the appearance of large sunspot groups eastward from the central meridian, and conversely the decrease of rotation velocity is connected with the appearance of large sunspot groups westward from the central meridian. Have you found such effects? (2) Can you explain why at your slide the rotation velocity (2.3) is 20% less than the standard one (2.8)?

Howard: I have examined only the data shown in these slides, which represent a short interval. For them I found no such effects. (2) These data represent high latitudes where the angular rotation rate is lower.

Giovanelli: The problem of interpreting velocities from line profiles is well known – under some conditions, one can even get velocities of the wrong sign! Can you be confident for such small velocities as reported there may not be errors introduced by variations in conditions in the line-forming large areas of the surface, e.g., plages distribution of magnetic points across quiet areas which may affect the low chromosphere or upper photosphere!

Howard: I agree that this is a serious worry. In the past we have examined this possibility by observing with $g = 0$ lines or lines that are not weakened in magnetic filaments. Such observations are quite similar to the $\lambda 5250$ observations.

DIFFERENTIAL ROTATION AND GIANT CELL
CIRCULATION OF THE SOLAR Ca⁺-NETWORK

E. H. SCHRÖTER and H. WÖHL

University Observatory Göttingen, West Germany

High precision computer controlled tracings of bright Ca⁺-mottles were performed during 1974 and 1975 at the Locarno Observatory of Göttingen to study solar differential rotation and to search for giant cell circulation pattern. Details of the observing method and the results from observations during 1974 have been published very recently (Schröter and Wöhl, 1975). Our method consists of measuring the position of 5–15 bright Ca⁺-mottles with respect to the center of the solar disc

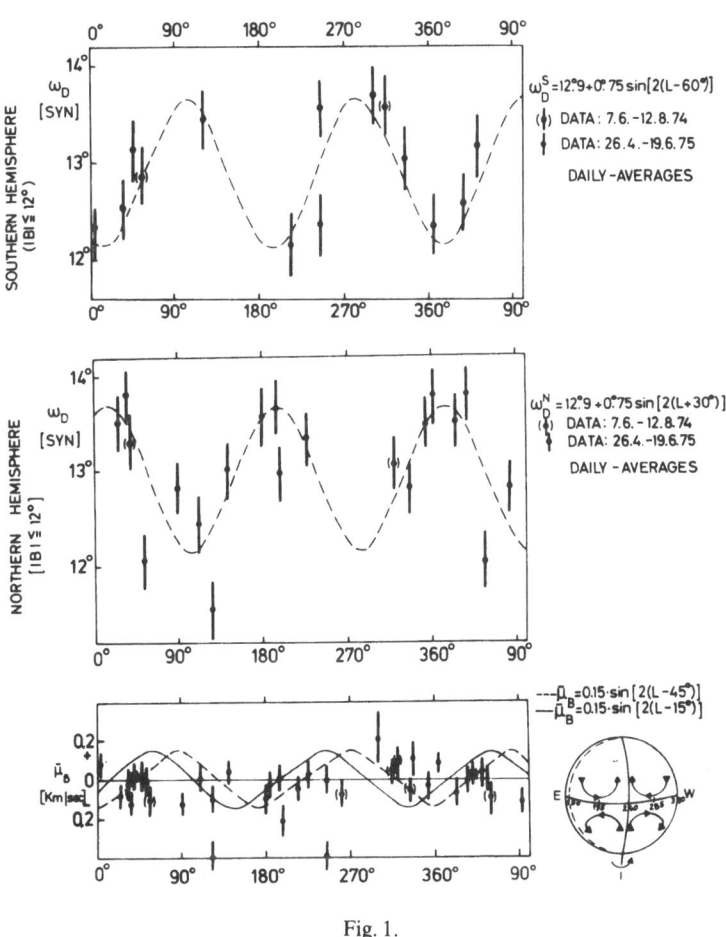

Fig. 1.

Bumba and Kleczek (eds.), Basic Mechanisms of Solar Activity, 37–40. All Rights Reserved.

every 10 to 15 min during 4 h every day. A computer controlled program determines first the center of the disc and immediately afterwards the right ascension and declination of a Ca^+-mottle which was centered by the observer into a 1″-diameter spectrograph pinhole by watching a TV-monitor displaying a Ca^+-K-line solar image. From a linear least squares fit of the observed positions the solar latitude and longitude were computed for the beginning and the end of the daily 4 h observation period. From this the components in latitude and longitude of the proper motions were derived which result from the differential rotation, possible giant cell circulation and the small scale random walk of these features.

The main goal of the 1975 observing program was to search for a sector-like structure of the solar rotation within the equator belt as a result of giant cell circulation pattern. We traced 341 Ca^+-mottles in latitudes $\leq 12°$ from 1975, April 26 to June 19. The result of these observations are shown in Figure 1, where the synodic rotation angle per day (ω_D) is plotted as a function of the solar longitude for the northern and southern hemisphere separately. The points represent the average longitudinal motion component from the 5–15 bright mottles traced during one day. The length of the bars corresponds to the mean of the daily rms deviation. These rms deviations do not present the accuracy of measurements which is much better, but reflect mainly the random walk of the Ca^+-mottles superposed on the drift, due to rotation. Clearly, a straight line representation of the data, resulting in a longitude independent rotation rate, is rather unsatisfactory. A much better fit is obtained by a sinoidal-curve leading to two sectors with faster than average rotation and two sectors with slower than average rotation for each hemisphere. The deviation of $0.75\ day^{-1}$ corresponds to $\Delta v = 0.105\ km\ s^{-1}$. There is a 90° phase difference between the sectors of the northern and southern hemisphere. These results suggest

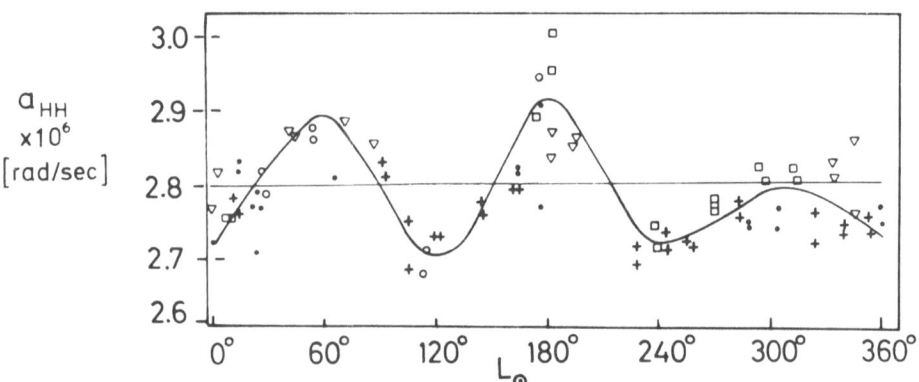

SECTOR – STRUCTURE in SOLAR ROTATION ?
TIME: 2.0369 – 2.3591 in H.H. time scale (4.5 rotations)

Howard and Harvey parameter $\underline{a}$ as function of longitude

Fig. 2.

very much a giant cell circulation pattern as indicated at the bottom of Figure 1. This coarse model may well be tested by the behavior of the latitudinal component of the proper motions. In longitudes $L = 60°$ and $240°$ a poleward (equatorward) drift is to be expected in the northern (southern) hemisphere whereas drifts with opposite directions should occur at longitudes $L = 150°$ and $330°$. We, therefore, converted the sign of the latitudinal motion component μ_B in the southern hemisphere and plotted μ_B from the northern and southern hemisphere as a function of the longitude ($\mu_B > 0$ means a poleward drift, units: km s^{-1}). The full-line curve corresponds to the circulation pattern model as indicated at the right, the dashed-line curve is a somewhat better fit of the data. However, both representations are not very convincing. We have to wait for the final reduction using more data (partly already obtained but not yet reduced) to check whether such a simple model of a giant cell circulation pattern is capable of representing well all observations.

Being aware of the fact that such a pattern, if existing, should also have been observed by previous measurements, we performed some sample probe compilations with the data published by Howard and Harvey (1970). We found several time periods in their Table II where the speed of the equator rotation clearly varied periodically for 2–4 solar rotations. For the sake of space we present here only one example. Figure 2 shows the equatorial rotation speed a_{HH} (units: radians per second) from Table II of Howard's and Harvey's paper as function of the solar latitude for the time interval 1967, January 14 to April 15. The different symbols represent successive solar rotations. This time, a 6-sector structure of roughly 60° extension in longitude is indicated with no phase difference between the northern and southern hemisphere. The amplitude of deviations is of the same order of magnitude as for our observations. However, Howard and Harvey reduced their data in a fashion being rather unsuitable for the detection of such sector structures, since full disc (or full hemisphere) observations were used to derive the rotation rate of the equator zone. Hence, the fluctuation of a_{HH} in Figure 2 is a kind of a net effect resulting from an uncomplete smoothing of the sector structure by the reduction procedure of Howard and Harvey.

After having reduced in final form all available data we shall submit a more detailed presentation to *Solar Physics*.

Acknowledgement

We thank Mr D. Soltau for the help in numerical reductions. The Locarno Solar Observatory is operated by the Deutsche Forschungsgemeinschaft.

References

Howard, R. and Harvey, J.: 1970, *Solar Phys.* **12**, 23.
Schröter, E. H. and Wöhl, H.: 1975, *Solar Phys.* **42**, 3.

DISCUSSION

Stenflo: There seems to be some contradiction between your results and those presented by Howard in his review, which showed coherent, simultaneous changes in the rotation velocity in the northern and southern hemispheres. Would you like to comment on this?

Schröter: I forgot to mention this point. Yes, what puzzles me, is the fact that during our observational period 1975, April 24–June 19, the large scale circulation cells of the northern hemisphere showed a circulation sense opposite to that of the southern hemisphere, whereas in those periods I picked out from Howard's and Harvey's Table II, apparently the cells in both hemispheres showed the same sense of circulation. At the moment I can only guess the reasons for this discrepancy: (a) Howard's and Harvey's data we used, have already been reduced according to a certain procedure and it is very hard to overlook to what extent such a large scale circulation pattern has been smoothed out by this procedure. One should look into the original data and reduce them in a more problem-oriented way. (b) Our observations cover a single period of 2.5 rotations only. We cannot exclude that such an antisymmetric mode of large scale circulation is not a common but a rare phenomenon occurring from time to time only. We are continuing our programme and shall be able to answer this question more definitely at the end of this year.

Howard: We plan to rereduce all the older data with the new reduction technique described in our previous paper. I would like to point out that for the two days from last summer for which we showed results in the previous paper, the southern hemisphere rotated slightly faster than the northern hemisphere. These data were from last summer (1974).

Schröter: We found in the period 1974, June–August 12, the southern equator belt to rotate faster than the northern belt in agreement with your findings. However, in comparing both results one should not forget, that our results refer to Ca^+-chromospheric layers.

Gilman: (1) Please draw your postulated circulation pattern on the board.

(2) *Comment:* This is the opposite of the dominant symmetry suggested by non-axisymmetric convection theory.

Schröter: To (1): I did so. (2) Since our observations cover a single period of 2.5 solar rotation only this type of large scale circulation may not necessarily be a common, but rather a rare phenomenon. But there is no doubt that the opposite sense of circulation in both hemispheres is far out of observational errors.

ROTATIONAL CHARACTERISTICS OF CORONAL HOLES

WILLIAM J. WAGNER

Sacramento Peak Observatory, AFCRL, Sunspot, N.M. 88349, U.S.A.

Abstract. From May 1972 to October 1973, daily measures were obtained of EUV coronal hole areas appearing at the central meridian. Autocorrelations of these coronal hole area time series provide synodic rotation periods which indicate an almost rigid rotation by such features for lag lengths as short as one rotation. The rotation periods of coronal holes at high latitudes best compare with inferred interplanetary field rotation periods.

In view of the complexity of the rotational characteristics of solar phenomena and because of the possible link between coronal holes and large-scale magnetic field structure, it seems important to determine the rotational properties of coronal holes.

Daily spectroheliograms were obtained in λ 284 of Fe XV by R. D. Chapman, W. M. Neupert, and R. J. Thomas of the Goddard Space Flight Center using OSO 7. These appear as intensity isophote maps in *Solar-Geophysical Data*, issued by the Environmental Data Service of NOAA, and serve as the basic data used in this work. I have arbitrarily chosen an EUV count level of less than 10 as a brightness level which defines a coronal hole. This threshold was allowed to drop to 5 or rise to 20 depending on the ratio of total EUV counts to 10-cm radio flux for each particular day (Chapman and Neupert, 1974). No figure of merit was applied for the 'depth' of the coronal hole: rather, a binary division of the spectroheliogram was made into either hole or non-hole areas.

Seven latitude zones were studied: 80°–60° N, 60°–40° N, 40°–20° N, 20° N–20° S, 20°–40° S, 40°–60° S, 60°–80° S. Figure 1a shows one typical λ 284 isophote map

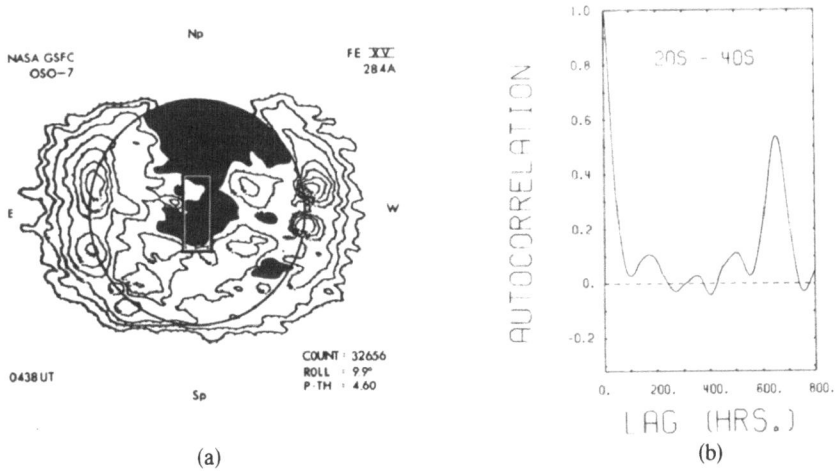

(a) (b)

Figs. 1a–b. The data from which rotation periods were derived appears in *Solar-Geophysical Data of NOAA*. (a) On daily OSO-7 isophote maps, coronal hole regions were identified (dark shading). Apertures of 14° width in each latitude zone were used in measuring coronal hole area at the central meridian. (b) The autocorrelation function for the 20°–40° S coronal hole area series. The location of the peak near 650 h marks the synodic rotation period. Isophote map from World Data Center A, NOAA.

Bumba and Kleczek (eds.), Basic Mechanisms of Solar Activity, 41–43. All Rights Reserved.

with an imaginary aperture one rotation day (14°) in width straddling the central meridian. The coronal hole area in each of the seven latitude zones was recorded with these apertures each day from 1 May 1972, through 31 October 1973.

Using such time series of coronal hole area, autocorrelation functions were generated for each of the seven latitude zones, as shown in Figure 1b. The major peak near a lag of 650 h marks the recurrence of those coronal hole features which persisted for at least one synodic rotation period. The zonal synodic rotation periods thus derived are shown in Figure 2. The error bars represent standard deviation from the mean of similar rotation periods obtained by subdividing the 18-month time series into four data subsets of 4.5 months each.

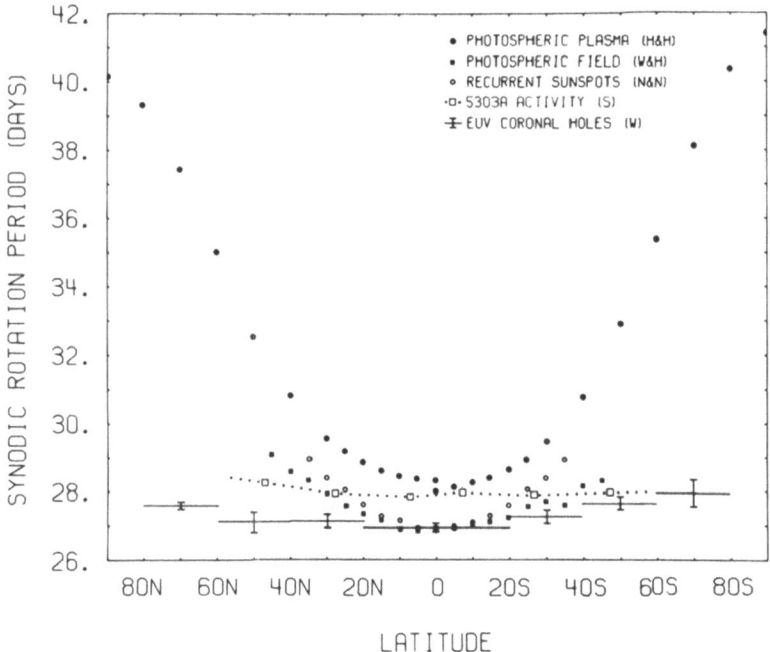

Fig. 2.   EUV coronal hole rotation rates from the present work. Low-emissivity coronal holes and active green line coronal structures (Sykora, 1971) perform almost rigid rotation when contrasted to: photospheric plasma (Howard and Harvey, 1970); photospheric magnetic fields (Wilcox and Howard, 1970); and the classical differential rotation (Newton and Nunn, 1951) of recurrent sunspots (with alterations from Wagner, 1975).

In Figure 2, it may be seen that coronal holes do not participate in the differential rotation exhibited by sunspots (Newton and Nunn, 1951), photospheric magnetic fields (Wilcox and Howard, 1970), or photospheric plasma (Howard and Harvey, 1970). While equatorial rotation periods are generally comparable, at higher latitudes considerable slippage of fields and plasma occurs with respect to the coronal holes. This near-rigid rotation of these low-emissivity coronal hole features is quite similar to that reported earlier by Sykora (1971) for bright coronal green line structure up to latitudes ±57°.5. I find that, for the corona in 1972–1973, the northern hemisphere rotated marginally faster than the southern, contrary to the 1947–1968 results of Sykora (1971).

Now, if bright coronal features correspond to centers of activity, then coronal holes may be said to represent inactivity. Analogously, during these 18 months, a rather consistent pattern was observed in these data of 'inactive longitudes'. This pattern was especially evident in the low ($\leq 40°$) latitude regions, and consisted of four orthogonal longitudes which were consistently favored by coronal holes.

Altschuler *et al.* (1972) have provided a description of coronal holes as regions of diverging magnetic field open to interplanetary space. From May 1972 to October 1973, the mean rotation period of the interplanetary field patterns (with field directions inferred by the method of Svalgaard, 1972) was somewhat greater than 28 days. If coronal holes are to be related to interplanetary fields at 1 AU, the results of the present work imply that the polar region coronal holes with 28-day rotation periods are more pertinent features than low latitude holes with 27-day periods.

In conclusion, coronal holes show almost rigid rotation. At low latitudes during this period, four orthogonal 'inactive longitudes' existed which showed coronal holes. Coronal hole rotation periods at high latitudes best compare with inferred interplanetary field rotation periods.

References

Altschuler, M. D., Trotter, D. E., and Orrall, F. Q.: 1972, *Solar Phys.* **26**, 354.
Chapman, R. D. and Neupert, W. M.: 1974, *J. Geophys. Res.* **79**, 4138.
Howard, R. and Harvey, J.: 1970, *Solar Phys.* **12**, 23.
Newton, H. W. and Nunn, M. L.: 1951, *Monthly Notices Roy. Astron. Soc.* **111**, 413.
Svalgaard, L.: 1972, *J. Geophys. Res.* **77**, 4027.
Sykora, J.: 1971, *Solar Phys.* **18**, 72.
Wagner, W. J.: 1975, *Astrophys. J. Letters* **198**, L141.
Wilcox, J. M. and Howard, R.: 1970, *Solar Phys.* **13**, 251.

DISCUSSION

Stix: Which latitude of the coronal hole exactly corresponds to the rotation of the interplanetary sector structure?

Wagner: I would have to say latitudes higher than about ±50 deg. Certainly not the equatorial or low latitudes.

Stenflo: The result that coronal holes appear to rotate rigidly agrees with the observation that the background magnetic fields have a rotation more similar to rigid rotation than fields in active regions. As magnetic fields reflect rotation rates in deeper layers, and as they rotate faster than the photospheric plasma, the angular velocity of rotation should *increase* with depth.

ON DIFFERENCES IN DIFFERENTIAL ROTATION

W. VAN TEND and C. ZWAAN

Astronomical Institute, Utrecht, The Netherlands

Summary. We investigated the reality of the differences in rotation rates found from various features. We only used published observations including almost forgotten work of more than half a century ago.

Here we summarize some conclusions which seem interesting in the present context. A more detailed account will be submitted to *Solar Physics* where also the references will be given.

(1) It is well known that larger rotation rates are found from sunspots and faculae than from photospheric Doppler shifts. However, the faculae and the rare spots occurring outside the activity belt indicate that the differences diminish for latitudes $|\phi| > 40°$ (see Figure 1).

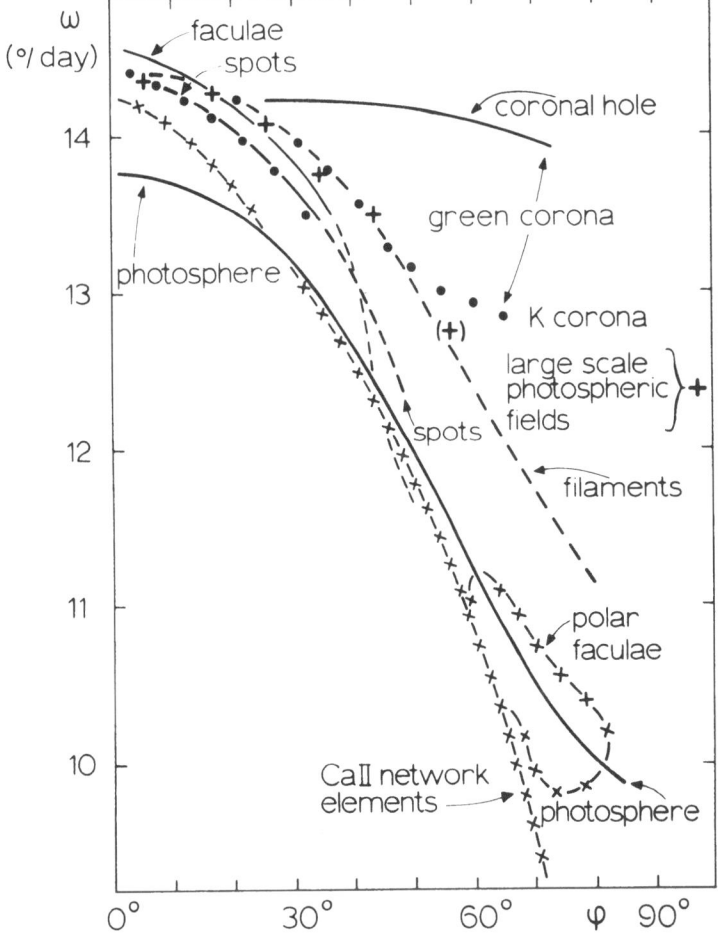

Fig. 1.   Mean sidereal rotation rates ω as a function of heliographic latitude ϕ.

Bumba and Kleczek (eds.), Basic Mechanisms of Solar Activity, 45–46. *All Rights Reserved.*
Copyright © 1976 by the IAU.

(2) There are no significant differences between the rotation rates determined from photospheric Doppler shifts, from white-light polar faculae, and from Ca II K 3 network elements outside active regions (Milošević), at least for latitudes $|\phi| > 20°$. In other words, magnetic flux tubes not (or, no longer?) belonging to active regions corotate with the photospheric plasma.

(3) There is no difference between the rotation rates derived from filaments and from autocorrelation analysis of long series of magnetograph data. This is not surprising since autocorrelation analysis yields mean rotation rates for large magnetic complexes of long duration (many months), whereas filaments mark the boundaries between magnetic complexes of opposite polarity.

(4) However, the intriguing result is that, at least for latitudes $|\phi| > 20°$, the large complexes rotate faster than all known elements which constitute those complexes: sunspots, faculae and network elements.

(5) Autocorrelation analyses of coronal features yield rotation rates larger than the rotation rates for large photospheric magnetic complexes. More specifically, the K corona and the short-lived condensations observed in the green line rotate about equally fast or slightly faster than the photospheric complexes whereas the very long-lived quiescent condensations rotate about as fast as the almost rigidly rotating coronal holes.

DISCUSSION

Howard: I wish to point out that you and other investigators have shown the results, that Harvey and I obtained some years ago in comparison with other rotation data. Often our results are shown as a solid line as if they were well established and invariant. In fact, they are neither. In particular the average rotation rate has been increasing in recent years, and is now early up to the sunspot rate.

Antonucci: At high altitudes the rotation period of the green corona appears to be dependent on the solar cycle phase, in the last two solar cycles. Before solar minimum the rotation period at high latitudes approaches the equatorial one (before 1954 the rotation period was exactly the same at 40°–60° and at 20°–0°), near activity maximum the differential rotation curves of the green corona and of the photospheric magnetic fields are in fairly good agreement. Therefore, the behaviour of the green corona could be better described by two different rotation curves corresponding to the period before solar minimum and to solar maximum.

Zwaan: I agree, your results are included in our graph.

LARGE-SCALE SOLAR MAGNETIC FIELDS

V. BUMBA

Astronomical Institute of the Czechoslovak Academy of Sciences,
Observatory Ondřejov, Czechoslovakia

Abstract. The characteristics of the large-scale distribution of the solar magnetic fields on the basis of a series of solar magnetic synoptic charts covering more than 15 years of observations are given. The major part of our information concerns the morphology and only some results deal with the kinematics of the field distribution. Results of averaged solar magnetic field fluxes and polarity reversal studies as well as of preliminary investigation of the very-low angular resolution magnetic measurements are given. The regular zonal and sectoral distribution of photospheric background fields, the different role or visibility of structures in both polarities is discussed. The reflection of both main types of the longitudinal distribution of large-scale solar background magnetic fields (the 27-day, the 28–29-day successions, the 'supergiant' structures) in the interplanetary magnetic field distribution is also considered.

1. Introduction

Existence of the solar magnetic fields outside sunspots has been demonstrated and discussed for the first time by H. W. and H. D. Babcock (1955). They showed the reality of a polar field and two types of the fields at lower heliographic latitudes: Bipolar Magnetic Regions (BMR's) and Unipolar Magnetic Regions (UMR's). Thanks to synoptic charts drawn from the daily magnetograms of the whole solar disk obtained regularly since August 1959 (Howard *et al.*, 1967) at the Mt. Wilson Observatory with a new magnetograph we have learned some more characteristics of the solar background magnetic fields distribution. The technique of the construction of the synoptic maps which fix the magnetic field situation around its central meridian passage once per solar rotation makes them suitable sources of our knowledge of the large-scale distribution of solar magnetic fields having appropriate long life-times.

One of the most important facts we have to pay attention to during the study of these large-scale fields is the angular resolution of the observational instrument. For example, the older Babcock's magnetograms made during 1953–1954 with an angular resolution of about $38'' \times 70''$, although they have not yet been fully evaluated, seem to demonstrate – at least in several up to now utilized cases – a different conception of the weak background magnetic field distribution from what we used to receive on the daily maps or synoptic charts published in the Mt. Wilson Atlas of Solar Magnetic Fields (Howard *et al.*, 1967) having the increased $23'' \times 23''$ resolution or even on today's magnetograms and charts obtained with the still increased angular resolution of about $12'' \times 10''$. With the growing resolution the large unipolar areas of BMR's or UMR's disintegrate into smaller mixed polarity islands and we are getting more and more confused by the complexity of the figure.

During the last decade a rich information concerning the distribution of the solar large-scale magnetic fields has been obtained from direct measurements, as well as from the distribution of the interplanetary magnetic fields reflecting the organization of the solar large-scale fields in a surprising manner, and from the investigation of various activity phenomena spacing on the solar surface.

Bumba and Kleczek (eds.), Basic Mechanisms of Solar Activity, 47–67. All Rights Reserved.
Copyright © 1976 by the IAU.

In this paper we would like to summarize the most characteristic features of the large-scale solar magnetic fields as they have been found during the recent few years. The large-scale in our case means that the dimensions of the investigated fields are greater than the scale of one active region.

2. Average Solar Magnetic Field Fluxes and Polarity Reversal

To understand better the rules following which the large-scale fields are distributed we may start with the problem of the Sun seen as a star. Howard (1974a, b, c) just recently examined the magnetic flux data, average magnetic field strengths and changes of the field polarity from the Mount Wilson magnetographic data over the time interval 1967–1973. He found that the fact that on the whole the activity in the northern solar hemisphere was greater than that in the south during the mentioned interval, is reflected in the 7% difference in total flux values between the two hemispheres in favour of the northern one. This N–S percentage difference is greater at the highest heliographic latitudes, where above 70° there is about one third more flux in the north than in the south. But in regard to the total amount of the solar magnetic flux, it seems, that about 95% of it is connected with latitudes below 40° in both hemispheres and that the flux above heliographic latitude 60° represents less than 2% of the total solar magnetic flux.

Activity-cycle-related increase in the total magnetic flux may be seen on both hemispheres and it is strongest in the activity zones but it may be seen at the higher latitudes as well. The negative flux is stronger than the positive magnetic flux at all latitudes around the time of solar maximum. This effect is proportionally strongest poleward of 40° where a 'wave' of the negative flux may be seen moving poleward.

It is known to observers and Howard (1974a, b) has demonstrated it again, that at times the fields appear, their polarity behaves in the same way over the entire solar disk or at least in a large portion of it. He brings two nice examples of it – the large positive 'spike' in the middle of 1972 caused by only a 'handful of days' observations with strong positive bias and the negative surge early in 1967. It is interesting that these biases in polarity of the solar magnetic field are reflected in biases of the same polarity in the interplanetary magnetic field (Wilcox, 1972). Thus the 'monopole' behaviour of the solar magnetic fields observed since the earliest magnetographic observations (H. W. and H. D. Babcock, 1955) seems to be a real effect as yet without explanation.

As in the polar field polarity reversal during the maximum activity of the previous solar cycle (Babcock, 1959), during the present cycle the south polar fields changed sign significantly earlier than did the north polar fields. But that reversal was a weaker one. The polar fields ($\varphi > 70°$) reversed in about September, 1969, and the 40–50° zone reversed in March, 1968. The most striking polarity reversal, that in the north, occurred approximately two years after the activity maximum. The north polar fields changed sign from weakly negative to rather strongly positive in August, 1971. This high-latitude reversal was accompanied by a rather general positive surge in field strengths over the whole solar disk. The fields at successively lower latitudes in the north changed sign earlier, as if there were a positive 'wave' starting in the

40–50° zone and reaching the pole in about one year. No evidence has been found for a related reversal in the equatorial zone. On the other hand Wilcox and Scherrer (1972) have shown from interplanetary field data inferred by Svalgaard (1972) from polar geomagnetic data that a change in phase of the predominant polarity of the interplanetary field occurs on the average 2.75 yr after activity maximum.

The average fields poleward of 40° in each hemisphere have over the whole period investigated by Howard (1974a) the sign of the follower spots, and the fields equatorward of 40° have the sign of the leader spots.

3. Large-Scale Distribution of Magnetic Fields on the Solar Disk Obtained with Very Low Angular Resolution

As a next step to make a transition from the magnetically wholly integrated Sun to its solar field distribution I would like to present as an example a few daily magnetic maps constructed from the Babcocks' low resolution (38″ × 70″) magnetograms obtained during the period of minimum activity in 1953. One of the advantages of those magnetograms seems to me the fact that they make it possible to estimate the boundary between the opposite polarities. Thus they visualized with a high degree of integration the distribution of very weak, highly integrated background magnetic fields.

The pictures available up to now display the three typical situations of the field distribution we may observe on such maps:

(a) Practically each solar hemisphere is occupied by one polarity field (the northern hemisphere by the positive polarity field, the southern hemisphere by the negative one) (Figure 1a).

(b) The leading magnetic fields of active regions are formed as gulfs of an opposite polarity reaching the given hemisphere from the opposite hemisphere. In other words, during 1953 the active regions developed in the northern hemisphere from the declining 18th cycle of activity have the polarity of their following portions the same as the background magnetic field of this hemisphere, that is the positive polarity. The leading parts of these active regions have the negative polarity and are formed from gulfs of this polarity reaching the place at the northern hemisphere from the other side of the equator as a prolongation of the southern negative polarity background field. The same is true but vice versa about the active regions on the southern hemisphere (Figure 1b, c).

(c) In this still more complicated situation the northern polar fields are positive, the southern polar fields are negative, but in activity zones the polarities of the background fields are reversed in such a way that the northern activity zone is negative and the southern one is positive. This is practically the same situation as mentioned above when speaking about Howard's results for 1967–1973: the Sun has four parallel zones with alternating polarities, two of them are positive and two of them are negative (Figure 1d). In heliographic longitudes with active regions the situation is, of course, much more complicated.

I do not mean to say that the mentioned three types of situations are the three evolutionary stages of the activity development. But I would like to underline the

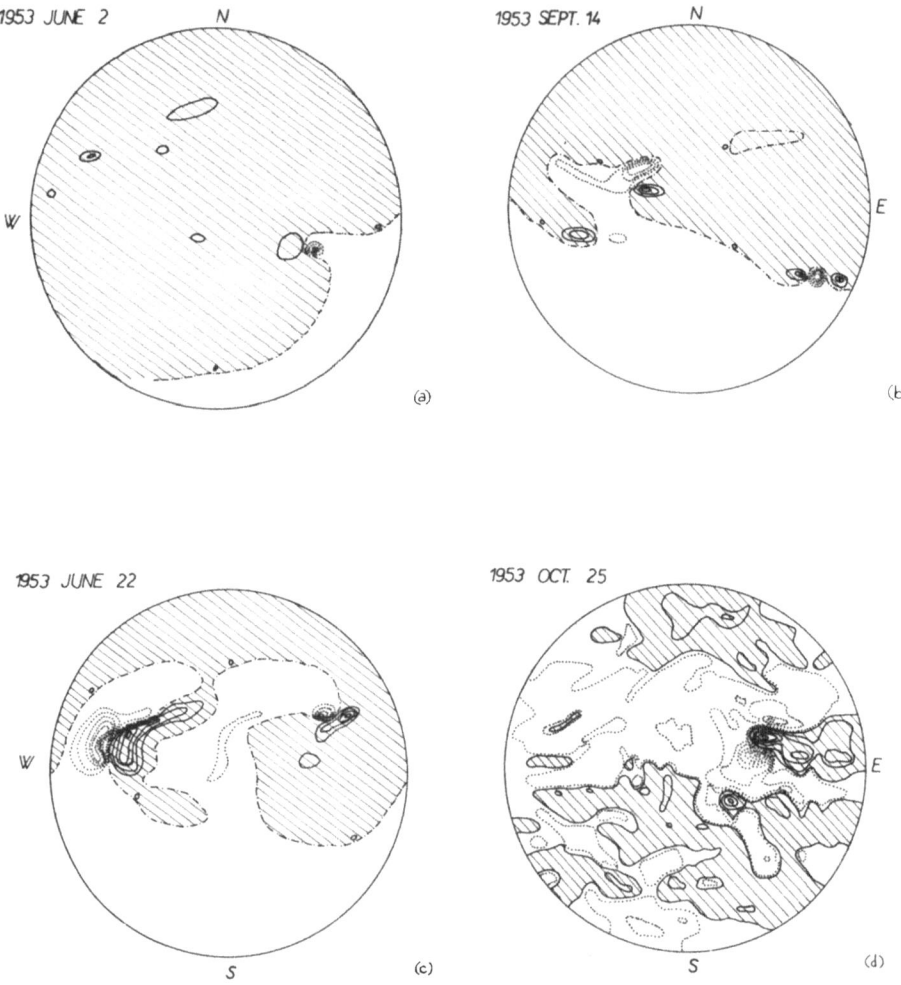

Fig. 1.   Four daily magnetic maps constructed from the Babcocks' low resolution (38″ × 70″) magneto-grams obtained during the period of minimum activity in 1953. The distribution of highly integrated background photospheric magnetic fields is visualized. (Negative polarity is indicated by dotted lines and shadowed areas, the positive one by full lines and parallel hatching.)

(a) 1953, June 2; one of the most simple situations.

(b) 1953, Sept. 14; one gulf of leading polarity in the northern hemisphere has been formed from the negative polarity main field of the southern hemisphere.

(c) 1953, June 22;

(d) 1953, October 25; the Sun has four parallel zones with alternating polarities, two of them are positive and two of them are negative.

interesting fact of the identity of the following magnetic field with that of the background field of the given hemisphere. This may be, for example, the reason why the active regions start to develop from their following portions first (Bumba, and Howard, 1965; Knoška, 1976) etc. A lot of work is needed to investigate further this question. But I believe that more observational material with very high degree of integration, i.e. with low angular resolution, is desirable.

4. Longitudinal Distribution of Solar Magnetic Fields

More than five years ago (Bumba and Howard, 1969) certain new patterns in the large-scale field distribution were characterized: *Sections* are the features of one polarity developing in activity zones from the magnetic fields of one to several active regions. They extend generally across the equator and they may be followed often for at least 10 rotations. Series of sections of one polarity on consecutive synoptic charts mounted in chronological order form *rows*. Two or more rows of one or both polarities may constitute a nearly continuous *stream*.

It has also been shown that the sections in the equatorial zone ($\varphi = \pm 20°$) have their recurrence period near to 27 days which means that they are slightly shifted to the west from one rotation to the following one, if drawn in Carrington's coordinate network, and the rows they form are then slightly inclined. This inclination practically disappears if drawn in Barthel's coordinates. Rows from opposite polarities differ at the given moment by various widths and level of concentration of their fields, sometimes also by their inclination. Sometimes the rows of the positive polarity, at other times those of the negative polarity are more pronounced.

In higher heliographic latitudes ($\varphi > \pm 20°$) the rows drawn in Carrington's coordinates are inclined in the opposite direction if compared with the equatorial rows and their inclination angle is greater. This inclination seems to be connected with the 28–29 day synodic rotational period of the sections developed in those latitudes.

The same characteristic patterns with the same properties described above may be now found during the whole material of the synoptic charts as yet available (1959–1975) (Bumba, 1976), (Figures 2 and 3).

During the descending part of the previous cycle as well as during the whole present cycle the evolution of the rows demonstrates a typical behaviour: a row starts in the form of a narrow section. The width of the section grows from one rotation to the next until it reaches its maximum of development. This maximal phase is usually well defined and visible in between the other background fields, due to the high degree of the field concentration and to the relatively large area occupied by the section. Afterwards the section of such a row narrows rather fast and disappears again in the noise of the background field. What is more important, the individual rows reach their maxima successively in such a way that with the decreasing heliographic longitude of the row the development of its maximum occurred later. The succession of consecutively developed maxima of one polarity rows forms a second type of 'row' inclination which in Carrington's coordinates goes parallel to those we may see in higher heliographic latitudes ($\varphi > \pm 20°$). This indicates that the synodic period of the rotation of these 'secondary rows' is also close to 28–29 days. These 'secondary equatorial rows' which are as a rule more pronounced in the negative polarity fields during both cycles of the activity for which we have the observational material develop simultaneously with the higher latitudinal zone rows (Figures 4, 2 and 3).

It may be underlined that the mentioned two basic inclinations of the rows or two basic periods of the synodic rotation of the sections in the large-scale distribution of the solar magnetic fields have been already demonstrated in several manners of their

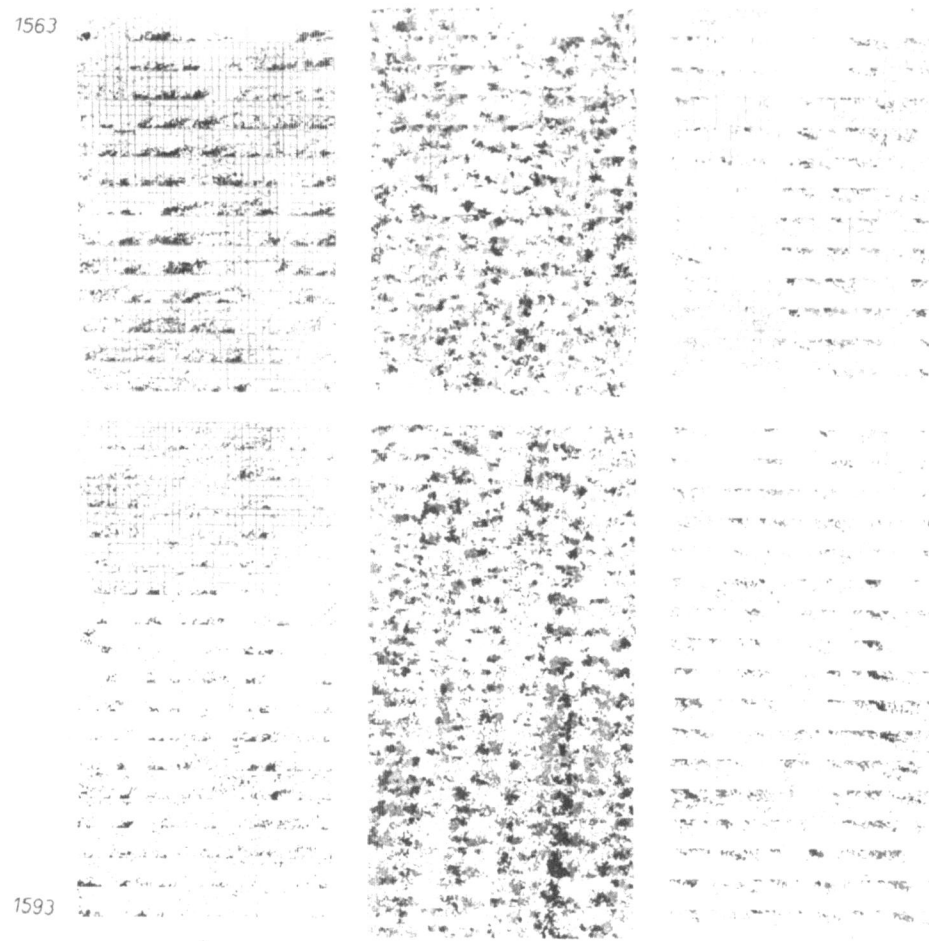

Fig. 2. Magnetic synoptic charts cut into strips representing two zones of latitude (±20°, ±20°–±40°) and mounted in chronological order during the maximum phase of the present cycle of solar activity (Rotations Nos. 1563–1593). The middle line represents the consecutive equatorial strips of the individual synoptic charts, the adjoining data represents the strips of higher latitudes on the Northern (left) and Southern (right) hemispheres. The dark regions represent the minus polarity areas.

manifestation (Ambrož *et al.*, 1971; Bumba, 1970; Bumba *et al.*, 1968; etc.). It seems even that these two main types of the rows (or streams) determine in the mutual interaction the morphology and may be also the dynamics of the background magnetic field distribution: the most interesting structures develop in the place and time when these two successions intersect.

Just recently Svalgaard and Wilcox (1974) investigated the same large-scale structures of the solar magnetic field using the polarity of the interplanetary magnetic field as inferred from polar geomagnetic observations. They succeeded to study the material for the past five sunspot cycles. I believe their results are very often in a good agreement with those obtained from the direct research of the solar magnetic fields. But since Dr Wilcox will tell us the main topics of their study in a contributed paper, I will not touch this question any more.

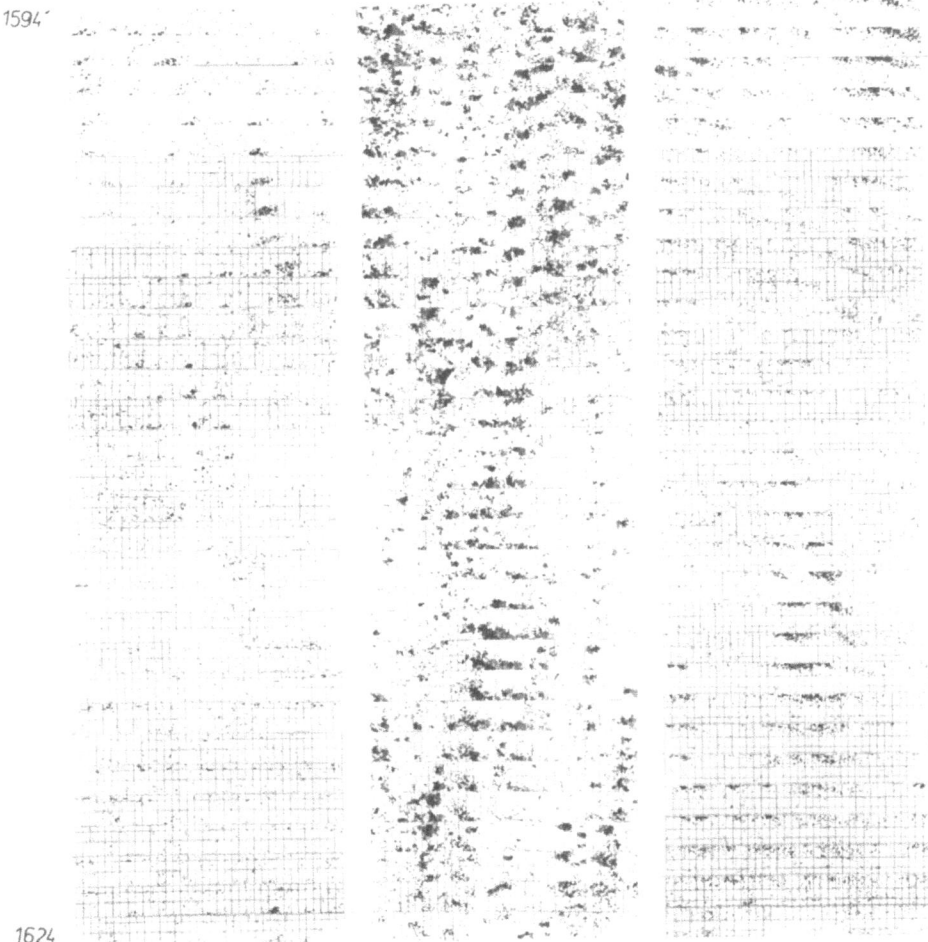

Fig. 3. Magnetic synoptic charts of solar photospheric fields organized as in Figure 2 during the descending part of the present cycle of solar activity (Rotations Nos. 1594–1624).

During the recent cycle of the activity two in longitude practically opposite complex magnetic streams used to exist on the solar surface, in each of which the opposite polarity seemed to be predominant. During the present activity cycle this regularity in distribution seems to lose its evident character one or two years after the maximum of the solar cycle.

Longitudinal distribution of the large-scale solar magnetic fields has also been many times studied in connection with the problem of the so-called 'Active Longitudes' (Vitinskij, 1969; 1973). Just recently Vitinskij (1974) using also the Mt. Wilson Observatory synoptic maps as initial data and applying his methods of isoline charts found several such active longitudes or concentrations of photospheric background magnetic fields but only one of these active longitudes is retained for each field polarity during the 19th and 20th cycle of activity. The other active longitudes either disappeared or are shifted in longitude. The conclusion made from

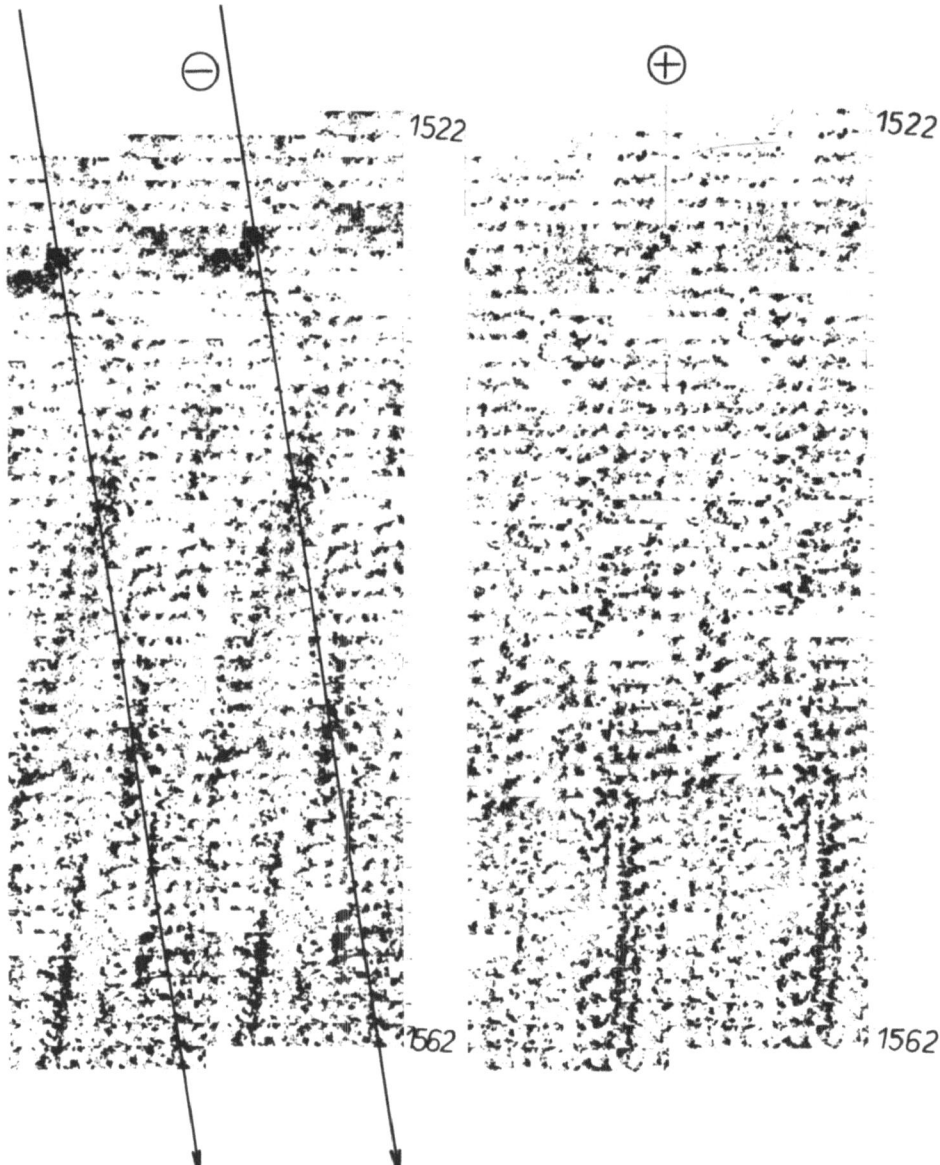

Fig. 4. Magnetic synoptic charts cut into equatorial strips (±20°) and mounted in chronological order (each strip of synoptic chart is drawn twice) for Rotations Nos. 1522–1562. Minus polarity is on the left, plus on the right of the figure. The main inclination of 27 days can be seen as well as on the minus polarity drawings the intensification of subsequent individual magnetic rows, starting in the upper left-hand corner and going in a succession forming the 28–29 day secondary row.

these observations by the author is as follows: the latitudinal-longitudinal distribution of photospheric background magnetic fields is not very stable in time (the time-scale here is one activity cycle). Therefore, we cannot be sure that they are the basis of active longitudes found from other indices of activity.

5. Regularities in the Large-Scale Distribution
of Solar Magnetic Fields

For a long time we have been trying to visualize in the large-scale distribution of the solar magnetic fields the so-called 'supergiant' regular structures (Ambrož et al., 1971) and to learn more about the dynamics of their growth, their relation to the development of the sections, rows and streams. As it has been said, their occurrence seems to be related to those heliographic longitudes and times in which the 27-day and the 28–29 day rows intersect or in which the equatorial streams have their evolutionary maxima. The same has been shown in proton-flare activity complexes which are parts of the 'supergiant' structures (Švestka, 1968). The development of subpolar maxima of green coronal emission seems also to be connected with these intersections (Bumba and Růžičková-Topolová, 1969). Both main inclinations may be seen in the shifts of the western (27-day) and eastern (28–29-day) boundaries during the development of the normal complexes of activity (Bumba and Howard, 1965b; Bumba et al., 1968) or even of 'supergiant' structures (Bumba, 1976) if drawn in Carrington's coordinates.

The study of such a magnetic field distribution on the whole synoptic charts may be simplified by drawing the charts in separated polarities. During the decreasing phase of the previous as well as of the present activity cycle we may observe these 'supergiant' regular or semi-regular structures – usually better visible in the negative polarity fields – as they form together with the positive polarity fields complex patterns in which both polarities are complementary (Ambrož et al., 1971; Bumba, 1972b; Bumba and Sýkora, 1973).

During the period of the minimum activity they are not seen, probably because of the low concentration of background fields, although the main knots in field distribution formed from magnetic fields of young active regions used to have the same distances as the main knots of the 'supergiant' structures. Sometimes even something like halves of the 'supergiant' structures either in the northern or in the southern hemisphere are observable. During the maximum of the present cycle of activity the huge regular structures seem to start to be apparent again. But probably because of a still relatively low background field concentration, before all in the equatorial zone, their portions in this zone are recognizable with difficulty. It seems also that during this phase of activity cycle their life-time is shorter if compared with that during the decreasing part of the recent cycle when the greater magnetic field concentration due to the higher activity level may be the reason why they disappear more slowly below the level of detection. The same is true about the visibility and duration of the 'supergiant' structures if we compare the decreasing parts of the present and recent cycle. Thus the level of background field concentration plays probably an important role in their formation. The different level of activity in both solar hemispheres during the declining phase of the present cycle influences the situation as well.

Examining the time development of large-scale field structures in relation to the role of the equator on it, relatively often we may see that the activity develops first in higher latitudes of one hemisphere, usually the northern one, than its center of gravity, and the resulting magnetic fields shift slowly – if followed in successive rotations – to the lower latitudes, cross the equator and continue to shift to the higher latitudes of the opposite hemisphere and vice versa. Such kind of 'latitudinal activity oscillations' may be sometimes seen repeated several times.

We have still to say a few words about the morphology and dynamics of development of the 'supergiant' structures. They are drop-shaped with the tail in higher latitudes. The drop with its head to the west, formed from older activity regions, more stable and regular, is usually about 90°–100° long. The center of gravity of the major solar activity is located at the eastern part of the feature, where the activity changes more rapidly, just below the root of the tail. The area of positive polarity forms a 'mirror' image to that of the negative polarity with its tail stretched out to the higher latitudes of the opposite hemisphere and with its body fitting into the drop of the negative polarity feature (Figure 5). The development of the described magnetic patterns is a very complex process, because morphologically the structures of both magnetic polarities do not develop simultaneously and in phase with regard to time and heliographic longitude. The life-time of such structures is of the order of one year although they can be best observed only for a few rotations (Figure 6).

The main drop-shaped body has also its own internal structure in which we may recognize the 'giant' – as we believe granular-like – structures during the periods of best visibility. They have a diameter of the order of 30°–35° and form clusters. Several years ago we tried to demonstrate that these 'giant' granules belong to a hierarchical system of convective motions (granulation, supergranulation and giant cells). They are better visible in the positive polarity distribution, which speaks in favour of a different depth of magnetic fields with different polarities and of a different interference of both polarities with old fields and motions during an active region formation (Bumba, 1970; 1967).

6. Averaged Synoptic Charts and Spherical Harmonics of the Large-Scale Photospheric Field

Although all the above discussed patterns of the large-scale distribution of the background solar magnetic fields are evident on the Mt. Wilson daily magnetic maps and especially on synoptic charts representing the whole solar surface, it is very difficult to describe the patterns in a more objective way.

Stenflo (1972) used the digitized Mt. Wilson data (Altschuler *et al.*, 1971) covering an 11-year period (August 1959–May 1970) to average each of the four existing synoptic charts into one so that a series of averaged magnetic synoptic charts might be studied. Then he constructed a butterfly diagram of the averaged magnetic fields and a sector diagram of the photospheric magnetic field averaged either over all the heliographic latitudes or for a latitudinal interval between ±8° only for the whole 11-year period.

Altschuler *et al.* (1971, 1974) divided the large-scale photospheric field into spherical harmonics. This procedure provides a description of the distribution of solar fields in terms of a set of objective parameters which can be visualized in certain geometrical configurations. The paper describes the photospheric field for a period of 13 years (August 1959–January 1973).

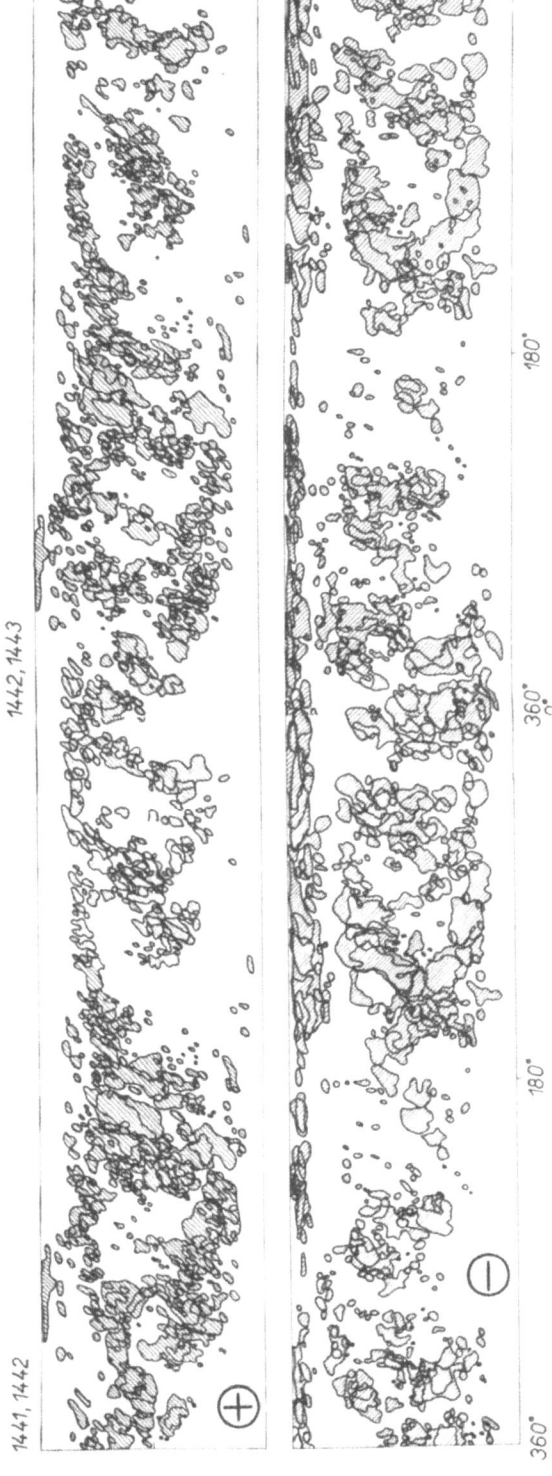

Fig. 5. The best example of 'supergiant' regular structures as yet observed, demonstrated on magnetic synoptic charts drawn for Rotations Nos. 1441–1443 in separated polarities (plus to the top, minus below). For integration two consecutive maps, one of which is repeated, are overlapped. The maximum of development of 'supergiant' patterns in practically opposite solar hemispheres is shown, each of them occupying one half of the Sun in both main active longitudes, formed from opposite polarities. The complementary role of both polarities may be also well seen.

V. BUMBA

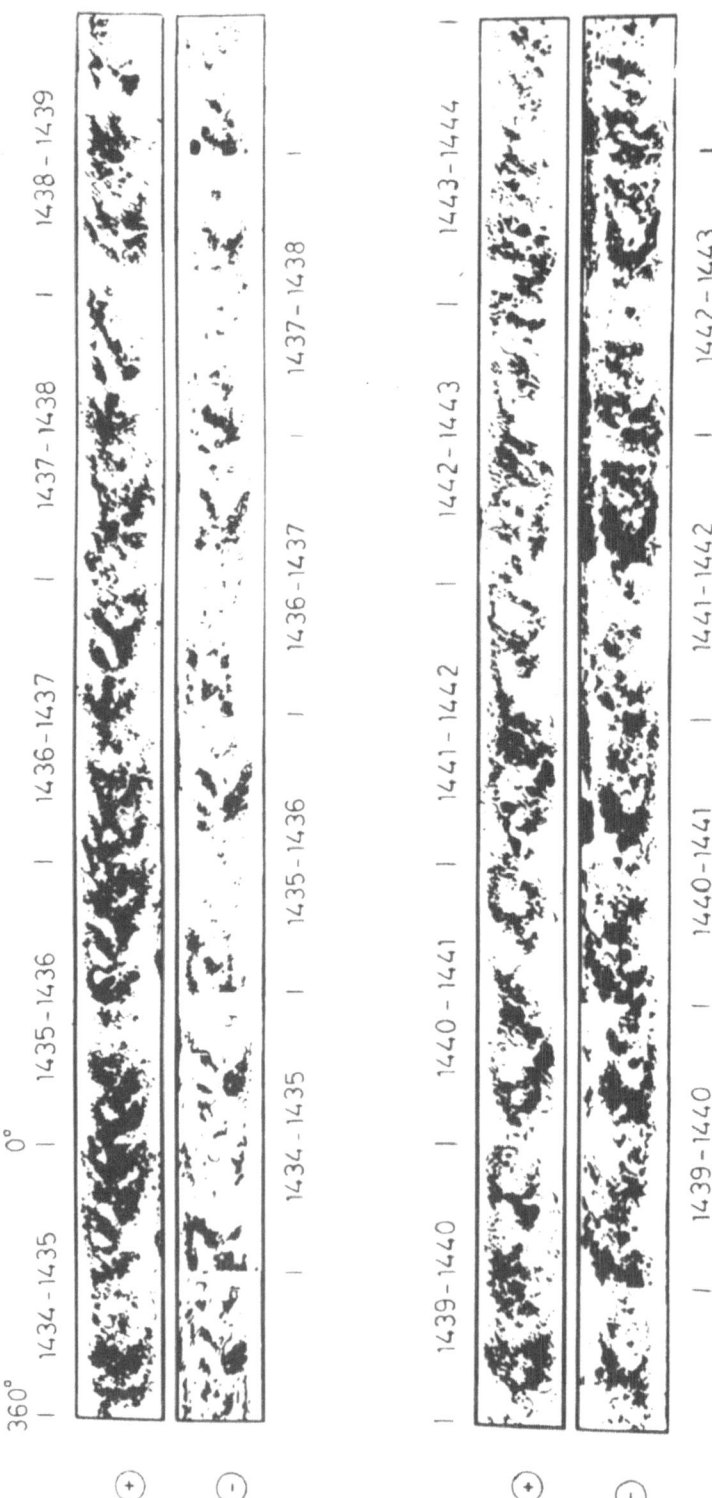

Fig. 6. Series of magnetic synoptic charts demonstrating the time-development of the best visible 'supergiant' regular structures shown in separated polarities (plus in the top line, minus in the lower line). In each line the succession of 10 solar rotations (Nos. 1434–1444) is presented. For integration again two consecutive maps are overlapped. The phase shift in the moment of maximum development of regular structure in each polarity may be seen as well as the decrease of plus magnetic flux and increase of the minus magnetic flux during the studied time interval. Hence the change of unbalance in both fluxes is clearly visible.

If we compare the determined dominant and important harmonics of the global photospheric field and the changes in the relative rankings of the harmonics over the time interval during which the material has been investigated by all the above mentioned methods, in other words if we compare the data represented on Figure 3 from Altschuler *et al.* (1974) paper with Stenflo's (1972) averaged figures or with our consecutively mounted integrated whole synoptic charts or their parts, we may see a good agreement of all the 'objective' as well as 'subjective' parameters representing the background field distribution. For example, investigating Stenflo's butterfly diagram (Figure 2; Stenflo, 1972) and the time sequence of important sector and zonal harmonics (Figure 3; Altschuler *et al.* 1972) for the period of the last activity minimum and for the increasing stage of the present activity cycle, we may see not only the prevailing dipole character of the solar magnetic field on both figures during the years 1963–1964 and the increasing importance of the zonal field component during the years 1965 till 1967; but comparing these data with the old Babcock's low resolution magnetic maps from the activity minimum in 1953–1954 – which we have demonstrated in the third section in the form of three typical magnetic field distributions – we may recognize in both materials again the same situation.

The same is true if we look at Stenflo's sector diagrams, especially at the one with averaged fields closer to the equator ($\pm 8°$; Stenflo, 1972) and if we compare it with the time variations of zonal and sector harmonics and with our diagram of longitudinal distribution of background magnetic fields from equatorial zones ($\pm 20°$) or with Stenflo's averaged synoptic charts. We may mention as an example the four sector character of the field in 1962, 1966 etc. Even the dominant behaviour of the $n = m = 5$ harmonic preceding the development of August 1972 proton-flare region may be well seen on our figure of the equatorial rows distribution as well as in the development of a 'supergiant' structure in which the proton-flare region characteristically developed (Figure 7).

7. Large-Scale Distribution of Solar Magnetic Fields
and Solar Activity Phenomena

It has been demonstrated several times that the distribution of solar background magnetic fields is the result of the expansion, weakening, stretching by differential rotation of magnetic fields of the old active regions and their interaction with neighbouring fields, and of the continuing development of magnetic fields of new regions within the patterns (Bumba and Howard, 1965a). Thus the dynamics of background magnetic field development is in a close relationship with the distribution of the active regions and vice versa. Consequently the same is true about the other activity phenomena linked to active regions. It has been shown, for example, that the active regions producing the particle-emitting flares occur in certain well defined places of regular 'supergiant' structures (Bumba, 1972a; Bumba and Sýkora, 1973; Bumba and Sýkora, 1974). Correlating the synoptic charts of the green (λ 5303 Å) coronal emission drawn in the form of isophotes from coronal data which were reduced to a unified photometric scale (Sýkora, 1971; 1973) with the position of 'supergiant' structures, we may see that the big elliptical features formed from an

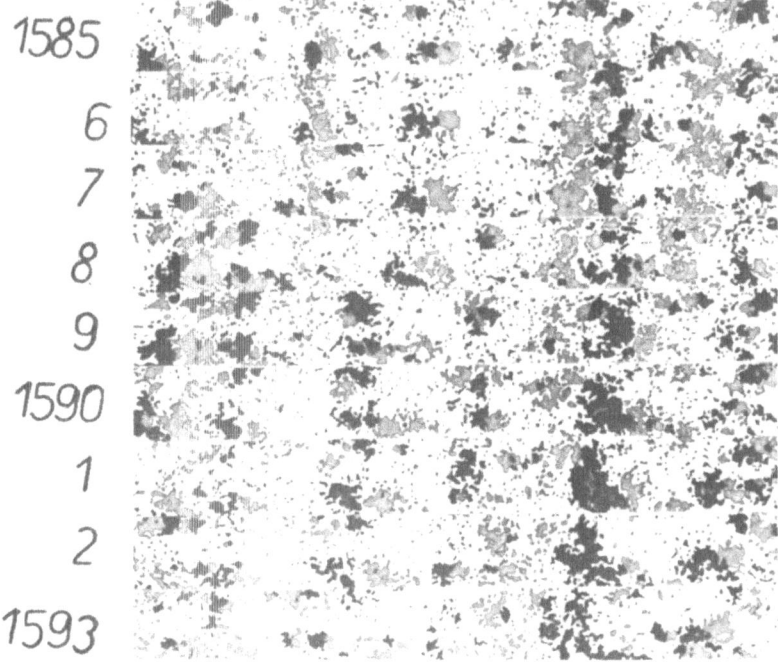

Fig. 7.   Magnetic synoptic charts cut into equatorial strips ($\pm 20°$) and mounted in chronological order for the time period (Rotations Nos. 1585–1593) when the dominant behaviours of the $n = m = 5$ harmonic preceding the development of the August 1972 proton-flares may be seen.

enhanced coronal emission practically conform with the 'supergiant' body of the magnetic field (Bumba and Sýkora, 1974).

The longitudinal distribution of the green coronal emission is closely related to the longitudinal distribution of individual polarities of the background magnetic field. Even the presence of coronal holes is demonstrated in the longitudinal distribution of the green coronal emission when its minima occur in the longitudinal intervals; they are minimally occupied by negative polarity fields. But on the other hand they agree with the position of the positive 'supergiant' body of the solar field (Bumba and Sýkora, 1973; 1974).

Using Sýkora's (1973) basic coronal data a lot of papers have appeared recently correlating the coronal structures with the distribution of interplanetary magnetic field and therefrom assessing the situation in the position of solar field sector boundary (Antonucci, 1974; Antonucci and Svalgaard, 1974). There are other papers demonstrating the situation in the corona around the magnetic field sector boundary, as, for example, by Wilcox and Svalgaard (1974) and Howard and Koomen (1974).

8. Interplanetary Magnetic Field Sector Structure and Large-Scale Distribution of Solar Magnetic Fields

The interplanetary magnetic field often shows a remarkably simple pattern (Wilcox and Ness, 1965) which correlates with the large-scale-averaged photospheric background magnetic field. (Ness and Wilcox, 1966; Wilcox, 1968; Schatten *et al.*, 1968; Wilcox and Howard, 1968). This is the reason why many authors use the satellite interplanetary magnetic field measurements as the source of information concerning the solar magnetic field distribution. But it is very difficult to visualize in the distribution of solar background fields the most apparent structure of the inter-planetary field – its regular alternation of sectors with opposite polarity. The only exception is the mean photospheric magnetic field of the Sun seen as a star – where each change in polarity of the mean solar field is followed about 4.5 days later by a change in polarity of the interplanetary field (Wilcox *et al.*, 1969; Severny *et al.*, 1970). That is why many papers trying to correlate the position of the interplanetary sector boundary with the distribution of various activity phenomena on the Sun were published during the recent years. Not only a specific situation in the occurrence of active regions toward the position of such a sector boundary has been formed (Wilcox, 1968; Schatten *et al.*, 1968); but the higher frequency of solar flares in connection with this boundary (Bumba and Obridko, 1969), the location of coronal condensations to the west of this boundary (Martres *et al.*, 1970), the influence of interplanetary field sector polarity on the orientation of radio emission polarization (Krüger *et al.*, 1968) etc. have been demonstrated. Several papers investigating the relation of coronal magnetic structures to the sector boundary showed that there is a certain difference between the boundaries changing the sector field polarity from the plus to the minus and from the minus to the plus. (Antonucci and Svalgaard, 1974). The same is true for the flares – they demonstrate a marked increase of occurrence in connection with the minus/plus sector boundary (Dittmer, 1974).

But in what relation is the position of the sector boundary to the regular 'supergiant' structures? Just recently (Bumba, 1976b) we succeeded in finding that the negative polarity interplanetary sectors coincide with the main body of the negative 'supergiant' structure (Figure 8). The minus/plus sector boundary lies close to the east end of this elliptical body which means that it precedes slightly the region of high activity occurrence. The plus/minus sector boundary separates the head of the negative field structure from the old positive polarity field. Even the longitudinal shift of the sector boundaries if examined in successive rotations goes very often parallel to that of the solar magnetic field rows and streams on equatorial ($\pm 20°$) strips of synoptic charts mounted successively, which means that the synodic rotation of sector boundaries is often close to 27 days. At the same time the centers of gravity of successive interplanetary field sectors (with decreasing heliographic longitude) are shifted to the east with the synodic rotation close to 28–29 days. Thus both main types of the longitudinal distribution of large-scale solar background magnetic fields (the 27-day and 28–29-day successions) are reflected in the interplanetary magnetic field distribution (Figures 9 and 10).

9. Conclusion

The main purpose of this talk was to show what we know up to now about the large-scale distribution of the solar magnetic fields on the basis of a series of solar magnetic synoptic charts covering more than 15 years of observations. The major

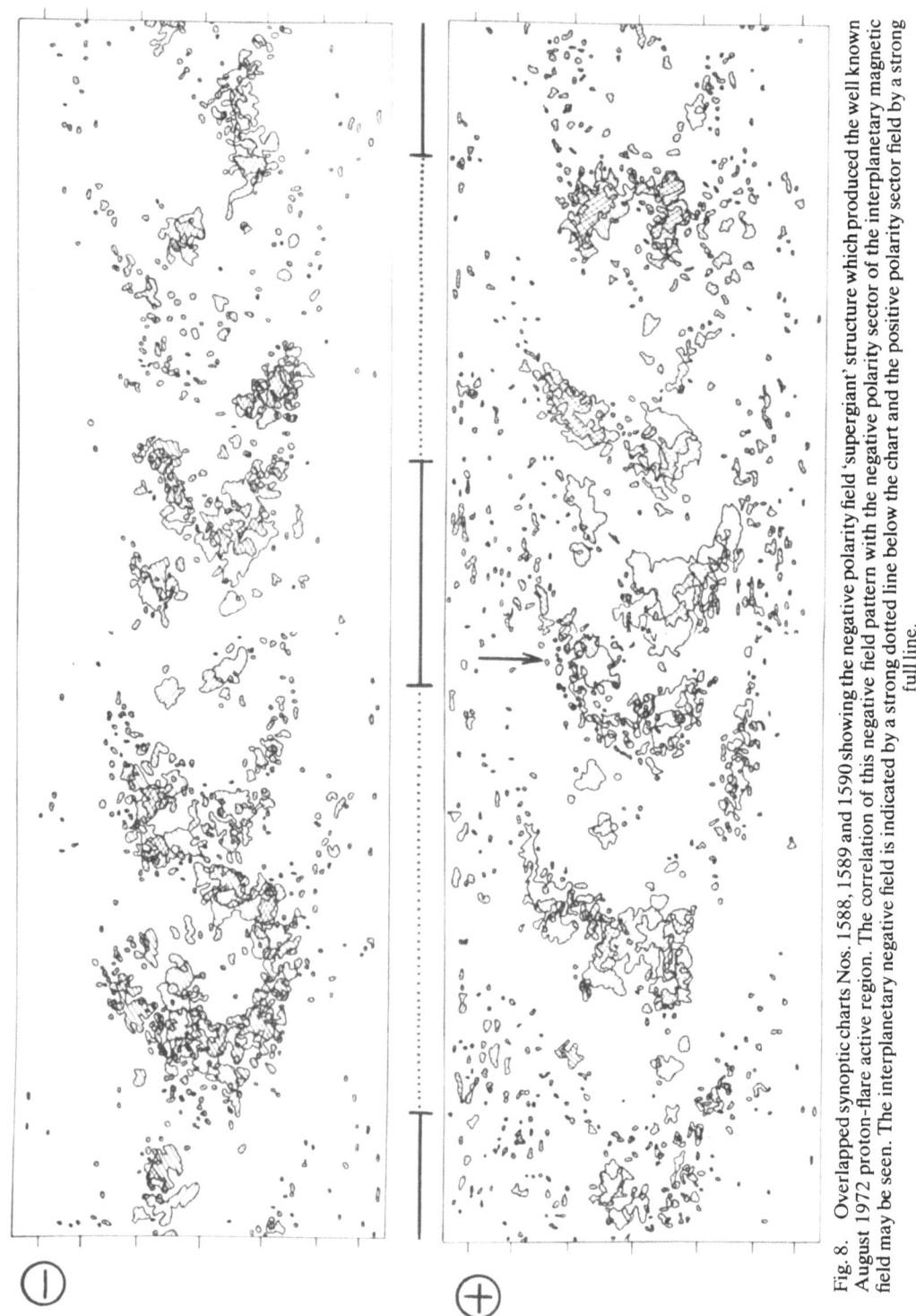

Fig. 8.   Overlapped synoptic charts Nos. 1588, 1589 and 1590 showing the negative polarity field 'supergiant' structure which produced the well known August 1972 proton-flare active region. The correlation of this negative field pattern with the negative polarity sector of the interplanetary magnetic field may be seen. The interplanetary negative field is indicated by a strong dotted line below the chart and the positive polarity sector field by a strong full line.

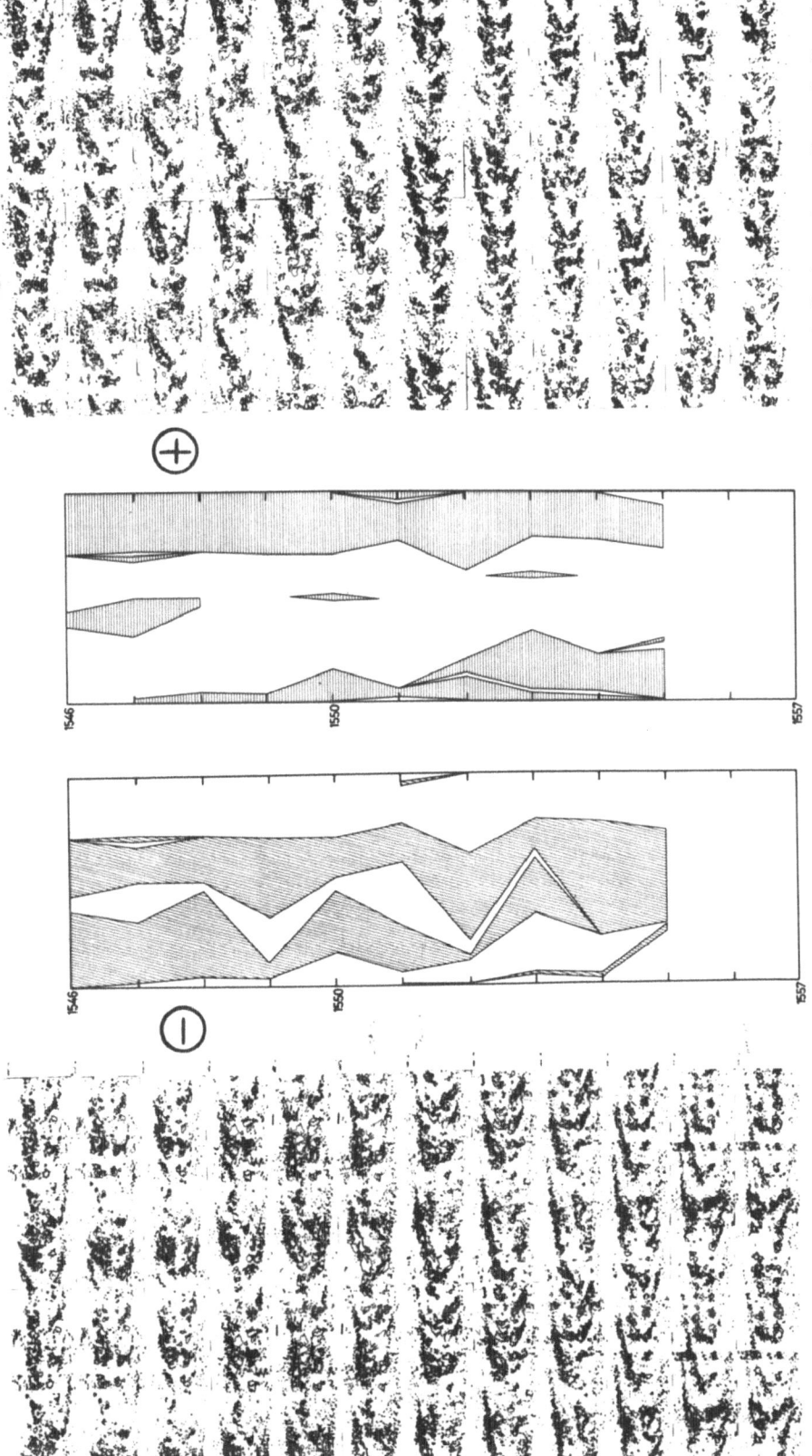

Fig. 9. Correlation of the large-scale distribution of solar magnetic fields with the interplanetary magnetic field sector structure drawn in Carrington's coordinates for Rotations Nos. 1546–1555. On the left side of the figure successively mounted magnetic synoptic charts of the negative polarity, on the right side of the positive polarity, distribution are shown. In the middle part of each figure the negative sector fields (to the left) and the positive sector fields (to the right) are drawn.

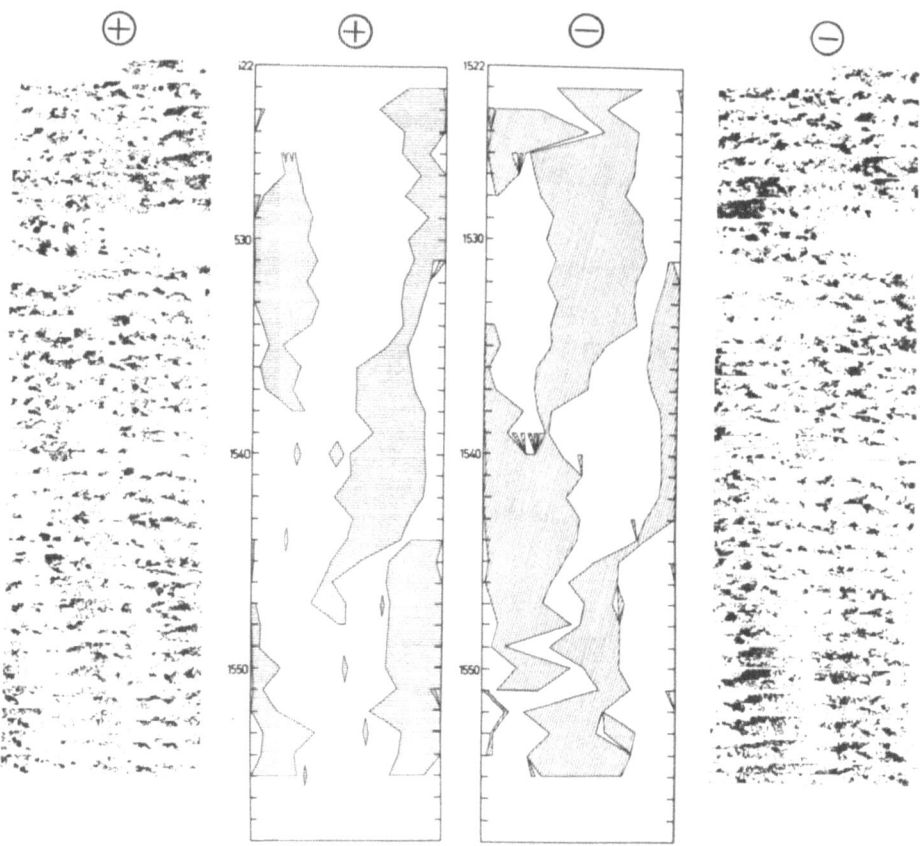

Fig. 10. Correlation of the secondary rows and streams (28–29 day synodic rotation) of solar magnetic fields from higher latitudinal zones ($\pm 20°$–$\pm 40°$) with the interplanetary magnetic field sector structure is shown. The negative polarity fields are demonstrated to the right, the positive polarity fields to the left of the figure. The interplanetary fields are drawn in Carrington's coordinates. The shift of centers of gravity of individual interplanetary field sectors coinciding with the shift of 28–29 day successions of rows of the solar fields as well as practically no shift of sector's boundaries in Carrington's network may be seen.

part of our information concerns the morphology and only some findings deal with the kinematics of the field distribution. Although there exist some regularities and recurrences, without a theoretical model of the process leading to such a distribution it is difficult to understand it. We spoke about the role of some agents influencing that distribution, especially about the differential rotation, but we did not discuss the effect of other important influences such as for example the convection and motions in general, although it seems that motions, before all the convective motions may play a very important role.

It is very difficult to explain the observed distribution regularities using the known semi-empirical models of the magnetic field distribution changes depending on the phase of the solar activity cycle (Babcock, 1961; Leighton, 1964). Especially difficult to explain is the regular zonal as well as sectoral distribution of background fields, the different role of both polarities, such as, for example, the better visibility of 'supergiant' structures in the negative polarity fields, the close relation of these negative polarity fields to the young solar activity regions and to the 28–29-day rows or successions of sections and the better visibility of 'giant' granules in the positive polarity fields, their relation to the remainders of old activity which seem to be the source of the solar wind, and their better evidence of the 27-day rows in their distribution. The same is true about the mutual interaction of both types of rows and the development of huge regularities occupying half of the solar surface and connected with the interplanetary field sector structure.

The major part of recent attempts to construct a model of observed regularities in field and activity distribution uses the interplanetary magnetic field sector structure as a starting point. We may quote as an example the papers by Svalgaard *et al.* (1974), Antonucci (1974). But there are also theoretical papers trying to explain some of the observed behaviours of the large-scale distribution of solar magnetic fields (Stix, 1971; Yoshimura, 1973; Wolff, 1974). Let us hope that the present Symposium will stimulate an increased interest in these problems.

Acknowledgements

I wish to thank Dr R. Howard from the Hale Observatories, Carnegie Institution of Washington, California Institute of Technology, Pasadena, for supplying me with the synoptic charts of the photospheric magnetic fields constructed from the daily observations of the Mt. Wilson Observatory prior to their publication.

References

Altschuler, M. D., Newkirk, G. Jr., Trotter, D. E., and Howard, R.: 1971, in R. Howard (ed.), 'Solar Magnetic Fields', *IAU Symp.* **43**, 588.
Altschuler, M. D., Trotter, D. E., Newkirk, G. Jr., and Howard, R.: 1974, *Solar Phys.* **39**, 3.
Ambrož, P., Bumba, V., Howard, R., and Sýkora, J.: 1971, in R. Howard (ed.), 'Solar Magnetic Fields', *IAU Symp.* **43**, 696.
Antonucci, E.: 1974a, *Solar Phys.* **34**, 471.
Antonucci, E.: 1974b, SUIPR Report No. 570, Inst. for Plasma Research, Stanford Univ.
Antonucci, E. and Svalgaard, L.: 1974, *Solar Phys.* **36**, 115.
Babcock, H. D.: 1959, *Astrophys. J.* **130**, 364.
Babcock, H. W.: 1961, *Astrophys. J.* **133**, 572.
Babcock, H. W. and Babcock, H. D.: 1955, *Astrophys. J.* **121**, 349.
Bumba, V.: 1967, *Hierarchy of Solar Magnetic Field Distribution*, Moscow.
Bumba, V.: 1970a, in E. R. Dyer (ed.), *Solar Terrestrial Physics*, Part I, Dordrecht, p. 21.
Bumba, V.: 1970b, *Solar Phys.* **14**, 80.
Bumba, V.: 1972a, in C. P. Sonett *et al.* (eds.), *Solar Wind*, Washington, p. 31.
Bumba, V.: 1972b, in C. P. Sonett *et al.* (eds.), *Solar Wind*, Washington, p. 151.
Bumba, V.: 1976a, *Bull. Astron. Inst. Czech.* **27** (in press).
Bumba, V.: 1976b, *Bull. Astron. Inst. Czech.* **27** (in press).
Bumba, V. and Howard, R.: 1956a, *Astrophys. J.* **141**, 1492.
Bumba, V. and Howard, R.: 1965b, *Astrophys. J.* **141**, 1502.
Bumba, V. and Howard, R.: 1969, *Solar Phys.* **7**, 28.
Bumba, V. and Obridko, V. N.: 1969, *Solar Phys.* **6**, 104.
Bumba, V. and Růžičková-Topolová, B.: 1969, *Bull. Astron. Inst. Czech.* **20**, 63.
Bumba, V. and Sýkora, J.: 1973, in M. J. Rycroft and S. K. Runcorn (eds.), *COSPAR Space Research*, Vol. XIII, Berlin, p. 1973.
Bumba, V. and Sýkora, J.: 1974, in G. Newkirk, Jr. (ed.), 'Coronal Disturbances', *IAU Symp.* **57**, 73.

Bumba, V., Howard, R., Kopecký, M., and Kuklin, G. V.: 1969, *Bull. Astron. Inst. Czech.* **20**, 18.
Bumba, V., Howard, R., Martres, M. J., and Soru-Iscovici, I.: 1968, in K. O. Kiepenheuer (ed.), 'Structure and Development of Solar Active Regions', *IAU Symp.* **35**, 13.
Dittmer, P. H.: 1974, SUIPR Report No. 597, Prepr. Inst. for Plasma Research, Stanford Univ.
Howard, R.: 1974a, *Solar Phys.* **38**, 283.
Howard, R.: 1974b, *Solar Phys.* **38**, 59.
Howard, R.: 1974c, *Solar Phys.* **39**, 275.
Howard, R. and Koomen, M. J.: 1974, *Solar Phys.* **37**, 469.
Howard, R., Bumba, V., and Smith, S. F.: 1967, Carnegie Inst. of Washington Publ. No. 626, Washington.
Knoška, S.: 1976, *Bull. Astron. Inst. Czech.* **27** (in press).
Krüger, A., Bumba, V., Howard, R., and Kleczek, J.: 1968, *Bull. Astron. Inst. Czech.* **19**, 180.
Leighton, R. B.: 1964, *Astrophys. J.* **140**, 1547.
Martres, M. J., Pick, M., and Parks, G. K.: 1970, *Solar Phys.* **15**, 48.
Ness, N. F. and Wilcox, J. M.: 1966, *Astrophys. J.* **143**, 23.
Severny, A. B., Wilcox, J. M., Scherrer, P. H., and Colburn, D. S.: 1970, *Solar Phys.* **15**, 3.
Schatten, K. H., Ness, N. F., and Wilcox, J. M.: 1968, *Solar Phys.* **5**, 240.
Stenflo, J. O.: 1972, *Solar Phys.* **23**, 307.
Stix, M.: 1971, *Astron. Astrophys.* **13**, 203.
Svalgaard, L.: 1972, *J. Geophys. Res.* **77**, 4027.
Svalgaard, L. and Wilcox, J. M.: 1974, SUIPR Report No. 605, Prepr. Inst. for Plasma Research, Stanford Univ.
Svalgaard, L., Wilcox, J. M., and Duvell, T. L.: 1974, SUIPR Report No. 537, Inst. for Plasma Research, Stanford Univ.
Sýkora, J.: 1971, *Bull. Astron. Inst. Czech.* **22**, 12.
Sýkora, J.: 1973, *Contr. Astron. Obs. Skalnaté Pleso* **5**, 5.
Švestka, Z.: 1968, *Solar Phys.* **4**, 18.
Vitinskij, Yu. I.: 1969, *Solar Phys.* **7**, 210.
Vitinskij, Yu. I.: 1973, *Ciklichnost i prognosy solnechnoj aktivnosti*, Nauka, Leningrad.
Vitinskij, Yu. I.: 1974, *Bull. Astron. Inst. Czech.* **25**, 222.
Wilcox, J. M.: 1968, *Space Sci. Rev.* **8**, 258.
Wilcox, J. M.: 1972, *Comments Astrophys. Space Phys.* **4**, 141.
Wilcox, J. M. and Howard, R.: 1968, *Phys. Rev. Letters* **20**, 1252.
Wilcox, J. M. and Ness, N. F.: 1965, *J. Geophys. Res.* **70**, 5793.
Wilcox, J. M. and Scherrer, P. H.: 1972, *J. Geophys. Res.* **77**, 5385.
Wilcox, J. M. and Svalgaard, L.: 1974, *Solar Phys.* **34**, 461.
Wilcox, J. M., Severny, A. B., and Colburn, D. S.: 1969, *Nature* **224**, 353.
Wolff, Ch. L.: 1974, *Astrophys. J.* **194**, 489.
Yoshimura, H.: 1973, *Solar Phys.* **33**, 131.

DISCUSSION

Krause: I would like to come back to the point, where you speak about the behaviour of the Sun as a magnetic monopole. I reject the assumption that the magnetic field is not source-free. Then this behaviour of the magnetic field indicates electric currents in higher levels than the photosphere. My question is are there other indications of subphotospheric currents at times, when the magnetic field behaves like a 'magnetic monopole'?

Bumba: I do not know such observations that might answer your question. As far as I know such indications are not observed, although the 'monopole' behaviour of the solar magnetic field is observed relatively often.

Stix: The identification of interplanetary sector boundaries with photospheric sector boundaries is difficult for two reasons: Firstly, *streams* in the solar wind may influence the position of sector boundaries at 1 AU, and secondly, *closed field* configurations near the Sun cause large deviations from the simple spiral field pattern.

Bumba: Nevertheless any 'supergiant' regular structure that I could correlate with the interplanetary sector structure shows the dependence of the position of its sector boundary on the situation I mentioned in my talk.

Vandakurov: May we consider the extended magnetic field with the largest scale to be a superposition of a rigidly rotating simple sector structure and a differentially rotating sector structure?

Bumba: As far as I know such a model of superimposed two kinds of magnetic field distribution has been suggested by Dr Wilcox. Maybe Dr Wilcox wishes to say a few words concerning this model himself.

Wilcox: There may exist a sector structure somewhere below the photosphere. These field lines have rigid rotation. When they rise to the photosphere (perhaps by magnetic buoyancy) the magnetic lines are distorted by differential rotation. Thus the observed photospheric field can display both rigid rotation (on a long time scale, i.e. several rotations) and differential rotation (on a shorter time scale of days).

SMALL-SCALE SOLAR MAGNETIC FIELDS

J. O. STENFLO

Lund Observatory, Lund, Sweden

Abstract. The observed properties of small-scale solar magnetic fields are reviewed. Most of the magnetic flux in the photosphere is in the form of strong fields of about $100-200\ mT$ $(1-2\ \text{kG})$, which have remarkably similar properties regardless of whether they occur in active or quiet regions. These fields are associated with strong atmospheric heating. Flux concentrations decay at a rate of about $10^7\ \text{Wb s}^{-1}$, independent of the amount of flux in the decaying structure. The decay occurs by smaller flux fragments breaking loose from the larger ones, i.e. a transfer of magnetic flux from smaller to larger Fourier wave numbers, into the wave-number regime where ohmic diffusion becomes significant. This takes place in a time-scale much shorter than the length of the solar cycle.

The field amplification occurs mainly below the solar surface, since very little magnetic flux appears in diffuse form in the photosphere, and the life-time of the smallest flux elements is very short. The observations further suggest that most of the magnetic flux in quiet regions is supplied directly from below the solar surface rather than being the result of turbulent diffusion of active-region magnetic fields.

1. Introduction

Solar magnetic fields are of fundamental importance for most solar phenomena. The solar cycle itself is a manifestation of the interaction between solar rotation, convection, and magnetic fields. It is thus not surprising that most review papers in solar physics contain discussions of magnetic fields, including their small-scale properties. Recent reviews with fairly extensive discussion of small-scale fields have been given by Harvey (1971, 1974), Stenflo (1971), Severny (1972), Mullan (1974), and Deubner (1975). The emphasis in many of these papers has however been on observed structures substantially larger than one arcsec, whereas the true sizes of the basic magnetic elements in the network and plages appear to be considerably smaller. Most values of the field strengths that have been quoted have been of limited value since the relation between the observed and true field strengths has not been clear.

During the past few years a new picture of solar magnetic fields has gradually emerged due to advances in observational techniques and in the theoretical understanding of the Zeeman effect in the solar atmosphere. According to this picture, there are hardly any weak magnetic fields in the photosphere (where the magnetic field is measured). Most of the net magnetic flux is carried by fields of $100-200\ mT$ $(1-2\ \text{kG})$, even when the average field strength is much less than $1\ mT$ $(10\ \text{G})$. The corresponding magnetic elements occupy only a tiny fraction of the solar surface and have sizes in quiet regions generally less than 200 km. Much of the discussion in the present review will be focused on the properties of the basic subarcsec magnetic elements, i.e. on scales comparable to the size of Bohemia or less. A spatial resolution of 100 km is considered to be exceedingly poor and completely inadequate in Earth studies, while such a resolution has never been achieved for the Sun. The best spatial resolution in direct magnetograph observations has been about 1000 km so far.

Bumba and Kleczek (eds.), Basic Mechanisms of Solar Activity, 69–99. All Rights Reserved.

The unresolved fine structure in the magnetic field has introduced errors in determinations of the average field strengths from magnetograph observations. This error has earlier often been considered to be a nuisance and an obstacle that should be removed to allow the magnetic-field structure to be determined more accurately. We now realize, however, that the error in itself is a powerful diagnostic tool and even seems to contain more information about the true structure of the magnetic field than the direct observations.

The deduced magnetic pressure in the magnetic elements apparently exceeds the ambient gas pressure in the photosphere and the dynamic pressure due to motions of the solar plasma. The generation of such an extreme fine structure poses a fundamental plasma physics problem. Further, as we shall see, the small-scale nature of solar magnetic fields affects our views on the dynamics and energy balance of the solar atmosphere, as well as our understanding of differential rotation and the solar cycle.

2. Observed Properties of Small-Scale Solar Magnetic Fields

2.1. TRUE FIELD STRENGTHS

The signal from a solar magnetograph recording the circular polarization in a line wing is a measure of the magnetic flux, *not* of the field strength. The apparent flux values need to be corrected to represent the true net flux through the scanning aperture of a magnetograph, but this correction is generally less than a factor of two (Howard and Stenflo, 1972). By dividing the net flux through the aperture by the aperture area, we arrive at an *apparent* field strength, $B_{\rm obs}$. As most of the magnetic flux through the solar surface appears to be channelled through subarcsec elements (see Section 2.3), which have never been spatially resolved by any magnetograph, the directly recorded field strengths $B_{\rm obs}$ alone do not provide much information on the true field strengths except that they may indicate some lower limits. In this sense the 'direct' methods of mapping solar magnetic fields do not represent the most direct ways of determining the field strengths. For this purpose the 'line-ratio' method is more adequate (see Section 2.1.2). Both methods will be reviewed in the following sections.

A typical example of the best spatial resolution presently achieved in magnetograph observations is given in Figure 1, due to Vrabec (1974).

2.1.1. *Direct Observations*

Various approaches to determine the true magnetic field strengths directly have been tried. Stenflo (1966, 1968a) and Severny (1967) made repeated scans of the same regions with different sizes of the scanning aperture. The results showed that the apparent field strengths increased rapidly with decreasing aperture size down to the smallest apertures that could be used then, 7 (arcsec)^2. It was however not clear how far this increase would continue if one could go down to infinitely small aperture sizes.

Many authors have tried to measure the apparent field strengths with the highest spatial resolution available at the best seeing conditions. Sheeley (1966, 1967) found field strengths in the quiet region network in excess of 30 mT (300 G). The diameters

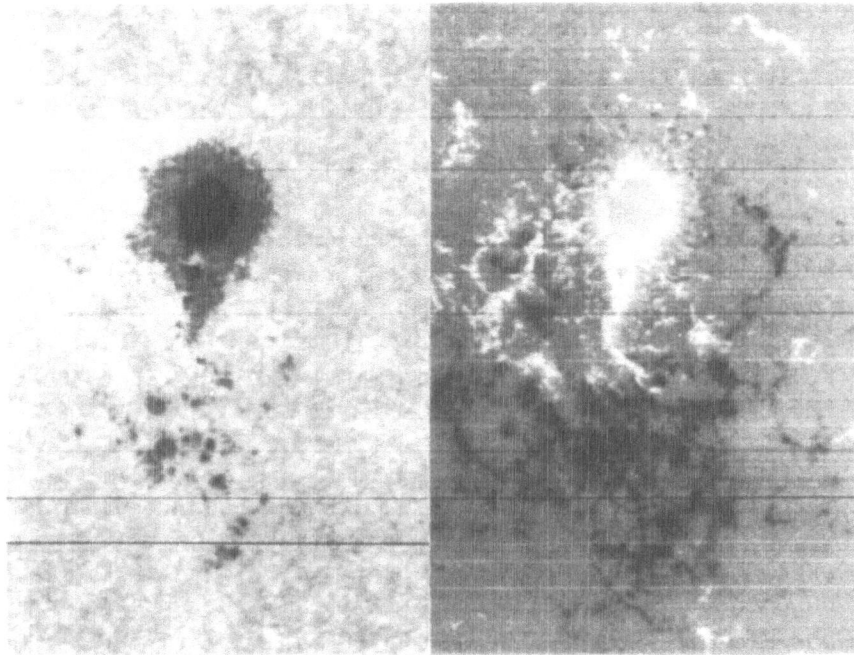

Fig. 1.   Magnetogram (right) of the active region shown to the left, obtained on January 9, 1971 at the Aerospace San Fernando Solar Observatory by D. Vrabec and T. Janssens (Vrabec, 1974). This magnetogram gives an impression of the highest spatial resolution that can be achieved in present magnetograph observations. It shows a complex pattern of flux elements imbedded in a virtually field-free photosphere.

of the flux concentrations were often less than one arcsec. In active regions Beckers and Schröter (1968a) found numerous flux concentrations with typical diameters of 1000 km and observed field strengths in the range 60–140 mT. These strong-field regions appeared to carry most of the magnetic flux occurring outside the sunspots. Similar results, including the evolution and dynamics of the flux concentrations, have been presented by Vrabec (1971, 1974), who finds that the small flux elements are imbedded in a virtually field-free photosphere.

Filtergram observations generally allow a higher spatial resolution than magnetograph observations due to the shorter integration times (which reduces the seeing effects). As there is a one-to-one correspondence between brightness enhancements and concentrations of magnetic flux (see Section 2.8), one can measure the sizes of the bright points and compare with magnetic flux recordings, to get an estimate of the field strengths. This requires however that one can resolve the individual bright points and assume that the brightness structures and magnetic elements are of the same size, which is far from obvious. Simon and Zirker (1974) tried to measure the sizes of the magnetic structures observed in high-resolution spectrograms and found typical apparent dimensions of 1–2″. As this is much larger than the sizes of the 'filigree' structures earlier recorded in $H\alpha$ filtergrams by Dunn and Zirker (1973), Simon and Zirker (1974) concluded that the magnetic elements are generally larger than the brightness elements. Model calculations of Stenflo

(1974b, 1975) indicate on the other hand that the brightness enhancements are spread out over a somewhat larger area than the associated magnetic flux (see Section 2.8).

Mehltretter (1974) counted the number of resolved bright points in filtergrams and compared with the observed magnetic flux in the region. This procedure gave a flux of 4×10^9 Wb per element. As the typical size of a bright point was about the same as or less than the resolution of the telescope (Sacramento Peak Observatory vacuum tower) and microphotometer, i.e. 140 km, Mehltretter estimated the typical field strength of a bright point to be 200–250 mT (2–2.5 kG).

As the Zeeman splitting is proportional to the square of the wavelength, it is advantageous to use infrared lines, which may be completely split in the flux elements, while spectral lines in the visible are only partially split (overlapping sigma components). By using the line Fe I λ 1.5648 μm, Harvey and Hall (1975) obtained evidence that the flux elements have field strengths of typically 150–200 mT (1.5–2 kG) inside as well as outside active regions.

A further indication that the true field strengths are quite high is obtained from observations of pores and the umbrae of small sunspots. Steshenko (1967) found $B_{obs} \approx 140$ mT (1.4 kG) for the smallest pores he could observe (1.5–2″). This seemed to be a lower limit for the field strength in sunspots. Bumba (1967) found $B_{obs} \approx 160$–210 mT in pores and spots of diameter 1000–3000 km. Both Steshenko and Bumba noted that the observed field strength increases only slowly with increasing sunspot area.

2.1.2. *The Line-Ratio Method*

Calibrating a solar magnetograph means to find the relation between the circular polarization recorded at a fixed position in the line wing and the magnetic flux through the scanning aperture. This relation (the calibration curve) is a linear one for small field strengths, but deviates strongly from linearity (Zeeman saturation) when the Zeeman splitting becomes comparable to the line width. Even in the case of weak fields, the magnetograph signal is sensitive to the shape of the line profile. When a magnetograph is calibrated directly, the observed *average* line profile is used. Due to the pronounced concentration of magnetic flux to small regions, the average profiles are typical for the non-magnetic regions. We know that there are indeed large changes in the line profiles in magnetic regions, visible as line 'gaps' (Sheeley, 1967) in spectrograms (see Section 2.8).

Straightforward interpretations of magnetograms using calibration curves based on average line profiles are generally in error due to the combined effects of Zeeman saturation, line weakenings in magnetic regions, and uncompensated Doppler shifts (occurring on a scale smaller than the spatial resolution). Unfortunately it has not been possible to calibrate magnetographs using the line profiles in magnetic regions, since the magnetic elements are far too small to be resolved by magnetographs or in spectrograms. Many people have considered this as a serious principle limitation of magnetograph observations. Instead it turns out that the 'error' in the flux determination contains information about the unresolved small-scale magnetic structures. This source of information can be tapped using the line-ratio method without the need for achieving extremely high spatial resolution or working at excellent seeing conditions.

The line-ratio method uses simultaneous magnetograph recordings in two or more properly selected spectral lines. Since the scanning aperture, telescope and spectrograph are the same for the different lines, and the observations are made simultaneously, smearing effects due to seeing and instrument resolution will be the same for the pair of lines chosen. The apparent field strengths recorded in the two lines will be different due to difference in Zeeman saturation, line weakenings, uncompensated Doppler shifts, and height of line formation. A comparison between apparent field strengths in two spectral lines can thus give information on these parameters. By a careful choice of pairs of spectral lines, the various effects mentioned can be untangled.

The methods of extracting information from the magnetograph 'calibration error' were developed by Stenflo (1968b) and have since been improved when applied to various sets of data (Harvey and Livingston, 1969; Howard and Stenflo, 1972; Frazier and Stenflo, 1972; Gopasyuk et al., 1973; Caccin et al., 1974; Frazier, 1974; Stenflo, 1971, 1973, 1974b, 1975).

Before the nature of the magnetic fine structure was known, it was natural to assume that there should be all kinds of magnetic elements in the photosphere, of all sizes and field strengths from zero to several hundred mT. Similarly a continuous distribution of temperature enhancements and Doppler shifts in the magnetic elements would be expected. As a general approach to such an inhomogeneous atmosphere, Stenflo (1968b) introduced a 'multi-component' model, each component occupying a certain fraction of the scanning aperture and having a given field strength distribution (Zeeman saturation), temperature enhancement (line weakening), and velocity (uncompensated Doppler shift). Following suggestions by Alfvén (1967), a filamentary model was considered as a special case of this general approach. We will presently consider such a 'two-component' model in some detail, since it has proven to represent the situation in the photosphere remarkably well and will be repeatedly referred to in our discussions of observational data.

Accordingly, we assume that there are two types of regions on the Sun: strong and weak-field regions. If Φ is the net longitudinal magnetic flux through the scanning aperture, we can write

$$\Phi = \Phi_s + \Phi_w, \tag{1}$$

Φ_s being the flux in strong–field form, and Φ_w the corresponding weak-field flux. The magnetograph calibration is correct for the weak-field flux, since the calibration refers to average line profiles, and the weak fields occupy the major fraction of the solar surface. For the strong-field regions, the magnetograph calibration is in error by a factor δ, determined by Zeeman saturation, line weakenings, and uncompensated Doppler shifts.

The observed line-of-sight magnetic field is thus

$$B_{obs} = (\delta \Phi_s + \Phi_w)/A, \tag{2}$$

A being the aperture area, while the true average field is

$$\bar{B} = \Phi/A. \tag{3}$$

If we define the average strength $\bar{B}_w$ of the weak-field component

$$\bar{B}_w = \Phi_w/A, \tag{4}$$

we find, using (1)–(4),

$$B_{obs} = \delta\bar{B} + (1-\delta)\bar{B}_w. \tag{5}$$

The commonly used line for magnetograph observations, Fe I λ 525.02 nm, is quite susceptible to Zeeman saturation, temperature enhancements, and Doppler shifts, causing δ to be considerably smaller than unity. The Fe I λ 523.3 nm line is on the other hand insensitive to these effects (except in the line core), being much broader and having considerably smaller Landé factor than the 525.0 nm line. As δ for the 523.3 nm line is thus approximately unity, which has been verified by more detailed model calculations (Stenflo, 1975), $B_{523.3}$ represents a good approximation for the average field $\bar{B}$, and (5) can be written

$$B_{525.0} = \delta_{525.0}B_{523.3} + (1-\delta_{525.0})\bar{B}_w. \tag{6}$$

An example of a scatter-plot diagram of $B_{525.0}$ vs. $B_{523.3}$ is shown in Figure 2, taken from Frazier and Stenflo (1972). The difference in intercepts of the lines fitted to the positive and negative points provides information on $\bar{B}_w$ (see Section 2.3).

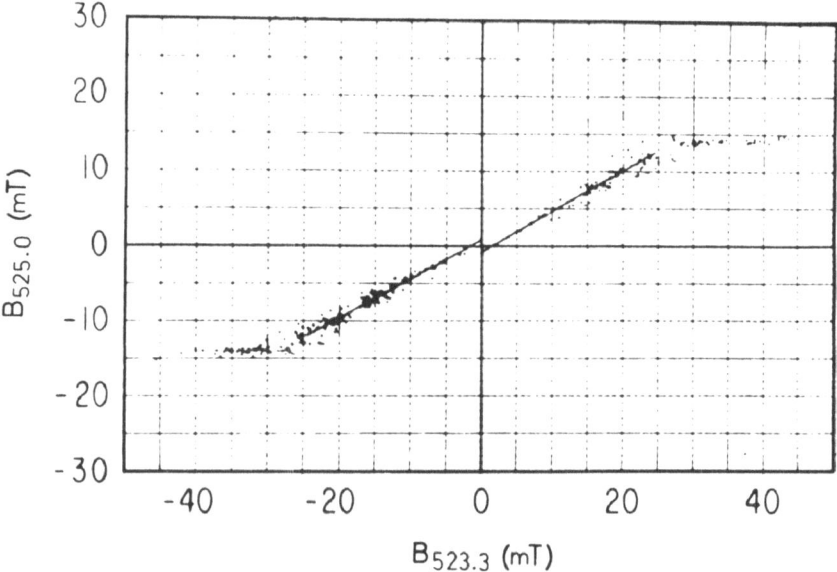

Fig. 2. Scatter-plot diagram of $B_{525.0}$ vs. $B_{523.3}$ for the brighter plage structures in an active region. Two straight lines have been fitted to the positive and negative points. Points with $|B_{523.3}| > 25$ mT contain an instrumental saturation in the $B_{525.0}$ channel and have not been used when fitting the straight lines. From Frazier and Stenflo (1972).

If the two-component model were too idealized, the real Sun containing a large range of various types of magnetic elements, we would have a broad distribution of values of δ. The points in a diagram giving $B_{525.0}$ vs. $B_{523.3}$ would then not fall along a straight line representing a single δ value, but there would be a large scatter in the diagram. The absence of such a scatter in excess of the instrumental noise provides empirical justification for the simple two-component model (see Section 2.2).

The effect of Zeeman saturation on the value of δ contains the information on the true field strengths in the flux concentrations. To extract this information we must find a method to separate out Zeeman saturation from the other factors affecting the slope in the scatter-plot diagrams. This is done by choosing a pair of lines which should behave identically in all respects except for the sensitivity to Zeeman saturation. Two such lines are Fe I λ 525.02 nm and Fe I λ 524.71 nm. Both belong to the same multiplet, have practically the same excitation potential of the lower level, the same oscillator strength and equivalent width. The only essential difference between the lines is their effective Landé factors, 3.0 for the 525.0 nm line, and 2.0 for the 524.7 nm line. Simultaneous recordings in these two lines have been made by Stenflo (1973) for a quiet region at disk center. If $\bar{B}_w$ in (5) is small compared with B_{obs}, it may be neglected when determining the slope k in the $B_{525.0} - B_{524.7}$ scatter-plot diagram. In this case

$$k \approx \delta_{525.0}/\delta_{524.7}. \tag{7}$$

k can differ from unity only if there are strong, unresolved magnetic fields inside the scanning aperture. As the 525.0 nm line is more sensitive to Zeeman saturation, we expect that k should be less than unity and be smallest when the magnetograph exit slits are closest to the line core, increasing towards unity when the exit slits are moved further out in the wings. This effect is observed, supporting the validity of the interpretation.

The longitudinal magnetic field in a flux element can be described by giving the peak value of the field strength at the center of the element and the shape of the magnetic cross-section. The relation between the slope k and the peak field strength is sensitive to the assumed shape of the cross-section. Assuming a cylindrically-symmetric magnetic element, four different magnetic profiles have been considered:

(1) Gaussian cross-section

$$B = B_0 \, e^{-(r/r_0)^2}. \tag{8}$$

(2) The profile

$$B = B_0[1 - (r/r_0)^2]^2. \tag{9}$$

(3) The profile

$$B = B_0[1 - (r/r_0)^2]. \tag{10}$$

(4) Single-valued field $B = B_0$ throughout the element.

The values of B_0 derived from the observed slope k were 230, 186, 167, and 110 mT, respectively, for the four types of magnetic cross-sections mentioned above (Stenflo, 1973, 1975). The value 110 mT represents a definite lower limit. This result should be contrasted with the fact that most of the field strengths $B_{525.0}$ observed with the $2.4 \times 2.4''$ scanning aperture were considerably smaller than 1 mT, not a single of the 1943 points in the raster having an apparent field in excess of 5 mT (Stenflo, 1973). This apparent contradiction is understood only if the sizes of the magnetic elements are very small compared with the size of the scanning aperture (see Section 2.5).

The derivation of B_0 from k is not model dependent except for the assumed magnetic cross-section. Various atmosphere models (Milne-Eddington, HSRA, etc.) give practically identical results (Stenflo, 1975), because the Zeeman saturation is a simple manifestation of the fact that the separation between the σ components becomes comparable to the line width. The total half-width of the lines we consider is about 10 pm (100 mÅ), and does not vary greatly with atmospheric model. The relation between Zeeman splitting and field strength depends in a well-known way on atomic parameters.

Harvey *et al.* (1972) made recordings of the complete line profiles of the Stokes I and V parameters for the 525.0 nm line and used these observations to derive average field strengths of 50 mT (500 G) in the unresolved magnetic elements, but needed to introduce a dispersion in the field strengths of about 100 mT. If magnetic profiles like (8)–(10) are used instead of characterizing the magnetic elements by an average field strength plus a dispersion, practically the same field strengths as those derived from the slope in the 525.0–524.7 scatter-plot diagram are obtained (Stenflo, 1973).

Wiehr and Wittmann (1975) have recently used a pair of spectral lines belonging to a different multiplet to isolate the effect of Zeeman saturation, and have found approximately the same high values of the field strengths in the magnetic elements.

Frazier (1974) combined Stenflo's (1973) observations with velocity data and showed that the maximum field strength B_0 within a flux concentration must be between 130 and 260 mT in order to satisfy both the magnetic and velocity data (see Section 2.7).

2.2. UNIQUENESS OF THE FIELD STRENGTH DISTRIBUTION

The flux elements appear to have remarkably unique and universal properties. This conclusion is mainly based on the following observations:

(1) The points in the 525.0–523.3 scatter-plot diagram fall along a straight line. No systematic deviation from linearity is observed for $B_{523.3} < 20$ mT (Harvey and Livingston, 1969; Frazier and Stenflo, 1972).

(2) The scatter of the points around this straight line does not appreciably exceed the scatter due to random instrumental noise, which was 0.1–0.2 mT for the quiet-region scan analysed (Frazier and Stenflo, 1972).

(3) The straight line has practically the same slope in quiet as in active regions (excluding sunspots) (Frazier and Stenflo, 1972).

Argument (1) shows that $\delta_{525.0}$ is independent of the average field strength in the scanning aperture, at least as long as $B_{523.3} \lesssim 20$ mT. The amount of Zeeman saturation, line weakenings, and uncompensated Doppler shifts are thus independent of the amount of magnetic flux within the aperture. This indicates that there is no intrinsic change in the physical parameters between different regions when we measure apparently different magnetic fluxes and enhancements in the photospheric brightness. The conclusion that the physical parameters are independent of the amount of magnetic flux is supported by argument (3) above, according to which the Zeeman saturation plus other effects appear to be the same in quiet regions as in active-region plages.

Argument (2) above indicates that the intrinsic scatter of the physical parameters characterizing the magnetic elements must be small compared with the average values ($\Delta B_0/\bar{B}_0 \ll 1$, etc.).

If the temperature enhancement in magnetic regions is independent of the amount of magnetic flux, we should expect a linear relation between apparent brightness enhancement $\Delta I/\bar{I}$ and apparent field strength $B_{523.3}$. The observed relations (Frazier, 1971) are shown in Figure 3 for brightness enhancements in the continuum as well as in the core of Fe I λ 525.0 nm. The scatter of the points is large, particularly for the continuum due to the superposed non-magnetic brightness fluctuations in the granulation, but average curves can be fitted to the good statistical sample of points.

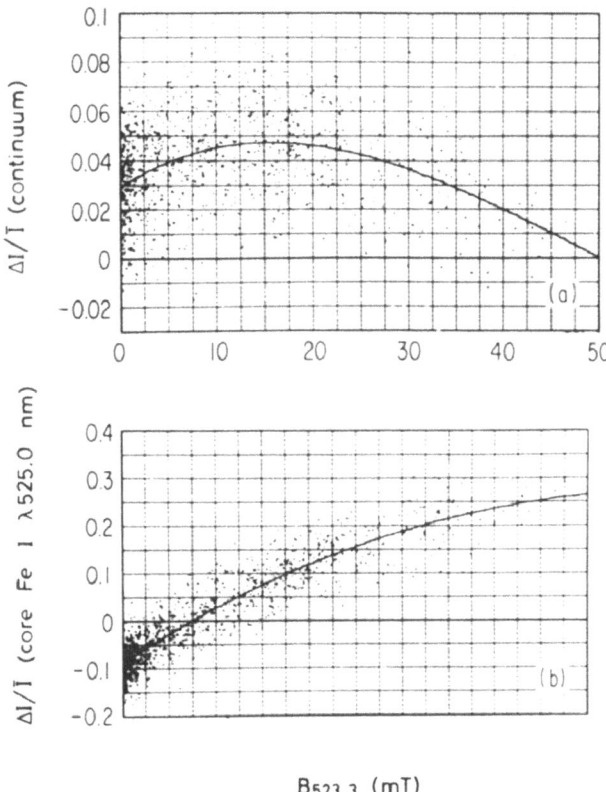

Fig. 3.   Relations between observed brightness enhancement $\Delta I/\bar{I}$ and observed magnetic field $B_{523.3}$.
(a) Continuum brightness. (b) Core brightness for Fe I λ 525.0 nm. From Frazier (1971).

The relation between observed magnetic field and brightness enhancement in the continuum is linear only up to $B_{523.3} \approx 5\ mT$ (possibly 10 mT). The brightness enhancement reaches a maximum around $B_{523.3} \approx 15\ mT$ and then decreases. This decline corresponds to a gradual transition to pores and sunspots as more magnetic flux occurs within the aperture (Frazier, 1971; Stenflo, 1975).

The enhancement in core brightness does not show such a decline for large values of the observed magnetic flux, but only a slight deviation from linearity when the observed field strength $B_{523.3}$ exceeds 30 mT. This degree of linearity is consistent with the observed linear relation between $B_{525.0}$ and $B_{523.3}$ for $B_{523.3} \lesssim 20\ mT$.

The magnetograph signal is fairly insensitive to variations in the continuum brightness, which explains why deviations from linearity in the continuum enhancement are not seen in the $B_{525.0} - B_{523.3}$ scatter-plot diagram. The behaviour of the continuum intensity shows that intrinsic changes in temperature and density in the lower photosphere are not independent of the amount of magnetic flux in a given region. It is remarkable, however, that the magnetic field strengths and line weakenings (temperatures and densities in the *upper* photosphere) appear to remain almost entirely independent of the amount of magnetic flux, at least for $B_{523.3} < 20\ mT$.

Chapman (1974) has pointed out another, more qualitative argument in favor of the uniqueness of the field-strength distribution. He observes that there is no consistent difference in appearance of dark $H\alpha$ fibrils lying on or near the magnetic features, although the apparent field strength of the features varies between 1 and 40 mT.

2.3. PROPORTION OF MAGNETIC FLUX IN STRONG-FIELD FORM

Howard and Stenflo (1972) and Frazier and Stenflo (1972) concluded from an analysis of simultaneous magnetograph recordings in the two lines Fe I $\lambda\lambda$ 525.0 and 523.3 nm that more than 90% of the total magnetic flux through the solar surface recorded by the magnetographs is chanelled through the strong-field regions. As this conclusion has quite far-reaching theoretical implications, we will explain the arguments behind it in some detail.

The aperture sizes used in the two investigations were $17 \times 17''$ and $2.4 \times 2.4''$, respectively, so it is not excluded that a considerable amount of flux is carried by some kind of 'turbulent' magnetic field, which is so tangled that opposite-polarity fluxes almost average out over an area of $2.4 \times 2.4''$. The magnetograph will hardly detect this type of flux unless the resolution is considerably improved (see Section 2.10).

To better understand the nature of the fluxes involved, it is useful to distinguish between two types of weak fields:

(1) Part of the weak-field flux is statistically unrelated to the strong fields.

(2) The remaining part of the weak-field flux is closely correlated with the strong fields, with a polarity either the same or opposite to that of the adjacent strong fields.

In the first case the weak fields will contribute to the scatter around the straight lines in the $B_{525.0} - B_{523.3}$ scatter-plot diagram. If there are no strong-field regions within the scanning aperture, the weak-field points will fall along a straight line with a slope of unity, and will be concentrated near the origin in the diagram. If there are strong fields inside the aperture, the weak-field flux will shift the points from the line of slope $\delta_{525.0}$ towards a line of slope unity, thus contributing to the scatter. In the case of the quiet-region scan of Frazier and Stenflo (1972), for which the highest magnetograph sensitivity was used, the scatter around the straight line of slope $\delta_{525.0}$ did not exceed the instrumental scatter. As the rms instrumental noise was 0.1–0.2 mT, the rms of the random component of $\bar{B}_w$ in (6) should be less than about

0.1 mT. This limit is less stringent for the active-region scans, for which the rms instrumental noise was about 0.8 mT.

The second kind of weak-field flux which is systematically related to the strong fields, is revealed as it causes a systematic displacement of the straight line. Since the sign of $\bar{B}_w$ depends on the sign of the strong fields, the second term in (6) determining the intercept of the straight line should change sign when going from the positive to the negative side of the scatter-plot diagram. If we fit two separate straight lines to the positive and negative points as in Figure 2 and determine the two intercepts l_+ and l_-, we can use (6) to obtain an estimate of the average value of $\bar{B}_w$:

$$\langle \bar{B}_w \rangle = \frac{l_+ - l_-}{2(1 - \delta_{525.0})}. \tag{11}$$

A positive value means that $\bar{B}_w$ is systematically of the same sign as the neighbouring strong fields.

$\langle \bar{B}_w \rangle$ has been found to be $\approx -0.08 \pm 0.04\ mT$ for the quiet-region scan and fluctuates between -0.2 and $-0.5\ mT$ for the active-region scans (Frazier and Stenflo, 1972).

As the straight lines fitted to the points in the scatter-plot diagram are mainly determined by points with $|B_{523.3}| \gg |\bar{B}_w|$, the weak fields we are talking about are located immediately adjacent to the strong fields, within the same aperture area. With no strong fields inside the aperture, the points would fall close to the origin in the scatter-plot diagram (or along a line of slope unity, which is not observed), where they do not contribute significantly to the measured value of $\langle \bar{B}_w \rangle$.

The negative values of $\langle \bar{B}_w \rangle$ indicate that the strong fields are immediately surrounded by weak fields of *opposite* polarity. Frazier and Stenflo (1972) interpreted this result in terms of a model of a strong-field region with diverging, 'mushroom'-shaped field lines, part of the magnetic flux returning back to the adjacent photosphere within a distance of a few arcsec.

To estimate the fraction of magnetic flux in strong-field form, the absolute values of the fluxes measured at each of the N aperture positions in a raster are summed up. The ratio ρ between the total fluxes measured in the 525.0 and 523.3 nm lines is

$$\rho = \sum_{i=1}^{N} |\Phi_{525.0}| \bigg/ \sum_{i=1}^{N} |\Phi_{523.3}|. \tag{12}$$

As we can assume that the true flux is measured in the 523.3 nm line (cf. the derivation of (6) in Section 2.1.2),

$$\rho = \sum_{i=1}^{N} |\delta \Phi_{s_i} + \Phi_{w_i}| \bigg/ \sum_{i=1}^{N} |\Phi_i|, \tag{13}$$

where the notations in (1) have been used.

Let us denote the fraction of the flux in strong-field form by R_s, i.e.

$$R_s = \sum_{i=1}^{N} |\Phi_{s_i}| \bigg/ \sum_{i=1}^{N} |\Phi_i|. \tag{14}$$

When strong-field elements are inside the scanning aperture, we can assume that $|\Phi_w| < \delta|\Phi_s|$, so that

$$|\delta\Phi_s + \Phi_w| = \delta|\Phi_s| + \varepsilon|\Phi_w|, \tag{15}$$

where $\varepsilon = +1$ or -1 depending on whether Φ_s and Φ_w are of equal or opposite signs. Away from strong-field elements $\Phi_s = 0$, and (15) is valid with $\varepsilon = 1$. Accordingly

$$\rho = \left(\delta \sum_{i=1}^{N} |\Phi_{s_i}| + \sum_{i=1}^{N} \varepsilon_i|\Phi_{w_i}|\right) \Big/ \sum_{i=1}^{N} |\Phi_i|. \tag{16}$$

Using (14), we have

$$\rho = \delta R_s + \sum_{i=1}^{N} \varepsilon_i|\Phi_{w_i}| \Big/ \sum_{i=1}^{N} |\Phi_i|. \tag{17}$$

Similarly, as $\delta = 1$ implies $\rho = 1$ according to (13) and (1),

$$1 = R_s + \sum_{i=1}^{N} \varepsilon_i|\Phi_{w_i}| \Big/ \sum_{i=1}^{N} |\Phi_i|. \tag{18}$$

Subtraction of (17) and (18) gives

$$R_s = \frac{1-\rho}{1-\delta}. \tag{19}$$

Observations (Frazier and Stenflo, 1972) give for a quiet region at disk center $\rho \approx 0.43$ and $\delta \approx 0.45$, resulting in $R_s \approx 104\%$. The explanation why R_s exceeds 100% is that Φ_s is not a directly observed quantity; for every aperture position weak-field fluxes Φ_w of *opposite* polarity always occur, cancelling out a small fraction of Φ_s. In the definition of R_s in (14), we have used the *observed* total flux $\sum |\Phi|$ in the denominator, which includes the flux cancelling effect inside an aperture area.

It thus appears that the magnetic flux through the solar photosphere in strong-field form is even *larger* than the total flux seen by a magnetograph with a scanning aperture of $2.4 \times 2.4''$ or larger. Part (at least 4%) of the flux seems to return to the solar surface within a distance of $2.4''$ from the strong-field regions. If the magnetograph resolution were infinitely high, the cancelling effect within the aperture would vanish, and R_s should be $\leq 100\%$. Similar results were obtained for the magnetic fields in active regions (Frazier and Stenflo, 1972).

2.4. QUESTION OF FLUX QUANTIZATION

As there is strong evidence for a unique field-strength distribution characterizing all magnetic elements in the quiet-region network as well as in active-region plages (see Section 2.2), the question naturally arises if there is also a unique amount of magnetic flux assigned to each of the smallest flux elements, in other words if there is flux quantization. Livingston and Harvey (1969) tried to determine the existence of such a flux quantum using a technique similar to the Millikan oil drop method for determining the charge of the electron. They reported some evidence for such a quantization, giving the value 2.8×10^{10} Wb for the flux quantum. This very difficult measurement was based on rather poor statistics, and later work has shown that the flux quantum Φ_0, if it exists at all, must be much smaller.

From an analysis of Stokesmeter data, Harvey *et al.* (1972) tried to derive sizes and field strengths of unresolved magnetic elements. According to their results, the average flux of an element was about 7×10^9 Wb, but both smaller and larger fluxes were found. Although this value is 4 times less than that found by Livingston and Harvey (1969), it should be regarded as an upper limit, since it is not known to what extent the flux concentrations studied represent aggregates of smaller flux elements.

Stenflo (1973) used a different approach to estimate an upper limit for Φ_0 from his $B_{525.0}$–$B_{524.7}$ data. The total observed magnetic flux divided by the area of the scanned quiet region corresponded to a field strength of 0.4 *mT*. Multiplying by the area of a supergranular cell, the average flux per cell is approximately 3×10^{11} Wb. If the average number of separate flux elements per supergranular cell is n, $\Phi_0 = 3 \times 10^{11}/n$ Wb. If we can set a lower limit for n, we obtain an upper limit for Φ_0.

As there is a one-to-one correspondence between flux concentrations and bright network points (see Section 2.8), we expect that n should equal the number of bright points making up the intricate filigree pattern of the network around one supergranular cell, as it would appear if seen with infinitely high spatial resolution. To judge from high resolution filtergrams, this number should be of the order of 100, maybe more, which also agrees with estimates of the number of spicules per supergranulation cell (Beckers, 1972; Lynch *et al.*, 1973). If the value of Φ_0 by Livingston and Harvey (1969) were correct, n would only be 10, which is clearly much too small. We estimate that $n \gtrsim 100$ and accordingly $\Phi_0 \lesssim 3 \times 10^9$ Wb.

Mehltretter (1974) compared the number density of bright facular points in a high-resolution filtergram with the magnetic flux in the same region recorded almost simultaneously at Kitt Peak with a $2.4 \times 2.4''$ aperture. He estimated the magnetic flux associated with a single facular point to be 4.4×10^9 Wb. This should probably be regarded as an upper limit, since clumps of bright points may not have been resolved, leading to an underestimate of the number density of bright points.

A further estimate, which is much less precise but leads to consistent results, is obtained by comparing large-scale flux measurements with counts of spicules or dark fine mottles, assuming that each such feature is associated with a flux element. The total magnetic flux through the Sun was typically 10^{15} Wb during the period 1967–1973 (Howard, 1974). The total number of dark fine mottles on the Sun is estimated to be about 300 000 (Michard, 1974). All dark fine mottles are likely to be associated with flux elements. The reverse may not be true; there need not be a mottle at every flux element. The total flux divided by the number of dark fine mottles should then be an upper limit for Φ_0, the flux per magnetic element. Accordingly, $\Phi_0 \lesssim 3 \times 10^9$ Wb.

Considering all these arguments together, we conclude that $\Phi_0 \lesssim 5 \times 10^9$ Wb, which is a rather conservative estimate of the upper limit. As we will see below (e.g. Section 2.6), it is highly doubtful that the concept of flux quantization is at all' physically meaningful, and it should better be abandoned.

2.5. TRUE SIZES OF MAGNETIC ELEMENTS

Let us define the size d of a magnetic element by the equation

$$\Phi = B_0 d^2, \tag{20}$$

Φ being the magnetic flux and B_0 the maximum field strength within an element. As B_0 must be in the interval 130 to 260 mT according to Section 2.1.2, and individual flux elements in the network should typically have $\Phi \leq 5 \times 10^9$ Wb according to Section 2.4, we obtain a corresponding upper limit for d by using $B_0 = 130$ mT and $\Phi = 5 \times 10^9$ Wb in (20). The result is

$$d \leq 200 \text{ km,} \tag{21}$$

or $d \leq 0.3''$.

It is not quite clear whether the associated brightness structures are of the same size as the magnetic elements, although more detailed data analysis (Stenflo, 1975) indicates that the brightness structures are slightly larger (see Section 2.8). Dunn and Zirker (1973) recorded at excellent seeing conditions an intricate small-scale brightness pattern in the far wings of Hα, which they called 'solar filigree'. Typical sizes of elements in the filigree pattern were 200 km. As this was about the same as their instrumental resolution, true sizes could not be derived.

Mehltretter (1974), using broad and narrow-band filters in the near UV, recorded photospheric faculae at the center of the solar disk with high spatial resolution. His results suggest that 'solar filigree' are identical with faculae and represent concentrations of magnetic flux. The full half-width of telescope + microphotometer was 140 km. Mehltretter (1974) found that the true half-width of the facular points, after correction for smearing, was about the same as or smaller than the instrumental half-width of 140 km. This result is consistent with our upper limit for d, the size of the magnetic elements. We recall that our estimate of d did not depend on observations with high spatial resolution but was mainly based on field strengths derived using the line-ratio method.

Let us stress that we have been discussing only the smallest flux elements in this section. Larger flux aggregates with correspondingly larger sizes often cannot be regarded as the mere sum of smaller flux fragments but may act cohesively as one single unit. This will be clarified more in the next section.

2.6. LIFE-TIMES

The life-times of flux concentrations and bright points depend on the size of the feature studied, which explains why various authors have reported highly different results. Mehltretter (1974) finds that his bright facular points with diameters of about 150 km live for 5–15 min. Associations of bright points and 'micropores' have life-times of 10–30 min (Mehltretter, 1974), which agrees with results of Bray and Loughhead (1961), Beckers and Schröter (1968a), and Sheeley (1969). Using the line-ratio method, Howard and Stenflo (1972) found evidence that the life-time of the basic magnetic elements is less than two hours, without being able to set any lower limit.

Assuming that the life-time of the bright elements also represents the life-time of the associated magnetic structures, a decay rate

$$d\Phi/dt \approx -10^7 \text{ Wb s}^{-1} \tag{22}$$

appears to account for most of the observational results remarkably well. An element with $\Phi = 4 \times 10^9$ Wb, typical for the facular points observed by Mehltretter (1974), would decay in a time of about 7 min according to (22), in agreement with Mehltretter's (1974) estimate of 5–15 min. Moving magnetic features (MMF) around sunspots have typical fluxes of 10^{11} Wb each, and seem to endure for hours (Harvey and Harvey, 1973; Vrabec, 1974). According to (22), an element with $\Phi = 10^{11}$ Wb will live for 2–3 h, consistent with the above results. Harvey and Harvey (1973) estimate that the net flux loss from a sunspot through MMFs is $d\Phi/dt \approx -10^{11}$ Wb h^{-1}, i.e. -3×10^7 Wb s^{-1}, which is slightly larger than (22) but of the same order of magnitude. Bumba (1963) and Gokhale and Zwaan (1972) observe that the slowest flux loss from a sunspot, during its most stable phase of decay, is $-(1.2–1.8) \times 10^7$ Wb s^{-1} independent of the sunspot area, in excellent agreement with (22).

The remarkably universal applicability of the decay rate expressed by the empirical law (22) indicates that sunspots, pores, 'magnetic knots', MMFs, and bright points are not merely to be regarded as an association of individual flux quanta, since the life-time of the aggregate would in that case be about the same as the life-time of its individual elements. Instead the flux aggregate should have some cohesiveness and act as a single physical unit, similar to the flux ropes described by Piddington (1975a, c).

The diffusion time τ_d for an element of size d is

$$\tau_d = \mu_0 \sigma d^2, \tag{23}$$

σ being the electrical conductivity. As the flux Φ is $B_0 d^2$, B_0 being the maximum field strength (cf. (20)), dispersion of the field lines occurs at a characteristic rate of

$$\Phi/\tau_d = B_0/(\mu_0 \sigma). \tag{24}$$

Using the photospheric values of $B_0 = 0.2$ T (see Section 2.1.2) and $\sigma = 10$ mho m^{-1} (Kopecký, 1971), we obtain $\Phi/\tau_d \approx 1.6 \times 10^4$ Wb s^{-1}, which is about three orders of magnitude slower than the decay rate given by our empirical law (22).

It thus appears necessary to invoke some kind of plasma instability occurring on a small scale, to explain the observed fast decay rate of solar magnetic fields. Larger flux concentrations must break up into smaller flux elements, which retain the strong values of the field strength. Otherwise a considerable fraction of the flux would appear in diluted, weak-field form, which is not observed (see Section 2.3). In the Fourier domain the flux moves from small to large wave numbers, where rapid decay may occur. This picture supports the recent suggestion by Piddington (1975a, c) that flux ropes may decay by untwisting and subsequent 'fraying' or separation of flux fibres by the flute instability.

Implications of the decay rate for theories of the solar cycle and for heating of the solar atmosphere will be discussed in Section 3.

2.7 RELATION TO VELOCITY FIELDS

Flux concentrations always appear to be correlated with downflow of matter, regardless of whether they occur in the quiet-region network (Simon and Leighton, 1964; Tannenbaum et al., 1969; Frazier, 1970) or in active-region plages (Servajean,

1961; Beckers and Schröter, 1968a; Giovanelli and Ramsay, 1971; Sheeley, 1971; Howard, 1971, 1972).The magnitude of these downdrafts is of the order of 0.5 km s^{-1} as observed with a spatial resolution of a few arcsec.

The next question that arises is whether the velocity field has as pronounced a fine structure as the magnetic field. Skumanich *et al.* (1975) found a proportionality between the apparent velocity and the apparent magnetic field: $v_{obs}/B_{obs} \approx 16$ m s^{-1} mT^{-1}. If the true maximum field strength in the magnetic elements is $B_0 = 200$ mT, the peak velocity would be $v_0 = 3.2$ km s^{-1} if the velocity is strictly proportional to the magnetic field at the smallest scale. Such a high value of v_0 seems to be ruled out by stokesmeter observations of Harvey *et al.* (1972) and line-ratio data by Stenflo (1973), indicating that the average downdraft velocity in the magnetic elements is less than 1 km s^{-1}. This apparent contradiction would be resolved if the velocity profile is broader than the corresponding magnetic profile, with a substantial fraction of the downward mass flux occurring immediately outside the magnetic structures. A similar conclusion was reached by Dravins (1974) from a comparison of the visibility of the bright network in the blue and red wings of Hα. Howard (1972) noted a slight tendency for the downward motions to be concentrated over the parts of the flux regions with high gradient of the apparent magnetic field, which is consistent with the above interpretation.

Frazier (1974) employed the line-ratio method to derive information on the velocity profile of the magnetic elements. The apparent downdraft velocities recorded simultaneously in the two lines Fe I $\lambda\lambda$ 525.0 and 523.3 nm are systematically different from each other. Frazier (1974) combined his velocity data with the $B_{525.0}-B_{524.7}$ results of Stenflo (1973), and assumed that the profiles of the true line-of-sight velocities $v(r)$ and magnetic fields $B(r)$ are related by

$$v(r) = \alpha B(r)^{\beta} \tag{26}$$

in the case of a cylindrically-symmetric element. r is the distance from the axis of the element. The constant α is fixed by the requirement that the true average downdraft velocity should be 0.5 km s^{-1}, as suggested by Harvey *et al.* (1972) and Stenflo (1973). β is the remaining free parameter to be determined. If β is less than unity, the velocity structures are coarser than the magnetic-field structures.

The value of β was found to be sensitive to the *shape* of the magnetic-field profile, which was not known. The observational data could however only be satisfied for profiles with the peak field strength B_0 in the interval 130 to 260 mT. The value of β decreased with increasing B_0 and was found to be smaller than unity for $B_0 \gtrsim 170$ mT. For the Gaussian magnetic-field profile ($B_0 = 230$ mT), $\beta \approx 0.25$. As indirect evidence supports the view that the velocity structures are more extended than the associated magnetic-field structures, i.e. $\beta < 1$, we might tentatively narrow down the allowed interval for B_0 to be between 170 and 260 mT. Let us however add a word of caution: The different levels of height of formation of the 525.0 and 523.3 nm lines were not taken into account by Frazier but may affect the results.

Finally it should be mentioned that the bright network or facular points show transverse motions, as if being shuffled around by the solar granulation over distances of the order of 1000 km (Sheeley, 1969, 1971; Dravins, 1974; Mehltretter, 1974).

2.8. RELATION TO TEMPERATURE AND DENSITY VARIATIONS

The flux concentrations are always accompanied by line weakenings, both in the quiet-region network (Sheeley, 1967; Chapman and Sheeley, 1968; Harvey and Livingston, 1969; Frazier, 1970; Harvey *et al.*, 1972; Stenflo, 1973, 1974b, 1975) and in active-region plages (Schmahl, 1967; Stellmacher and Wiehr, 1971, 1973; Stenflo, 1974b, 1975). Due to the unique and universal properties of the magnetic elements (see Section 2.2), their temperature and density structure should be essentially the same in quiet as in active regions (if the average observed field strength $B_{523.3} \lesssim 10\ mT$). A facular model should thus be identical with a model of a magnetic element or a model of the quiet-region network. A number of facular models have been constructed in the past, based on continuum and line-profile data (Kuz'minykh, 1964; Schmahl, 1967; Chapman, 1970; Wilson, 1971; Stellmacher and Wiehr, 1971, 1973; Shine and Linsky, 1974). All of these models have however been hampered by the finite spatial resolution of the observations, which have only been able to indicate lower limits to the true brightness enhancements. This serious limitation was avoided by Stenflo (1974b, 1975), by combining continuum and line profile data with magnetograph observations using the line-ratio method. The zero point of the geometrical height scale, which could not be obtained from the observational data available, was chosen by Stenflo (1974b, 1975) in such a way that the temperature difference between the inside and outside of the flux tube is zero at large depths (more precisely at $\tau_0 = 25$, τ_0 being the continuum optical depth at 500 nm). The justification for this is that the narrow flux tube is almost optically thin making it difficult to maintain large temperature differences in the lower layers, where the radiative relaxation time is short (cf. the discussion in Section 3.1).

The resulting model, which satisfies the observational data including the magnetograph observations, shows no temperature deficit in the lower layers. The temperature excess starts at a height of about 180 km above the level where $\tau_0 = 1$ in the Harvard-Smithsonian Reference Atmosphere (HSRA), and increases rapidly with height. It gives line weakenings concentrated to the line core (Stellmacher and Wiehr, 1971, 1973), the neutral lines being considerably more weakened than the ionized lines, as observed by Chapman and Sheeley (1968). It further gives a positive continuum contrast at disk center, in agreement with observations of Schmahl (1967), Frazier (1971), Dunn and Zirker (1973), and Mehltretter (1974). The higher layers appear to be efficiently heated in the flux elements, presumably by MHD waves. This heating may be of fundamental importance for the whole energy-balance of the solar atmosphere (see Section 3.2).

Let us now face the interesting question whether the brightness structures seen in the continuum are coarser than the magnetic-field structures or not. Analogously to the case of magnetic fields and velocities (see Section 2.7), Skumanich *et al.* (1975) found a proportionality between the apparent continuum brightness enhancement at disk center and the apparent magnetic field B_{obs} for $B_{obs} < 12\ mT$: $(\Delta I_c / \bar{I}_c)_{obs} / B_{obs} \approx 9.3 \times 10^{-4}\ mT^{-1}$. If there were strict proportionality between the true brightness enhancement and the true magnetic field at the smallest scale, a peak magnetic field B_0 of 200 mT would correspond to a peak brightness contrast at disk center $\Delta I_c / \bar{I}_c$ of as much as 19%. According to the model calculations of Stenflo (1974b, 1975), the observational data could not be satisfied if the true average continuum contrast across a magnetic element exceeds 8%. The most probable model had an average contrast of about 6%. This indicates that the diameter of the brightness elements slightly exceeds that of the magnetic elements. Applying the line-ratio method on combined brightness, velocity, and magnetic-field data, a future determination of the shapes of the magnetic, brightness, and velocity profiles should be possible, as indicated in Section 2.7 when discussing the velocity profiles.

Dunn and Zirker (1973) and Mehltretter (1974) found that the bright network and facular points are preferentially located in the intergranular lanes and are associated with abnormal granulation. As the facular material fills the darker intergranular lanes with emission, the contrast in the granulation is washed out. The presence of facular emission at disk center in the continuum can only be discerned in photographs with spatial resolution better than 0.5" or by sensitive photoelectric recordings, revealing average enhancements in the continuum signal of the order of one percent.

As described in Section 2.2 and by Figure 3, the increase of the continuum brightness turns into a decrease and darkening as larger and larger flux concentrations are considered. This shows again that a large flux concentration is physically different from the sum of a number of smaller flux elements with the same total flux, a conclusion reached in Section 2.6 when studying the life-times of flux aggregates with different amounts of flux. Active cooling seems to be able to reduce the temperatures in the lower layers only when the flux concentration is sufficiently large and optically thick, so that radiative relaxation is slow enough, and the large flux may effectively interact with convection (Stenflo, 1975). According to Figure 3, deviations from a linear relation between continuum brightness flux and magnetic flux start to become significant for $B_{523.3} \gtrsim 10 \ mT$, i.e. when more than about 10% of the $2.4 \times 2.4"$ aperture area is occupied by strong-field magnetic flux. Only when more than 30% of the aperture area is covered by the strong fields, the region is darker in the continuum.

Finally we note the close relation between bright network points and spicules. Spicule counts (Beckers, 1972; Lynch et al., 1973) indicate satisfactory agreement with estimates of the number of flux elements, as was noted in Section 2.4. The spicule life-time of 5–15 min (Beckers, 1972) is about the same as the life-time of bright points (see Section 2.6). Each bright point cannot give rise to more than *one* spicule, since it has disappeared before a new spicule can occur. Successive spicules that appear to be ejected from one and the same bright area thus represent an apparent effect caused by the limited spatial resolution.

2.9. HEIGHT VARIATION OF THE MAGNETIC FIELD

Very little is known about the height variation of the field. Straightforward comparison between magnetograph recordings made in lines formed at different heights is not possible, because the generally dominating effects of Zeeman saturation, line weakenings, etc., affect various lines differently. Further, the magnetograph signal is mainly proportional to the flux, not the field strength. Averaged over an area of a few arcsec, the flux from a 0.2" magnetic element may be the same at different heights, although the field lines are diverging and the field strength decreases with height.

The most direct way to measure the height variation therefore seems to be to use the line-ratio method. The field strengths derived by Stenflo (1973) from his $B_{525.0} - B_{524.7}$ observations refer to a height of approximately $+100$ km relative to the height where $\tau_0 = 1$ in HSRA (Stenflo, 1975). By using other pairs of lines formed at different heights, the true height variation may be found. Another way is to study the center-to-limb variation of the $B_{525.0} - B_{524.7}$ and $B_{525.0} - B_{523.3}$ scatter-plot diagrams. The center-to-limb variation of the ratio $B_{525.0}/B_{523.3}$ indicates that the field lines diverge rapidly with height (Howard and Stenflo, 1972). The conclusion that part of the strong-field flux returns in the form of weak-field opposite-polarity flux adjacent to the magnetic elements (Frazier and Stenflo, 1972) as described in Section 2.3 is a further indication of diverging field lines. No quantitative evaluation of the height gradient has however yet been made.

Other indirect evidence for rapid spreading of the field lines with height is obtained from the observation that the brightness structures become coarser higher up in the atmosphere (Hale and Ellerman, 1903; Simon and Noyes, 1971; Dunn and Zirker, 1973). It is not clear, however, whether the diameter of a bright element is strictly proportional to the diameter of the flux tube at all heights. Beckers (1963, 1968) finds that the inclination angle to the vertical for spicules and dark fine mottles has a very wide distribution, with an average inclination of about 20° for both phenomena. It is natural to expect that the field lines should show similar inclinations. Deubner (1975) has recently obtained evidence from magnetograph observations that the field vectors have a nearly isotropic distribution in quiet regions.

Mehltretter (1974) finds that some of his bright 'points' are slightly elongated. The wide and narrow-band filtergrams, representing two height ranges in the solar atmosphere, show a relative displacement of the bright points in the direction of elongation, indicating a tilt against the normal. Strong tilts of facular elements have also been reported by Stoyanova (1970) and Krat (1973).

2.10. NON-NETWORK MAGNETIC FIELDS AND POSSIBLE MAGNETIC TURBULENCE

In Section 2.3 we came to the conclusion that practically all the magnetic flux through the Sun seen by a magnetograph aperture of $2.4 \times 2.4''$ or larger is in the form of strong fields associated with line weakenings. This magnetic flux is located in the supergranulation network when it occurs outside active regions. This conclusion does not exclude the possibility that large amounts of flux and strong fields can occur away from the network. The strong non-network field, if it exists, must however be so tangled that opposite-polarity fluxes cancel out so effectively over an area of $2.4 \times 2.4''$ that the net flux recorded by the magnetograph aperture is small compared with the net flux in the network.

Livingston and Harvey (1971) recorded weak fields inside supergranular cells, which had an apparent strength of $0.2-0.3$ mT. Recently the fields in the cell interiors have been recorded with considerably improved spatial resolution (Livingston and Harvey, 1975). The polarities were found to be mixed and independent of the surrounding, dominating network-field polarity. Many elements appeared bipolar, although the flux of the two polarities seldom balanced. The apparent fluxes involved were generally quite small.

Such a tangled field is similar to a 'turbulent' magnetic field. If there exists a strong 'turbulent' field on a small scale, which does not contribute significantly to the net flux recorded with the smallest apertures that can presently be used, it will nevertheless reveal itself by its line-broadening effect. The contribution to the line width from magnetic 'turbulence' is proportional to the Landé factor of the spectral line. Unno (1959) tried to measure B_{turb} by comparing the widths of lines with different Landé factors and obtained an upper limit for B_{turb} of about 30 mT. Howard and Bhatnagar (1970) studied if the difference in line width between granular and intergranular regions varied with Landé factor, and concluded that the difference in B_{turb} between granules and intergranular lanes is less than 3.5 mT. Other authors (Steshenko, 1960; Semel, 1962; Leighton, 1965; Livingston, 1968; Beckers and Schröter, 1968b) have tried to find a magnetic field associated with the granulation in the interior of supergranular cells and have been able to place an upper limit of a few mT for such a field as observed with a resolution of the order of one arcsec.

3. Implications of the Small-Scale Magnetic Structure

3.1. ORIGIN OF THE OBSERVED STRONG FIELDS

There are essentially three different types of forces that may be invoked to explain how locally strong solar magnetic fields can be confined.
(1) Gas pressure.
(2) Dynamic forces.
(3) Electromagnetic forces.

Figure 4, taken from Stenflo (1975), illustrates the difficulties encountered by gas pressure and dynamic pressure in confining the field. The average height of formation of the Fe I $\lambda\lambda$ 525.0 and 524.7 nm line wings, which are used to determine the strength of the field, is about 100 km above the level where $\tau_0 = 1$ (Lites, 1972). At this level the ambient gas pressure equals the magnetic pressure of a field of about 120 mT (1.2 kG). This is the maximum field strength that can be confined by gas pressure if the flux tube is completely evacuated at this level (zero internal gas pressure). The field strengths deduced from observations exceed this value. Still Spruit (1975) and Parker (1976) manage to confine 200 mT fields by means of gas pressure effects. This is done by assuming active cooling inside the flux concentrations (Parker, 1976), similar to the cooling mechanism in sunspots (Parker, 1974d, e). The cooling makes the flux region more transparent, producing a Wilson depression of about 200 km (Spruit, 1976). The wings of the spectral lines are accordingly formed deeper down in the magnetic regions, where the external gas pressure is large enough to confine the strong fields.

An argument against this idea is that it is difficult to maintain a large temperature difference between interior and exterior when the flux tube is narrow and close to being optically thin ($d \lesssim 200$ km as shown in Section 2.5). The reason is the short radiative relaxation time for an optically thin plasma, 29 s at the 0 km level, and 12 s

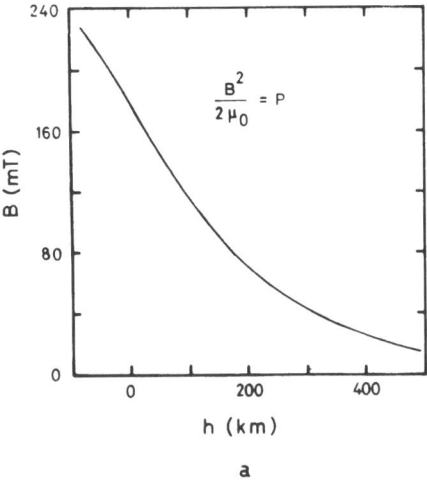
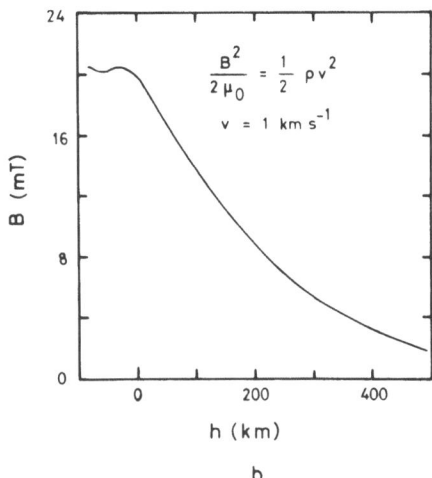

Fig. 4. Variation with height of the field strength that has a magnetic pressure equal to (a) the gas pressure, (b) the dynamic pressure from a flow of 1 km s^{-1}. Pressures and densities are from the facular model of Stenflo (1975).

at the -100 km level (Ulmschneider, 1971). The Wilson depression is expected to increase with optical thickness of the flux tube. The observations of Frazier (1971) show (see Figure 3) that the small isolated flux tubes are physically different from small sunspots in the deeper layers. The temperature deficit in the lower layers in a sunspot is revealed by a corresponding brightness reduction in the continuum. In contrast, the isolated flux elements in plages and in the network are *brighter* in the continuum at disk center (cf. Figure 3 and Section 2.8).

In spite of these difficulties, Parker (1976) believes active cooling to be a more likely cause of the field enhancement than any other mechanism he has considered, i.e. longitudinal and transverse aerodynamic drag (Parker, 1974a, 1976), the Bernoulli effect (Parker, 1974b), turbulence at great depth, viscous stresses in a downdraft, and force-free field configurations.

As shown by Figure 4b, the supergranular flow with velocities of $0.3-0.5$ km s^{-1} has a kinetic energy density equal to the magnetic energy density of a magnetic field of less than 10 mT. Even in the ordinary granulation, where velocities of a few km s^{-1} occur, the kinetic energy density is much less than the observed magnetic energy density. Peckover and Weiss (1972) have however shown that convection may be able to amplify the field to peak magnetic energy densities considerably exceeding the kinetic energy density in the convective flow. Accordingly it is not ruled out that the granular flow may contribute significantly to maintaining the field confined.

There remains the possibility that the magnetic flux may be confined by internal rather than external forces, by the constricting, pinching forces of longitudinal electric currents (corresponding to a twisted flux-rope), as advocated by Stenflo (1974b, 1975) and Piddington (1975a, c). Indications that twisted fields are a common phenomenon on the Sun come from a variety of observations. The Hα fibrils around sunspots often show a vortex pattern (Hale, 1908, 1927; Richardson, 1941). Observations of transverse magnetic fields (e.g. Severny, 1965; Kotov, 1970) show strong vertical electric currents in active regions. The time sequence of orientation angles for evolving arch filament systems (Frazier, 1972; Weart, 1972) indicate strong twists. Vorphal (1974) measured 2–3 twists in a length of $(2-3) \times 10^4$ km of an emerging filament. All these observations relate however to relatively large-scale structures. There is presently no direct observational evidence relevant to the question whether all the subarcsec flux tubes are twisted or not.

Parker has rejected force-free fields in his discussion of field confinement, because he believes that twisting cannot enhance the field (Parker, 1976). On the other hand Sreenivasan (1973a, b) and Sreenivasan and Thompson (1974) have shown that certain types of flows, so-called hydromagnetic Beltrami flows, preserve the force-free character of the magnetic field in a resistive medium, and can amplify the fields far beyond equipartition between magnetic and kinetic energy. Part of the reason for this apparent discrepancy between Parker and Sreenivasan appears to be that Parker considers force-free fields in hydrostatic equilibrium through the equation

$$(\mathbf{\nabla} \times \mathbf{B}) \times \mathbf{B} = \mathbf{0}, \tag{26}$$

while Sreenivasan treats the general case of dynamic force-free fields with velocity fields in a resistive medium, including the evolution of the field through the equation

$$\frac{\partial \mathbf{B}}{\partial t} = \mathbf{\nabla} \times (\mathbf{v} \times \mathbf{B}) + \frac{1}{\mu_0 \sigma} \mathbf{\nabla}^2 \mathbf{B}. \tag{27}$$

Stenflo (1974b, 1975) has suggested that Sreenivasan's mechanism can contribute to amplification of the small-scale solar magnetic fields through vorticity flow around the downdrafts (hydromagnetic tornado effect). The Beltrami type of flow can be similar to the whirl in a bath-tub when water goes down the drain. The flux concentrations on the Sun are always cospatial with sinks in the flow (see Section 2.7), located in the intergranular lanes at the supergranular cell boundaries. Vorticity around these sinks may be generated by colliding pressure fronts, likely to occur in a dynamic, inhomogeneous atmosphere, by turbulent shear, and to some extent by Coriolis forces.

So far we have only discussed how the magnetic field may be confined in quasi-stationary flux tubes in the photosphere, and how the field may be amplified in or near the surface layers of the Sun. Parker (1976) has argued that all the amplification from about 10 to 200 mT must in fact take place at the surface or immediately below, since flux tubes are buoyant, and strong fields rise to the surface at nearly the Alfvén speed (Parker, 1955; 1974c). Accordingly fields substantially stronger than 10 mT can hardly be retained in the convection zone.

Contrary to this Piddington (1975a, c) believes that all amplification takes place in the convection zone by differential rotation producing toroidal magnetic flux, which is subsequently rolled into helically twisted flux ropes. Piddington suggests that the typical field strength in the flux ropes in the convection zone is 400 mT.

If not all the field amplification takes place in the convection zone, it is clear that most of the preamplification must occur there. Otherwise the flux would first emerge in a diffuse form, contrary to observations (see Section 2.3). Let x be the fraction of surface magnetic flux, which at any given time occurs in the form of fields stronger than 100 mT, and let τ be the average life-time of such fields. Let us further assume that the magnetic flux first emerges in the form of fields weaker than 100 mT, and that it takes an average time τ_a to amplify them to fields stronger than 100 mT. Assuming that the magnetic flux is eliminated when the strong flux elements die, and that this flux is replaced by new flux emerging from below, we have in a stationary situation

$$x = \frac{\tau}{\tau_a + \tau},$$ (28)

or

$$\tau_a = \frac{1-x}{x}\,\tau.$$ (29)

According to Sections 2.3 and 2.6, $1-x$ may be estimated to be of the order of 10% or less, while τ is about 10 min, assuming that the magnetic flux is destroyed when the cospatial bright points disappear. With these values we get an amplification time $\tau_a \approx 1$ min which is extremely short. In this time an Alfvén wave would travel of the order of 200 km only, and granular motions even less. The observation that only a very small fraction of the total magnetic flux is seen in diffuse form in the photosphere strongly supports the idea of Piddington (1975a, c) that in fact *all* the amplification of the field must occur in the convection zone before the field appears in the photosphere, in spite of Parker's arguments about the buoyancy of flux tubes. Some of the effects we have been discussing, like gas and dynamic pressure effects, vorticity, etc., might play some role down in the convection zone as well, although their contribution to field amplification in the photosphere is likely to be marginal.

When the smallest magnetic elements disintegrate at the end of their life-time τ, it really seems that their flux is completely eliminated from the photosphere, because a mere dispersion of the field lines would be observed as a diffuse field. Such a fast flux elimination is however presently not physically understood.

Finally a word of caution again: The conclusion that all the field amplification occurs below the photosphere rests on the order-of-magnitude correctness of our estimates of $1-x$ and τ in (29), and on the assumption that the flux disappears after time τ, which in turn is justified by our estimated small value of $1-x$. The estimate of τ rests mainly on Mehltretter's (1974) study of the life-time of bright facular points, and on the observation that the flux concentrations always appear to be associated with such bright points, as discussed in Section 2.8. If the strong fields were considerably more long-lived than the bright points, a substantial fraction of the strong-field flux as observed at any given time would be unrelated to bright points or line weakenings, which seems to contradict observations. Nevertheless our estimates of x and τ need verification by new observations.

3.2. Heating of the Solar Atmosphere

Magnetic fields play a role in the energy balance of the solar atmosphere in a variety of ways. Some of the most important are:

(1) Dissipation of magnetic energy, related to the elimination of magnetic flux.
(2) Generation and guiding of MHD waves, which heat the higher layers.
(3) Channelling the heat conduction from the corona.
(4) Causing spicules.

For a given magnetic flux Φ through a surface area A, the lowest energy state corresponds to the situation when the flux is evenly spread out over the area. In that case the average magnetic energy per unit volume in the area, $\bar{W}_{M,min}$, is

$$\bar{W}_{M,min} = \bar{B}^2/2\mu_0, \tag{30}$$

$\bar{B} = \Phi/A$ being the average field strength. If we instead have an inhomogeneous distribution, with fields of strength B_0 covering a fraction α of the area, while the rest of the area has zero field, we get

$$\bar{W}_M = \alpha B_0^2/2\mu_0. \tag{31}$$

As

$$\bar{B} = \alpha B_0 \tag{32}$$

we can eliminate α from (31), yielding

$$\bar{W}_M = \bar{B}B_0/2\mu_0, \tag{33}$$

or, using (30),

$$\bar{W}_M = \frac{B_0}{\bar{B}} W_{M,min}. \tag{34}$$

In quiet regions, with a typical value for $\bar{B}$ of $0.4\ mT$, and with B_0 in the strong-field elements estimated at $120\ mT$, the average magnetic energy density exceeds the energy density in a uniformly spread out field with the same magnetic flux by a factor of about 300.

A flux rope with a radius of 100 km and a field strength of $100\ mT$ has a magnetic energy stored within 1000 km along the rope of about 10^{20} J if it is untwisted, more if it is twisted. A twisted, long flux rope may easily contain a magnetic energy of 10^{21} J in the solar atmosphere, which equals the total energy released during a small solar flare, although it carries only a very small amount of magnetic flux. If the annihilation or untwisting of such a flux filament were the cause of a solar flare, a solar magnetograph would hardly notice any changes in the longitudinal magnetic flux. Instead, fictitious flux changes may be recorded, related to line weakenings caused by flare heating (cf. Frazier and Stenflo, 1972; Stenflo, 1974b).

The total number of small flux elements on the Sun was estimated at about 300 000 or more in Section 2.4. This would correspond to a magnetic energy in the solar atmosphere of 10^{26}–10^{27} J. If the magnetic energy in each element is eliminated in as short a time as about 10 min indicated by Mehltretter's (1974) observations (see Section 2.6), the rate of energy dissipation averaged over the solar surface is of the order of $10^5\ J\,m^{-2}\,s^{-1}$. This must however be regarded as an upper limit, since far from all the magnetic flux is in the form of the smallest elements corresponding to bright points, and larger elements have longer life-times (see Section 2.6). Still it is interesting to note that this upper limit is about two orders of magnitude more than is needed to heat the outer layers of the solar atmosphere. The net radiative energy loss from the chromosphere is $(2-6) \times 10^3 \, \text{J m}^{-2} \, \text{s}^{-1}$, while the energy loss from the corona due to radiation and solar wind but ignoring heat conduction downwards to the chromosphere is $55-350 \, \text{J m}^{-2} \, \text{s}^{-1}$ according to Bray and Loughhead (1974), who give further references. The heating of the solar atmosphere might thus be accounted for, at least partially, by the dissipation of magnetic energy. This energy transformation may occur by dissipation of magnetic flux, or by unwinding of helically twisted fields.

The flux concentrations will also provide heating by MHD waves travelling up along the field lines, generated by the interaction between convection and the magnetic field (Piddington, 1973; Stenflo, 1975). The height variation of the temperature excess in magnetic elements (see Section 2.8) resembles what would be expected by MHD wave heating (Osterbrock, 1961; Milkey, 1970). It is not clear, however, how much this heating contributes to the overall heating of the solar corona.

The relation between flux concentrations and spicules has been touched upon in Section 2.8. It was pointed out that only one spicule should be ejected from each flux element, because the life-time of the field concentration does not seem to exceed the spicule life-time. The excess gas pressure due to efficient heating by MHD waves may be released by outward expansion of the over-heated plasma along the field lines. If these field lines are twisted, the ejected plasma will get a spinning motion (Stenflo, 1974b, 1975). The inclined emission lines of spicules have been interpreted in terms of spinning (Michard, 1956; Beckers et al., 1966; Pasachoff et al., 1968), with rotational velocities at the periphery of the spicules of as much as $30 \, \text{km s}^{-1}$, comparable in magnitude with the velocity along the spicule axis.

In small bipolar regions with field-lines in the form of closed loops, the over-heated plasma cannot escape and accumulates at the top of the loop. This may explain the X-ray bright points (Golub et al., 1974), which are cospatial with ephemeral active regions (Harvey and Martin, 1973; Harvey et al., 1974).

3.3. DIFFERENTIAL ROTATION

Two important results obtained from observations of solar rotation are:

(1) The magnetic fields rotate faster by about 5%, i.e. $0.1 \, \text{km s}^{-1}$ at the equator, than the photospheric plasma (Howard and Harvey, 1970; Wilcox and Howard, 1970; Wilcox et al., 1970; Stenflo, 1974a).

(2) The fields outside active regions (background fields) show a latitude variation of angular velocity more similar to rigid rotation than the active-region fields (Wilcox and Howard, 1970; Wilcox et al., 1970; Stenflo, 1974a). The angular velocity of the various fields agree at the equator, where the maximum angular velocity occurs.

The difference in rotation between the magnetic field and the plasma implies that the magnetic flux cannot be smoothly spread out over the solar surface. The high electrical conductivity in the photosphere prevents slippage of the field lines over a large scale at the required rate. As it is now known that most of the magnetic flux seen by solar magnetographs is in the form of strong fields occupying only a small fraction of the solar surface, the discrepancy between the rotational velocities can be understood without invoking field-line slippage. The observations of Doppler shifts refer to average line profiles determined mainly by non-magnetic regions (non-magnetic in the sense that they do not contribute significantly to the flux recorded by a magnetograph). The plasma within the strong-field regions, which carry most of the magnetic flux, may well be approximately at rest with respect to the field lines, although the surrounding photospheric plasma may stream with a velocity of about 0.1 km s^{-1} in the eastward direction relative to the flux tubes (Foukal, 1972; Stenflo, 1974a).

The flux tubes can be regarded as rigid columns, exerting a viscous drag on the surrounding photospheric plasma. The effect of this drag is smaller, the smaller the cross-section of the tube. If the flux tubes are anchored in deeper, faster-rotating layers, their drag will tend to speed up the photospheric layers. Foukal and Jokipii (1975) estimate that the spin-up time t due to viscous drag of the flux tubes is

$$t \approx d/(\alpha v), \tag{35}$$

where d is the diameter of the flux tube, v is the relative velocity between the tube and the surrounding plasma, and α is the fraction of the solar surface covered by flux tubes. For d we can use 150 km (see Section 2.5). $v \approx 0.1$ km s^{-1}. α can be estimated in the following way: The total magnetic flux through the solar surface is typically 10^{15} Wb as observed with a $17 \times 17''$ magnetograph aperture (Howard, 1974). This corresponds to an average field of 10^{15} Wb$/4\pi R_\odot^2 \approx 0.16$ mT. Assuming that the true average field strength (not the peak) is 100 mT (see Section 2.1.2.) in the flux tubes, and that they carry most of the magnetic flux, $\alpha \approx 0.16\%$. With these values, the spin-up time becomes $t \approx 10^6$ $s \approx 10$ days. Foukal and Jokipii (1975) arrive at the same value for t by using $d = 1000$ km and $\alpha = 1\%$.

In spite of this extremely rapid spin-up, the velocity difference between the magnetic regions and the rest of the photosphere is observed to be maintained. Foukal and Jokipii (1975) explain this in terms of convection, which transports angular momentum. They point out that the spin-up by the flux tubes can be effective only if it is fast compared with the convective period, and note that the life-time of supergranular cells is only about 10^5 s, much shorter than t. An alternative explanation is given by the life-time of the magnetic structures, which seems to be as small as 10 min for the smallest structures (see Section 2.6). During this time no appreciable drag effect can be effective.

Finally a few words about the difference in rotation rates between the 'background' magnetic field and the field in active regions. The rotation of surface magnetic fields should reflect the rotation rate of deeper layers in which the fields are anchored or from which they are expelled. As the surface fields rotate faster than most of the photosphere, we can conclude that the angular velocity increases with depth. The angular velocity of the background or quiet-region magnetic fields is systematically higher than for active-region fields, indicating that the background fields are related to layers *deeper* than those associated with active-region fields (Stenflo, 1974a). This means that the background fields can hardly be due to the dispersion or turbulent diffusion of the magnetic flux in active regions, in contradiction with the solar cycle model of Leighton (1964, 1969). The short life-times of the magnetic elements (see Section 2.6) is another strong argument against the background fields being formed by turbulent diffusion. Instead they appear to be more directly linked to the subsurface sources, which all the time supply new magnetic flux to the solar surface to replace the flux that has been dissipated. This direct link with subsurface sources may account for the rigid rotation properties of magnetic sector structures (Schatten *et al.*, 1969) including the associated coronal holes (Timothy *et al.*, 1975).

3.4. THEORIES OF THE SOLAR CYCLE

Several ways in which our new picture of small-scale magnetic fields influence solar cycle theories have already been touched upon in the present paper. In this concluding section we will comment on the implications of the small-scale fields for:
    (1) Mean field electrodynamics.
    (2) Relation between active and quiet region magnetic fields.
    (3) Role of supergranulation.
    (4) Dissipation of magnetic flux.
    Modern dynamo theories of the solar cycle are based on the concept of mean field electrodynamics and turbulent diffusion of the field (e.g. Steenbeck and Krause, 1969; Leighton, 1969; Parker, 1970; Stix, 1974). Observations indicate that more than about 90% of the magnetic flux observed in the photosphere is channelled through strong-field elements which occupy only of the order of 0.2% of the solar surface (see Section 2.3). This makes the concept of a 'mean field' fairly meaningless in the photosphere. The field seems to be concentrated in this form before it emerges in the solar atmosphere (see Section 3.1). Hence it is doubtful that the mean-field approach will be much more valid deeper in the convection zone (cf. Piddington, 1975c).
    Leighton (1964) and Bumba and Howard (1965) suggest that all the background magnetic fields originate in active regions and are dispersed over the Sun by turbulent diffusion, mainly due to the flow in supergranular cells. This dispersion of the field is one of the main ingredients in Leighton's (1969) solar cycle theory. There are two main arguments against the background fields being formed in this way:
    (a) The life-time of magnetic elements (see Section 2.6) appears to be very short in comparison with the time it takes to move an element an appreciable distance across the solar surface.
    (b) The more rigid-rotation properties of the background fields indicate that they are generally not associated with active-region magnetic fields (see Section 3.3).
    There is also observational evidence for new magnetic flux emerging in quiet regions directly from below (e.g. Sheeley, 1969).
    Supergranulation has long been believed to be responsible for the enhancement of the field at the cell boundaries (Leighton *et al.*, 1962; Parker, 1963; Simon and Leighton, 1964). As shown in Section 3.1, the flux should however be in a concentrated, strong-field form already when it emerges from below into the photosphere. Bumba and Howard (1965) conclude that new magnetic flux emerges at the inferred edges of supergranules. More recent observations (Harvey and Martin, 1973; Zirin and Tanaka, 1973) lead to the opposite conclusion: New magnetic flux emerges at random with respect to the inferred cell boundaries.

Piddington (1972, 1975a, b, c) has criticized current dynamo theories because they do not account for the removal of the large amounts of flux generated each solar cycle. Turbulent diffusion does not annihilate the field, it only moves the field lines around. Field annihilation is only possible through true ohmic diffusion, which requires length scales less than about 200 km to be sufficiently effective during a solar cycle. This type of elimination of surplus flux is not understood in terms of the 'mean-field' theories.

Observations indicate however that the magnetic flux is destroyed soon after it has appeared in the photosphere (see Section 2.6). The life-time of a flux concentration is roughly proportional to the amount of flux it contains. The largest flux concentrations, the sunspots, may live for months, while the smallest, individual bright network points, live of the order of 10 min. A larger flux concentration is dissolved by breaking apart (fraying) into smaller flux elements, with a flux loss rate of about $10^7 \, \mathrm{Wb \, s^{-1}}$ independent of the amount of flux in the decaying structure (see Section 2.6). This is the rate at which the magnetic flux is transferred from smaller to larger Fourier wave numbers. The time it takes to reach wave numbers sufficiently large for ohmic dissipation to be effective is much smaller than the length of the solar cycle.

References

Alfvén, H.: 1967, in Hindmarsh, Lowes, Roberts, and Runcorn (eds.), *Magnetism and the Cosmos*, Oliver & Boyd, Edinburgh/London, p. 246.
Beckers, J. M.: 1963, *Astrophys. J.* **138**, 648.
Beckers, J. M.: 1968, *Solar Phys.* **3**, 367.
Beckers, J. M.: 1972, *Ann. Rev. Astron. Astrophys.* **10**, 73.
Beckers, J. M., Noyes, R. W., and Pasachoff, J. M.: 1966, *Astron. J.* **71**, 155.
Beckers, J. M. and Schröter, E. H.: 1968a, *Solar Phys.* **4**, 142.
Beckers, J. M. and Schröter, E. H.: 1968b, *Solar Phys.* **4**, 165.
Bray, R. J. and Loughhead, R. E.: 1961, *Australian J. Phys.* **14**, 14.
Bray, R. J. and Loughhead, R. E.: 1974, *The Solar Chromosphere*, Chapman and Hall, London.
Bumba, V.: 1963, *Bull. Astron. Inst. Czech.* **19**, 91.
Bumba, V.: 1967, *Solar Phys.* **1**, 371.
Bumba, V. and Howard, R.: *Astrophys. J.* **141**, 1502.
Caccin, B., Falciani, R., and Donati-Falchi, A.: 1974, *Solar Phys.* **35**, 31.
Chapman, G. A.: 1970, *Solar Phys.* **14**, 315.
Chapman, G. A.: 1974, *Astrophys. J.* **191**, 255.
Chapman, G. A. and Sheeley, N. R. Jr.: 1968, *Solar Phys.* **5**, 442.
Deubner, F. L.: 1975, in C. Chiuderi, M. Landini, and A. Righini (eds.), 'First European Solar Meeting (Florence, February 25–27, 1975)', *Osservazioni e Memorie Osservatorio di Arcetri* **105**, 39.
Dravins, D.: 1974, in R. G. Athay (ed.), 'Chromospheric Fine Structure', *IAU Symp.* **56**, 257.
Dunn, R. B. and Zirker, J. B.: 1973, *Solar Phys.* **33**, 281.
Foukal, P.: 1972, *Astrophys. J.* **173**, 439.
Foukal, P. and Jokipii, J. R.: 1975, *Astrophys. J.* **199**, L71.
Frazier, E. N.: 1970, *Solar Phys.* **14**, 89.
Frazier, E. N.: 1971, *Solar Phys.* **21**, 42.
Frazier, E. N.: 1972, *Solar Phys.* **26**, 130.
Frazier, E. N.: 1974, *Solar Phys.* **38**, 69.
Frazier, E. N. and Stenflo, J. O.: 1972, *Solar Phys.* **27**, 330.
Giovanelli, R. G. and Ramsay, J. V.: 1971, in R. Howard (ed.), 'Solar Magnetic Fields', *IAU Symp.* **43**, 293.
Gokhale, M. H. and Zwaan, C.: 1972, *Solar Phys.* **26**, 52.
Golub, L., Krieger, A. S., Silk, J. K., Timothy, A. F., and Vaiana, G. S.: 1974, *Astrophys. J.* **189**, L93.
Gopasyuk, S. I., Kotov, V. A., Severny, A. B., and Tsap, T. T.: 1973, *Solar Phys.* **31**, 307.

Hale, G. E.: 1908, *Astrophys. J.* **28**, 100.

Hale, G. E.: 1927, Nature **119**, 708.

Hale, G. E. and Ellerman, F.: 1903, *Publ. Yerkes Obs.* **3**, 1.

Harvey, J. W.: 1971, *Publ. Astron. Soc. Pacific* **83**, 539.

Harvey, J. W.: 1974, *Flare Related Magnetic Field Dynamics*, NCAR, Boulder, Colorado, p. 1.

Harvey, J. W. and Hall, D.: 1975, Abstract of Report at the 146th AAS meeting, San Diego, California, 17–20 August, 1975.

Harvey, J. W. and Livingston, W.: 1969, *Solar Phys.* **10**, 283.

Harvey, J., Livingston, W., and Slaughter, C.: 1972, *Line Formation in a Magnetic Field*, NCAR, Boulder, Colorado, p. 227.

Harvey, K. and Harvey, J.: 1973, *Solar Phys.* **28**, 61.

Harvey, K. L. and Martin, S. F.: 1973, *Solar Phys.* **32**, 389.

Harvey, K. L., Harvey, J. W., and Martin, S. F.: 1975, *Solar Phys.* **40**, 87.

Howard, R.: 1971, *Solar Phys.* **16**, 21.

Howard, R.: 1972, *Solar Phys.* **24**, 123.

Howard, R.: 1974, *Solar Phys.* **38**, 59.

Howard, R. and Bhatnagar, A.: 1969, *Solar Phys.* **10**, 245.

Howard, R. and Harvey, J.: 1970, *Solar Phys.* **12**, 23.

Howard, R. and Stenflo, J. O.: 1972, *Solar Phys.* **22**, 402.

Kopecký, M.: 1971, *Bull. Astron. Inst. Czech.* **22**, 343.

Kotov, V. A.: 1970, *Izv. Krymsk. Astrofiz. Obs.* **41**, 67.

Krat, V. A.: 1973, *Solar Phys.* **32**, 307.

Kuz'minykh, V. D.: 1964, *Astron. Zh.* **41**, 692 (*Soviet Astron.* **8**, 551).

Leighton, R. B.: 1964, *Astrophys. J.* **140**, 1547.

Leighton, R. B.: 1965, in R. Lüst (ed.), 'Stellar and Solar Magnetic Fields', *IAU Symp.* **22**, 158.

Leighton, R. B.: 1969 *Astrophys. J.* **156**, 1.

Leighton, R. B., Noyes, R. W., and Simon, G.: 1962, *Astrophys. J.* **135**, 474.

Lites, B. W.: 1972, *NCAR Cooperative Thesis No. 28*, University of Colorado and High Altitude Observatory, NCAR, Boulder, Colorado.

Livingston, W.: 1968, *Astrophys. J.* **153**, 929.

Livingston, W. and Harvey, J.: 1969, *Solar Phys.* **10**, 294.

Livingston, W. and Harvey, J.: 1971, in R. Howard (ed.), 'Solar Magnetic Fields', *IAU Symp.* **43**, 51.

Livingston, W. and Harvey, J.: 1975, *Bull. Amer. Astron. Soc.* **7**, 346 (Abstract).

Lynch, D. K., Beckers, J. M., and Dunn, R. B.: 1973, *Solar Phys.* **30**, 63.

Mehltretter, J. P.: 1974, *Solar Phys.* **38**, 43.

Michard, R.: 1956, *Ann. Astrophys.* **19**, 1.

Michard, R.: 1974, in R. G. Athay (ed.), 'Chromospheric Fine Structure', *IAU Symp.* **56**, 3.

Milkey, R. W.: 1970, *Solar Phys.* **14**, 62.

Mullan, D. J.: 1974, preprint.

Osterbrock, D. E.: 1961, *Astrophys. J.* **134**, 347.

Parker, E. N.: 1955, *Astrophys. J.* **121**, 491.

Parker, E. N.: 1963, *Astrophys. J.* **138**, 552.

Parker, E. N.: 1970, *Ann. Rev. Astron. Astrophys.* **8**, 1.

Parker, E. N.: 1974a, *Astrophys. J.* **189**, 563.

Parker, E. N.: 1974b, *Astrophys. J.* **190**, 429.

Parker, E. N.: 1974c, *Astrophys. Space Sci.* **22**, 279.

Parker, E. N.: 1974d, *Solar Phys.* **36**, 249.

Parker, E. N.: 1974e, *Solar Phys.* **37**, 127.

Parker, E. N.: 1976, *Astrophys. J.* (in press).

Pasachoff, J. M., Noyes, R. W., and Beckers, J. M.: 1968, *Solar Phys.* **5**, 131.

Peckover, R. S. and Weiss, N. O.: 1972, *Computer Phys. Communications* **4**, 339.

Piddington, J. H.: 1972, *Solar Phys.* **22**, 3.

Piddington, J. H.: 1973, *Solar Phys.* **33**, 363.

Piddington, J. H.: 1975a, *Astrophys. Space Sci.* **34**, 347.

Piddington, J. H.: 1975b, *Astrophys. Space Sci.* **35**, 269.

Piddington, J. H.: 1975c, *Astrophys. Space Sci.* **38**, 157.

Richardson, R. S.: 1941, *Astrophys. J.* **93**, 24.

Schatten, K. H., Wilcox, J. M., and Ness, N. F.: 1969, *Solar Phys.* **6**, 442.

Schmahl, G.: 1967, *Z. Astrophys.* **66**, 81.

Semel, M.: 1962, *Compt. Rend. Acad. Sci. Paris* **254**, 3978.

Servajean, R.: 1961, *Ann. Astrophys.* **24**, 1.

Severny, A. B.: 1965, *Astron. Zh.* **42**, 217.
Severny, A. B.: 1967, *Izv. Krymsk. Astrofiz. Obs.* **38**, 3.
Severny, A. B.: 1972, in C. de Jager (ed.), *Astrophysics and Space Science Library*, Vol. 29, 38.
Sheeley, N. R. Jr.: 1966, *Astrophys. J.* **144**, 723.
Sheeley, N. R. Jr.: 1967, *Solar Phys.* **1**, 171.
Sheeley, N. R. Jr.: 1969, *Solar Phys.* **9**, 347.
Sheeley, N. R. Jr.: 1971, in R. Howard (ed.), 'Solar Magnetic Fields', *IAU Symp.* **43**, 310.
Shine, R. A. and Linsky, J. J.: 1974, *Solar Phys.* **37**, 145.
Simon, G. W. and Leighton, R. B.: 1964, *Astrophys. J.* **140**, 1120.
Simon, G. W. and Noyes, R. W.: 1971, in R. Howard (ed.), 'Solar Magnetic Fields', *IAU Symp.* **43**, 663.
Simon, G. W. and Zirker, J. B.: 1974, *Solar Phys.* **35**, 331.
Skumanich, A., Smythe, C., and Frazier, E. N.: 1975, *Astrophys. J.* **200**, 747.
Spruit, H. C.: 1976, to be published.
Sreenivasan, S. R.: 1973a, *Physica* **67**, 323.
Sreenivasan, S. R.: 1973b, *Physica* **67**, 330.
Sreenivasan, S. R. and Thompson, D. L.: 1974, *Physica* **78**, 321.
Steenbeck, M. and Krause, F.: 1969, *Astron. Nachr.* **291**, 49.
Stellmacher, G. and Wiehr, E.: 1971, *Solar Phys.* **18**, 220.
Stellmacher, G. and Wiehr, E.: 1973, *Astron. Astrophys.* **29**, 13.
Stenflo, J. O.: 1966, *Arkiv. Astron.* **4**, 173.
Stenflo, J. O.: 1968a, *Acta Univ. Lund.* II No. 1 (= *Medd. Lunds Astron. Obs.* Ser. II No. 152).
Stenflo, J. O.: 1968b, *Acta Univ. Lund.* II No. 2 (= *Medd. Lunds Astron. Obs.* Ser. II No. 153).
Stenflo, J. O.: 1971, in R. Howard (ed.), 'Solar Magnetic Fields', *IAU Symp.* **43**, 101.
Stenflo, J. O.: 1973, *Solar Phys.* **32**, 41.
Stenflo, J. O.: 1974a, *Solar Phys.* **36**, 495.
Stenflo, J. O.: 1974b, *Flare Related Magnetic Field Dynamics*, NCAR, Boulder, Colorado, p. 153.
Stenflo, J. O.: 1975, *Solar Phys.* **42**, 79.
Steshenko, N. V.: 1960, *Izv. Krymsk. Astrofiz. Obs.* **22**, 49.
Steshenko, N. V.: 1967, *Izv. Krymsk. Astrofiz. Obs.* **37**, 21.
Stix, M.: 1974, *Astron. Astrophys.* **37**, 121.
Stoyanova, M. N.: 1970, *Solar Phys.* **15**, 349.
Tanenbaum, A. S., Wilcox, J. M., Frazier, E. N., and Howard, R.: 1969, *Solar Phys.* **9**, 328.
Timothy, A. F., Krieger, A. S., and Vaiana, G. S.: 1975, *Solar Phys.* **42**, 135.
Ulmschneider, P.: 1971, *Astron. Astrophys.* **14**, 275.
Unno, W.: 1959, *Astrophys. J.* **129**, 375.
Vorpahl, J.: 1974, in R. G. Athay (ed.), 'Chromospheric Fine Structure', *IAU Symp.* **56**, 197.
Vrabec, D.: 1971, in R. Howard (ed.), 'Solar Magnetic Fields', *IAU Symp.* **43**, 329.
Vrabec, D.: 1974, in R. G. Athay (ed.), 'Chromospheric Fine Structure', *IAU Symp.* **56**, 201.
Weart, S.: 1972, *Astrophys. J.* **177**, 271.
Wiehr, E. and Wittmann, A.: 1975, private communication.
Wilcox, J. M. and Howard, R.: 1970, *Solar Phys.* **13**, 251.
Wilcox, J. M., Schatten, K. H., Tanenbaum, A. S., and Howard, R.: 1970, *Solar Phys.* **14**, 255.
Wilson, P. R.: 1969, *Solar Phys.* **6**, 364.
Zirin, H. and Tanaka, K.: 1973, *Solar Phys.* **32**, 173.

DISCUSSION

Schröter: I have a comment regarding your 'line-ratio' method and a question regarding your new facular model.

(a) It would be much more convincing if you applied this method to three or more lines of the same multiplet and compared the measured circular polarizations in all lines with 'saturation curves' computed accordingly by solving the Stokes-parameter equations numerically. Why did you not do this? We (Dr Wiehr) at Göttingen intend to follow this line.

(b) You observe apparent fields of ≤10 G and arrive after reduction at 2 kG (factor of 200!!). Why did you not perform your measurements by positioning your magnetograph-entrance pinhole on bright Ca^+ fine mottles where the directly observed magnetic field strengths are already 50–200 G? You would have avoided such large corrections!

Stenflo: It is irrelevant to speak of large correction factors, because no such factors are derived or used. The value of about 2 kG for the field strength is determined only from the *slope* of the straight line in the scatter-plot diagram, regardless of the values of the apparent field strengths. My observations were made in a quiet region at disk center in different parts of the wings of the 5250 and 5247 Å lines. Frazier and I are presently working on data for the same lines, obtained in active regions, where the signal-to-noise is better. These new data indicate the same high values of the field strength. We intend to use other combinations of spectral lines in the future.

Zwaan: Spruit has continued his calculations on models of flux tubes. One of these results is that indeed a model explaining intensity and magnetograph data can be found if a flux tube with temperatures lower than the surroundings is embedded in the convection zone and photosphere. Lateral influx affects the structure down to depths over $\tau = 1$ but no longer at substantially larger depths. Of course, as Dr Parker has pointed out, the assumed lower temperature deep down is a dilemma.

Deubner: The figure given by yourself at the blackboard for the flux dissipation rate raises a serious problem as I pointed out already at the Florence meeting: If you take this number at face value, i.e. if the flux was really dissipated or destroyed, it had to be replaced either from below or by random walk processes from sunspot regions at a rate of 10^3 times higher than the average rate of production of active regions. We are led to assume, therefore, that rather than being destroyed the magnetic flux elements are squeezed and juggled around by granular motions. This conclusion is in fact corroborated by observations of white light faculae at high angular resolution, as e.g. published by Mehltretter.

Stenflo: The life-time of the bright points observed by Mehltretter was 5–15 min. If the associated flux tubes were substantially more long-lived, we find that a large fraction of the magnetic flux at any given time is unrelated to brightness enhancements, which seems to contradict observations. Such flux would correspond to a straight line of different slope in the scatter-plot diagram of magnetic fields observed in two different lines.

Parker: I would like to ask what you had in mind with your statement that the 2000 G flux tubes can be confined by currents, twisting and dynamical amplification, etc. The fact is that no matter what the origin of these flux tubes, the concentration to 2000 G is a mechanical one, involving compression, to a pressure $B^2/8\pi$. I did not understand how you hope to accomplish the compression. Only a force can compress a field. What force do you propose?

Stenflo: My general conclusion was that the production of the strong fields must occur in the convection zone, before they emerge in the photosphere. No significant further amplification should occur at the surface. When discussing confinement mechanisms, I pointed out that electromagnetic confinement (force-free fields) is not ruled out, as Sreenivasan has shown by studying the evolution of force-free fields in a resistive medium.

Giovanelli: There seems to be no problem with confinement of the magnetic field in small magnetic elements. A much smaller pressure is needed in the magnetic tube for pressure equilibrium, and this reduces the opacity greatly, so that we see deeper in the magnetic element than in the magnetic-free photosphere. The external pressure at this depth may be substantially greater than at the normal $\tau = 1$ level. Whether the element appears bright or dark depends on the magnetic flux and the internal pressure (and hence the diameter and opacity); theoretically the smaller fluxes are associated with brighter structures, the larger ones with darker structures in qualitative agreement with Frazier's observations.

Krause: You deny that on the Sun there is a mean magnetic field which is generated by a dynamo mechanism. However, I don't see any possibility to gain magnetic energy in the convection zone of the Sun apart from that by a dynamo mechanism feeding energy in the large scale magnetic field, from where it is transported by the interaction with the convective motion into the small scales. Do you see another mechanism which provides for the magnetic fields in the Sun?

Stenflo: I do not reject dynamo theories in general, I conclude that the approach of a 'mean-field electrodynamics' is not a very meaningful approximation. Further, if we can presently only conceive of one way of generating the magnetic fields, this does not necessarily imply that it must be the correct one.

INVESTIGATION OF CORONAL ROTATION BY THE
SPECTROSCOPIC METHOD

V. E. STEPANOV and N. F. TJAGUN

Sibizmiran, Irkutsk, U.S.S.R.

Abstract. Coronal rotation is different in the northern and southern hemispheres of the Sun and varies with the cycle of solar activity. A sharp decrease in linear velocity by 0.4 km s^{-1} on the latitude and its relative constancy in the latitude interval $35° < \varphi < 50°$ are the most typical.

In the southern polar region an intensive stream of coronal gas directed oppositely to the solar rotation was detected during the time interval 1969–1972. In the greater part of the latitude range, on average, the corona rotates more slowly in comparison with the solar atmosphere at the photospheric level.

1. Introduction

In our earlier paper [1] the solar corona rotation rate was inferred using the Doppler shift of the Fe X 6374 Å line. In this paper there were used only 525 line profiles obtained at Sayn Mountain Observatory during the period March 1968 to September 1969. However, an analysis of errors showed that for more exact determination of the coronal rotation rate at higher latitudes more numerous observations are needed. Therefore, the program of observations has been enlarged and the results of this paper are based on the analysis of the 'gravity center' displacement of 4976 profiles of the Fe X 6374 Å line, obtained during the period 1969 to 1972.

The number of observed profiles used for determination of the rotation rate during different time intervals, is given in Table I. In 1971 and 1972 we paid special

TABLE I

Date		Number of profiles observed	Number of profiles for hemisphere N	Number of profiles for hemisphere S
Year	Month			
1969	6–9	215	79	136
1970	6, 7, 9	1624	902	722
1971	5–7	1191	550	641
1972	8, 9	1946	810	1136
Total number of profiles		4976	2341	2635

attention to the region of latitude within $20° < \varphi < 85°$. The same lines as in [1] were taken as standard lines. We used their wavelengths as determined in [2]. The coronal spectra were obtained with the help of a coronograph with an objective diameter of 53 cm using grating in the 2nd order as in [1]. The linear dispersion is equal to 1 Å mm^{-1}. Spectrogram measurements were performed with a step equal to 0°9 of latitude. The processing of measured data and the computations were made on the computer BESM-4. The software of both automatic processing of spectra and

Bumba and Kleczek (eds.), Basic Mechanisms of Solar Activity, 101–106. All Rights Reserved.

computation of mean radial velocities for different heliographic latitudes is described in [3].

2. Observational Results

The Doppler velocities in the corona, which are averaged over the whole observational period as well as over 10° latitude interval and reduced to one quadrant, are presented in Figure 1.

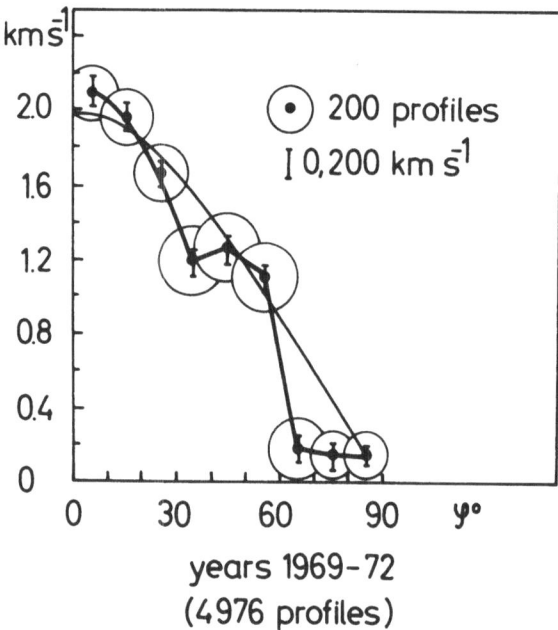

Fig. 1.   The curve of coronal rotation, on the average, for the period from 1969 to 1972. The heavy full line is the rotation of the corona according to our determinations. The areas of circles correspond to the number of observations, the verticals correspond to the mean square error. The light line is the curve of the photosphere rotation from sunspots [4], extrapolated for higher latitudes.

Within the interval of heliographic latitudes from 0° to 15° the velocity of coronal rotation exceeds the rotation velocity inferred from sunspots by 100 m s^{-1}. Further, with the increase of latitude up to 35° the velocity decreases sharply.

At the latitude of 35° the coronal rotation velocity appears to be less than the photospheric rotation by 400 m s^{-1}. In the latitude interval from 35° to 55° the velocity remains constant. Then, a second new sharp decrease of velocity at the latitude of 65° takes place. At this latitude the coronal rotation rate is by 500 m s^{-1} less than that of the photospheric rotation rate obtained by extrapolating the Newton-Nunn curve [4] at higher latitudes. Within the latitude interval from 65° to 85° the velocity of coronal rotation remains constant.

Hence, the corona on the whole rotates a little slower than the photosphere; the difference of velocity rotation being most appreciable at the latitude of 35° and 65°.

We can distinguish three zones in the character of dependence of the coronal rotation velocity on heliographic latitudes.

The first one appears to be a zone of active latitudes where the angular momentum transfer to the equator occurs; the second zone contains the latitudes from 35° to 55° where an increase of angular velocity of rotation occurs, i.e. in the second zone the coronal rotation velocity is being restored up to the value of the photospheric rotation velocity. The third zone appears to be a zone at high latitudes, where the character of the motions, as we shall see below, is very complicated.

In Figure 2 the velocity curves are given for the northern (full line) and southern (dashed line) hemispheres. In general, the coronal rotation in different hemispheres looks like the pattern of the rotation for one quadrant.

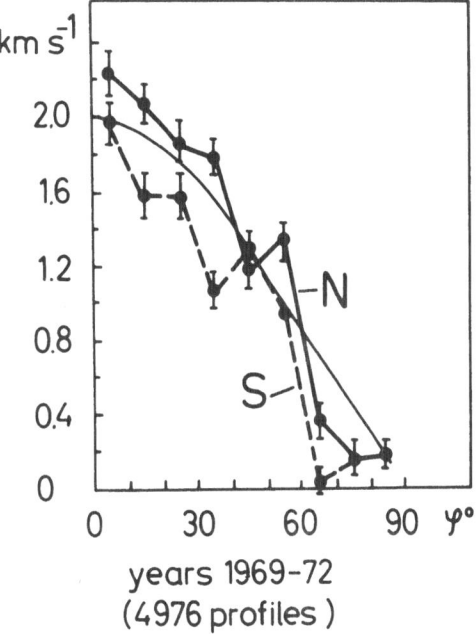

Fig. 2. Coronal rotation in different hemispheres for the period from 1969 to 1972. The heavy full line is for the northern hemisphere in the corona. The dashed line is the southern hemisphere. The light line is the rotation of the photosphere from sunspots [4].

However, the essential differences exist too. Firstly, during the whole observational period the corona in the southern hemisphere rotates, on average, slower than in the northern hemisphere. Secondly, a clear difference in the character of decrease of rotational velocity at the latitudes $\varphi > 25°$ takes place for both hemispheres – the relative minimum of velocity in the southern hemisphere occurs at the latitude of 35°, while in the northern hemisphere that minimum appears to be shifted 10° towards higher latitudes.

Finally, very low velocities are observed in the southern hemisphere at the latitude of 65°, that can be treated as the presence of a strong zonal motion along the direction opposite to that of the solar rotation.

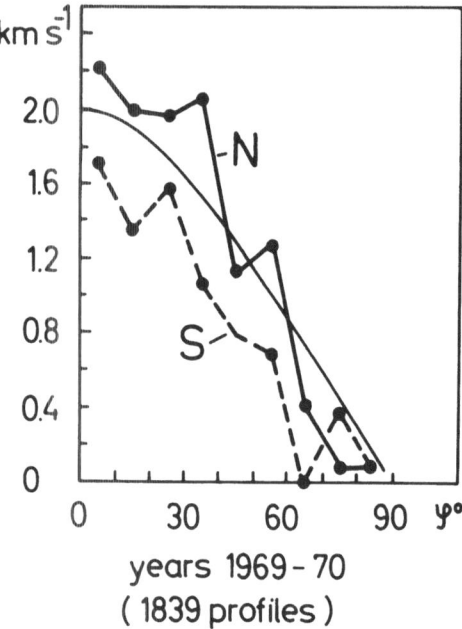

Fig. 3.   Coronal rotation from 1969 to 1970. The heavy full line is the northern hemisphere in the corona. The dashed line is the southern hemisphere. The light line is the rotation of the photosphere from sunspots [4].

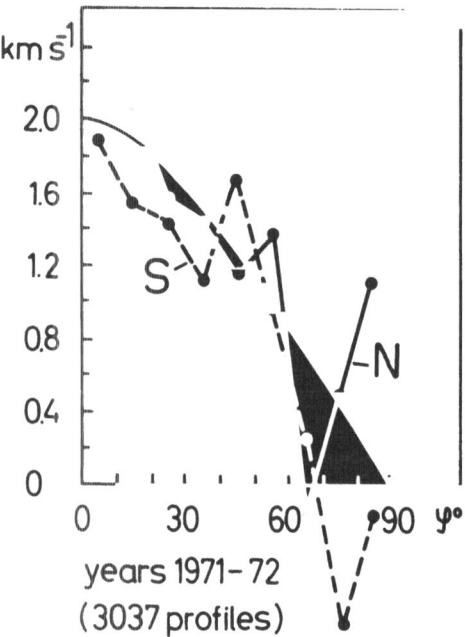

Fig. 4.   Coronal rotation from 1971 to 1972. The heavy full line is for the northern hemisphere in the corona. The dashed line is the southern hemisphere. The light line is the rotation of the photosphere from sunspots [4].

In Figures 3 and 4, the velocity curves are given separately for the periods 1969 to 1970 and 1971 to 1972 accordingly. It is seen that the velocities in the first period of observations (1969–1970) are larger in the northern hemisphere than in the southern zone. In the second period (1971–1972) the character of coronal rotation has changed; at a latitude of 45° in the southern hemisphere the velocity is larger than in the northern hemisphere.

From Figures 3 and 4 we can draw the conclusion that the location of the stream, which moved oppositely to the direction of solar rotation, had shifted towards higher latitudes.

3. Discussion

Up to now the passage of typical coronal formations across the eastern and western limbs of the solar disk (the method of 'tracers' [5–15]) was used for investigation of coronal rotation rates. This method yields fairly high accuracy but it does not supply sufficiently comprehensive information on the variation of the rotation rate with the heliographic latitude. Besides, it can be applied only to the study of motions of distinctly visible coronal formations.

The method of the coronal rotation study by the Doppler shift of spectral lines allows, in principle, to obtain the rotation velocities of the bright formations as well as that of the diffuse background of the corona. However, the spectroscopic method has its disadvantage, which consists in summarizing emission along the line of sight. The structures with the different brightness contribute unequally to the formation of the spectral line. This circumstance causes a large dispersion of the observed velocities. Therefore, the determination of the coronal rotation velocities with the proper accuracy needs a large number of observations.

How can we explain the difference between the observational results obtained by the tracers method and our results? It is obvious that the lines with the largest equivalent width belong to the brightest coronal formations and that the lines with the small equivalent width belong to the quiet coronal background. Thus we must expect in accordance with the results of many authors [5–15] that the lines with largest equivalent widths should possess the largest velocities. Our investigations have shown that this regularity is not observed. The lines with large equivalent widths exhibit minimal velocities and give the smallest velocity dispersion [1].

The same lines have also the smallest line width [16]. Therefore, observations by using the method of tracers do not represent real gas motion in the corona. They give the displacement, a disturbed region, the position of which depends on the source of disturbance located at the deep level of the solar atmosphere.

The questions how to distinguish between rotation of a bright coronal object and coronal background as well as how to reveal the local motions in the vicinity of active regions will be considered by us in the future.

4. Conclusions

The results of this study are the following:

(1) The difference between rotation rates of corona and photosphere at almost all heliographic latitudes was found. The sharp decrease of the coronal rotation rate at latitudes of 35° and 65° appears to be the most remarkable peculiarity.

(2) The difference in coronal rotation rates for the northern and southern hemispheres was also found.

(3) In the southern polar region the intensive long-lived zonal streams were observed. These streams are directed oppositely to the solar rotation direction.

(4) The coronal rotation rate has been studied by us on the descending branch of the solar activity cycle. Nevertheless, using that material, one can conclude certainly that the character of coronal rotation varies with the phase of the solar cycle.

References

1. Stepanov, V. E. and Tjagun, N. F.: 1971, *IAU Symp.* **43**, 667.
2. Pierce, A. K. and Breckinridge, J. B.: 1973, The Kitt Peak Table of Photographic Solar Spectrum Wavelengths.
3. Tjagun, N. F., Zhukov, V. D., Stepanov, V. E., and Katz, I. O.: 1973, *Issled. Geomagnetizmu, Aeronomii i fizike Solntsa* **28**, 118.
4. Newton, H. W. and Nunn, M. L.: 1952, *Monthly Notices Roy. Astron. Soc.* **111**, 413.
5. Livingston, W. C.: 1969, *Solar Phys.* **7**, 144.
6. Livingston, W. C.: 1969, *Solar Phys.* **9**, 448.
7. Trellis, M.: 1950, *Ann. Astrophys. Suppl. Ser.* **5**, 81.
8. Cooper, R. H. and Billings, D. E.: 1962, *Z. Astrophys.* **55**, 28.
9. Waldmeier, M.: 1950, *Z. Astrophys.* **43**, 29.
10. Hansen, R. T., Garcia, C. J., Hansen, S. F., and Loomis, H. G.: 1969, *Solar Phys.* **7**, 417.
11. Mohamed el-Raey and Scherrer, P. H.: 1972, *Solar Phys.* **26**, 15.
12. Ward, F.: 1966, *Astrophys. J.* **145**, 416.
13. Dupree, A. K. and Henze, W.: 1972, *Solar Phys.* **27**, 271.
14. Henze, W. and Dupree, A. K.: 1973, *Solar Phys.* **33**, 425.
15. Antonucci, E. and Svalgaard, L.: 1974, *Solar Phys.* **34**, 3.
16. Tjagun, N. F. and Stepanov, V. E.: 1975, *Solar Data* **2**, 56.

FAST VARIATIONS OF THE SOLAR ROTATION

V. F. CHISTYAKOV

Service of Sun, Ussurijsk, U.S.S.R.

The rotation law of the Sun is often described by the Faye formula:

$$\xi(\varphi) = a - b \sin^2 \varphi,$$

where φ is a heliographic latitude and a and b are constants. For a long time researchers have been studying the problem in what way the constants a and b change in time and in dependence on a phase of 11-year cycles.

The results of solar rotation observations in cycle No. 19 (1955–1963) and No. 20 (1965–1973) are presented in this report.

The photoheliograms of sunspots received during a standard program were used for these investigations. The diameter of Sun in the picture was 75 mm. The measurement of φ and λ coordinates was carried out using orthographic charts 25 mm in diameter. Measuring accuracy of φ and λ coordinates was equal to $\pm 0°.5$. If during the time T the spot moved along the parallel to $\Delta\lambda$, and if the measuring error $\Delta\lambda$ does not exceed $1°$, the error in measuring the rotation velocity is equal to $\Delta\xi \leq 1° \, T^{-1}$. For these investigations the sunspots in the central zone were selected. The relative distance from the centre of the disc for spots was $r/R \leq 0.7$. The interval between observations was $T = 6$ days. Consequently, the measuring accuracy of rotation velocity $\xi(\varphi)$ was about $0°.17$. Real rotation velocity of spots deviated from the average one within the limits $\pm 2°$. It greatly exceeds error limits. The coefficients a and b were determined by the method of minimal squares.

General statistics for No. 19 cycle included 619 sunspots. 361 spots were situated in the northern hemisphere and 258 in the southern. General formulae of the rotation law for the cycle are as follows:

the northern hemisphere:

$$\xi(\varphi) = 14°.583 \pm 2°.708 \sin^2 \varphi$$

$$+ 0°.092 \pm 0°.496$$

the southern hemisphere:

$$\xi(\varphi) = 14°.583 - 2°.638 \sin^2 \varphi$$

$$\pm 0°.072 \pm 0°.623$$

Statistics of the cycle No. 20 included 854 sunspots. 505 spots were in the northern hemisphere and 349 in the southern.

The following formulae were obtained for this cycle:

the northern hemisphere:

$$\xi(\varphi) = 14°.610 - 1°.485 \sin^2 \varphi$$

$$\pm 0°.117 \pm 0°.989$$

Bumba and Kleczek (eds.), Basic Mechanisms of Solar Activity, 107–111. All Rights Reserved.

the southern hemisphere:

$$\xi(\varphi) = 14°563 - 1°480 \sin^2 \varphi$$

$$\pm 0°119 \pm 1°245$$

The coefficients a_N and a_S, b_N and b_S coincide in the two cycles.

But we observe another picture when we take short time intervals, for example, 0.5–1 yr. Then a and especially b change significantly. These changes are not conditioned by the low number of annual selections but they have a real solar origin, because a and b variations are determined by the series of objective laws in both hemispheres.

Figure 1 shows the time course of Δb, of deviation coefficient b from the average for the cycle No. 20: a curve Δb_N for the northern hemisphere and a curve Δb_S for the south. The curve Δb_S is represented by a mirror image. It equals multiplication of Δb_S by (-1). The parallel course of curves Δb_N and Δb_S shows that the variations of rotation velocity of spots in the northern and southern hemispheres are in the opposite phase. If the rotation velocity of spots in the northern hemisphere increases, it decreases in the southern and vice versa.

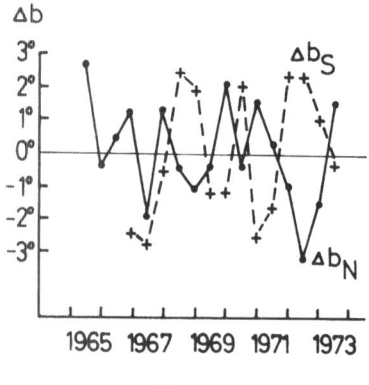

Fig. 1.

Let us use a dimensionless parameter $\Delta\xi/\bar{\xi}$, where $\bar{\xi}$ is the average rotation velocity of the spots in the latitude for the cycle (in agreement with the Faye formulae), and $\Delta\xi$ is the deviation from the average velocity. Thus we shall be able to obtain the characteristics of solar rotation independent from the latitude for the short intervals of time, for example a month.

A time course $\Delta\xi/\bar{\xi}$ for the spots of the northern (the solid line) and the southern (dashed line) hemispheres is given in Figure 2. A curve $\Delta\xi/\bar{\xi}$ for the southern hemisphere is turned round the horizontal axis. The mean value of its coordinate was multiplied by (-1).

Figure 2 shows sharp variations of the solar rotation in an annual period on the ascending branch of cycle No. 20 (1965–1969). After the cycle maximum (at the beginning of 1970) the variation $\Delta\xi/\bar{\xi}$ had the smallest amplitudes and has lost its regular, almost harmonic character. The parallel course of both the curves in Figure 2

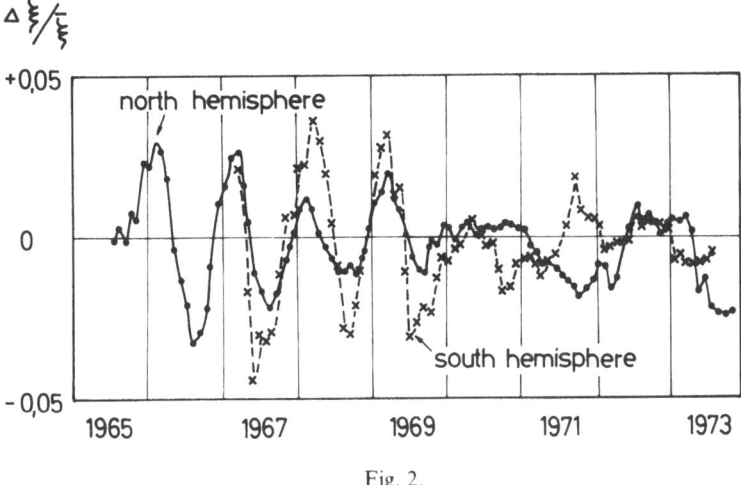

Fig. 2.

indicates that there are variations of parameter $\Delta\xi/\bar{\xi}$ for the northern and southern hemispheres in opposite phase. In other words, this result shows the alternation of acceleration and retardation of the rotation in the northern and southern hemispheres. Investigations show that the rapid variations of rotation velocity of the spots are accompanied by rapid variations of meridional velocity of motion. These are shown in Figure 3. Here, the continuous curve shows time variations of the average velocity of the meridional motions in the northern hemisphere and the dashed curve in the south. The curve $\Delta\varphi/\Delta t$ for the northern hemisphere is turned round the horizontal axis. The parallel course of both the curves on Figure 3 shows that the motions of the spots along the latitude have a global character. At a given moment the spots on the solar surface are moving in the direction of one of the solar poles. This situation changes in time.

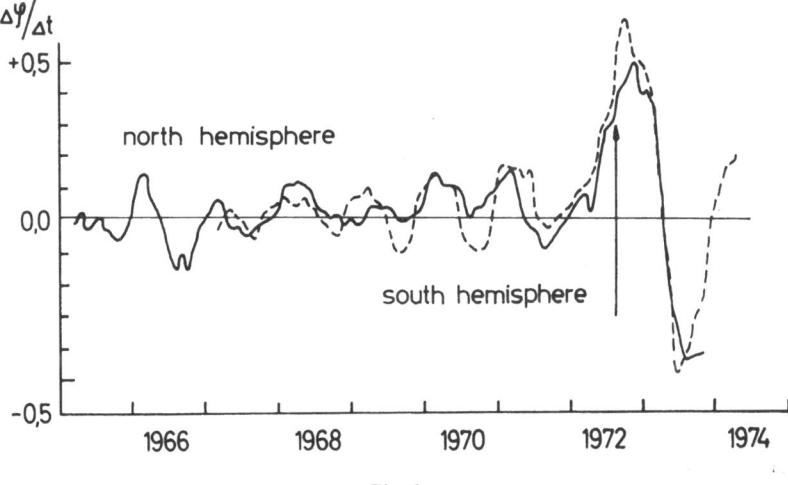

Fig. 3.

The following general picture of motions is observed on the Sun. If the spots of the northern hemisphere rotate faster than the average velocity, the meridional streams are directed to the south, and vice versa. The connection of the meridional motions of the spots with the velocity of their rotation shows that there is a large-scale circulation in subphotospheric layers. The alternation of rotation acceleration of spots in the north and the south indicates that there is an interchange of motion quantity between these hemispheres.

Such a kind of rotation variations reminds us of torsional variations.

Since there are variable meridional streams, whose direction alternately changes to the north and to the south, the coefficient a in the Faye formula formally indicating the velocity of equatorial rotation, should be considered as a kind of kinematic characteristics of the sunspots zone. A diagram of dependence of coefficients a and b for the cycle No. 19 (points) and for the cycle No. 20 (crosses) is given on Figure 4.

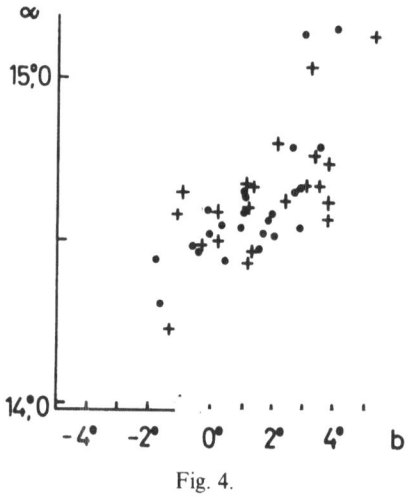

Fig. 4.

We can see the proportionality of a and b values. It has a simple sense: increasing rotation velocity in the equatorial belt is accompanied by decreasing rotation velocity in higher latitudes. The larger scatter of spots on the diagram indicates the influence of other factors.

One of these factors may be the connection of a with a phase of an 11-year cycle. For cycle No. 19 values a were the biggest at the beginning and at the end of the cycle and smallest in the period of the cycle maximum. Variations a_S may have developed with the retardation by 1.5 yr in respect to the variations a_N in the sunspots cycle No. 20. The same retardation by 1.5 yr was observed in spots formation of the southern hemisphere in comparison with the spots formation in the northern hemisphere.

In conclusion we would like to point out one interesting possibility to investigate the solar rotation. Spectrographic methods permit us to register the rotation velocity of the surface layers of the solar atmosphere. As a rule we assume that spots originate in deeper layers. Consequently, the rotation of the sunspots should reflect the peculiarities of these deeper layers. For the large time interval the law of the solar rotation determined by motion of the sunspots and by the spectral methods is described by Faye formulae. But in this case the rapid rotation variations are not explained. That is why it would be interesting to find the differences in the rapid variations of the rotation between the surface and the deeper layers.

DISCUSSION

Gilman: The units on your figure are degrees per day. The magnitudes of the coefficient b seem very large.

Chistyakov: The unit along the vertical axis is 1° per day. The rotational velocity of sunspots changes from +3° (per day) to −3°. Those deviations are averaged over a short time interval, 0.5–1 yr.

Howard: Do I understand from your first slide that the $\Delta\beta$ is the average at times for a whole year larger than b, so that for negative $\Delta\beta$ there is an equatorial deceleration instead of acceleration?

Chistyakov: Yes. In practice we meet the following occurrence: Δb is positive and Δb is negative. If Δb is negative then the rotation law is described by the formula:

$$\xi = a + b \sin^2 \varphi.$$

Howard: I have measured rotational velocities from sunspots only for a two-year interval, but I am surprised at the magnitude of the effect.

Chistyakov: We established such a situation for the two 11-year cycles, i.e. No. 19 and No. 20.

ABOUT THE RELATION BETWEEN THE LIMB EFFECT
OF THE REDSHIFT ON THE SUN AND THE LARGE-SCALE
DISTRIBUTION OF SOLAR ACTIVITY

P. AMBROŽ

Astronomical Institute of the Czechoslovak Academy of Sciences, Ondřejov

Abstract. The measurement of the magnitude of the limb effect was homogenized in time and a recurrent period of maxima of 27.8 days was found. A relation was found between the maximum values of the limb effect of the redshift, the boundaries of polarities of the interplanetary magnetic field, the characteristic large-scale distribution of the background magnetic fields and the complex of solar activity.

Systematic measurements of radial velocities on the Sun carried out at the Mt. Wilson Observatory in the years 1966–1968 and processed and published by Howard and Harvey (1970) indicate that the difference of the redshift between the limb and the centre of the solar disc usually changes day by day. The theoretical models as proposed by Schröter (1957) and Hart (1974) do not, however, give a satisfactory explanation of these fast changes. As follows from the study by Howard (1971) series of values indicating the magnitude of the limb effect show, during the autocorrelation analysis, long-term persistent regular recurrences with a period which is practically identical with the period of solar synodic rotation, i.e. about 27–28 days. We can, therefore, expect a characteristic relation between the typical changes of the limb-effect magnitude and the regularities in the large-scale distribution of solar activity, i.e. the photospheric background magnetic fields, the typical incidence of filaments, the structure of the boundary of interplanetary magnetic fields polarities etc. If such a relation could be found, it would be very interesting, especially since Howard interprets the changes of the limb-effect magnitude as a consequence of nearly horizontal large-scale velocity fields in the solar photosphere.

As the material of Howard and Harvey (1970) contains – for the chosen time interval – sequences of days when observations could not be realized, it is not homogeneous in time and thus not convenient for direct comparison.

The primary assumption for the elaboration of the method for processing the measured data is based on Howard's finding of persistent recurrences in the change of the limb-effect magnitude. The aim of processing was to homogenize in time the series of the measured limb-effect magnitudes.

The measured values were therefore ranged in the 'solar calendar' with a recurrence of 27 days. The material thus covers 31 Bartels rotations from rotation No. 1821 to rotation No. 1852. In that way we have obtained a table where the columns were denoted by the index $i \sim 0, 1, 2, \ldots, 26$. The line index j had values (number of rotations) $j \sim 1821, 1822, \ldots 1852$. Each measured value of the limb effect can, therefore, be denoted by two indexes e_{ji}. For the purpose of time homogenization we have used the following algorithm:

$$e_{kl} = \frac{1}{X} \sum_{i=l}^{l+2} \sum_{j=k}^{k+3} e_{ji},$$

Bumba and Kleczek (eds.), Basic Mechanisms of Solar Activity, 113–118. All Rights Reserved.
Copyright © 1976 by the IAU.

where k reaches the values $1821, 1822, \ldots 1849$ and l reaches the values $0, 3, 6, 9, \ldots 24$. The value e_{kl} indicates the group average value of the limb effect and X is the number of measurements in one group. At the same time the value X indicates the relative statistical weight of the value e_{kl} homogenized in time.

As a result of this arrangement we obtain a matrix with 9 columns and 29 lines containing homogeneous in time and equidistant average values of the limb effect, while not all the values have the same statistical weight. The result of this procedure is graphically represented in Figure 1, where two subsequent 27-day intervals are

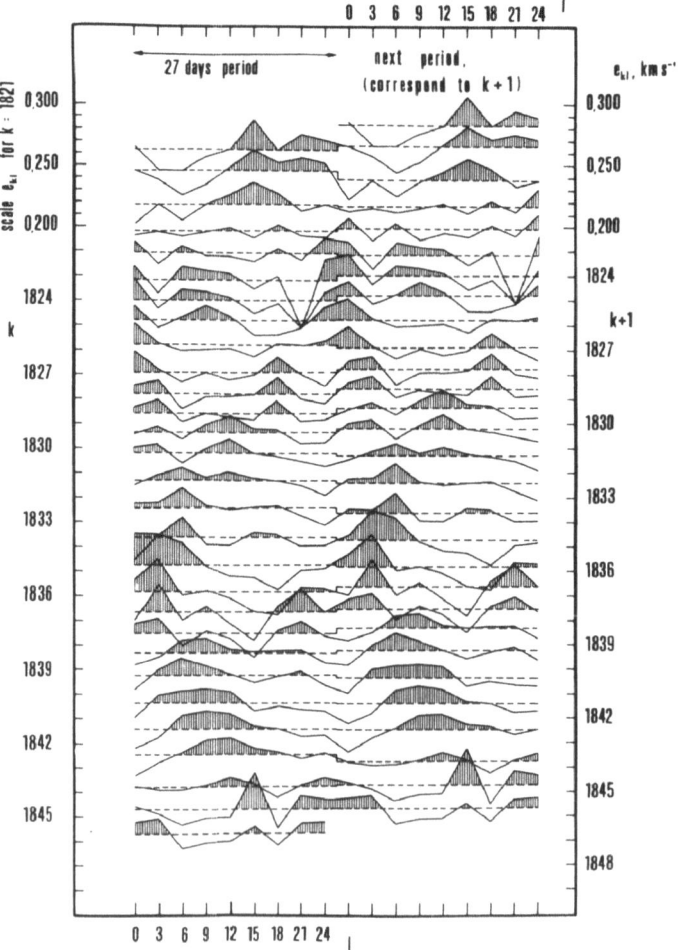

Fig. 1.   Graphical representation of the shape of the e_{kl} value. Sections illustrating the 27-day interval are drawn side by side. Values above the average for 27 days are hatched.

always plotted side by side. The values larger than the average over 27-day intervals are hatched. We can see from the figure that the averaged value of the limb effect shows a marked variation and the above average values are ranged in a vertical stripe slightly inclined towards the vertical axis. The inclination corresponds to the recurrence of 27.8 days. Thus we might conclude that the changes of the parameter e_{kl} occur approximately in heliographic latitudes from 20° to 40°.

Figure 2 shows the distribution of polarities and of the boundaries of the interplanetary magnetic field sectors ranged, according to Bartels. Rotations as measured by satellites (Wilcox and Colburn, 1970) are hatched (positive polarity is vertically hatched, the negative polarity is horizontally hatched). The dark circles

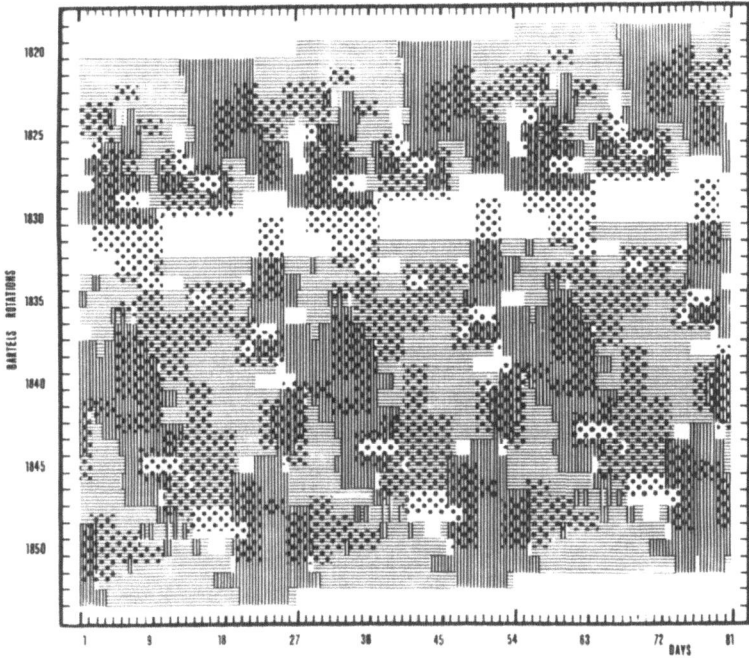

Fig. 2.   Chart illustrating the relation between the above average values e_{kl} (the zones in Figure 1 are hatched, here they are represented by dark circles) and the distribution of polarities of the interplanetary magnetic field measured by satellites. The positive polarity is horizontally hatched, the negative one vertically.

indicate the distribution of the above average values of the averaged limb-effect magnitude. The plot of the limb-effect values is shifted by 4.5 days in the time direction to reduce the time difference of the solar wind flight from the Sun to the Earth. The two plots are mutually superposed and are drawn three times side by side so that three rotations in one line are in a chronological sequence. In the figure we can clearly see the systematic incidence of the above average limb-effect magnitude in the close vicinity of the polarities boundary during the transition from (+) polarity to (−) polarity. In the transition from (−) polarity to (+) polarity such a relation is far less marked.

In this connection we should quote the studies by Ambrož et al. (1971) and by Bumba (1972a, b) dealing with the super-giant regular structures of the magnetic field whose characteristic dimension is comparable with the diameter of the solar disc. These structures contain both magnetic polarities, one of which is always prevalent. Bumba pointed out the relation between the structures with a prevalent negative polarity and with an increased incidence of the solar activity including large flares. Regular structures with the prevalent positive polarity are well correlated with the increased geomagnetic activity.

Experience acquired by the processing of the limb-effect values shows that on the Sun another measurable large dimensional physical parameter exists which has the dimension of velocity and behaves differently in relation to the type of transition between average polarities all over the solar disc.

In the last part of this study the variation of group average limb-effect values e_{kl} was compared with the large dimensional structure distribution of sunspots, filaments, and photospheric magnetic fields. Synoptic charts drawn in the Carrington's coordinate system were used for this comparison.

In Figure 3 we have illustrated a typical and relatively simple and synoptic situation in the time interval including Carrington's rotations No. 1537–1540. In the upper part we have a synoptic map of the sunspots created by the superposition of four Zürich heliographic maps. In a similar way the map of filaments on the basis of Meudon material was designed. The map of magnetic fields was designed by Stenflo (1972) as a sum of magnetic fields intensities in relation to their polarity after four rotations. The resulting positive polarity is vertically hatched, the negative one is horizontally hatched. In the graph 3d the above average values of the homogenized limb-effect magnitude are illustrated by hatching; we can observe one marked maximum. In map 3b a large typical filament wake was created in the same region of heliographic longitudes. This area is characterized by an increased production of filaments and by their intense development from the equator up to the latitudes $\pm 50°$. The incidence of sunspots is illustrated in Figure 3a. The distribution and the structure of the sunspot groups in the given area of longitudes are not, however, marked. The situation describing the distribution of magnetic fields can be found in map 3c. In the longitudinal region with an above average limb-effect value we can observe a typical configuration of the magnetic field. We observe an exceptionally enlarged surface of the average magnetic fields with intensities higher than ± 2 G. These background fields reach up to the polar areas and as a rule fill even the area around the equator.

We have designed altogether about 7 of such series for the period covered by the measurement of the limb-effect magnitude. The relation found between the different expressions of the solar activity is a typical one and in fact it can be found in a more or less marked way during all the studied period.

The results of this preliminary study are therefore the following ones:

(a) The recurrence determined by the autocorrelation method was found, also using the 'solar calendar' method; the recurrence period can be more precisely determined as being 27.8 days.

(b) The recurrent maxima in the course of the limb-effect magnitude are topologically linked to solar surface phenomena and the recurrence is given by the solar rotation in the latitudes from 20 to 40 heliographic degrees.

(c) The longitudinal areas of the maximum limb-effect values are linked to those areas on the Sun where there occurs the boundary of polarities of the total average value of the magnetic field strength on the disc and consequently of the interplanet-

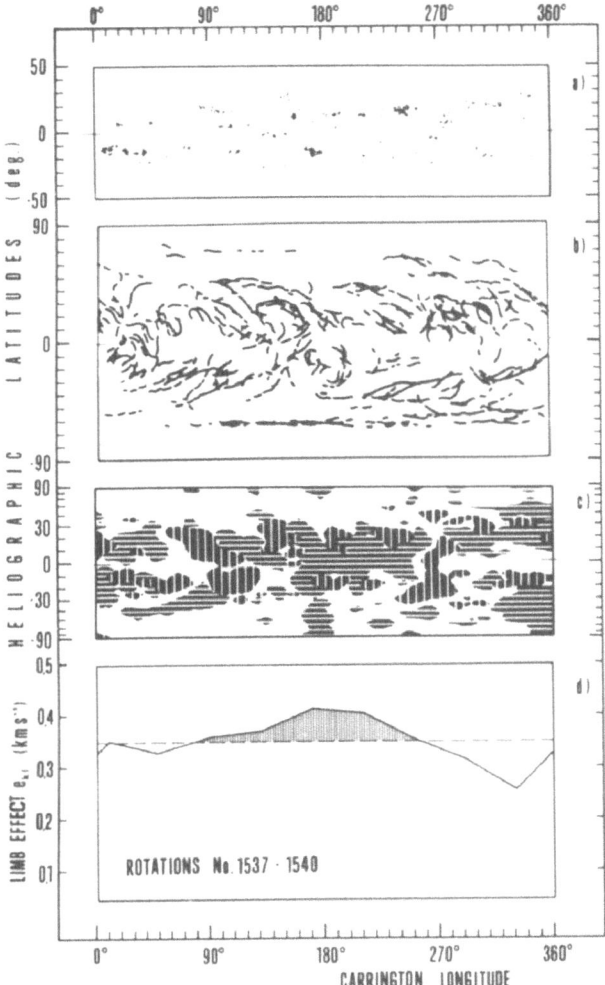

Fig. 3.   Illustration of the relation between the sunspots distribution (map a), the filaments (map b) and the photospheric magnetic fields (map c). Values of the e_{kl} limb effect parameter above the average are hatched (plot d). In the figure the data from Carrington's rotations 1537 to 1540 are superposed.

ary magnetic field. The effect is much more marked at the boundary of the type $(+ -)$ than at the boundary of the type $(- +)$.

(d) On the solar surface such areas are characterized by an extensive complex of activity whose typical feature is especially a complex structure of the photospheric background magnetic fields reaching not only the active zone but also the polar and equatorial areas. As a rule one polarity then passes from one hemisphere to the other one. The complex magnetic structure is accompanied by the incidence of active regions and especially by the filament formation creating usually filamentary systems having a long life and development.

(e) The maximum average values of the limb effect have, therefore, a close relation to the large-scale distribution of the magnetic field and of the solar activity.

The chosen method is not, however, sufficient for the physical interpretation of this relation. A study oriented in this direction is under way and will be published in the *BAC*.

References

Ambrož, P., Bumba, V., Howard, R., and Sýkora, J.: 1971, *IAU Symp.* **43**, 696.
Bumba, V.: 1972a, in *Solar Wind* (NASA, Asilomar conference, ed. by C. P. Sonett, J. P. Coleman, J. M. Wilcox), p. 31.
Bumba, V.: 1972b, in *Solar Wind* (NASA Asilomar conference, ed. by C. P. Sonett, J. P. Coleman, J. M. Wilcox), p. 151.
Hart, M. H.: 1974, *Astrophys. J.* **187**, 393.
Howard, R. and Harvey, J.: 1970, *Solar Phys.* **12**, 23.
Howard, R.: 1971, *Solar Phys.* **16**, 21.
Schröter, E. H.: 1957, *Z. Astrophysik*, **41**, 141.
Stenflo, J. O.: 1972, *Solar Phys.* **23**, 307.
Wilcox, J. M. and Colburn, D. S.: 1970, *J. Geophys. Res.* **75**, 6366.

DISCUSSION

Deubner: How can you distinguish changes of the limb effect from effects due to systematic vertical motions connected with active regions? Did you check whether an explanation of the observed changes in these terms would also be applicable?

Ambrož: The answer to the first part of your question is included in the paper by Howard and Harvey. It is very difficult to explain the physical origin of the limb effect. It seems very probable that the value of the limb effect is a super-position of the constant and variation term. But I cannot say more about the ratio of the horizontal and vertical component of the motion.

Karabin: How did you separate the limb effect from other velocities especially near the centre of the disc where the limb effect is small?

Ambrož: This separation was made by Howard and Harvey.

DRIFT OF SUNSPOTS IN LATITUDE

JAAKKO TUOMINEN
University of Helsinki, Finland

I am going to show two slides which should clearly prove that sunspots are drifting very regularly in the north–south direction on the solar surface.

Slide 1 is from a paper published in 1941. On the x-axis is the heliographic latitude. The y-axis gives the average proper motion of sunspots in latitude, expressed in degrees per day. The material is composed of long-lived spots, i.e. spots which have been observed at least at two rotations of the Sun. The dotted curve gives northern and southern latitude spots treated separately. The continuous curve has been drawn using averages of the values on the northern and southern latitudes. The diagram shows clearly that, on the average, between the parallels $\pm 16°$ sunspots are drifting towards the equator, while outside these parallels they are drifting towards the poles.

In the year 1954 Udo Becker showed that the drift curve changes when spots around sunspot minima and maxima are separated. And what is interesting, the errors decrease from the values obtained when all spots are combined. This happens in spite of the fact that in the separated material the number of values is much smaller than in the case when all the spots are combined. It clearly proves that the drift curve around minimum is different from that around maximum.

Slide 2. In this slide the dotted curve represents the values derived by Becker. Here the latitude is on the y-axis and the mean daily proper motion in latitude on the x-axis. Becker's material was practically the same as that of the first slide, i.e. long-lived spots only. As the number of spots in the minimum phase is small compared to that in the maximum phase, the curve of Slide 1 practically corresponds to the curve of the maximum phase, which can be seen easily.

Unfortunately, Becker did not treat northern and southern latitudes separately, but combined the two hemispheres. But, in order to obtain a check of the result, in 1960 I made an exactly similar study to that of Becker, this time based entirely on short-lived spots. Then the continuous line on the slide was obtained. Its accuracy is not inferior to that obtained from long-lived spots, because the number of short-lived spots is much greater than that of long-lived ones. It is seen that the two entirely different materials give practically the same result. It can be asked why there is such an oscillation in the north–south drift of sunspots.

References

Becker, U. 1954, *Z. Astrophys.* **34**, 129.
Tuominen, J.: 1941, *Z. Astrophys.* **30**, 96.
Tuominen, J.: 1960, *Z. Astrophys.* **51**, 91.

Bumba and Kleczek (eds.), Basic Mechanisms of Solar Activity, 119. *All Rights Reserved.*
Copyright © 1976 by the IAU.

VELOCITIES OBSERVED IN SUPERGRANULES

S. P. WORDEN and G. W. SIMON

Sacramento Peak Observatory
Air Force Cambridge Research Laboratories, Sunspot, N.M. 88349, U.S.A.

Abstract. The evolution of the velocity and magnetic fields associated with supergranulation has been investigated using the Sacramento Peak Observatory Diode Array Magnetograph. The observations consist of time sequences of simultaneous velocity, magnetic field, and chromospheric network measurements. From these data it appears that the supergranular velocity cells have lifetimes in excess of 30 h. Magnetic field motions associated with supergranulation were infrequent and seem to be accompanied by changes in the velocity field. More prevalent was the slow dissipation and diffusion of stationary flux points. These observations suggest that surface motions do not exhibit the detailed flux redistribution expected in the random-walk diffusion of magnetic fields. It is suggested that the surface motions are only the reflection of magnetic field-convective motion interactions which occur deeper in the convection zone.

1. Introduction

Following Leighton *et al.* (1962) the term 'supergranulation' refers only to a horizontal velocity phenomenon within the solar photosphere. Any study involving descriptive parameters, such as lifetimes and sizes, must consequently deal with the velocities directly. The development of fast photoelectric magnetographs makes it possible to study the supergranular velocity flow, rather than secondary effects such as the location of chromospheric emission regions. Several questions concerning supergranulation can therefore be investigated.

Supergranulation has been interpreted as a convective flow pattern (Simon and Leighton, 1964; Leighton, 1964, 1969; Simon and Weiss, 1968) and consequently used in discussions of convective theory. An important parameter in these discussions is the lifetime of the convective 'supergranular cell', since convective theory can provide estimates of the velocity and temperature structure within a convection cell if the lifetime is known. Several studies of supergranular lifetimes have been undertaken. In the studies of Simon and Leighton (1964) and Rogers (1970) the chromospheric emission network as observed in strong absorption lines was used to define supergranular boundaries. Lifetimes close to 24 h were obtained in this indirect manner. However, detailed studies of magnetic field elements by Smithson (1973) revealed that these elements do not change significantly in their positions over periods of approximately 36 h. If magnetic field elements delineate supergranular boundaries as presumed (Simon and Leighton, 1964) this observation would lead to supergranular lifetime estimates somewhat longer than 24 h. Additional evidence for this hypothesis is provided from Livingston and Orrall's (1964) observation of long-lived magnetic features with supergranular appearance. These 'cells' had lifetimes of 3–5 days and occurred within active regions.

Smithson (1973) suggested that the shorter lifetime derived from emission network studies was misleading due to the manner in which it was derived. Lifetimes in the previous studies were derived from mathematical cross-correlations. However, since the emission network is a 'thin' system defining only cell boundaries, and not the cell itself, small changes in the shape of a cell will cause a large decrease in

Bumba and Kleczek (eds.), Basic Mechanism of Solar Activity, 121–134. All Rights Reserved.

cross-correlation coefficients. This may produce a spuriously short lifetime compared to the real lifetime of the velocity cell. For this reason supergranule lifetimes are more appropriately derived from measurements of the horizontal velocity flow which extend over virtually the entire cell.

Supergranule cell lifetimes are important for another reason. Leighton (1964, 1969) in his theory of the solar activity cycle used supergranular flows to disperse magnetic fields over the solar surface in a random-walk process. However, if the supergranule lifetimes are significantly longer than 24 h as discussed by Smithson (1973), the motions due to supergranular flow are too small to provide for the observed magnetic field dispersal. Moreover, field motions have never been correlated with any material motions, supergranular or any other. Smithson (1973) observed an occasional rapid movement of magnetic flux elements; however, the frequency of these occurrences is insufficient to explain observed changes in the magnetic field pattern. His data appear more consistent with the total disappearance of magnetic field elements. Clearly it is important to determine whether the rapid movement and disappearance of magnetic elements is associated with changes in the velocity flow.

In order to determine mass flow rates within the supergranule, knowledge of the vertical velocity field is needed. Simon and Leighton (1964), Frazier (1970), Deubner (1972), and Musman and Rust (1970) have reported vertical flows associated with supergranulation. However, only the vertical downdrafts associated with magnetic field elements appear well confirmed. A corresponding vertical upflow in cell centers has not yet been shown convincingly. Accurate photoelectric velocity observations are needed to investigate these questions.

This paper reports a time series of observations including simultaneous magnetic, velocity, and emission network information obtained with the Sacramento Peak Observatory Diode Array Magnetograph (Dunn et al., 1974). The results of this study will be discussed in light of the problems mentioned above.

Our use of the strongly magnetically sensitive line Fe I 8468 Å presented difficulties in observing velocities in magnetic regions (this problem has recently been discussed by Frazier (1974)). Recent observations of magnetic field strengths by Harvey and Hall (1974) in the infrared indicate solar magnetic fields may be as strong as 2000 G. At such high field strengths Fe I 8468 Å becomes Zeeman split to such a degree that the components are completely separated and serious errors occur in the velocity measurements. Consequently, a group of magnetically insensitive lines ($g = 0$) were also used to obtain velocity observations. These lines were chosen to represent a range of heights within the solar atmosphere. Data concerning these lines are presented in Table I.

2. Analysis and Results

Leighton et al. (1962) reported a vertical oscillatory velocity in the photosphere with a period about 300 s and with an amplitude of approximately 0.5 km s^{-1}. This phenomenon is of similar velocity amplitude to the supergranular field. As it constitutes an interference to direct observation of the non-oscillatory flow pattern of supergranules it is necessary to remove the 300-sec oscillations from these observations. This was accomplished by averaging together the signals from a sequence of individual scans (each taking 48 seconds of time) over one or more 300-sec periods.

The digital nature of the two-dimensional data made it possible to use two-dimensional Fourier analysis. A two-dimensional Fast Fourier Transform (FFT) computer algorithm was written for this purpose. Since Fourier transforms have the property of separating in frequency space information on differing size scales they are ideal for supergranular studies. The various velocity phenomena, granulation, supergranulation, and 300-sec oscillations have well-defined and different size scales so the information on each of these velocity fields is separated within the frequency space of the Fourier transform. Thus to isolate supergranular effects the data are transformed and all frequencies representing size scales significantly smaller than supergranular flow (30 000 km) are removed; then the transform is inverted producing a filtered picture. However, these filtered pictures suffer from the disadvantage that small scale effects which may be associated with supergranulation, such as the downflows at cell boundaries, may also be removed. In conjunction with conventional means of analysis, such as cinematography of the time sequences obtained, the lifetime of the supergranular velocity flow, the transport of magnetic field, and the vertical velocity structure of the supergranule were studied.

The Lifetime of Supergranulation

Several sequences of observations covered roughly 10 h each on a single region. While this time was less than the presumed lifetimes (20–40 h) changes in some supergranules within the observed area may be expected. The data from 1974, March 5 covered $9\frac{1}{2}$ h at radius vector $\rho = 0.6$. During that run the seeing remained consistently good, so the March 5 data were chosen for detailed analysis.

A movie was made from the original data. The only processing in addition to magnetic and velocity reductions involved time averaging to remove the 300-sec oscillation effects. Each frame in the movie consisted of one 288-sec average of six 48-sec observations. The movie gives the impression that few changes occurred in the supergranular flow pattern during the 9-hour observation period. However, granular velocities were strong and interfered with the definition of supergranular cells. Additionally, when the seeing became slightly variable, as in the late afternoon, leakage of the 300-sec oscillation was present due to the inconsistent seeing during the 288-sec average. These problems were reduced using the two-dimensional Fourier filtering scheme described earlier to remove granulation and 300-sec leakage. Figure 1 shows the unfiltered data and corresponding filtered data. Each frame consists of an average over four 300-sec oscillations (20 min) or 24 observations. These observations cover nine hours on 1974, March 5. Averages over periods longer than 20 min were impractical due to imperfect guiding in the telescope.

From the time sequence in Figure 1 the impression of minimal change in the supergranular pattern derived from the movies is strengthened. Over the 9 h covered by the data a significant change can be observed in only one of the approximately 9

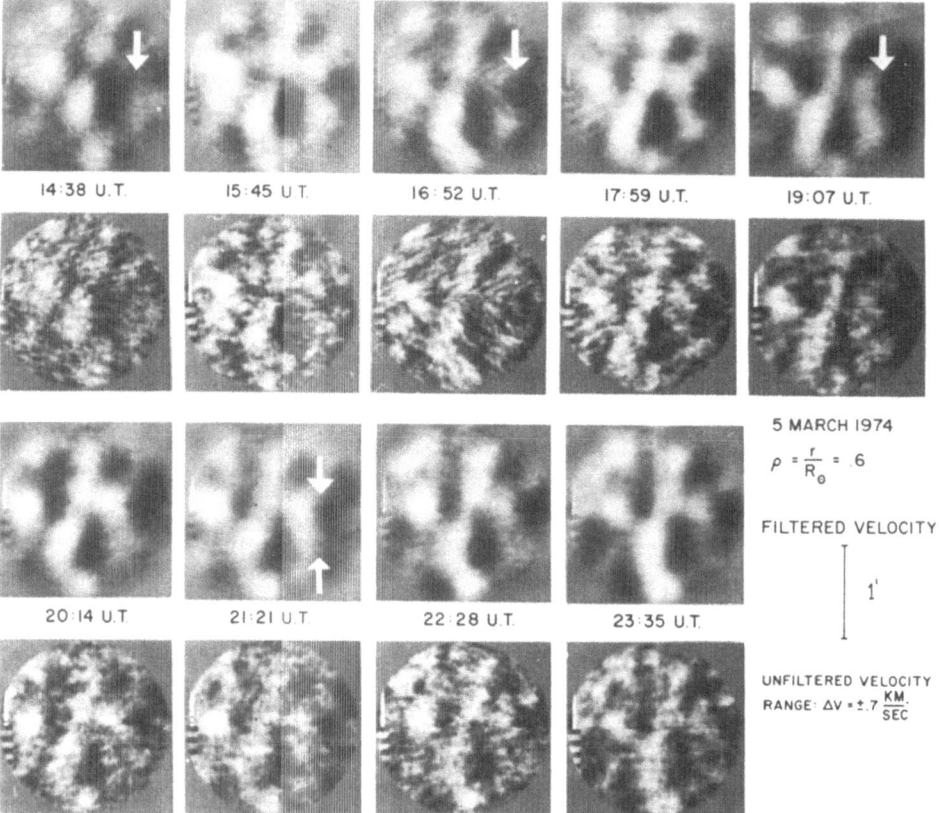

Fig. 1.   All day filtered and unfiltered velocity observations for 1974, March 5.

supergranules present. In the early frames the supergranular flow in the right central part of the picture appears weak; however, later in the day the flow has strengthened. Similar behavior was observed in the data of 1974, March 11 and 1974, March 16; however, only 1–2 supergranules in the region under observation appeared to change significantly in the course of the day's observations. The relatively infrequent changes observed are inconsistent with a mean lifetime of approximately 24 h. If the lifetimes were that short, roughly 1/3 or 3–4 supergranules per day, could be expected to change dramatically in an area covering 9–10 supergranules.

The lifetime of the supergranular flow can be derived from the two-dimensional cross-correlation function. This function is given by:

$$XC(\Delta x, \Delta y) = \int_{-\infty}^{\infty} \int_{-\infty}^{\infty} f(x, y) g(x + \Delta x, y + \Delta y) \, dx \, dy$$

$$= \int_{-\infty}^{\infty} \int_{-\infty}^{\infty} F(u, v) G^*(u, v) \, e^{i 2\pi (u \Delta x + v \Delta y)} \, du \, dv. \tag{1}$$

The latter function is the inverse Fourier transform of the cross product of the Fourier transforms of the data fields f and g under study, and Δx and Δy are the shifts in x and y of the two data arrays, f and g, relative to one another. The cross-correlation coefficient at zero shift ($\Delta x = 0$; $\Delta y = 0$) is a measure of differences between two sets of data when they are perfectly matched, as such it can be used to derive lifetimes of structures present in time series of data. The mean lifetime is defined as the time needed for the cross-correlation function to fall to $1/e$ of its original value at zero Δx and Δy shifts (Simon and Leighton, 1964). The two-dimensional FFT program was used to calculate these functions for the velocity data of 1974, March 5 shown in Figure 1. The results for zero spatial lag ($\Delta x = 0, \Delta y = 0$) are independent of signals due to granular velocity structure, since the first data for which cross-correlations were computed are separated by 1 h in time, during which the granular pattern should have completely changed. The correlation fell to 0.7 of its maximum after 9 h. A least squares fit to these points gave a slope of 0.017 ± 0.009 yielding a tentative mean lifetime of 36^{+70}_{-12} h. To verify this longer lifetime, data covering several days are required. These have been obtained at Sacramento Peak in the last two months and are being analyzed. However, the results discussed in this paper already suggest that velocity cells have lifetimes in excess of the 24-hour value derived from emission network studies.

3. Horizontal Transport of Magnetic Field Associated with Supergranulation

Several of the longer data runs showed evidence of magnetic field motions. Just as for the velocity data, casual inspection of the movies leaves the impression of relative inactivity; however, an occasional horizontal motion of magnetic flux points occurred. As with Smithson's (1973) observations these motions were of several forms. The most frequent form of motion appeared to be a slow (<1 km s^{-1}) motion of existing flux points in which part of the relatively stationary flux point splits off from the magnetic element and moves away. Figure 2 shows an example of this phenomenon. In most cases the moving flux-element moved less than 5000 km and then dissipated. In a few cases of larger velocity (0.5 km s$^{-1} < v < 1$ km s^{-1}) the daughter flux point moved 5000–10 000 km and remained visible for the remainder of the day.

A more dramatic form of magnetic field motion was associated with the emergence of new flux. In the few cases of this behavior, new flux emerged and moved rapidly (1-2 km s^{-1}) for a distance of 5–10 000 km. In the three single-day observations available no more than one, or in one case, two of these events occurred. Additionally, they always appeared in regions where changes in the velocity flow were underway. One such example is present in the data of 1974, March 5. As mentioned previously a velocity cell was observed to change during the day in the right central portion of Figure 1. During this period a small flux point appeared and rapidly moved to the boundary of the developing supergranule. Figure 3 shows this motion. Associated with this phenomenon changes in the chromospheric emission network occurred. In Figure 3 the λ 8542 intensity data for 1974, March 5 are also displayed to show these changes. In the data from early in that day the emission network in the region where the velocity cell changes appears chaotic. Later, when the velocity flow

20 APRIL 1974
CENTER OF DISK

1438 U.T.

1511 U.T.

1'

1643 U.T.

1745 U.T.

1902 U.T.

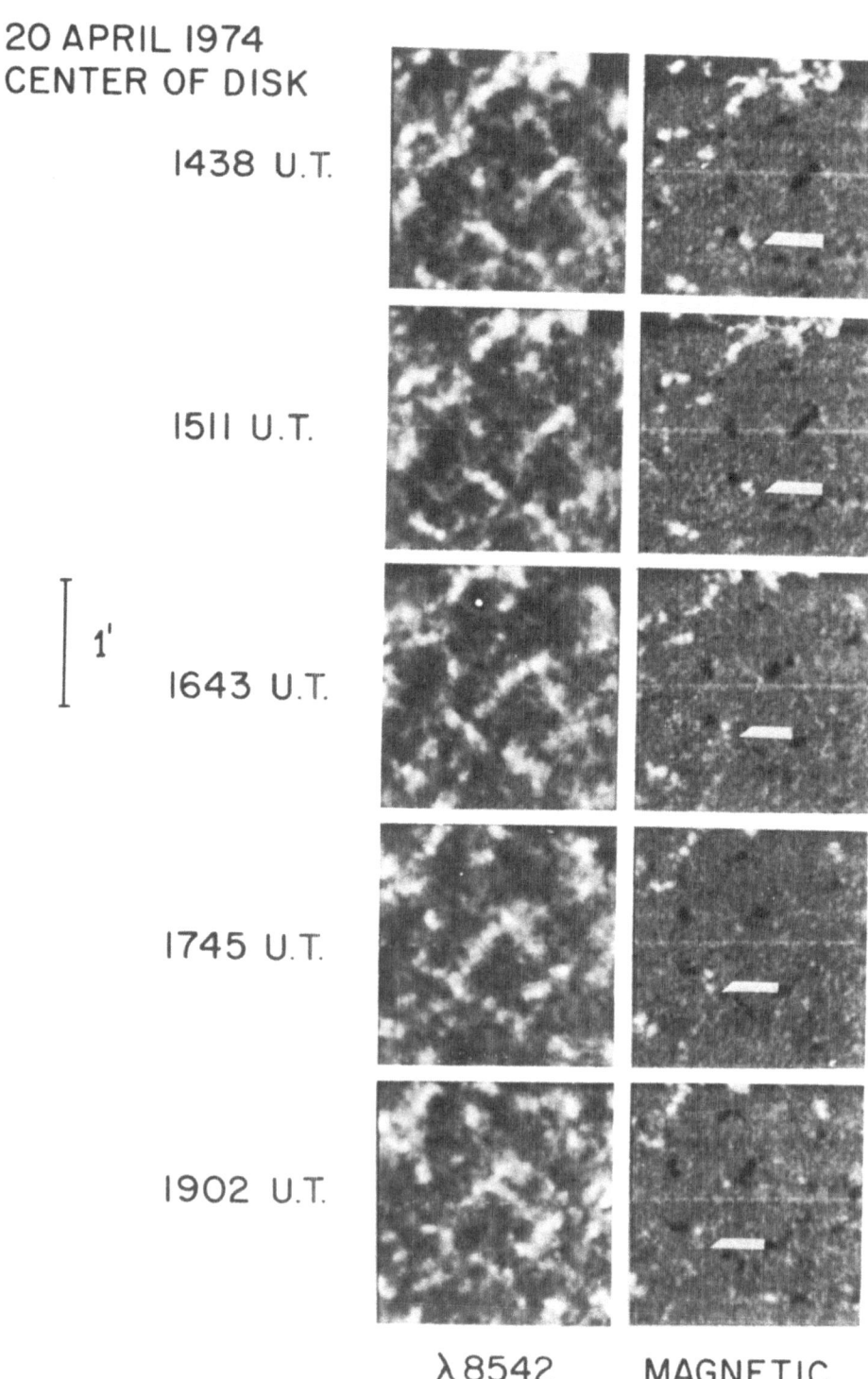

λ8542   MAGNETIC

Fig. 2.   Example of slow form of flux motion: splitting of existing flux.

5 MARCH '74

1628 U.T.

1716 U.T.

1804 U.T.

2311 U.T.

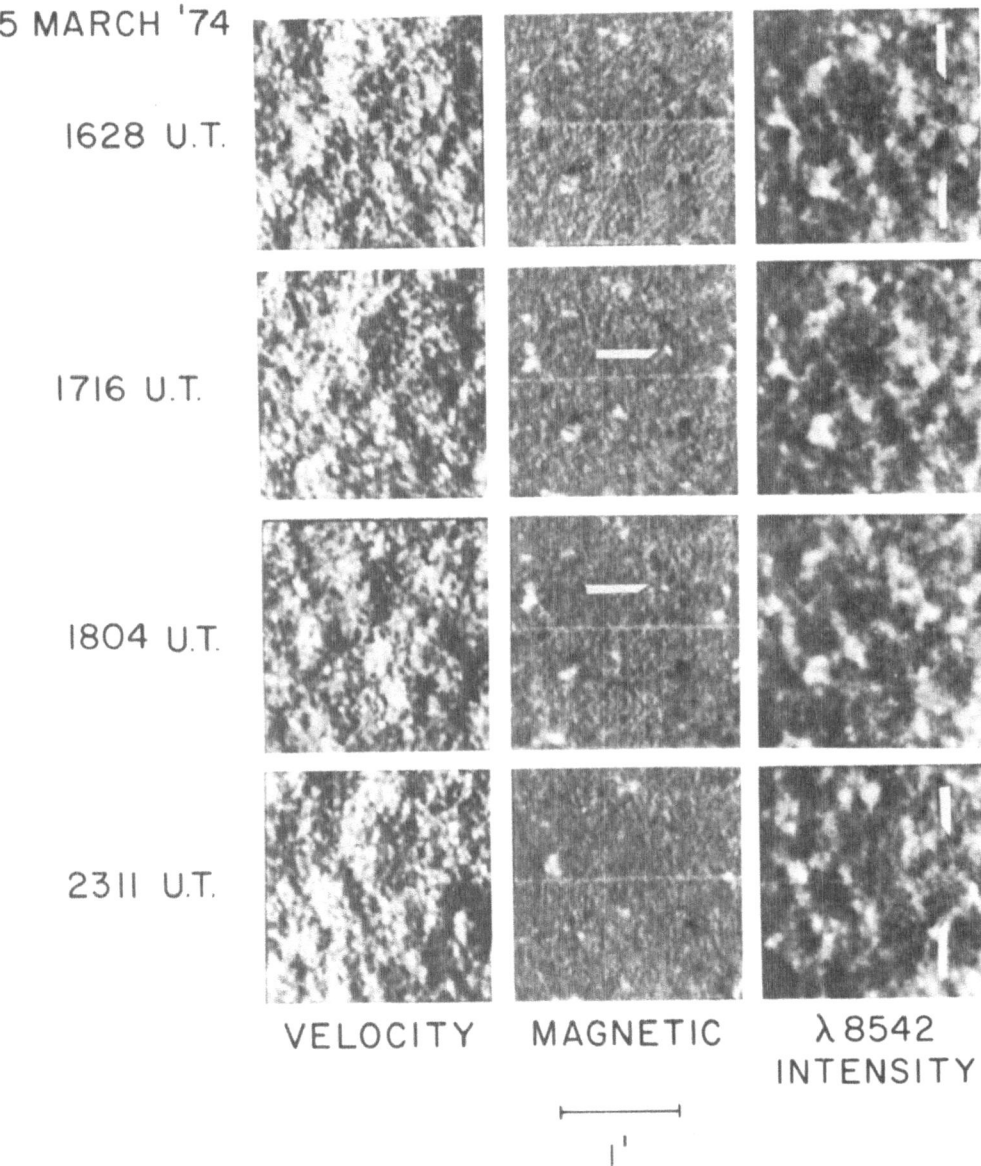

VELOCITY   MAGNETIC   λ 8542
                                   INTENSITY

I'

Fig. 3.   Example of fast form of flux motion: the appearance of new flux. The frame from 23 11 UT is
shifted slightly to the right relative to the other three frames.

has strengthened, the emission network has arranged itself into a well-defined
network 'cell'. During this period the emission network showed rapid and frequent
disappearances, motions, and reappearances. Similar behavior was observed in other
cases of cell development from the data on different days. In all three of these cases
the cell strengthening may well represent the formation of a new supergranule cell.

4. Vertical Velocity Flow within the Supergranule

Vertical velocities within supergranules have been reported by several investigators. Simon and Leighton (1964), Tannenbaum *et al.* (1969), Deubner (1972a), Musman and Rust (1970), and Frazier (1970) have detected what appear to be downdrafts at supergranular vertices. However since Frazier (1970) was only able to observe these downdrafts in certain lines there exists the possibility that the downdrafts may not be real, but may represent an artifact of the differing line formation, or the magnetic sensitivity of the lines used in the magnetic field regions (Frazier, 1974). Since the observed downflows occur only within the flux elements concentrated at supergranular boundaries this latter suggestion is a distinct possibility. The group of non-magnetically sensitive lines is ideal for a study of this problem, since they present a range of heights of formation, from levels where magnetic regions differ little in temperature structure from non-magnetic regions to levels where a large temperature differential exists.

TABLE I

Data for non-magnetic lines used in the vertical velocity investigation

Line	Spectrograph dispersion (Å mm^{-1})	Order	Range of formation heights for wings of the line (Altrock *et al.*, 1975)
Fe I 5123.730 Å	10.507	45	$-51 \rightarrow 286$ km
Fe I 5434.418 Å	9.368	42	$-47 \rightarrow 316$ km
Fe I 4065.388 Å	12.314	56	$-57 \rightarrow 122$ km

The disk center observations of 1974, July 16 were used for this portion of the study. During several hours of the morning during which these data were obtained, the seeing remained excellent (1 to 1.5″). As with the previous data 24 frame averages over 20 min were computed. A resulting 20-min average velocity map for the wavelengths λ 4065, λ 5123, λ 5435, and λ 8468 is shown in Figure 4. Also shown in this figure are the filtered images with the granular velocities removed, as well as the simultaneous magnetogram. In the velocity data a large scale pattern, which shows especially well in the filtered images, is apparent. While this pattern appears to have a supergranular size scale, the mean velocity signal is only slightly higher than the noise level. Several methods were used to sort out the true nature of this pattern, which does not appear to correlate well with the magnetic field structure.

To determine whether this pattern is related to the long-lived supergranular flow, observations obtained 20 min later during the same run were compared with the earlier data to see if the same velocity structures remained. While a similar large scale velocity pattern was evident, detailed agreement was only partial. Consequently other methods were attempted to study these results.

As mentioned previously, magnetic field elements have been observed to define the vertices of the horizontal supergranule pattern (Simon and Leighton, 1964). A comparison was therefore made between the magnetic field and velocity

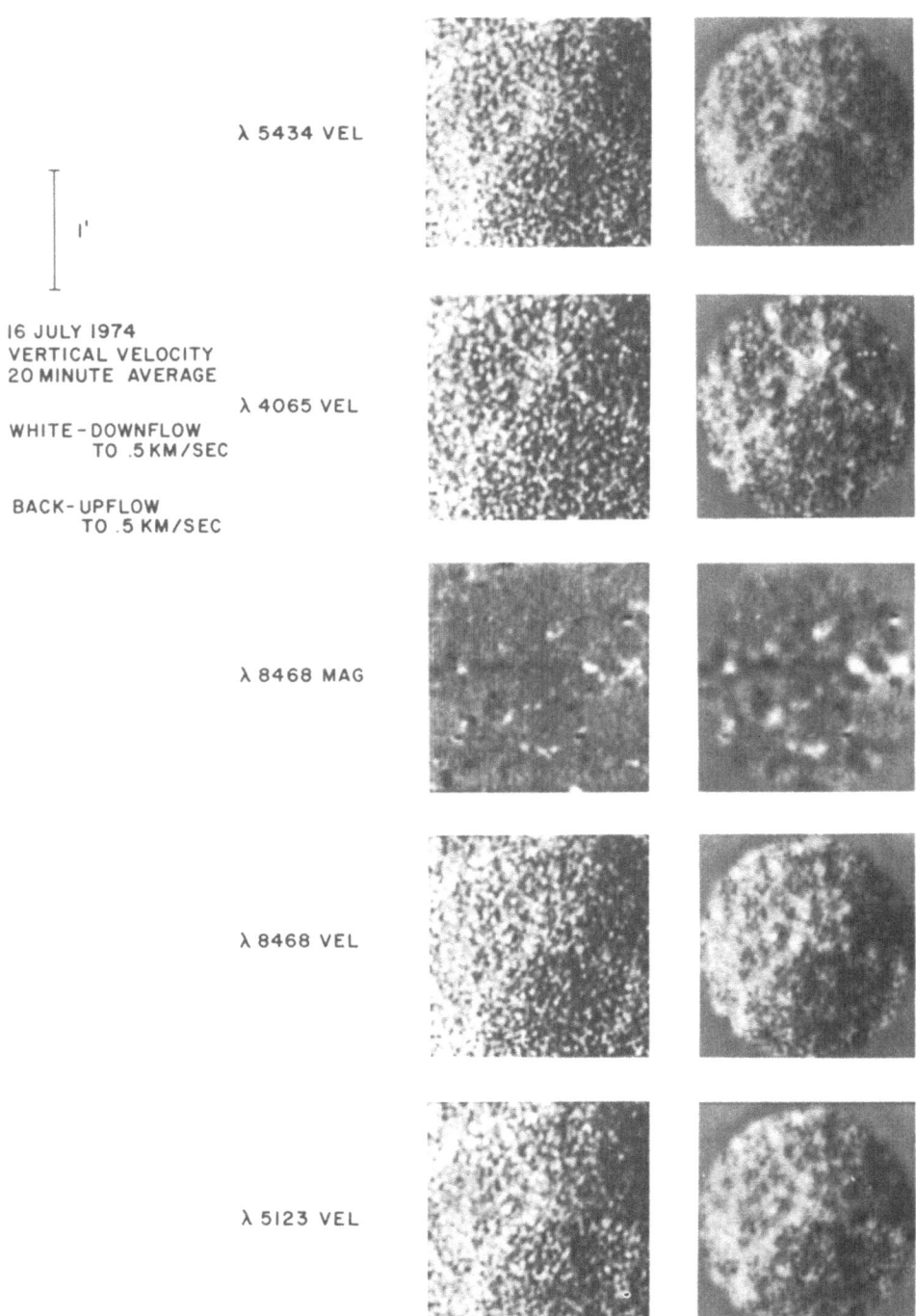

Fig. 4.   Filtered and unfiltered velocities in different lines from 1974, July 16.

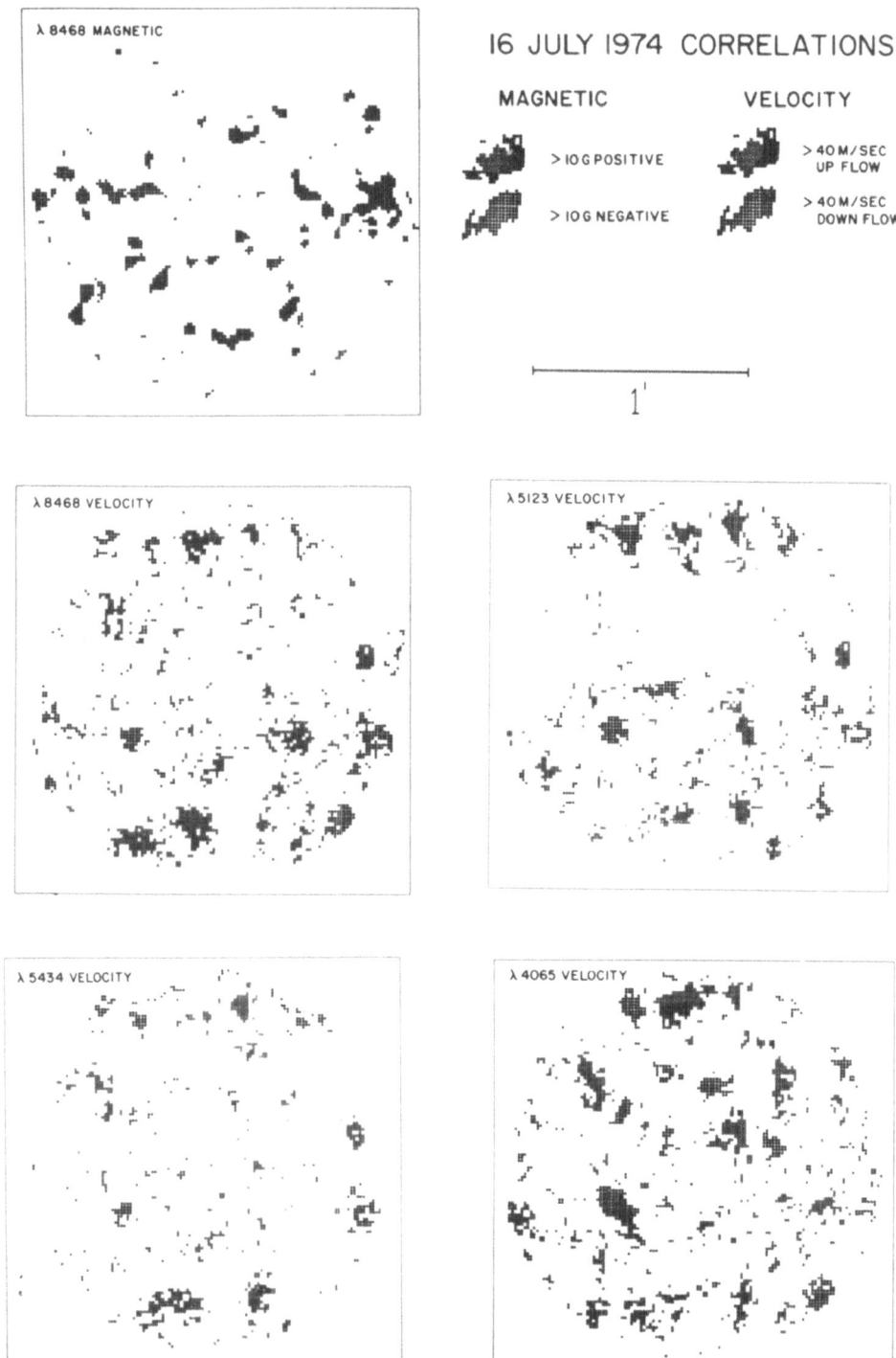

Fig. 5.   Velocities and magnetic field above a minimum threshold which correlate over forty minutes in center of the disk data.

observations. Only those velocity features which appeared on both consecutive 20-min time averages were used, since supergranular motions should persist over substantially longer periods. Figure 5 shows the result of this procedure, with all velocities less than a threshold of $40 \, \text{m s}^{-1}$ on both sets of data removed. For comparison the magnetic field structure is also displayed in Figure 5; as expected, the emission network is clearly outlined by the magnetic field. An upflow of roughly $50 \, \text{m s}^{-1}$ relative to the average velocity over the entire frame appears in the center of some emission network cells. However, downflows in magnetic field regions are apparent *only* in the λ 4065 data. Table II lists the velocities observed for each of the lines as a function of observed magnetic field strength. For small field strength λ 8468 seems to show downdrafts; however, as the field becomes progressively stronger this downflow diminishes in strength, until it disappears entirely. This behavior may be attributable to the magnetic splitting and distortion of this line profile as discussed previously.

TABLE II

Mean vertical velocities in magnetic field regions
(unfiltered data-downflows positive)

Magnetic field threshold	Number of 1″ points used	Velocity (m s^{-1})			
		λ 8468	λ 4065	λ 5123	λ 5434
5 G	3140	3 ± 2	14 ± 2	2 ± 1	0 ± 1
10 G	594	13 ± 4	49 ± 4	−3 ± 3	−4 ± 3
15 G	217	15 ± 5	85 ± 7	−16 ± 5	−7 ± 5
20 G	114	13 ± 8.2	105 ± 10	−33 ± 7	−6 ± 6
25 G	67	3 ± 10	116 ± 14	−45 ± 10	−15 ± 8
30 G	39	10 ± 14	153 ± 18	−70 ± 12	−23 ± 8
35 G	28	0 ± 14	174 ± 22	−84 ± 14	−29 ± 10
40 G	19	−9 ± 19	193 ± 26	−92 ± 18	−38 ± 13

Two-dimensional autocorrelation functions were used to verify these ideas. The two-dimensional autocorrelation of a data field $f(x, y)$ is given by:

$$AC(\Delta x, \Delta y) = \int_{-\infty}^{\infty} \int_{-\infty}^{\infty} f(x, y) f(x - \Delta x, y - \Delta y) \, dx \, dy$$

$$= \int_{-\infty}^{\infty} \int_{-\infty}^{\infty} |F(u, v)|^2 \, e^{i 2\pi (u \, \Delta x + v \, \Delta y)} \, du \, dv \tag{2}$$

which is identical to Equation (1) if $g(x)$ is replaced by $f(x)$. As with the cross-correlation function, this may be calculated by inverse Fourier transforming the amplitude (power spectrum) of the Fourier transform of the data under investigation. This two-dimensional function can be converted to a one-dimensional function by summing the two-dimensional function over radial annuli. The resulting one-dimensional function is a measure of the correlation of the data field with itself for radial displacements. The mean size scale in the data is given by the half width of the central autocorrelation peak. Highly periodic data will show secondary maxima at displacements representing the mean size of the periodic structures. The radial autocorrelation function summed over 20 min for each of the wavelengths is shown in Figure 6. The sharp fall-offs of the central maxima with full widths at half maximum (FWHM) of 1500 km are probably attributable to granular signal. However, the more gradual fall-off out to 20 000 km and the possible weak secondary maxima at 30 000 km are indications of velocity structure on a supergranular size scale. The magnetic field autocorrelation function shown in Figure 6 is interesting since its FWHM fall-off of 3500 km is indicative of a mean size of magnetic elements of 4–7″, significantly larger than the seeing size and in line with the visual appearance of the data.

5. Summary

The observational results of this study are summarized below:

(1) Supergranule velocity cells may have mean lifetimes considerably longer than the previously accepted value of 24 h, with probable values near 36 h. However, detailed several day observations are still needed to verify this lifetime.

(2) No realignments of magnetic field elements already present at supergranular boundaries in a manner expected from Leighton's 'random walk' mechanism were observed in any of the 10-hour runs which were studied. Since 10 h is only a small fraction of a supergranule lifetime this negative result does not preclude the operation of supergranule random walk processes. However, two forms of magnetic flux motions were observed. Often an existing field element will split and a portion of the flux move away and dissipate. A second form of flux motion occurred when new flux appeared and moved rapidly in regions of changing supergranular flow. The motions are large ($\sim \sim 10\,000$ km) and rapid ($v \sim 2$ km s^{-1}) in the latter process.

(3) Magnetic field elements have an apparent size of 5″ ($\sim$3500 km) in data with 1″ resolution. This is larger than the seeing size and is presumably real.

(4) Supergranules may exhibit an upflow of $\sim$50 m^{-1} in the center of each cell; however, this observation needs to be verified with higher accuracy data. A corresponding downflow of $\sim$200 m s^{-1} is observed in magnetic field regions at the boundaries of supergranules. However, this downflow is only observed in the most deeply formed line. The disappearance of this downflow when observed in strongly magnetically sensitive lines is consistent with the hypothesis that strong fields ($B \sim 1000$ G) are present within these regions.

The longer lifetimes of velocity supergranules indicated from this work match Smithson's (1973) values for lifetimes obtained from magnetic field elements. As shown by Smithson this value is too small by a factor of two to explain the observed diffusion of active region magnetic fields as a random walk process due to supergranular motions. The slow breakup of magnetic field elements is difficult to explain in terms of any surface motion. However, this behavior and the sudden emergence of new flux in a supergranule appears more consistent with a model similar to the 'flux-rope' model of Piddington (1975). It may well be true that strong subsurface magnetic fields dominate convective motions to a far deeper level than previously

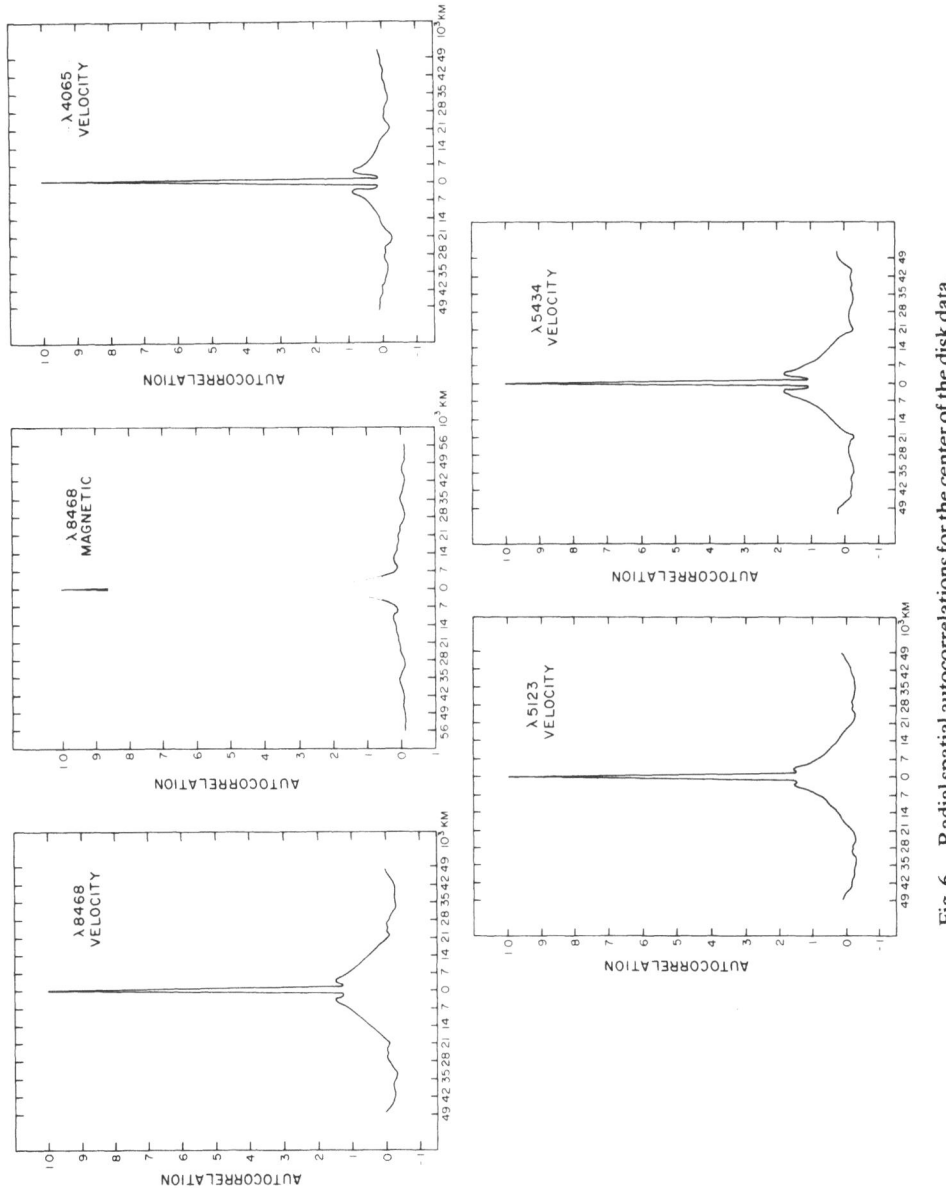

Fig. 6.   Radial spatial autocorrelations for the center of the disk data.

thought. If surface magnetic flux strengths are as large as 2000 G, subsurface fields may be concentrated to even larger strengths by convective motions. Thus, if the supergranule is roughly 10 000 km in depth (Mullan, 1971; Simon and Weiss, 1968) as predicted from theoretical considerations, the magnetic field might constrain the convective flow to a considerable degree as suggested by Wilson (1972) for sunspot regions. The observation of infrequent motion of existing flux elements would support the idea that magnetic fields are intimately tied to a single convective cell. Clearly, detailed convective analysis including the effects of strong magnetic fields is necessary.

Chapman (1974) and Frazier (1974) have suggested that the observed downflow at supergranule vertices may be due to line profile changes in the magnetic field regions caused by heating in the low chromosphere. Since the lines used in part of this study were chosen to represent heights where this effect shows a range of importance, this hypothesis was checked. The three non-magnetic lines were studied using an LTE computer program developed by R. W. Milkey for the KPNO CDC 6400 computer and it was found that the absence of vertical velocities in the emission regions is explainable as a line profile effect as suggested by Frazier (1974).

References

Altrock, R., November, L., Simon, G., Milkey, R., and Worden, S. P.: 1975, *Solar Phys.* **43**, 33.
Chapman, G. A.: 1974 (private communication).
Deubner, F. L.: 1972, *Solar Phys.* **17**, 6.
Dunn, R. B., Rust, D. M., and Spence, G. E.: 1974, *Proc. SPIE* **44**, 109.
Frazier, E. N.: 1970, *Solar Phys.* **14**, 89.
Frazier, E. N.: 1974, *Solar Phys.* **38**, 69.
Harvey, J. W., and Hall, D. N. B.: 1974 (private communication).
Leighton, R. B.: 1964, *Astrophys. J.* **140**, 1559.
Leighton, R. B.: 1969, *Astrophys. J.* **156**, 1.
Leighton, R. B., Noyes, R. W., and Simon, G. W.: 1962, *Astrophys. J.* **135**, 474.
Livingston, W. C. and Orall, F. Q.: 1974, *Solar Phys.* **39**, 301.
Mullan, D. F.: 1971, *Monthly Notices Roy. Astron. Soc.* **154**, 467.
Musman, S. and Rust, D. M.: 1970, *Solar Phys.* **13**, 261.
Piddington, E. H.: 1975 (preprint).
Rogers, E. H.: 1970, *Solar Phys.* **13**, 57.
Simon, G. W. and Leighton, R. B.: 1964, *Astrophys. J.* **140**, 1120.
Simon, G. W., and Weiss, N. O.: 1968, *Z. Astrophys.* **69**, 435.
Smithson, R. C.: 1973, *Solar Phys.* **29**, 365.
Tanenbaum, A. S., Wilcox, J. M., Frazier, E. N., and Howard, R.: 1969, *Solar Phys.* **9**, 328.
Wilson, P. R.: 1972, *Solar Phys.* **27**, 363.

LONG TERM EVOLUTION
OF SOLAR SECTOR STRUCTURE

LEIF SVALGAARD and JOHN M. WILCOX

Institute for Plasma Research, Stanford University, Stanford, Calif. 94305, U.S.A.

Abstract. The large-scale structure of the solar magnetic field during the past five sunspot cycles (representing by implication a much longer interval of time) has been investigated using the polarity (toward or away from the Sun) of the interplanetary magnetic field as inferred from polar geomagnetic observations. The polarity of the interplanetary magnetic field has previously been shown to be closely related to the polarity (into or out of the Sun) of the large-scale solar magnetic field. It appears that a solar structure with four sectors per rotation persisted through the past five sunspot cycles, with a synodic rotation period near 27.0 days, and a small relative westward drift during the first half of each sunspot cycle and a relative eastward drift during the second half of each cycle. Superposed on this four-sector structure there is another structure with inward field polarity, a width in solar longitude of about 100° and a synodic rotation period of about 28 to 29 days. This 28.5 day structure is usually most prominent during a few years near sunspot maximum. Some preliminary comparisons of these observed solar structures with theoretical considerations are given.

Full text of the paper submitted to *Solar Physics*.

Bumba and Kleczek (eds.), Basic Mechanism of Solar Activity, 135. *All Rights Reserved.*
Copyright © 1976 *by the IAU.*

SOLAR CYCLE EVOLUTION
OF THE GENERAL MAGNETIC FIELD

HIROKAZU YOSHIMURA

Hale Observatories, Carnegie Institution of Washington,
California Institute of Technology, Pasadena, Ca. 91101, U.S.A.

The data from magnetic field synoptic charts at Mt. Wilson for 16 years are separated into axisymmetric and nonaxisymmetric fields. The axisymmetric field derived simply by averaging over longitude corresponds to the general magnetic field and can be regarded as reflecting the poloidal (radial) field since bipolar magnetic fields which have been regarded as reflecting the toroidal field are cancelled out by the averaging. The evolution of pattern of the latitudinal distribution of this field shows a conspicuous appearance similar to the Butterfly Diagram of sunspots but having two branches of different polarity in each hemisphere. The two branches start from the middle latitudes, and one branch propagates towards the pole and the other toward the equator. This shows that the solar general magnetic field behaves like a quadrupole not a dipole as was previously believed. This feature is exactly what has been predicted by a numerical solar cycle model driven by the dynamo action of the global convection. Another axisymmetric field is also derived by averaging over longitude the absolute value of the magnetic field after subtracting the poloidal field. This field corresponds to the toroidal field since if this field is averaged over longitude it vanishes. The evolutionary pattern of the latitudinal distribution of this field shows a feature quite similar to the Butterfly Diagram of sunspots. These features of the two fields become conspicuous only after averaging over many rotations, e.g., over 27 rotations (2 yr): such a diagram averaged over a small number of rotations shows rather large noise.

The Butterfly Diagram of sunspots is drawn for the same period as the magnetic field data. There are slight differences between the Butterfly Diagram of sunspots and the Butterfly Diagram of the general toroidal magnetic field. That is, in the former case, the equatorial migration is quite clear but not so clear in the latter case. This can be interpreted as reflecting the state of the rotation in the upper and lower parts of the convection zone where the magnetic field observed at the surface and the sunspots originate respectively. These three kinds of information, i.e., the evolutions of the poloidal field, of the toroidal field, and of the sunspots should be considered as important indicative phenomena of the solar cycle to distinguish the validity of the various solar cycle models and to determine the basic mechanisms of the solar cycle.

References

Yoshimura, H.: 1975a, *Astrophys. J. Suppl. Series* **29**, 467.
Yoshimura, H.: 1975b, *Astrophys. J.* **201**, 740.
Yoshimura, H.: 1976, *Solar Phys.* (in press).

Bumba and Kleczek (eds.), Basic Mechanism of Solar Activity, 137–143. All Rights Reserved.
Copyright © 1976 by the IAU.

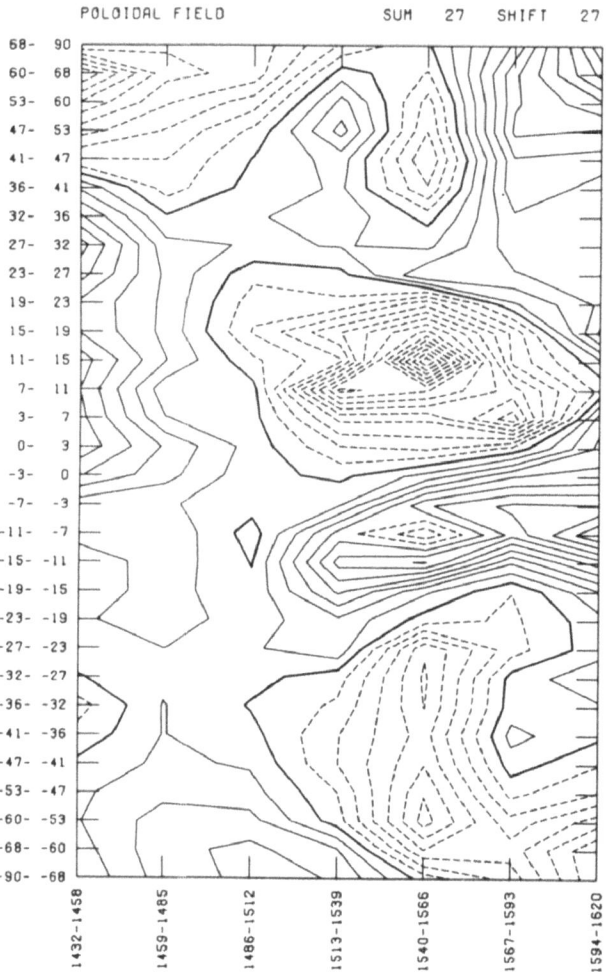

Fig. 1. The evolution of the latitudinal distribution of the *poloidal* general (axisymmetric) magnetic field which was derived by simply averaging the observed magnetic field over longitude, thus cancelling the magnetic field with bipolar structure. The abscissa is time from Carrington rotation number 1432 (1960 September) to 1620 (1974 October) and the coordinate is sin (latitude). The original data are stored in card form with 30 sections with equal interval in sin (latitude). The corresponding latitudes are shown in the figure. The broad lines designate zero lines; solid lines positive field, dotted lines negative except in the donut-like situations. Note that there are two branches with opposite polarities in each hemisphere so that the general magnetic field of the Sun behaves quadrupole-like not dipole-like. These features appear only if we average over many rotations (27 rotations) to cancel noises although there still are some fluctuations even in this figure. The maximum value in this field is 1.5 G.

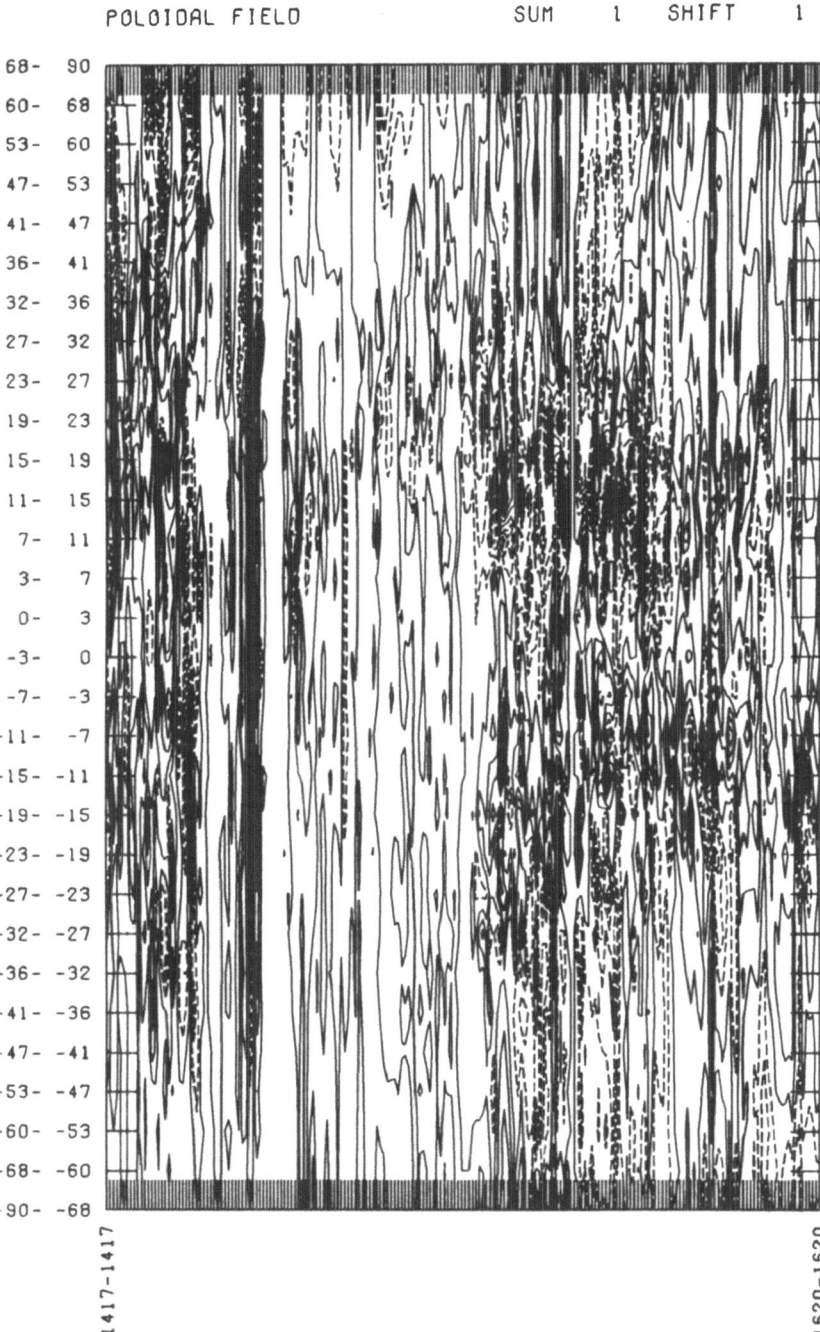

Fig. 2. The same as Figure 1 but the averaging over rotations is not done. Note that the noise is so large that clear features in Figure 1 cannot be seen.

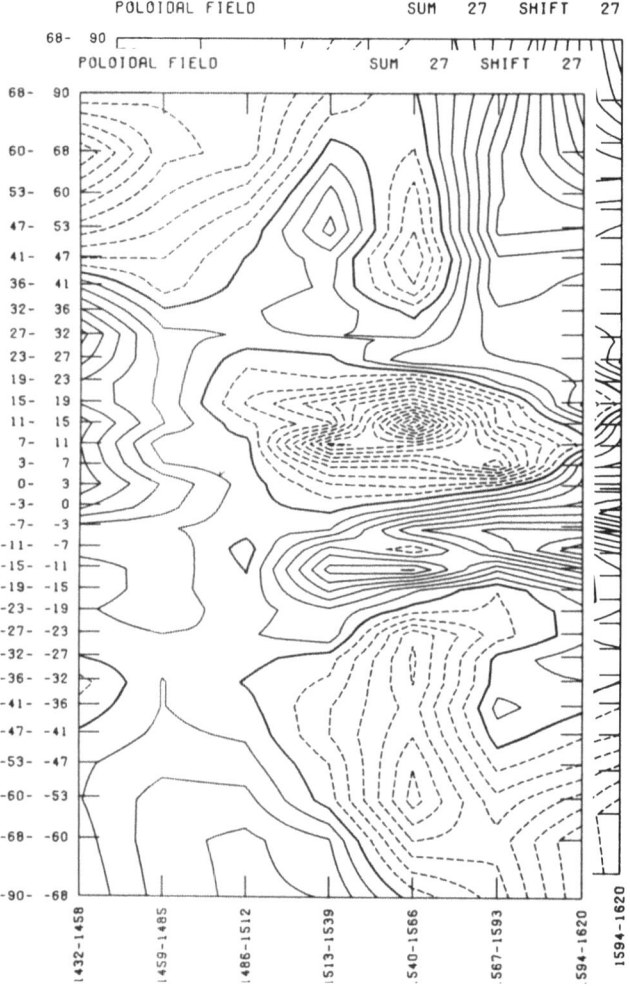

Fig. 3.   The same as Figure 1 but the coordinate is drawn with equal interval in latitude. Note that the
main branches are rather concentrated near the equator.

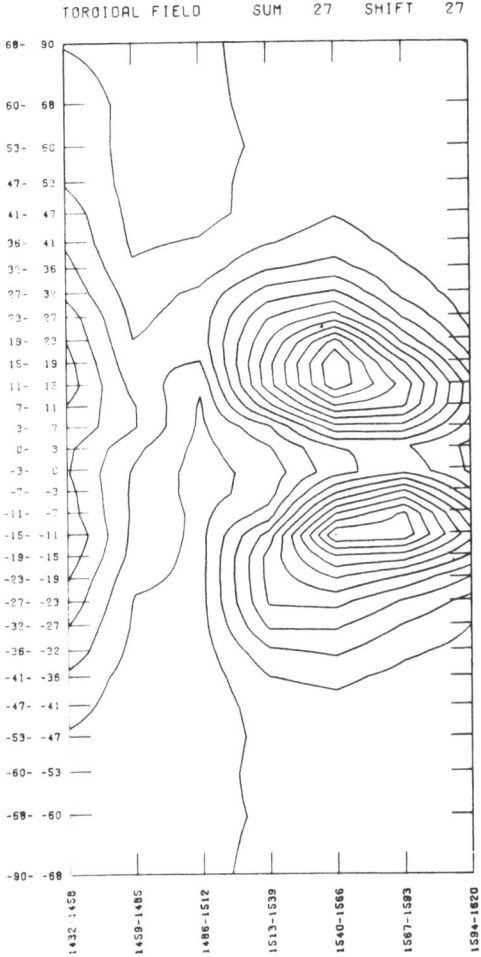

Fig. 4.   The evolution of the latitudinal distribution of the *toroidal* general (axisymmetric) magnetic field which was derived by averaging the absolute value of the magnetic field after subtracting the poloidal field. Thus this diagram shows the solar magnetic field which has bipolar structure. Note that this is similar to the butterfly diagram of sunspots but there seems to be some slight differences in the equatorial propagation of the wings (branches). Note also that, in order to compare it with the Butterfly Diagram of sunspots of the same period shown in Figure 6, the ratio of the scales of the abscissa and the coordinate is adjusted to that of Figure 6 where only the latitude region between −50 to +50 is shown. The maximum value on this figure is 6.9 G.

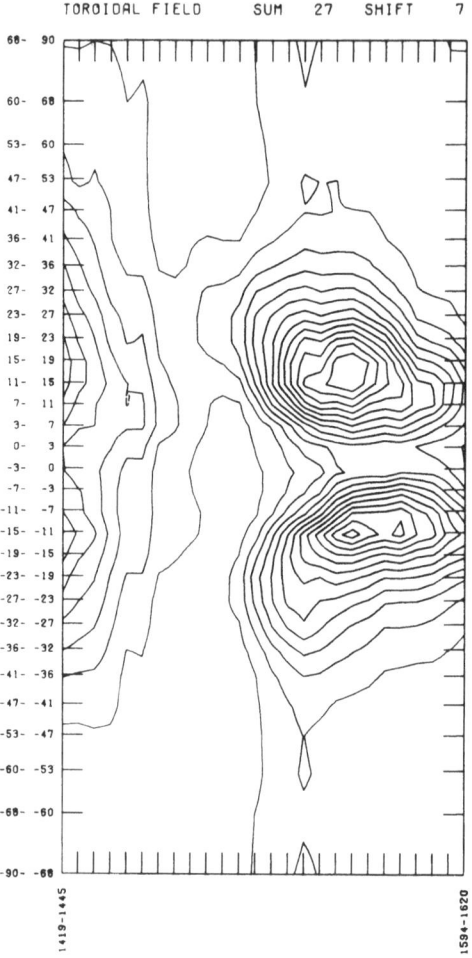

Fig. 5. The same as Figure 4, but the averaging over rotation is now running, the origin point of each averaging is shifted by 7 rotations not 27 rotations in order to show more details of the propagation of the wings. The propagation gives us important information about rotation law inside the Sun. The evolution of the distribution of the magnetic field shown above gives us information independent from that of the sunspot Butterfly Diagram since the zones from which the sunspots and surface magnetic field originate differ.

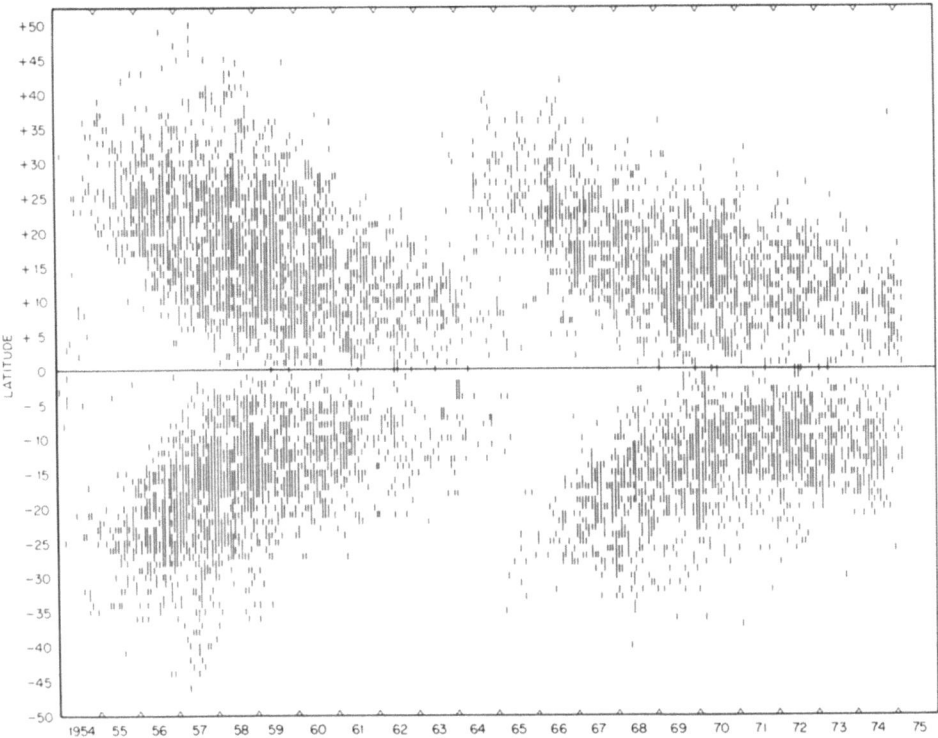

Fig. 6.   The Butterfly Diagram of sunspots of the period from 1954 to 1975. Note that in the most recent cycle 20, the equatorial propagation occurs first and then the wings go on rather parallel to the equator. This implies, according to Yoshimura (1975a) model, that the dynamo processes are weak. This can also be verified by the rather lower activity shown by the sunspot relative number curve. According to Yoshimura (1975a, b) model, this predicts longer period of the cycle 20. This figure was drawn by Dr Howard using Mt. Wilson sunspot data.

DISCUSSION

Stix: Could you determine a phase-shift between the poloidal and toroidal mean fields?

Yoshimura: Yes, I could. However it depends on how we interpret the phase-shift observed at the surface. According to my numerical model of the solar cycle, the dynamo waves propagate mainly radially. So, if we accept this model, we should not regard the observed phase-shift as the phase-shift of the dynamo waves along the propagation path. However, the observed phase-shift at the surface could be an important index of the solar cycle.

CORONAL BRIGHT POINTS

L. GOLUB and A. S. KRIEGER

American Science and Engineering, Inc., Cambridge, Mass. 02139, U.S.A.

and

G. S. VAIANA

Smithsonian Astrophysical Observatory, Cambridge, Mass. 02138, U.S.A.

Soft X-ray images of the inner corona obtained with the AS & E spectrographic telescope aboard Skylab revealed the presence of coronal bright points in far greater numbers than had previously been suspected. Bright points are associated with

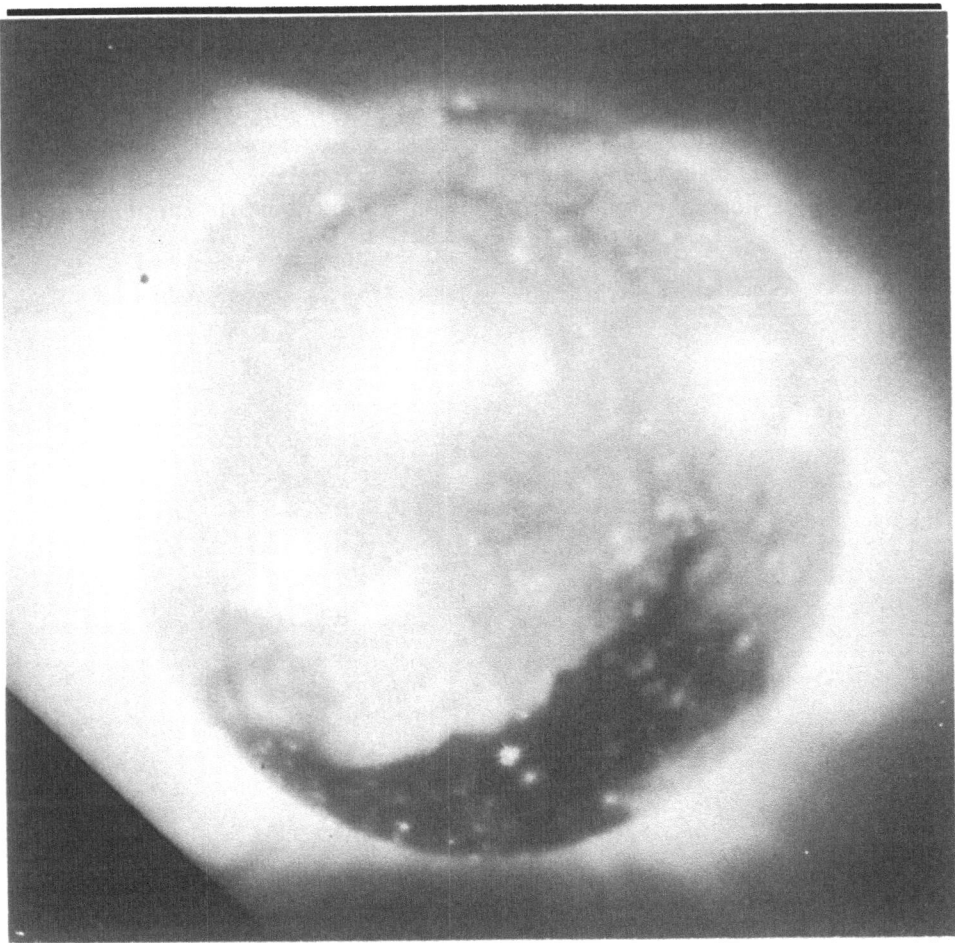

Fig. 1.   X-ray image of the solar corona in the wavelength range 2–32, 44–54 Å obtained 1973, June 12 at 05 10 UT. Over one hundred bright points are visible, corresponding to small regions of emerging bipolar magnetic flux. A bright point flare can be seen in the south coronal hole.

Bumba and Kleczek (eds.), Basic Mechanism of Solar Activity, 145–146. All Rights Reserved.
Copyright © 1976 by the IAU.

bipolar magnetic features with typical diameters of $1-2 \times 10^4$ km, mean lifetime of eight hours and magnetic flux $10^{19}-10^{20}$ Mx. Several thousand bright points emerge over the solar surface per day, thereby bringing up more magnetic flux than is contributed by the larger active regions during the period of observation, May 1973 to February 1974. Bright points identified in X-ray photographs are seen as small (5–10″) emission features in ground-based Hα and Ca K spectroheliograms as well as in transition region lines observed in other Skylab instruments. Typical bright point temperatures are $1.5-2 \times 10^6$ K and typical densities are $\sim 5 \times 10^9$ cm^{-3}.

Bright points are found at all latitudes on the Sun, including both poles. The distribution appears to have two components, with approximately half of the points distributed uniformly in both latitude and longitude. The remaining points have a distribution similar to that of active regions, being confined mostly to within $\pm 30°$ of the equator and showing statistically significant longitudinal variations. The peaks in the longitude distribution appear in both the northern and southern hemispheres, are persistent for several rotations and show a strong correlation with a major outbreak of activity in August 1973. However, the enhancement in bright points precedes the outbreak by at least three solar rotations. In addition, examples of occasional large fluctuations in full disk bright point number counts are shown which may indicate cooperative magnetic phenomena on a horizontal scale of 3×10^5 km or more.

CYCLICAL AND SECULAR VARIATIONS
OF SOLAR ACTIVITY

G. V. KUKLIN

Sibizmiran, Irkutsk, U.S.S.R.

It is difficult to prepare an invited paper on topics which are ABC for any specialist in Solar Physics. I considered my task as a systematization of a lot of information about cyclic and secular variations of solar activity rather than an account of new striking results in order to illustrate a certain concept. I beg your pardon in advance for possible defects in my report and for lack of time to quote all related papers.

1. Philosophy

1.1. Aspects, concepts, indices

One can speak about three aspects of the study of solar activity, as a whole, although usually no essential difference is made between them. The frequency-time aspect of solar activity study involves the consideration of a set of time variations in different solar activity indices and the determination of the power spectrum of quasiharmonic and cyclic variations. The spatial aspect implies the study of the spatial distribution of solar activity phenomena and, in particular, the clearing up of the degree of homogeneity or inhomogeneity in the distribution of various solar activity indices in latitude and longitude. The study of the significance of individual phenomena, which depends directly on the phenomena lifetime, reflects the importance of these investigations.

The basic concept crystallized from results of papers done in the past 20 yr, is associated with the division of various solar activity phenomena and indices characterizing the former into two large classes. This idea was proposed for the first time by Kopecký (1958). To the first class belong indices which predominantly reflect how frequently new formations appear on the Sun. A nearly uniform distribution in longitude and a relatively low concentration in latitude is typical for them. The main time variation is the 11-year cycle. Time variations in the spatial distribution pattern are influenced by the differential rotation. The indices pertaining to the second class contain information on phenomena importance. Their spatial distribution corresponds to a substantial inhomogeneity in longitude (the presence of so-called active longitudes), and the concentration in the low latitude region below 20°. Time variations of these indices mainly are represented by a secular (80–90 yr) cycle. It is remarkable that the spatial distribution of indices is not influenced by the differential rotation, but it rotates rigidly with a constant angular velocity. The presence of the two classes requires a different approach to indices which contain frequency and importance aspects of the phenomena. The mixing of these factors, as a rule, leads to uncertainty and sometimes even to contradictions in the results, depending strongly on data processing.

Bumba and Kleczek (eds.), Basic Mechanisms of Solar Activity, 147–190. All Rights Reserved.
Copyright © 1976 by the IAU.

All the wealth and diversity of the forms of solar activity phenomena are coded in a compact numerical form by various indices. These indices can be classified in different ways. First, we can subdivide them into basic physical indices and derived indices. To the former pertain numerical characteristics which are observed and measured directly: sunspot number, sunspot coordinates, sunspot area, flare brightness, radio emission level, and so forth. The latter imply all indices resulting from some processing or other of basic indices or their combination: Wolf numbers, flare indices, mean characteristics, fluctuation indices, asymmetry indices, and so on. Secondly, indices are subdivided into static and dynamic ones. Static indices, to which the majority of values pertain, are obtained by either averaging procedure, contain information about the level of the measured value. Dynamic indices determine the variability of either values. These are various measures of fluctuations, derivative evaluations, and similar constructions. Probably, essentially more emphasis than has been made so far, should be placed on dynamic indices due to their greater efficiency.

As an example let us consider a certain dynamic index representing the relative speed of variation $[(1/C)(dC/dt)]$ of a value C. The values of this index correspond to the inverse value of the characteristic-time variation of the initial parameter. Usually objects being studied represent a complicated open system with many couplings. One of the tasks of the functioning of such a system is the adaptation to variations of external and internal conditions. This process is characterized by a definite delay time τ. Until variations in the system occur during a time greater than τ, necessary recombinations take place without essential qualitative changes in the actual system. In a case where absolute values of the dynamic index are so great that the corresponding times are less than τ, disagreement takes place in the system which can qualitatively affect this system. For that reason, dynamic indices can prove to be highly effective for forecasting.

As mentioned above, indices can be divided into importance and frequency indices. Kopecký (1958, 1967) proposed the frequency of new formation phenomena on the Sun f_0 and their mean life-time T_0 to be considered as primary indices. Originally, these values were determined by him for sunspots. In this case the formation distribution density in respect to life-time $F(T)$ must play an essential role. For the case when these characteristics are time independent, corresponding mathematical methods were developed which made it possible to express the indices usually being used, with the help of the primary f_0 and T_0:

observed sunspot number $N = f_0 T_0,$

Wolf number $W = K f_0 T_0,$

sunspot mean area $\bar{S} = C f_0 T_0^k, \qquad k = 1$ or $2.$

Primary indices by themselves cannot be determined directly from observations because they are produced values. The methods to compute them are described by Kopecký (1958, 1967). In recent years Kopecký developed mathematical methods for the case when primary indices and required distribution functions are time dependent (1972, 1973, 1974, 1975). Naturally one expects that indices for which determination and computation areas, life-time, intensities were used will predominantly reflect the behaviour of importance indices; moreover, the frequency index effects can also reveal themselves, although to a lesser extent. Indices describing the number of observed formations, to a greater extent are similar frequency indices. Unfortunately, practically in all cases, the observed values cannot be reduced to primary indices because the corresponding expressions contain coefficients in an unknown manner depending on time and coordinates. Only by averaging for time intervals of order 1 yr and over coordinate intervals of several tens of degrees, can one attempt to make such reduction and in this case the statistical errors of the primary indices obtained, turn out rather large, so that an additional smoothing of the results is required.

Finally, indices used in various investigations can be divided into standard and non-standard ones. The former involve widely used indices determined according to standard programs and published in various issues as catalogues, Bulletin 'Solar Data' etc. Usually they do represent the basic material of the studies being carried out. Non-standard indices introduced in individual papers, are not broadly accessible and are not available in the form of long homogeneous series.

1.2. The frequency spectrum of variations

1.2. THE FREQUENCY SPECTRUM OF VARIATIONS

According to the approach accepted now, we shall take frequency spectra of the index time series as a basis of classification of solar activity variations. There seems to be no particular need to list a great number of papers in which either periods were chosen using different methods. A summary of these results, being far from complete, is given in Table I.

The overwhelming majority of authors have used Wolf number series as initial material, an index in common use, convenient due to the presence of relatively long series, but in a certain degree inappropriate because of its complicated nature. The Wolf number series of monthly values cover now more than 220 yr; for 11-year cycle minimum or maximum epochs these are followed back to 1610. The investigation of considerably longer periods is carried out using materials about phenomena which to some extent are associated with solar activity. This includes tree ring thickness measurements, deposit accumulation, chronicles in which evidences are noted on outstanding natural phenomena: drought, aurorae, bright comets, etc. On the basis of such materials one can roughly reconstruct the run of solar activity in a qualitative or even semi-quantitative scale, permitting an approximate reduction to Wolf numbers (Schove, 1955).

Data on sunspot group importance cover only 100 yr of regular observations. Evidences on sunspot flocculi, prominences, filaments, magnetic fields can be utilized practically since the beginning of the 20th century. Finally, fairly detailed and reliable evidence on flares, corona and solar radio emission are available from the beginning of the IGY, i.e., for about 2 cycles, although there are less reliable data which are accessible starting with the post-war years.

It is advisable to speak about three classes of periods which have been found: (a) periods of tens and hundreds of years; (b) periods within 7–25 yr; (c) periods less than

TABLE I

Periods and cycles of solar activity manifestations

Period	Author	Period	Author	Period	Author
1700–1900	Shnitnikov	22	Rima	4.47, 2.50	Pavelyev and Pavelyeva
900, 1000	Rubashev, Golubtsov, and Henkel	22.4	Cheng and In	5.57, 3.90	Rao ($\bar{S}$)
600, 700	Rubashev, Maksimov, Golubtsov, and Vitinsky	22.11	Zhukov and Muzalevsky	5.60, 4.67, 3.50	Rao (T_0)
		22.4, 20.3	Vasilyev and Kandaurova	3.53, 3.48	Rao (f_0)
400	Link	16	Rima	5.5, 3.6, 2.8, 2.2	Currie
350	Vasilyev and Kandaurova	17.75	Cheng and In	2.3, 4.1, 5.6	Rima
189	Predtechensky	17.8, 16.2	Vasilyev and Kandaurova	2.1	Shapiro and Ward
178.55	Jose	15.4, 14.1		2.22	Toman
176	Bonov and Chirkov	17.6, 16.2	Zhukov and Muzalevsky	1.47, 1.96, 3.1	Labrouste
178	Vasilyev and Kandaurova	11.13	Newcomb and Ol'	1.96, 4.79	Schuster
169	Anderson, Gabrovsky, Vasilyev and Vitinsky	11.2	Rima	6.0, 5.5	Vasilyev and Kandaurova
		10.45, 11.8, 11.9	Cole	4.6, 3.4	
		10.5	Covington	7–5.5	
79	Gleissberg	10.47, 10.33	Gabrovsky et al.	2.4–1.8	Kandaurova
80	Henkel	12.7, 11.9, 11.1	Vasilyev and Kandaurova	1.5–1.3	
75	Vasilyev and Vitinsky	10.6, 9.9, 9.3		1.21–1.09	
				0.8–0.6	
91.3	Vasilyev and Kandaurova	12.7, 10.7, 9.6	Vasilyev and Vitinsky	0.41, 0.64, 0.71	Kandaurova
65	Currie	10.79	Pavelyev and Pavelyeva	0.78, 1.05, 1.21	Kandaurova
55	Berdichevskaya	8.3	Rima	5.0	Vasilyev and Vitinsky
50	Trellis	8.9, 7.9	Vasilyev and Kandaurova		
35	Rima	7.9	Vasilyev and Vitinsky		
57.8, 44	Vasilyev and Kandaurova				
30, 26.2					

6 yr. The class is composed of supersecular and secular cycles, as well as a number of intermediate values of periods. Recurrence periods of various geophysical phenomena correspond to a great many of these values. The second class is very rich with spectral lines, of which the 22-year and 11-year cycles are the main. The frequency spectrum structure in this region is complicated and variable since many periodicities are not always present and are able to disappear during some time intervals. Otherwise, the spectral lines have finite width independent of the method by which they are detected. A greater variety of periods is observed in the third class. A special analysis by Kandaurova (1971) distinguished five subclasses of short periods, fairly clear-cut in time of Wolf numbers (roughly, frequency index) as well as of the sunspot area (greater contribution of the importance index). The last subclass with periods less than one year essentially exhibits effects of fluctuation of separate indices which, as considered now by most investigators, are caused by the existence of large individual active regions. It is typical that short periods correspond to non-stationary variations lasting about one–three 11-year cycles. It should be noted that some periodic components appear simultaneously, their periods being in whole number ratio between each other as well as with regard to the 11-year cycle, which qualified them to be considered as overtones of the basic 11-year cycle.

We have already mentioned that indices characterizing the phenomena importance manifest themselves most strikingly in the secular cycle, and for characteristics of new formation origin the 11-year cyclic periodicity is more typical. This statement is not absolute in the sense that a secular variation in f_0 and a 11-year variation in T_0 can be found. Kopecký and Kuklin (1971) showed the presence of an 11-year variation in T_0 and proposed to explain secondary 11-year cycle maxima, found by Gnevyshev (1963), through this very variation. Once one looks at the diagram of f_0 values, the 22-year component is found at once, but it would be irresponsible to speak about the presence of a secular cycle in f_0 using a series of such short length. Really, the autospectra of f_0 and T_0 contain significant components with a period of about 11 yr. The component with ~ 22-yr period is present in the spectrum of f_0, but it is negligible in the spectrum of T_0. Finally, the difference mentioned in secular variations is quite sharply illustrated by the diagrams of the cross spectrum modulus

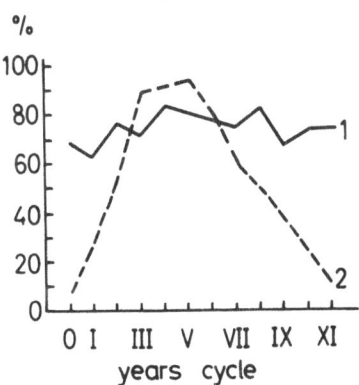

Fig. 1.   The variation of primary indices $T_0(1)$ and $f_0(2)$ during the 11-year cycle.

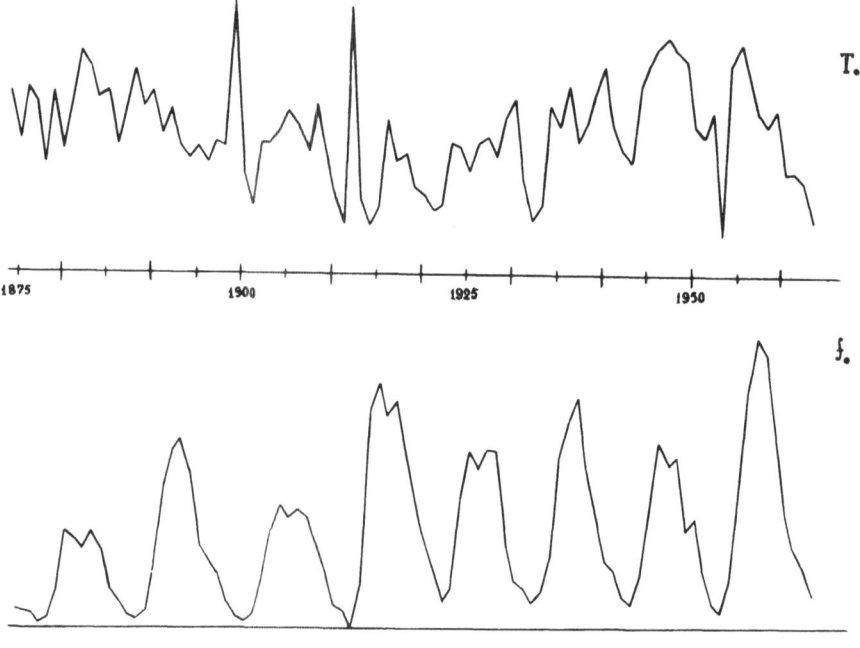

Fig. 2.   The time-series of f_0 and T_0.

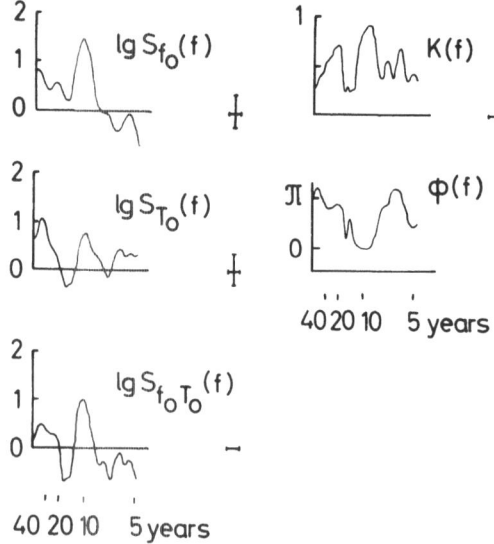

Fig. 3.   The autospectra of f_0 and T_0, for the cross-spectrum, the coherence index K, the phase spectrum for the time-series of f_0 and T_0.

and coherence index. The diagram of spectral phases shows that the basic harmonics of 11-year variations of f_0 and T_0 are synphased, but the secondary harmonics are already nearly in counterphase which generally means that maxima of 11-year variations of f_0 and T_0 do not coincide. However, the coherence index for the secondary harmonic cannot be considered to be significant. This agrees quite well with the fact that the 11-year variation in T_0 varies in shape when transiting from the descending branch of the secular cycle to the ascending branch (Kopecký and Kuklin, 1971).

It should be noted that Romanchuk and Sergeeva (1974) detected an inverse dependence between f_0 and T_0, qualitatively invariable, but quantitatively distinct at different phases of the 11-year cycle. This regularity essentially reflects peculiarities of f_0 and T_0 spectra.

An analysis made by Vasilyev and Vitinsky (1969) has shown that sunspot areas, Wolf numbers and sunspot weighted latitudes have close basic periodic components (10.7, 169, 9.6, 75, 12.7 yr). The components with periods 5.0, 7.9 and 16.1 yr present in the Wolf number spectrum, do not appear always in spectra of the indices to a greater degree.

There are curious attempts made by some authors following the idea by Anderson (1939), to convert the normal series of Wolf numbers into a sign-changing series, ascribing, e.g. to Wolf numbers minus in odd cycles on the basis of the Hale law. In this case the spectrum is strongly simplified and only three components with periods 22.1, 17.6 and 16.2 yr are left in it, which should be considered as 'basic' ones (Muzalevsky and Zhukov, 1968a, b). All variety of periods can be obtained as a result of the 'basic' oscillation interference and the assumption that the observed indices are the response of the Sun which is interpreted as a square detector. The invalidity of such an approach from the physical viewpoint and the absence of any gain in accuracy in the formal-mathematical sense were demonstrated by Vasilyev and Vitinsky (1969).

In connection with above, it is necessary to comment upon the problem of beats. The question is that many authors employing spectral or periodogram methods, obtained several 'basic' periods, the superposition of pairs of which provided all other periods, including even supersecular ones. More frequently data on a single index, e.g. Wolf numbers, were used. Pavelyev and Pavelyeva (1965) obtained three 'basic' periods of 10.8, 4.5 and 2.5 yr. Combining these, one can obtain cycles of 22, 44, 36 and 85 yr. *Combining the periods of 22.2 and 17.7 yr a secular cycle can be obtained* (Cheng and In, 1965; Muzalevsky and Zhukov, 1968c). Similar considerations were also given by Cole (1973), Cohen and Lintz (1974) and others. *However, a doubt inevitably arises as to reality of 'fictitious' secular and other cycles. In fact, the problem must be turned around. The secular cycle is a physical reality, but it has a length oscillating near a mean value of 80 yr. The result of its superposition with other physically real cycles (in simplest case, the amplitude modulation) does produce a 'fictitious' non-stationary fine structure of frequency spectra of many non-primary indices.*

1.3 MATHEMATICAL MODELS

Recently it has become popular to construct mathematical models of solar periodicity without going deeply into the physical nature. For example, it is postulated that the series of Wolf numbers can be considered as a random process, and characteristics of such process are analyzed. The construction of mathematical models implies the application of the apparatus of random process theory, but the known series of indices can be approximated neither by a determinated process nor by a stochastic one in the pure form. According to estimations by Vasilyev and Vitinsky (1969) about 90% of the process dispersion falls on the side of regular variations of solar activity, and the rest on the side of the random component. For this reason an analysis of the nature of each index oscillation by statistical methods must precede the construction of mathematical models.

According to Drozdov (1950) *all oscillations of either index may be classified as follows*: (1) *random oscillations*, (2) *disturbances*, (3) *rhythms*, (4) *periodic oscillations*, (5) *latent periodicity* (*or periodicity with disturbances*). Random oscillations can be rather well presented with the help of a model of quasi-white noise. Periodic oscillations have practically constant period, an amplitude and a phase. Rhythmic oscillations represent an interchange of disturbances (aperiodic deviations) of distinct signs with certain mean period and with an accumulation of phase difference, occurring in a gradual or stepped manner. Latent periodicity represents oscillations with random varying periods, amplitudes and phases near their mean values. In contrast to rhythms, no accumulation of phase difference takes place here. Investigations by Newcomb (1901) and Ol' (1960) have shown that 11-year cycles in these terms are an example of latent periodicity with a mean period of 11–13 yr. Analogous statements can be made with respect to secular cycles, the mean duration of which according to Gleissberg (1955) is 7.1 11-year cycles. However, deviations of observed epochs of 11-year cycle extrema from the calculated ones are not purely random: when smoothed over 4 cycles, deviations exhibit a regular interchange of deviation signs. This is an indication of the influence of a higher range cycle. Therefore, the totality of cycles of different lengths can represent some hierarchy (structural co-ordination). This idea was conceived by Eigenson (1963). Fluctuations of solar activity according to Vitinsky (1961b) must be referred to the random oscillation type, and even for this reason one can speak neither about a morphological likeness nor a hierarchic relation of fluctuations as an example of the 11-year cycle.

The superposition hypothesis was proposed by Wolf at the end of the last century, and it considers the curves of index variations of solar activity as a result of superposition of many periodic components. The formalism and unwarranted determination in earlier papers of this type have led to a striking failure of this approach. Recently, the superposition hypothesis has been revived on the basis of new achievements of mathematical statistics and application of computers. In the new variant, solar activity is considered as a polyperiodic process with a rather great random component. However, no physical interpretation is provided to the set of 'basic' frequencies, and the set itself is formed on the principle of the best approximation of observed series. Naturally, the meaning of the procedures utilized is lost in this case.

To counterbalance the superimposition hypothesis, Waldmeier (1935) proposed an 'eruptive' hypothesis according to which each 11-year cycle represents an independent explosion. It played a great role in the study of internal regularities of 11-year cycles but proved to be inconsistent in the sense that it does not take into account couplings of the statistical and physical character of the neighbouring 11-year cycles.

In a number of papers by Gudzenko and his colleagues, a model of the 'black box without input' was developed. On the basis of the series of Wolf numbers a model was constructed of the auto-oscillation system with one degree of freedom synchronized by a weak periodic signal. On one hand this model served as an additional argument in favor of the 'eruptive' hypothesis, as the physical interpretation of this model corresponds to the consideration of the Sun as a relaxation generator. On the other hand, using such an approach, physical links between cycles in a pair (Hale Law) and the presence of a secular cycle are neglected. So, this model possessed no vital capacity either.

If we consider more particular models of the formal kind, we should quote the results by Jakimiec (1969) who has shown that sunspot areas index deviation from smoothed values may be described as a stationary multiplicative random process, as well as the model of 'auto-regression – integrated sliding average' proposed by Phadke and Wu (1974) for the series of Wolf numbers. In the last model the value of 10.83 yr was obtained for the period. In both cases, mathematical models are formal and have no concrete physical interpretation.

Thus, mathematical models of solar cycle periodicity proposed hitherto predominantly bore formal character, were essentially functionally-adequate ones and were of practical interest only for the approximation of initial data series.

1.4. The problem of external effects on the sun

The finite aim of solar studies is to build a common theory of solar activity. Up to now, one cannot name any theory which could explain and describe the whole complicated pattern of spatial-time variations of solar activity, all hierarchy of the cycles of solar activity, the presence of a system of active longitudes, and the difference between indices of frequency and importance phenomena. Nearly all theories are limited by the 11- or 22-year cycle and Spoerer's law. A large amount of data still awaits corresponding theoretical constructions.

All theories of solar activity conceived up to now may be divided into three groups: (a) causes and sources of solar activity lies within the Sun itself; (b) external effects serve as a regulator of processes on the Sun; (c) external effects are the causes and sources of solar activity. Theories of the first group were intensively developed during the 20th century; they prefer mechanisms with direct contribution of magnetic fields, and they will be discussed in detail at this Symposium. For this reason, we will permit ourselves to pay some attention to the state of theories of the second and third groups.

Mostly, external effects imply the gravitational influence of planets upon solar processes. Recently, considerable interest has revived in this direction. Unfortunately, investigations very often are reduced only to a search of formal and statistical links between indices of solar activity and planetary configurations. The idea of planetary conditionality of phenomena on the Sun is very old (it was suggested as far back as Wolf). Supporters of this idea existed practically always, although nobody went into the question of what is the cycle of their number variation.

The accordance between the frequency spectrum of index variations of solar activity and the periods of planetary configuration recurrence, as a rule, serve as a basis for research in this direction. Particular attention is paid to Jupiter as the largest body of the solar system and as a planet with a rotation period close to the 11-year cycle length.

Mostly one considers tidal disturbances and variations in the solar center position with respect to the mass center of the whole solar system (baricentric motion of the Sun) as external influences, according to estimations by Anderson (1954), the total tidal solar gravitation force ratio is of order 5×10^{-12}. Trellis (1966a) has concluded that the tide height on the Sun does not exceed 1 mm, however, the mean daily work of tidal forces of about 10^{35} erg is comparable with the energy released by the most important flares $\sim 10^{33}$ erg (1966c). Displacements of the solar mass center in the baricentric coordinate system may exceed $2\,R_\odot$. It is supposed that such motions may serve as a trigger to produce disturbances in the unstable convective zone.

Let us list the results giving evidence in favor of planetary effects. Suda (1962) found in the Wolf number series periods and semi-periods of the Earth and Jupiter rotation by means of a periodogram analysis. The joint influence of Mercury, Venus, Earth and Jupiter also was studied (their mean relative contributions are $1.0 : 2.2 : 1.0 : 2.3$, and the influence of other planets can be neglected). A basic period of 178 yr was found which together with its harmonics must determine secular variations of solar activity. Maksimov and Smirnov (1967) found considerable correspondence between structures of frequency spectra of variations of solar indices and planetary tides. Jose (1965) found the period of 178.8 yr in the baricentric motion of the Sun and connected it with the same period in the Wolf number series (double secular cycle). The analysis of the Sun's position considering its baricentric motion, its velocity, its acceleration and acceleration change ('jerk'), taking account of all planets, led to a period of 11.08 yr (Wood and Wood, 1965). Bigg (1967) distinguished the variation of Wolf numbers with the period of Mercury. Dauvillier (1970) showed that the height of the 11-year cycles is proportional to the value of the Sun's displacement in the baricentric coordinate system. Takahashi (1967, 1968) found in variations of tidal forces a period of 22.4 yr and in addition distinguished a period of 11 yr with a 2-yr phase shift and a ~ 0.7 amplitude. A correspondence of periods was detected by Kolomeets *et al.* (1974). Trellis (1966b) found that the sunspot area and the number of new sunspot groups linearly depends on the tide height. Ambrož (1971) when comparing maps of calcium flocculi and the vertical component of tidal force came to the conclusion that there was a coupling between these characteristics in active longitudes.

Not a few empirical correlations were used with the aim of forecasting on the basis of planetary configuration analysis. In a number of papers Romanchuk conceived and applied a composite algorithm of prediction of solar activity depending on the planetary location with due regard for special empirical functions of action. A relation between 11-year cycle maxima and the Jupiter-Saturn quadrature was found here. Nemeth (1966) constructed a more simple algorithm utilizing the time interval between Venus-Earth and Jupiter-Earth conjunctions. This value is minimal during maximum epochs of 11-year cycles and exhibits a mean period of 2×11.2 yr. Shuvalov (1970) has found that planetary clustering in longitude corresponds to maxima of solar activity and blurring, conversely, to minima. Various combinations of planets lead to 11-year and 100-year periods. According to Liese (1971) the maxima are associated with Jupiter-Saturn quadratures and the minima with positions when the difference in longitudes of these planets is equal to 45° or 225°. Prokudina (1973a) has found that an activity increase in a given region of the Sun is associated with the crossing by this region of a direction determined by planetary configuration. Later (1973b), an attempt was made to explain these correlations through tidal forces and variations of the angular orbital moment of the Sun. Separate important events on the Sun were predicted using planetary configurations (Nelson, 1963; Blizard, 1965). Analyzing resonance correlations for the solar system (Molchanov, 1966) Kozelov (1972), has found a correspondence between secular variations of resonance longitudes and superlong cycles of solar activity. According to the same study, the Earth's position, when its longitude is close to the resonance one, corresponds to activity increases on the Sun.

TABLE II

The resonance relationships (Kozelov)

Longitudes	Periods	Possible cycles
$\lambda_1 = 62.28-1°84\ T$	196^a	
$\lambda_2 = 187.08-1.25\ T$	288	$T_2-T_1=92^a$
$\lambda_3 = 87.90-0.40\ T$	901	$T_1=196^a$
$\lambda_4 = 182.95-0.74\ T$	487	$T_1-T_3+T_5=178^a$
$\lambda_5 = 188.19-0.41\ T$	883	$T_5-T_4=396^a$
$\lambda_6 = 61.18-0.36\ T$	1017	$T_3=901^a$
$\lambda_7 = 75.28-0.08\ T$	4270	$T_3-T_2=613^a$
$\lambda_8 = 266.61-0.04\ T$	9747	

All the above correlations cause a mixed reaction. On the one hand they, to a certain extent, are convincing and stimulate further studies in this direction. On the other hand, there is no impressive physical explanation of the links found. Invalidity of all mechanisms in the energetic aspect was shown more than once (Ferris, 1969; Vlasov et al., 1974). The weak point of the results of correlations is that attempts are made to find a correspondence between frequency spectra of strictly periodic events (planetary configurations) and periodicities with disturbances (manifestations of solar activity). Even if the needed periods are revealed, either phase correlations often are absent (Dolginov et al., 1972) or it is impossible to explain irregular variations of momentary values of solar periods. Moreover, certain dissatisfaction is also caused by that fact that different authors succeed in describing the run of solar activity in the same time interval by means of the action of different planetary combinations. The negative result of the Schuster test application to verify the conditionality of flare activity by the action of Venus and Jupiter can be set off against the link of separate important phenomena of solar activity with planetary configurations (Dingle et al., 1973). Concrete physical mechanisms, apart from Dauvillier (1970, 1973), have not been proposed by anybody. Makarenko (1973) proved the possibility of the existence of a cybernetic model of a 'black box with input' in which the frequency trapping is realized, for the Sun, in an extremely abstract way but without indicating the physical mechanisms. This result is of purely academic interest only.

So, the hypothesis of the planetary effect on solar activity is not more than a working hypothesis, but a detailed and objective analysis of the reality and nature of the correlations, is needed.

One should dwell separately on the ideas developed in a number of papers by Vasil'eva and her colleagues. According to these studies, a predominant direction exists in space. It corresponds to projections at the ecliptic of the direction to the galactic center, and of the direction to the standard apex of the Sun's motion, and to the direction of the large-scale galactic magnetic field. The authors postulate the electromagnetic character of the interaction between the planets and the Sun with due regard for the magnetic field of the Galaxy. Hence, *the problem of planetary action on solar activity is transformed into the problem of the galactic action on solar activity, through the means of the planets.* One of the possible mechanisms is the connection with a variation of the EMF produced under the influence of the external galactic field in a closed conductive circuit formed by the Sun, a planet and the interplanetary medium plasma with the frozen-in regular magnetic field of solar origin. *However, all correlations of EMF changes with the variation of the solar activity level have again a conventional character due to a great number of assumptions, and in this sense they differ little from similar papers by other authors.*

1.5. SOME GENERAL CONSIDERATIONS

It is easy to calculate that the energy of solar rotation is 10^{38} erg (all estimations are made for 24-hour interval). The energy of convective motions in the subphotosphere layer of order 20 000 km thick, where the major part of magnetic energy associated with sunspots seems likely to be concentrated, is equal to 5×10^{36} erg. Of the same order is the energy of differential rotation. The estimation of the magnetic energy of sunspots has the form $8 \times 10^{31} f_0 T_0$ erg. Thus, magnetic energy from minimum to maximum of the 11-year cycle varies on the average, from 10^{31} to 14×10^{31} erg. If the magnetic energy of sunspots is assumed to be a result of mechanical energy transformation, for mechanisms related to either convective motions or differential rotation, the efficiency proves very small, of order 10^{-4}–10^{-3}. With respect to tidal forces it does not exceed 10^{-3}. We can consider the sources of mechanical energy to be inexhaustible. The losses in magnetic energy and mechanical energy carried away in the main by the solar wind, are correspondingly estimated by the values 10^{30} and 10^{32} erg. In the simplest form, the equations of energy balance have the form:

$$\frac{d}{dt} E_k = Q - F(E_k, E_m) - \lambda E_k,$$

$$\frac{d}{dt} E_m = F(E_k, E_m) - \mu E_m,$$

where Q is the source of mechanical energy, $F(E_k, E_m)$ is the term describing the transformation of mechanical energy into magnetic energy, and λE_k and μE_m are corresponding losses owing to the solar wind. The speed of magnetic energy changes is $8 \times 10^{31} f_0$ erg, varying during a cycle on the average from 10^{31} to 16×10^{31} erg per day. Unfortunately, estimations of the rate of change of mechanical energy are highly uncertain. Because a limit exists for sunspot dimensions and for the number of very large sunspots, it is evident that $F(E_k, E_m)$ is limited and has a single maximum.

All terms of our equations are positive, and a qualitative analysis shows that the system described by the equations must have a stable node. Really, the simplest auto-oscillation system with relaxation oscillations is described by Van der Pole's equations:

$$\dot{x} = y, \quad \dot{y} = -x + \varepsilon f(x, y), \qquad f(x, y) = y(1 - x^2)$$

different from our equations. In order to produce oscillations E_m it is necessary for either the system to be non-autonomous, i.e. subjected to an external influence, or phenomena of parametric resonance must be realized in it. Essentially, dynamo mechanisms precisely are one of the cases of parametric resonance manifestations. There is no doubt that the first equation 'does not' depend on E_m. Conversely, it is the governing equation. Thus even very small variations of the E_k speed change may cause considerable variations of E_m. It follows from the above that the problem of solar activity mechanism autonomy is not an artificial one and needs particular attention and critical analysis.

As regards the physical nature of the secular cycle, only the hypothesis by Kopecký (1964) has been suggested about the possible existence of variations of the sub-photospheric layer characteristics within the secular cycle where magnetic field tubes are formed. Dmitrieva et al. (1971) have shown that the sunspot distribution in respect to area varies regularly from the 15 to the 18th cycle, and the variations of some importance indices is caused by this. Kuklin (1973) has shown that the variations of a number of importance indices from cycle to cycle are linearly dependent. On can introduce a certain parameter, q, determined with an accuracy up to linear transformation on which many importance indices depend linearly. This parameter, naturally, varies with the secular cycle, although probably, it has a weak 11-year variation. The simplest interpretation suggests the existence of two relatively stable sunspot populations, the number ratio of which varies with the secular cycle. Population II contains predominantly small sunspots and sunspots of increased stability, the magnetic field in them being stronger. The fraction of Population II is larger during the minimum epoch of the secular cycle. Population I represents the common sunspot mass. One may interpret secular variations as cyclic variations in an original ecologic system where competition exists between two populations. The introduction of a 'universal' q importance parameter stimulates the need for a more detailed study of its significance, sense and properties.

2. Facts

2.1. SUPERSECULAR CYCLES

It is rather difficult to say how far the solar activity cycles hierarchy extends to long periods. A sufficiently detailed list of basic cycles of the principal phenomena in the atmosphere, hydrosphere and lithosphere in connection with solar activity is given by Eigenson (1963b). Analysis of the complex of geophysical, geological and geographical phenomena allows us to assert that a cycle of 1700–1900-year length

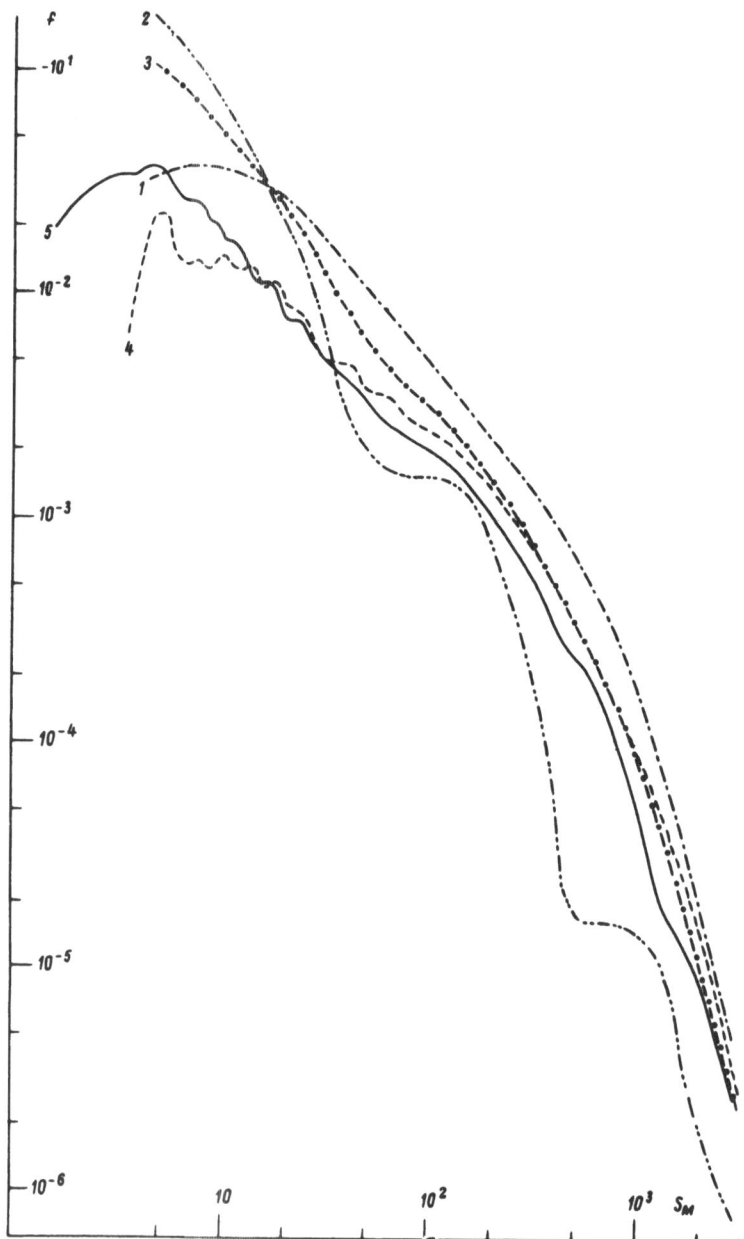

Fig. 4.   The sunspot distribution in respect to the maximal area S_m for the Population I (1) and II (2).

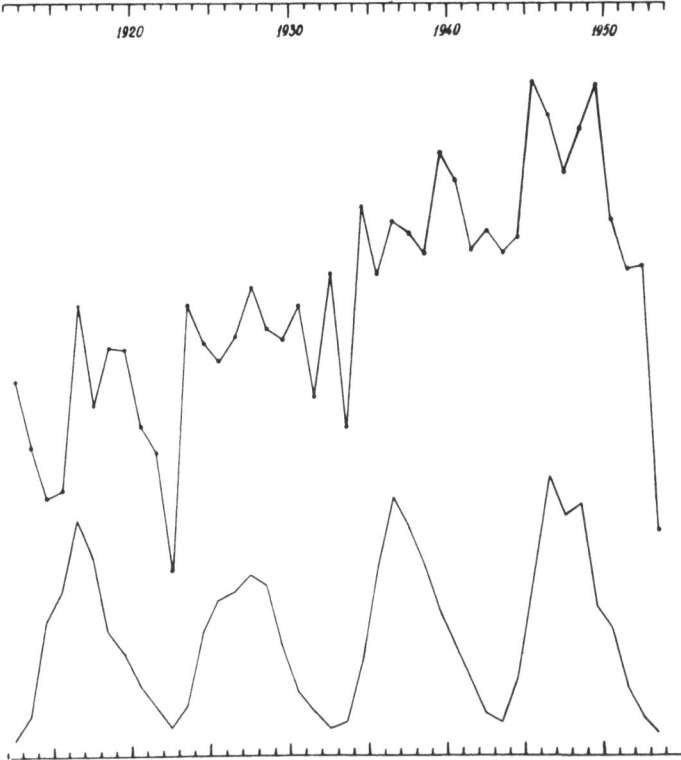

Fig. 5.   The time-series of q parameter values and of their relative weights n'.

exists, Shnitnikov (1951) considering changes of continent moistenings which, according to his opinion, are caused by solar-terrestrial relationships, concluded that such a cycle ought to be present in the solar activity level variations. Aside from questions about the existence of a significant influence of solar activity on terrestrial processes, we also can consider such a cycle to be quite real. However, in this case all information is reduced to a statement of the fact that the cycle of such duration exists.

Rubashev (1949) examined the catalogue of comets visible with the naked eye and concluded that a solar activity cycle of nearly 900-year duration exists. The study of 11-year cycle lengths on the basis of Schove catalogue allowed Henkel (1972) and earlier Golubtsov (1965) to discover this cycle. Also Gleissberg mentions the existence of this cycle as a result of study of the asymmetry index of 11-year cycles.

A solar activity cycle of nearly 600-year duration reveals itself by a study of comet number (Rubashev, 1949), by a study of 80-year cycle amplitudes using the width of tree rings (Maksimov, 1952), and by a study of the variability of the relationship between 11-year cycle height W_m and its ascending branch length T_A (Waldmeier, 1966; Vitinsky, 1969c). Golubtsov (1965) has obtained an estimation of 650 yr for a cycle consisting of two semicycles, 325 yr each. This cycle has proved itself by forecasts of activity (Vitinsky, 1973c). Besides, there are indications that this cycle manifests itself in hydrological, geological and climatic processes.

Link (1963) has revealed a cycle of 400-year duration using data on climate changes, the numbers of discovered comets, and aurorae, and the secular variations of geomagnetism. It was found that the 80-year cycle length is longer at its ascending branch.

The data on aurorae number and climatic changes are evidently in favour of the existence of a 300-year cycle but it is difficult to state direct arguments following from solar activity changes. However, Kolomiets has apparently found an 11-year cycle length variation within a period of 32 cycles (1969). It ought to be shown that if both the values of certain index and its change in speed (dynamical aspect) are important, then any cycle of duration T must give rise to cycles of duration $T/2$.

Predtechensky (1948) on the basis of a study of alternation regularities in the form of the cycle proposed a hypothesis on the existence of a 189-year period called 'indiktion' and containing 17 11-year cycles, but there are no serious reasons to take this hypothesis into consideration (Rubashev, 1969).

In a number of papers by Bonov (1957 and so on) a cycle of 176-year length (16 11-year cycles) was studied. It manifests itself in variations of a number of 11-year cycle pair characteristics: the length ratio even–odd, the sum of the ascending branch length etc. Bonov (1973) comes to the conclusion that such a cycle must begin from an even–odd pair and at the borders of it the solar activity level falls unevenly. Vertlib and Kuklin (1971c) studying the neighboring cycle pair links found the 176-year period also.

Vasilyev (1970) has revealed a 178-year period from the spectral analysis of Wolf numbers. According to Chirkov (1971) the total importance of the 22-year cycle W_M varies within 8×22 yr. One ought to note that doubts in these results were expressed since the initial data series is a little longer than the cycle length considered, although this cycle may be placed twice in the interval of the telescopic observations of the Sun.

An assumption on the existence of a cycle of 169-year length was suggested by Anderson (1954) and it must consist of two parts of 88- and 81-year duration. This cycle is traced in annual and monthly Wolf numbers. Djurkovic (1956) has discovered a cycle of 168-year length using data on the 11-year cycle extrema from 1610. It is necessary to note that the spectral analysis also revealed a period of 168 yr (Vasilyev and Vitinsky, 1969). The period of 169.0 ± 0.5 yr was found with the help of an autocorrelation function by Gabrovsky et al. (1967, 1968).

Apparently just now it is difficult to decide if the cycle of 169- or 176-year length is simply a double secular (80–90 yr) cycle or not but this is not important for the forecasting significance of these cycles. Particularly in the opinion of Eigenson and Mandrykina (1962b) when comparing the increasing branches, the current secular cycle is similar not to the preceding one but to the one before the preceding branch, which argues in favor of double secular cycle reality. However some authors suppose that these cycles are not independent. So Cohen and Lintz (1974) consider that the 179-year period arises as a result of beating at close frequencies $\sim 0.09\,\mathrm{yr}^{-1}$ and $0.1\,\mathrm{yr}^{-1}$ present in the spectrum. Analogously Cole (1973) has obtained the result that superposition of oscillations with derived periods of 10.45 and 11.8 yr has a phase variation with a period ~ 190 yr.

Thus, it is permissible to speak about the presence of supersecular cycles of 600- and 170–180-yr length which manifest themselves in a number of solar activity characteristics, and about the possible existence of other supersecular cycles.

2.2. THE SECULAR CYCLES

The secular cycle, or 80–90-year cycle, sometimes gives rise still to doubts by individual authors concerning its reality, but we don't consider them. Even the first investigators of solar cyclic variations suspected its existence. Contributions to its study were made by Wolf, Gleissberg, Eigenson, Waldmeier and others. In the simplest manner it reveals itself when smoothing procedures are employed or envelope curves are constructed.

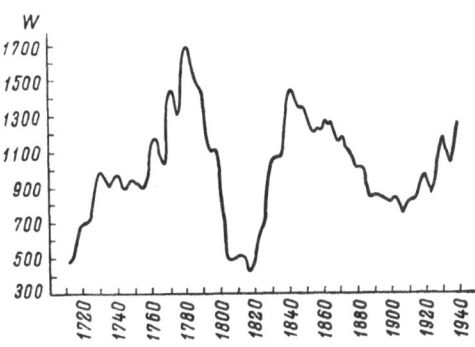

Fig. 6.   The secular cycle of Wolf numbers in case of 23-year sliding summation.

The properties of the secular cycle determined from Wolf numbers differ in great extent from the 11-year cycle properties. The length of the secular cycle fluctuates within broad limits from 20 to 130 yr with an average duration of about 80 yr. In different centuries the secular cycle is unequally sharp; Sometimes it does not reveal itself at all. Spectral analysis (Vasilyev and Kandaurova, 1968, 1970a, b) provides a 91.3-year estimate for the period. *According to Rubashev (1964) the correlation between secular cycle in principle represents a pseudo-harmonic variation, the first year after the* descending branch length also increases with the cycle importance, but this relationship is more weak. Correlations will be higher if the secular cycle importance is estimated by the difference in activity levels (Gleissberg, 1966). *The importance ratio in maximum and minimum is of order 3.* Vitinsky (1971a) *has shown that the secular cycle in principle represents a pseudo-harmonic variation, the first year after the 11-year cycle maximum being important for the secular cycle.* The extreme epochs of secular cycles are closer to those of the 11-year cycle maxima (Vitinsky, 1968a). *Chistyakov (1963) on the basis of a mirror symmetry of secular cycle branches found by him, believes that a secular cycle begins and ends with a maximum.*

The secular cycle reveals itself in variation of different characteristics of the 22-year cycles (Bonov, 1964), 11-year cycle asymmetry index (Eigenson and Mandrykina, 1962c), 11-year cycle length (Henkel, 1972; Kolomiets, 1969), jerk index (Vitinsky,

1968c). *According to Chistyakov's studies* (1973), *secular cycles reveal themselves more on those* 11-*year cycle phases when the mean sunspot zone latitude is close to* 16°. Afanasyev (1961) concluded that according to the 11-year cycle mean importance index $\sum W/T$ on the secular cycle ascending branch, pairs are formed in combination (low, odd, high, even), and on the descending branch in the reverse way. The length of the 22-year cycle determined with regard for superimposition varies with the secular cycle too (Chistyakov, 1961). Studying the break-points in 11-year cycle curves Chistyakov concluded that only points k_2 and t were associated with the secular cycle. As far as the same determined phases of the most important cycle manifestation, this once more supports the secular cycle interpretation as an importance cycle.

Actually, one can easily make sure of the latter as we shall show now in the consideration of other characteristics of solar activity. Because regular observations of sunspot areas cover only 100 yr, the reliability of determining the secular cycle length from phenomena importance indices is small, but it is easy to show their variations to have a secular cycle character. The difficulty of determining the secular cycle length from available importance indices series is illustrated by the results obtained by Vasilyev and Vitinsky (1969), where estimates are obtained within 55–95 yr. *According to Kopecký* (1967) *a great number of indices have such variations: mean sunspot group lifetime, mean sunspot group area, recurrence index, mean maximum sunspot group area, and maximum being more than* 1000×10^{-6}. *A secular*

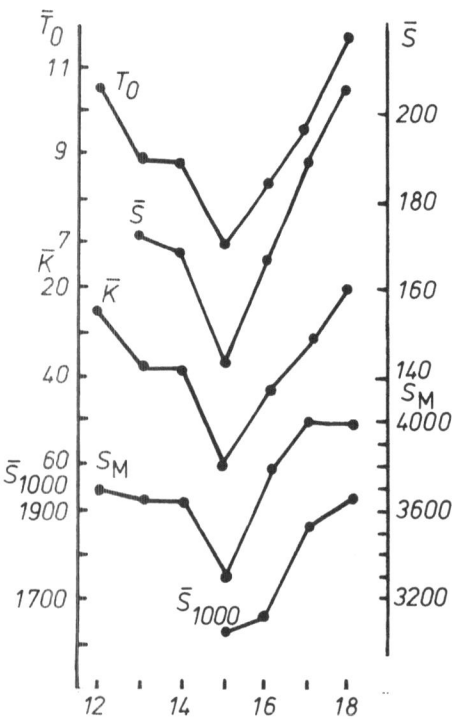

Fig. 7.   The secular cycle of importance indices according to Kopecký.

variation is present in the one-day sunspot group number – total sunspot group number ratio (Ringnes, 1962), *mean magnetic field strength* (Ringnes, 1965), *regression equation coefficients H – S* (Ringnes, 1965), *sunspot penumbra area – sunspot umbra area ratio* (Ringnes, 1965b), *mean sunspot group latitude* (Ringnes, 1968; Kuklin, 1971b), *high-latitude of sunspot formation* (Gleissberg, 1958), *high-latitude sunspot group number* (Kopecký, 1958; Waldmeier, 1966), *and fluctuation duration in the 11-year cycle descending branch* (Vitinsky and Ikhsanov, 1970).

The attempts to present the secular cycle as a result of oscillation beating with close periods using only the analysis of Wolf number series seem to be somewhat naive (Pavelyev and Pavelyeva, 1965; Cheng and In, 1965; Cole, 1973; Zhukov and Muzalevsky, 1969; Muzalevsky and Zhukov, 1968a, b). These results should be attributed to the category of erroneous ones. *It should also be noted that the parameter q, figuring in the concept of two populations of sunspot groups* (Kuklin, 1973) *since it is inferred from phenomena importance indices, also must vary with the secular cycle. In the framework of this concept the secular variation of precisely such characteristics does cause similar changes of all the above indices.*

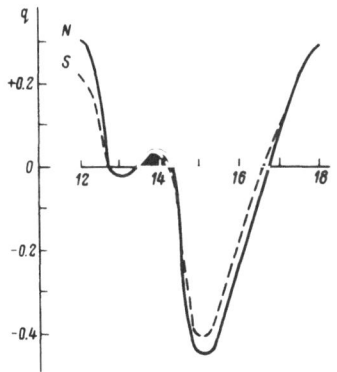

Fig. 8. The secular cycle of q parameter.

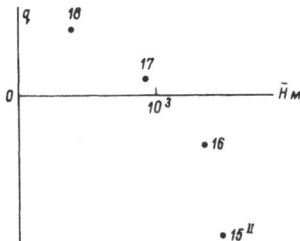

Fig. 9. The dependence of the mean magnetic field strength on the q parameter.

Asymmetry of the N and S hemispheres of the Sun manifests itself in the difference in shape between 11-year cycle curves and extremum epochs in the surplus of summarized area and sunspot group number in either hemisphere.

For a given characteristic C, separately determined in both hemispheres, one can introduce two nearly equivalent asymmetry indices

$$A_c = \frac{C_N}{C_S}, \qquad a_c = \frac{C_N - C_S}{C_N + C_S}.$$

Waldmeier (1957) investigated in detail the asymmetry between the hemispheres total sunspot group area. According to his results this asymmetry has a secular cycle (correct oscillation) with N hemisphere predominant in minimum and the S one in maximum. Ringnes (1968) considered the asymmetry of hemispheres using the mean sunspot latitude. The schemes of asymmetry manifestation according to Waldmeier and Ringnes are presented in Figures 10 and 11. It is curious to note that variations in

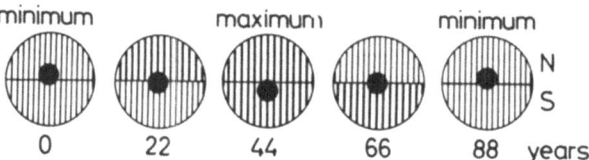

Fig. 10.   The scheme of the N–S asymmetry variation during the secular cycle according to Waldmeier.

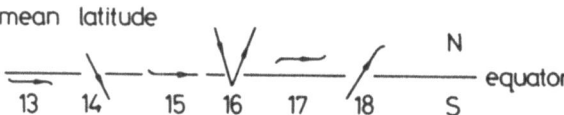

Fig. 11.   The scheme of the mean sunspot latitude variation during the secular cycle according to Ringnes.

mean latitude are opposite in high latitude sunspot groups in comparison with low latitude ones. However, one should pay attention to the fact that in cycles 19 and 20 a predominant development of activity was observed in the N hemisphere in disagreement with Waldmeier's scheme. The cause for such a marked and long-stable one-sign asymmetry is worthy of a special study.

Berdichevskaya (1967) considered the area-weighted latitude of the first sunspot groups at the beginning of 11-year cycles. These characteristics show a 55-year cycle periodicity, the N hemisphere being predominant in the minimum of the current secular cycle.

It should be noted that in a number of values the secular cycle appears as a cycle of about 55-year duration. The same period reveals itself in a spectral analysis (Vasilyev and Kandaurova, 1968). In connection with the secular cycle it is necessary to mention the study by Bezrukova (1968), where it is found that cyclic curves of sunspot group area sums averaged during their appearance in one and the same hemisphere are more similar every 4 cycles, i.e. every 44 yr. Besides, Trellis (1973) found that the duration of activity region displacements from mean positions varies with an approximately 55-year cycle.

Therefore, the secular cycle exists in reality, it is associated predominantly with important manifestations of solar activity and asymmetry effects and has properties different from those of the 11-year cycle.

2.3. THE 22-YEAR CYCLE

The 22-year cycle became recognized only after Hale's discovery of the law of sunspot magnetic field polarity changes, although earlier investigations of Wolf and Turner show indications of the existence of such a cycle. This cycle reveals itself more sharply in magnetic characteristics and therefore is more physical. However, it is harder to find using other indices, therefore it is more just to be considered in principle as a cycle of qualitative characteristics of solar activity.

Above all, the 22-year cycle manifests itself in forming 11-year cycle pairs according to the Gnevyshev-Ol' rule (1948). Where the statistical relationship between 11-year characteristics is more close in the even–odd combination. This refers to Wolf number sums (Gnevyshev and Ol', 1948), to the relationship between the length of the 22-year cycle and the interval between 11-year cycle maxima of a pair (Bonov, 1958), and 11-year cycle relative height in a pair (Rubashev, 1964). Essential deviations from obtained correlations fall on secular cycle extremum epochs (Chistyakov, 1959b).

Turner as long ago as 1925 emphasized that in odd 11-year cycles, the sunspot areas are larger and their latitude is 1° higher in comparison with even cycles. Usually an even cycle is lower and this regularity is more evident in the behavior of the new sunspot group appearance (Kopecký, 1958, 1967). The 22-year cycle length decreases with the growth of its importance (Bonov, 1958; Chistyakov, 1965; Chirkov, 1971). An analysis of cyclic curves (Vertlib and Kuklin, 1971c) supported the Gnevyshev-Ol' rule and showed a difference, in this sense, between even and odd 11-year cycles. Investigating the break points Chistyakov (1965) found that there is a rather close relationship in the pairs for values corresponding to the points k_2 and t. However, in the odd–even combination, characteristics of nearly all break points exhibit high correlation to the 22-year cycle length. These facts as well as the results by Afanasyev (1961) seem to speak well for the hypothesis that it is not important what the cycle order is in a pair, but interchange of properties is important in neighboring 11-year cycles (Chistyakov, 1973).

The study of cyclic curves of different indices in separate latitude zones (Vitinsky, 1965a) shows a weak 22-year cyclic periodicity. Vasilyev and Vitinsky (1959) did not succeed in clearly revealing the 22-year cyclic periodicity for phenomena frequency indices which to a certain degree contradicts the results by Kopecký (1958) and Figure 2.

The sign of the Sun's polar cap magnetic field changes with the 22-year length cycle and the same for the coupling character between increased brightness regions of the green corona and the sector structure of the IMF (Antonucci, 1974). Vasilyev and Rubashev (1972) have established that variations of solar radius occur with the 22-year cycle. The radius, in even cycles, increases with a velocity varying in parallel to the Wolf number run, and decreases in odd cycles. The coupling between neighboring 22-year cycles is reduced to that with increasing preceding odd cycle, the maximum of the next even cycles comes later, the length of the next 22-year cycle is longer and its importance is less (Chistyakov, 1959a). Many authors attach great importance to the 22-year cycle in view of the fact that the latter in its nature, is a magnetic cycle. For this reason, attempts are continued to operate with Wolf number sign-changing series (Anderson, 1938; papers by Muzalevsky and Zhukov). Vasilyev and Vitinsky (1969) showed that these formal operations in principle do not yield any new or better results, although it is tempting to explain rich spectra by the presence of a small number of basic periods. Nevertheless, periods close to 22 years, are found in almost all papers on frequency spectra of solar activity.

On the other hand, there is no sense in disregarding the reality of the 22-year cycle. For this reason, any physical and mathematical models of solar activity must lead to the basic period of order 22 yr. At the same time all variants of dynamo-mechanisms satisfy this. Apparently, this was sure to be the direction of models considered in a series of papers by Gudzenko and colleagues. From this viewpoint, the eruptive hypothesis of independent 11-year cycles raises objections.

The change of sunspot magnetic field polarities suggests an idea on the relationship between hemispheres in neighboring cycles. In particular, such a coupling follows from the Alfvén mechanism of solar activity (1945a, b). Recently (Fredga, 1965), the presence of such a coupling was supported when considering the similarity of sunspot distributions in latitude intervals at the beginning of each cycle. It is curious that on the 'latitude-time' diagram, single fluctuations with 1-year resolution form chains which, being extended into another hemisphere and another cycle, indicate either the beginning or maximum of similar chains.

An original concept of the magnetic cycle was suggested by Ol' (1972). He places emphasis on the coupling between geomagnetic recurrent disturbances on the descending branch of a previous cycle, and the characteristics of the oncoming cycle maximum. Taking into account the correlation for even–odd cycles, Ol' comes to the following scheme (Figure 12). The cycle of solar magnetic activity begins with the development of unipolar magnetic regions at the end of the old cycle which induce the development of unipolar and bipolar regions in the even cycle, and it ends with

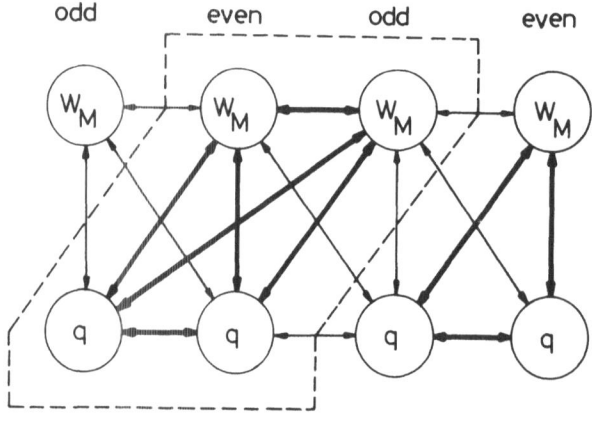

Fig. 12.   The cycle of magnetic activity according to Ol'.

the development of bipolar regions in the next even cycle. The length of such a cycle, with due regard for superposition, ought to be about 26 yr. Finally, if a cause-effect interaction between the laws of Hale and Gnevyshev-Ol' (physical and statistical aspects of 22-year cyclic periodicity) is postulated, then one should expect that violations of the second law caused by the secular and supersecular cycles, must correspond to violations of the first one. Otherwise, the expected violation of the Gnevyshev-Ol' law in the pair of cycles 20–21 must be accompanied by the absence of a magnetic field polarity change.

Hence, the 22-year cycle is mainly associated with the variation of magnetic sign characteristics and secondly it appears only in frequency or importance indices. Besides, N–S asymmetry does not seem to be related to the 22-year cycle (Vasilyev and Vitinsky, 1969).

2.4. THE 11-YEAR CYCLE

We shall not consider now the list of the main peculiarities of the well-known 11-year cycle of solar activity but in the first instance we shall focus our attention on its less known properties and manifestations.

It is known that the granule number on the Sun varies in parallel with the run of Wolf numbers (Macris, 1960). The intensity of Fraunhofer lines in the spectrum of the undisturbed photosphere also depends on the cycle phase: the central intensity is a minimum in years of maximum (Zhukova and Mitrofanova, 1973).

Somewhat contradictory information has been obtained using sunspot structure variations with the cycle phase. Deszö and Gerlie (1964a, b, 1965) found a cyclic variation of the penumbra-umbra area ratio. According to their data this index consecutively decreases from minimum to maximum of the cycle showing on the descending branch values greater than those on the ascending one. Highly detailed investigations of this index were performed by Antalova (1971) who came to the conclusion that it is an increasing function of Wolf number, i.e. it repeats, the shape of the cyclic curve. In the epoch of the secondary maximum the values of the index usually are less than in the epoch of basic maximum. The index is, essentially, a function of only cycle phase and sunspot area.

At the counter-phase polar plages (Sheeley, 1964), polar prominences (Godoli and Mazzuconi, 1967) as well as intensities of weak magnetic fields in polar caps (Howard, 1965) vary with the run of Wolf number during the cycle. It is considered that the background fields of a previous cycle with related plages and prominences reach polar regions with considerable delay of more than half a cycle and by that time magnetic fields of a new cycle have time to appear at intermediate latitudes. These data are an indication of the existence, at high latitudes $\varphi > 40$–$45°$, of another ring of meridional circulation directed towards the pole in the photosphere. It is of interest that in cycle 20 a secondary zone of polar prominences within 1968–1971 was observed (Waldmeier, 1973). Simultaneously, a discontinuous displacement of the sunspot production zone towards the poles was noted. Such incidents in operation of the cycle machine are interesting and need detailed study.

The coronal emission in the red line of 6374 Å does not undergo considerable cycle changes but the emission intensity of the green line of 5303 Å exhibits cyclic

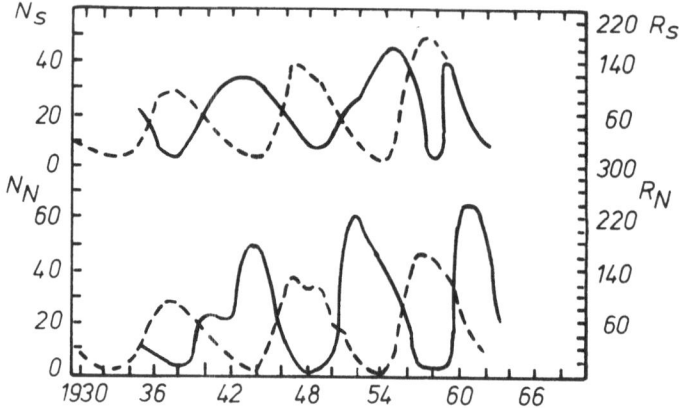

Fig. 13.   The cyclic curve of the polar faculae number according to Sheelly.

variation. Because the regions of increased coronal brightness coincide with solar activity phenomena, this allowed Gnevyshev to detect secondary maxima of the 11-year cycles (1963, 1966, 1967a). The N-S asymmetry index of green corona brightness also has two maxima and exhibits negative correlation with the sunspot number (Pathak, 1972). As has often been mentioned, the secondary maxima are caused by the increasing share of importance phenomena of solar activity. Therefore, the secondary maxima are found on cyclic curves of the number of type-IV radio bursts and the S-component level (Křivský and Krüger, 1966), of important calcium flocculi (Basu and Das Gupta, 1966), and of proton flares (Gnevyshev and Křivský, 1966). More correctly, the basic maxima of these cyclic curves lag 1–2 yr with respect to the cyclic maximum and coincide in time with the secondary Wolf number maxima.

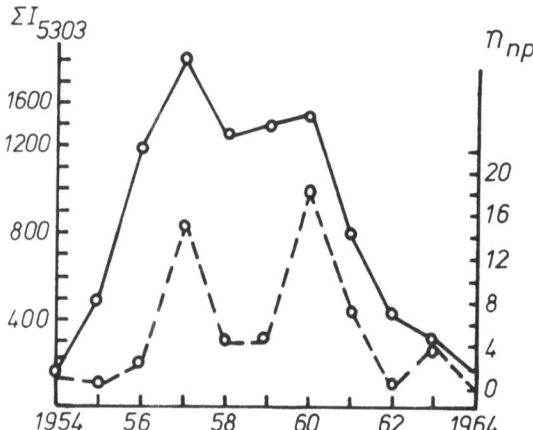

Fig. 14.   The cycle curve of the green corona emission (solid line) and of the proton flare number (dashed line) according to Gnevyshev.

If we operate with sunspot, plage, flocculi, flare, and prominence numbers irrespective of their importance, the corresponding cyclic curves differ little from Wolf number curves (Rubashev, 1964). Zabza (1962, 1964) has found that the following index combinations have conformal variations during a cycle: sunspot area – flare index – radio emission flux $\lambda = 10.7$ cm, red corona brightness – radio emission flux $\lambda = 10.7$ cm, red corona brightness – prominence index.

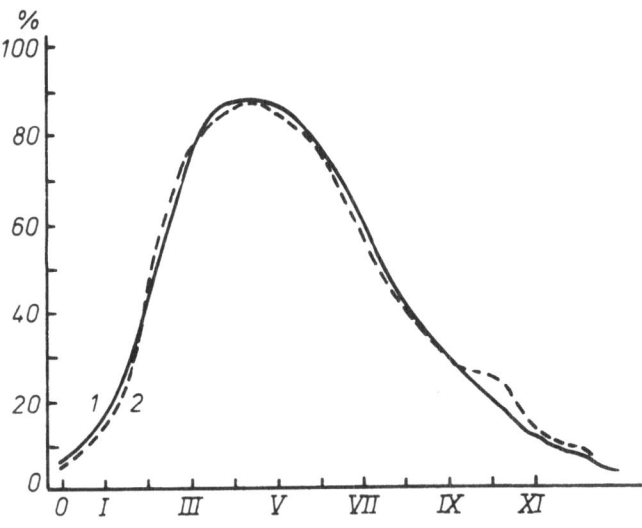

Fig. 15.   The cycle of Wolf numbers.

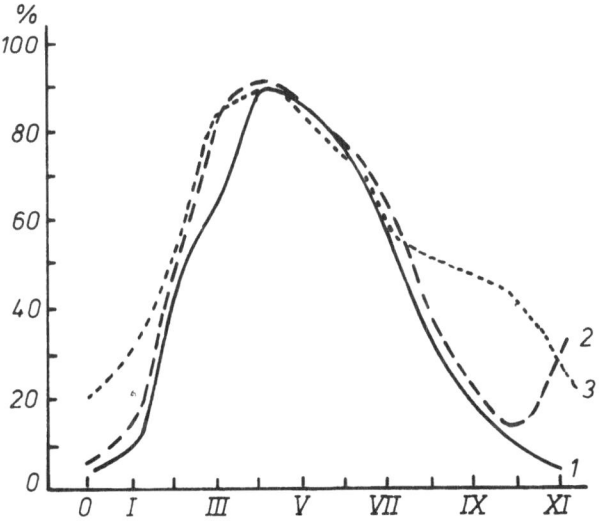

Fig. 16.   The cyclic curves of the sunspot area (1), the faculae areas (2), the prominences areas (3).

According to Fokker (1963), small microwave bursts are encountered equally often, independently of the cycle phase, but large bursts prefer to concentrate on the descending branch where they are encountered more often than in the cycle maximum.

As mentioned, the number of important proton flares has a maximum after 1–2 yr of the maximum but for proton flares of small importance (class 1 and 2) a simple increase of their part to minimum is typical (Levitsky, 1967b) as well as for flares accompanied by type IV bursts.

The fluxes of X-ray emission during the epoch of cycle maximum, in comparison with the epoch of minimum, increase 20 times in the 44–60 Å wavelength range and 200 times in the 8–20 Å wavelength range (Kreplin, 1970). It should be taken into account that estimates of magnetic energy, calculated for sunspots only vary by about 16 times.

Wilcox and Colburn (1069) emphasize that the IMF sector structure on the increasing branch is characterized by considerable instability.

According to Hakura (1974), geoefficiency of the Sun during the 11-year cycle three times reaches maximum values. The first maximum approximately coincides with the main cycle maximum or somewhat precedes it and is associated with the maximum of relatively small phenomena number on the Sun. The second maximum enters 1–2 yr later than the main one and corresponds to the maximum of important phenomena number on the Sun. Finally, the third maximum comes 4–5 yr after the main one when on the Sun there is a maximum of unipolar long-lived regions responsible for a great number of recurrent disturbances.

So, additional data emphasize the peculiarity of the 11-year cycle that indices of phenomena frequency with no regard to their importance have the usual cyclic curves, and importance indices of phenomena have less marked 11-year variation with displaced maxima.

It is typical for the 11-year cycle that the ascending branch, i.e. its length, is of prime significance. This was mentioned more than once by Eigenson and Mandrykina (1962a). In a number of papers by Xanthakis he succeeded in constructing a family of formulae, permitting him to express many cycle parameters through this value and to give empirical formulae of its computation, using the ordinal number of the cycle. Unfortunately, they are very bulky and formal in their nature. The determining significance of the ascending branch length means that mathematical models of the 11-year cycle ideally seem to be one-parametric ones, and cyclic curves are likely to be close to relaxation oscillations. Hence, the basic ideas of the papers by Gudzenko and colleagues cannot be simply discarded, there is still some value in them.

Although Vitinsky (1972) has come to a conclusion that it is extremely difficult to reveal among the two tens of known cycles a pair of analogous cycles, similar to the extent that one can make predictions, nevertheless, the search for similarity is of some interest for a more general interpretation. Vertlib and Kuklin (1971a, b) employed for these purposes both a generalization of the method of principal components and non-orthogonal expansions. It turned out that two factors play their part in the difference between cyclic curves: the cycle height and the curve shape.

One can reveal typical and individual cycles, using the latter factor. Typical cycles are divided into threes and such groups follow every 6 cycles.

Chistyakov (1965) has established a number of empirical correlations for 5 inflection points, found by him on the cyclic curve. These correlations, determining the shape of the cyclic curve, are helpful for predictions. It is important that the time interval between break points S and U is almost constant and is as large as 6 yr. With the help of this the cycle 'core' is detected, in which nearly all cycle importance is concentrated. The inflection points demarcate separate parts of the cyclic curve, within the limits of each part its own behaviour of fluctuations is established.

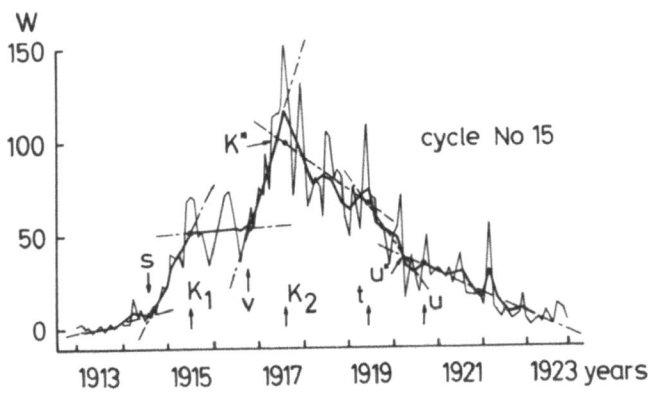

Fig. 17. The break points at the cyclic curve according to Chistyakov.

The result by Giovanelli (1964) is of interest not only for forecasting. It is shown that in the minimum epoch the curves of appearing sunspot group numbers are the same for all cycles when descending and ascending branches are examined separately. Szymansky (1974) has concluded that stabilization of cycle length takes place during several cycles as well as the correlation of intervals between cycles and their importance difference exists.

2.5. THE SEASONAL VARIATION

In recent years the problem of the reality of solar activity seasonal variation acquired new sharpness in connection with the fact that the presence of this variation is considered a nearly decisive argument in favour of gravitational planetary influence upon the Sun.

Suda (1962) and Romanchuk (1963) have concluded that the existence of seasonal solar activity variation is beyond doubt. Loewe (1973) has also found a secular variation of plage area with a maximum in July and a minimum in November. The results by Kozelov (1972, 1975) and those by Kozelov and Mingaleva (1975) are more impressive. The distribution of proton and important flares observed exhibits a

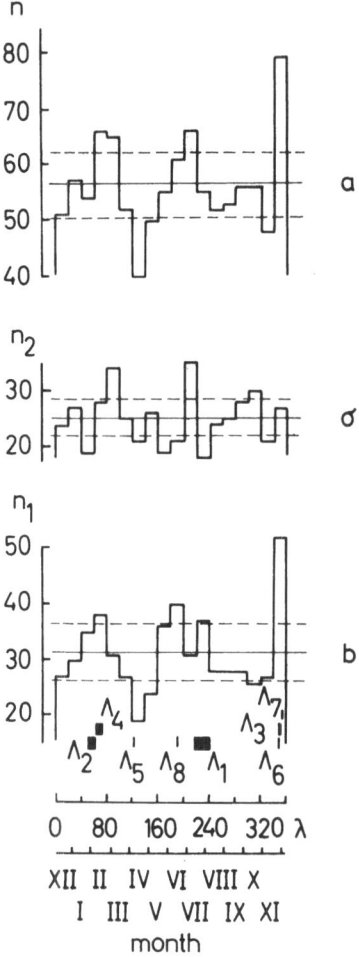

Fig. 18.   The seasonal variation of the flare number and the resonance longitude location according to
Kozelov.

non-random inhomogeneity with a probability more than 0.99. The extrema of this
polynodal distribution coincide with time intervals when the Earth has a longitude
corresponding to resonance directions. It is curious that such a distribution is
independent of activity level and is revealed from yearly repetition of geomagnetic
disturbances which is effectively taken into account by a number of geophysicists in
their planning of investigations. Finally, the seasonal variation of solar activity was
revealed by Vasilyeva *et al.* (1974), but in contrast to papers by other authors, this
variation turned out to be variable and dependent on cycle phase.

From papers of the opposite character one may quote the study by Vitinsky
(1973a) in which the absence of any statistical significance of the obtained seasonal
variation was proved after the analysis of the same material, used in the paper by
Vasilyeva *et al.* Besides, Ambrož (1973b) has shown that the seasonal variation of
Wolf numbers is unstable.

The above information gives no reason to draw a final conclusion on the reality of seasonal variations. The fact of instability itself is no reason to deny flatly the existence of such variation. However, the results by Kozelov require this question to be approached extremely seriously and critically.

2.6. FLUCTUATIONS

Usually, fluctuations of solar activity are considered as relatively short-term (less than a year) deviations of index values from their smoothed values. Because the distinguishing of fluctuations implies the presence of a long uniform series of indices, the study of fluctuations up to now was based on data of Wolf numbers and total sunspot areas. Physically, fluctuations are caused by the appearance of activity centers, the life-time of which is comparable with the fluctuation length. The activity centers are the totality of related phenomena of solar activity, covering nearly all layers of the solar atmosphere and concentrated in a limited part of the solar surface.

Fluctuation indices may be constructed as differences or ratios of unsmoothed and smoothed values. Zhukov and Muzalevsky (1969) have shown that fluctuations of Wolf numbers may be considered as a multiplicative random process. In this case, fluctuation indices of the 'ratio' type will be free of 11-year cycle influence, concentrated in smoothed values. 'Difference' type indices are more subject to the influence of smoothed values but, nevertheless, they also bear useful information.

Vitinsky (1961a) has investigated in detail the properties of Wolf number fluctuations. He proved that fluctuations have an aperiodic character, i.e. they show no marked periodicity. As already mentioned, the 11-year cycle seldom manifests itself in fluctuations but the secular cycle is followed in a number of fluctuation characteristics, especially in their average duration. Here, the fluctuation length implies the interval between two neighboring maxima. The fluctuation character on the ascending and descending branches of the 11-year cycle is different, their amplitude and duration behaving less regularly on the descending branch. The relationship between the relative fluctuation amplitude and the ascending and descending branch lengths is positive and weak.

Parallel with the secular cycle in the fluctuation length on the descending branch of an 11-year cycle, a longer (double secular) cycle for the same value is revealed on the ascending branch of the 11-year cycle (Vitinsky and Ikhsanov, 1970). According to data of this investigation, 80% of the fluctuations are 2–4 months, and they are different on the ascending and descending branches. More short fluctuations with greater amplitude appear on the descending branch.

Fluctuations in amplitude, exceeding the mean square deviation, are called strong fluctuations. Vitinsky has made a catalogue of strong fluctuations (1960). There are more strong positive fluctuations on the longer ascending branch of the 11-year cycle. On the descending branch the density of these fluctuations depends on the cycle length and height. Strong positive fluctuations prefer the 1st year before the maximum and the 3rd year after the maximum. Strong negative fluctuations are more often encountered in the 1st, 2nd and 6th years after the cycle maximum (Vitinsky, 1963b). They, to a lesser extent, enter on the long ascending cycle branch.

Chistyakov (1966) has found that 80% of the fluctuations in both hemispheres are synchronized. In the epoch of minima, fluctuations of the old and new cycle turn out to be synchronized to a considerable extent too (Chistyakov, 1968). A linear positive relationship exists between the fluctuation latitude of a new cycle and the fluctuation delay of the old cycle at low latitudes.

Dodson and Hedeman (1970) detected primary and secondary fluctuations of solar activity, related to activity centers. The mean time interval between primary fluctuations is about 15 rotations, and between the secondary ones – from 3 to 5 rotations. On the descending branch of the 11-year cycle 5–6 primary fluctuations are observed from which 2–3 are accompanied by the increased solar activity.

So, Wolf number fluctuations are independent random oscillations, manifesting themselves simultaneously in both hemispheres and possessing some regularities of lengths and amplitudes with respect to the 11-year cycle.

2.7. LATITUDINAL DISTRIBUTION OF SOLAR ACTIVITY AND ITS DYNAMICS

The distribution of solar activity phenomena in latitude is governed by Spoerer's law which says that during a cycle the zone of sunspot formation moves from a high latitude boundary to the equator. The character of the latitudinal drift, dependent upon the cycle phase, is of great importance for the construction of physical models of solar activity.

In the most general form Spoerer's law determines the variation of solar activity distribution in latitude with the cycle phase. An obvious idea of it is given by the 'latitude-time' diagram (Butterfly Diagram). Not a few papers are devoted to different aspects of the study of such diagrams.

The fine structure of the sunspot diagram was studied by Bell (1960) who concluded that the diagram represents a zonal pattern of sunspot formation ('cater-pillars') without any latitude drift. Vitinsky (1961b) has proved the inconsistency of such conclusions. In recent years the fine structure of an index distribution in latitude with a period of order $1°-2°$ revealed itself (Granova et al., 1972; Kozhevnikov, 1973), but these results provoke the same objections. Kuleshova (1962) has come to the conclusion that 'butterflies' are composed of impulse-chains existing for 15–20 months and slipping down to the equator. Dmitrieva (1965), by special methods, has supported the reality of impulse-chains and has proved that those depend on the secular cycle phase and the longitude. Minasyants and Obashev (1969) have revealed the same impulse-chains for flares with life-time of about 12 months and drift velocity of order 1.8 deg/month, associated with sunspot impulses.

Gnevyshev, when finding the existence of secondary maxima, questioned the reality of Spoerer's law. This conclusion was refuted by Vitinsky who considered cyclic curves for separate latitude intervals (1965a) and has shown, by means of the dispersion analysis, the predominant role of the latitude factor in comparison with the phase factor (1965b). Tschegoleva (1965) examined cyclic curves using separate latitude zones of the sunspot number index, supporting Spoerer's law, although the latitude drift turned out different in distinct cycles and hemispheres. Ramantan and Natarayan (1965) have constructed diagrams for different longitude intervals, in which there are no violations of Spoerer's law and the 'active latitude' is found,

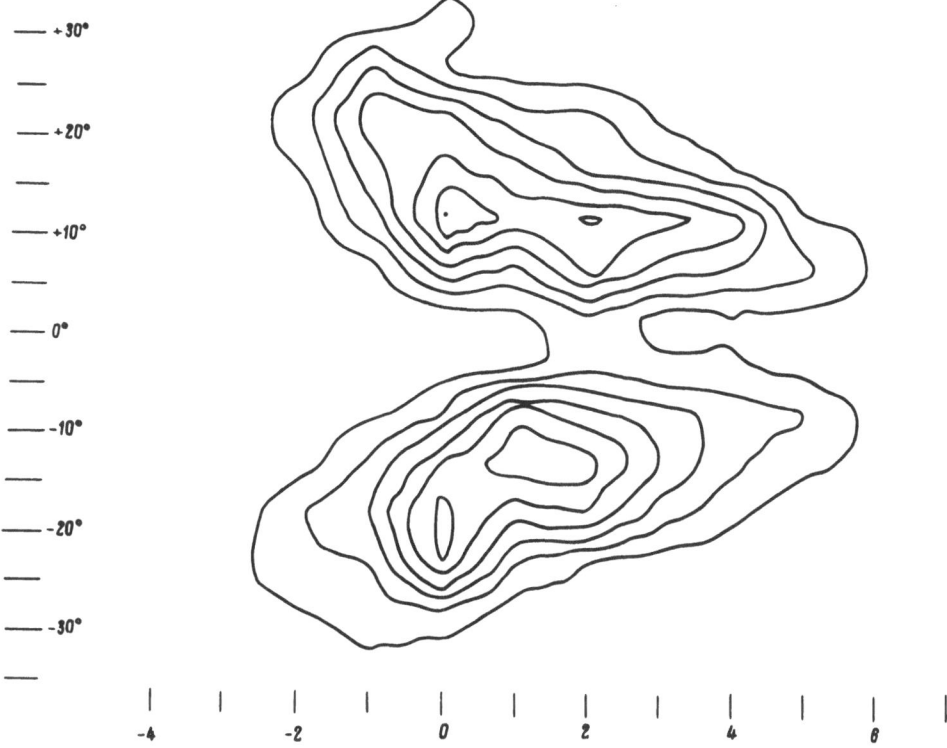

Fig. 19.   The latitude-time diagram of the total sunspot area index P.

invariable at different cycle phases. In order to construct an average diagram for 7 cycles Kuklin (1971a) used the same data on the total-area index as Antalova and Gnevyshev (1965) utilized to disprove Spoerer's law. The obtained diagram visually demonstrates the existence of a latitude drift with many peaks. Besides, the secondary Becker zone is revealed on it with poleward directed drift (1959) the reality of which was disputed by Kopecký (1962).

All dynamics of total sunspot-area latitude distribution reflecting the important group distribution of the E and F classes (Kopecký and Künzel, 1962), may be revealed by solving the reverse problem of the dynamic model (Kuklin, 1971d) in application to the diagram obtained by Kuklin. The distribution of the source function which corresponds predominantly to either birth or decay of large groups, shows that the cycle 'works energetically' no farther than the second year after the maximum. Approximately, in this time, according to Vitinsky and Tschegoleva (1971), the sunspot formation zone slipping down to the equator, ceases. The meridional circulation velocity field is divided into two zones: a zone of drift to the equator below ~25°, and a zone of drift to the pole above ~25°. This picture also agrees with the result by Becker and with the well-known pattern of the latitude drift, according to Tuominen (1942), with a difference in this case, with the behavior of sunspot formation centers. The velocities are of order 10^2–10^3 cm s^{-1}, the estimations of the diffusion coefficient are of 10^{12} cm^2 s^{-1}.

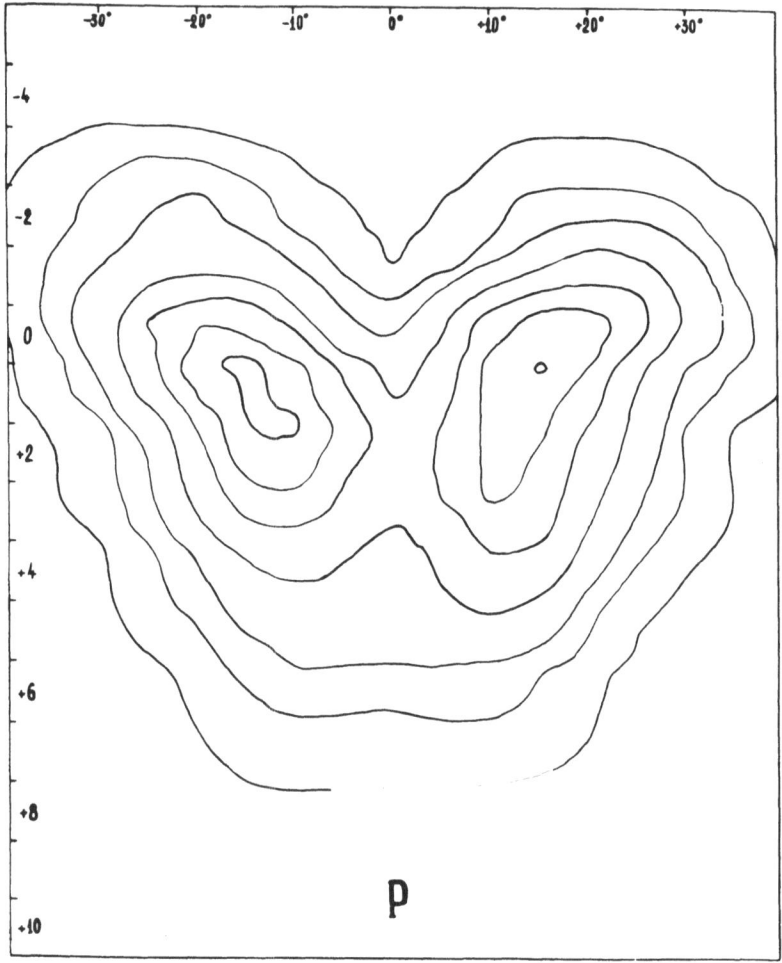

Fig. 20.   The smoothed -t diagram of the P index.

The sunspot formation zone width on the diagram increases towards high latitudes, during cycles with high activity level (Mursalimova, 1957; Schmidt, 1962). Morozov *et al.* (1973) have found that within the cycle the displacements of the high latitude border occur synchronously in both latitudes. According to Gleissberg (1968), the zonal activity at different latitudes lasts from three years at high latitudes to a practically continuous one at low latitudes, the cycle height having no influence on this. In each latitude interval, the activity (and the cycle) begins with the appearance of small groups which are followed by large groups after 1–2 yr (Szymansky, 1970). The activity decay takes place in the reverse order.

One should note that the classic 'latitude-time' diagram reflects mainly the properties of the behavior of the frequency index. Kopecký showed that the diagram loses its customary regular appearance and acquires a very fluctuating structure (1962) as we consider indices in greater degree presenting the phenomena importance. The index of conditional average sunspot area, proportional to T_0^2, turns out to

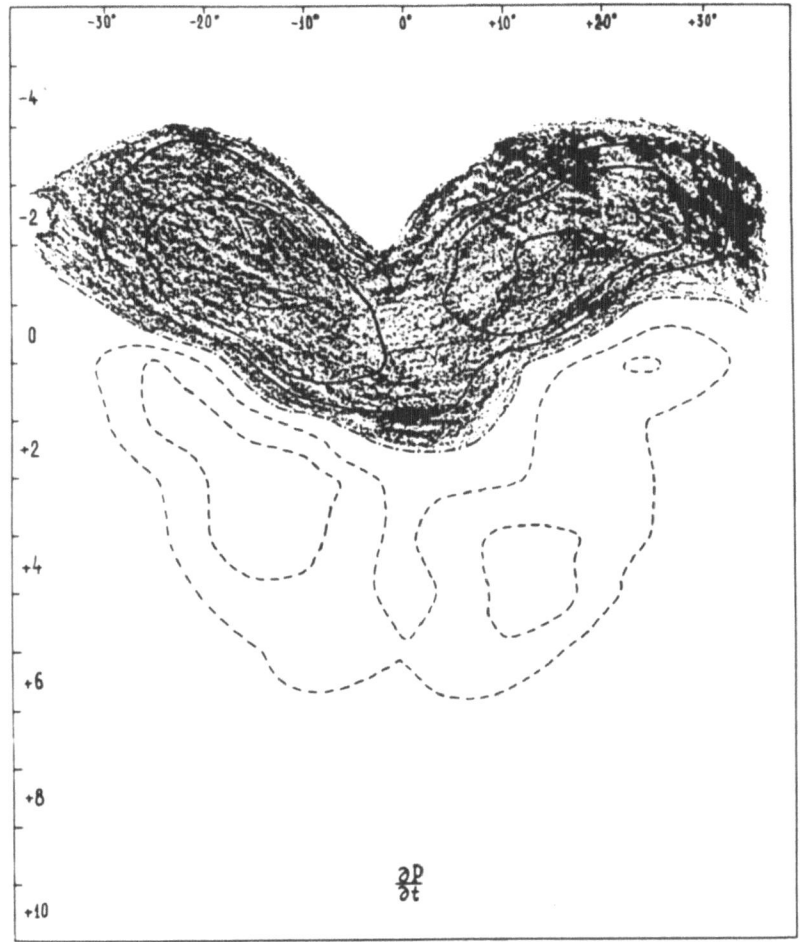

Fig. 21. The distribution of dP/dt (positive values are shaded).

be indifferent to both the phase factor and the latitude factor (Vitinsky and Kopecký, 1968). However, on the average, the mean latitude of sunspot groups decreases with the growth of their importance (Kashirin, 1962; Ringnes, 1968).

The 'latitude-time' diagram for flares was studied by Křivsky and Knoška (1967, 1968). The equatorial drift of the zone of the highest solar activity is revealed markedly, which may be disturbed by short displacements in the opposite direction. On the diagram, fluctuations ('jerks') are present, which in the 18 cycle were located more symmetrically with respect to the equator and synchronously than in the following one.

Certain information on the magnetic field geometry of active regions may be given by investigations of the flare position with respect to sunspots. The positions of flares in latitude, according to the first results (Greatrix, 1970), were independent of the cycle phase and flare importance, but they showed a relationship with the field structure according to magnetic classifications of the groups. Detailed studies by

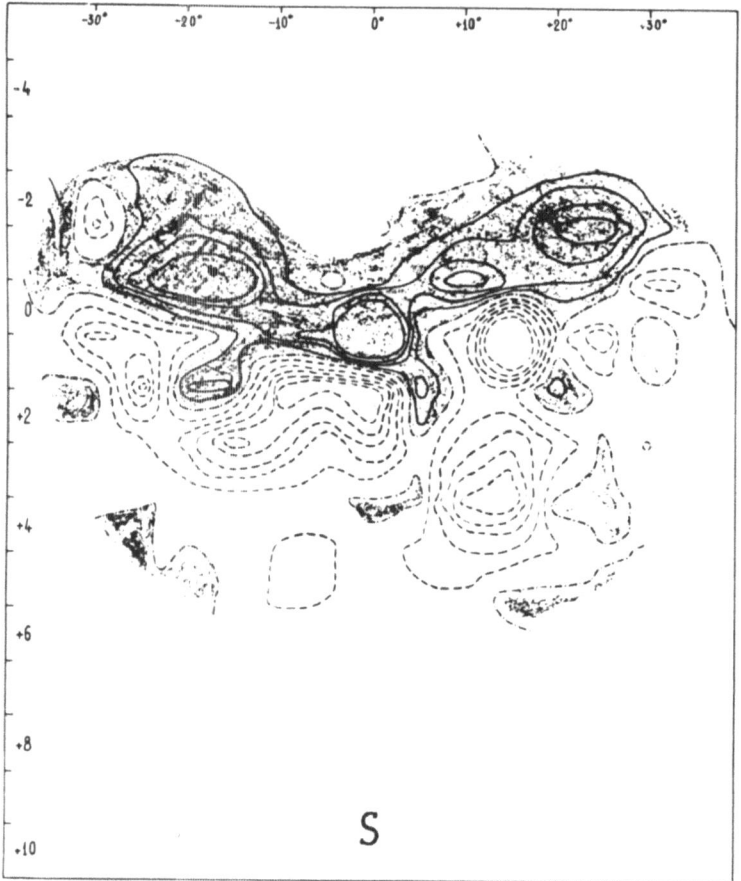

Fig. 22.   The distribution of sunspot source function S.

Kasinsky (1973a, b) have shown that the isoline of zero displacement in latitude during the cycle drifts to the equator with a velocity of order 1 deg per month, but this motion gets broken by fluctuations. The displacement direction in latitude changes its sign at least twice a cycle. At the beginning of the cycle, the displacement is directed towards the east and the pole, and at the end – towards the west and the equator. On the diagram boundaries, the displacements are directed inside the diagram.

So, the latitude distribution of indices reflecting the frequency, varies with cycle phase according to Spoerer's law, and the importance indices exhibit a less regular pattern with pronounced concentration at low latitudes within the cycle core. The analysis of index dynamics using the 'latitude-time' diagram, reveals a number of significant details.

2.8. ACTIVE LONGITUDES

In the longitude distribution of indices on the solar surface, inhomogeneity is expressed more significantly than in the latitude distribution, which involves the

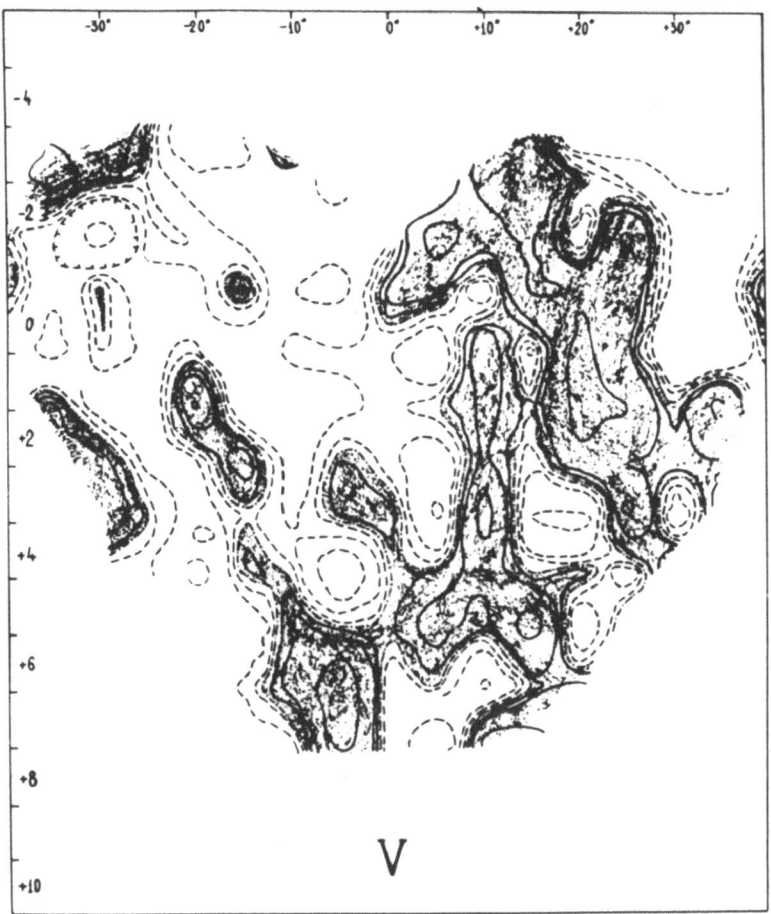

Fig. 23. The distribution of the meridional circulation velocity V (isolines $0.3, 1, 3, 10\,\mathrm{m\,s}^{-1}$).

content of the longitude distribution law of solar activity. This is expressed in terms of the existence of active longitudes, found at the beginning of the present century. The typical property of active longitudes is that they are not subjected to any effects of the differential rotation.

Active longitudes are very noticeable when data for relatively large time intervals of the order of several years or an entire cycle, are utilized. Active longitudes, as a rule, are 180° apart so that they form antipodal pairs with unequal activity level. The main population of active longitudes is composed of important recurrent sunspot groups. Relatively high stability of active longitudes is observed which may exist during several cycles (Vitinsky, 1958). They may, from cycle to cycle, irregularly displace in longitude at distances less than their own size. Besides active longitudes Vitinsky introduced the concept of cycle-like longitudes (1962a). The longitude interval within which the cyclic curve differs little from the cyclic curve for the Sun, as a whole, is called the cycle-like longitude. Usually, considerations are conducted separately for both hemispheres. Generally, there are more cycle-like longitudes

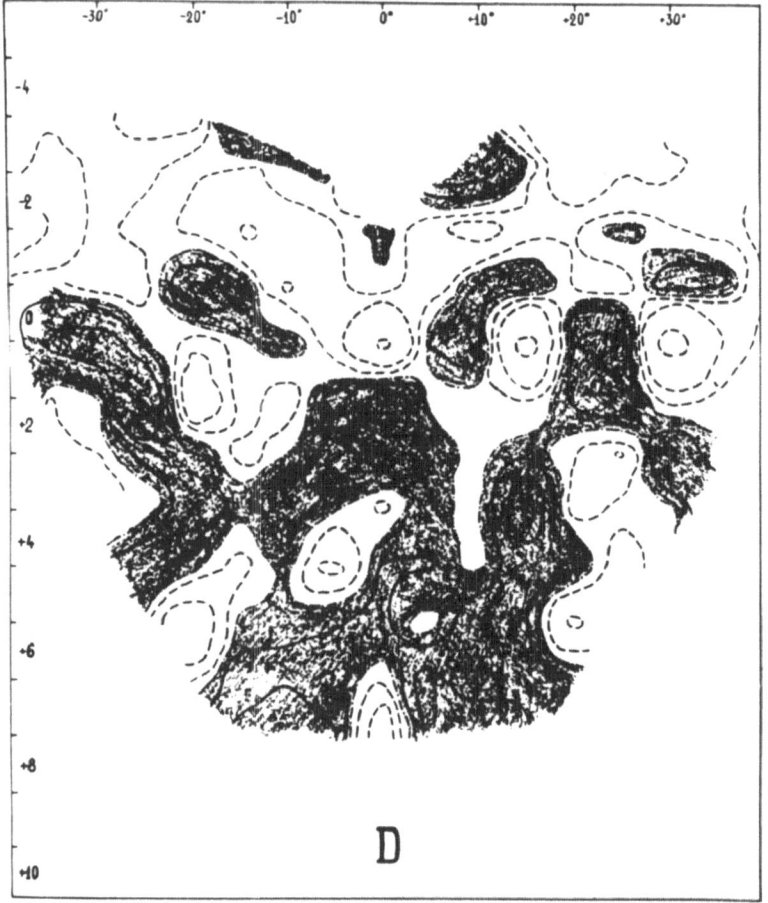

Fig. 24.   The distribution of the quasi-diffusion coefficient D (isolines 10^{12} and 3×10^{12} cm^2 s^{-1}).

than active ones, and only a part of them coincides with active ones. The cycle-like nature is expressed stronger for the sunspot number than for the sunspot area. Generally, active longitudes of the sunspot number are less stable, and the number of them is less than of active longitudes of sunspot areas (Vitinsky, 1967). In 50% of cycle-like longitudes, the solar activity level is below the average. Cycle-like longitudes remain during two or more cycles and practically determine the character of the cyclic curve for the Sun as a whole. Asynchronism in separate longitude intervals is expressed stronger than in hemispheres. Active longitudes are often characterized by an extremely fluctuated cycle evolution (Vitinsky, 1962a, b).

Asynchronism in sunspot formation was noted by Bezrukova (1963) with respect to antipodal active longitudes. Kuklin (1962) tried to distinguish active longitudes of high latitude sunspot groups. It turned out that they are not associated with the main active longitudes of the major sunspot formation zone but they may precede (1–2 yr) the secondary active longitudes of the major zone. According to Vitinsky (1968b) Spoerer's law in different longitude intervals manifests equally and the active zone width varies similarly. But if we pay attention to new sunspots in the beginning of a cycle, Berdichevskaya (1968) has shown that their active longitudes, from cycle to cycle, are considerably displaced, passing into the neighboring longitude quadrant in the solar rotation direction.

As was already mentioned, active longitudes may have displacements. According to Rodionov (1962) such oscillations are more typical for cycle extrema epochs and differ for active longitudes of small and large groups. Vitinsky (1963a) concluded that displacements in active longitudes increase from cycle to cycle with decreasing activity level in them. There is some evidence that in different cycles the Sun rotates differently. The problem of the existence of cyclic variations of solar rotation seems to be quite urgent (Shodo 1955, 1957).

Kozhevnikov (1970) has found active longitudes of the kinematic index, characterizing sunspot group motion velocities in latitude and longitude. They coincide with active longitudes of sunspots where, consequently, an increase in activity is caused by an increase in motions.

In the sunspot group longitude distribution periodicities were found which are different in both hemispheres (Morozov and Obashev 1969, Morozov et al. 1973). These periods vary synchronously with mean linear group dimensions but with a phase shift in different hemispheres. This effect is probably not linked directly with the existence of active longitudes but indicates structure of quite another time scale. Also Stanck (1972) has found that longitude distributions in both hemispheres are independent and are characterized by the presence of the main maxima at a distance of 90° in longitude and secondary maxima at a distance of 30°. It is curious that he has found long-lived regions with minimal sunspot number on whose boundaries activity is increased (sunspots and prominences). Such regions are most likely to correspond to coronal holes (Kasinsky and Tomozov, 1975).

Active longitudes were also distinguished using other phenomena of solar activity. Vitinsky (1969b) has distinguished active longitudes of weak background magnetic fields which for different polarities are situated in different hemispheres. In the northern hemisphere they coincide with active longitudes of other indices. The regions of magnetic field emergence, revealed from changes in the fine structure of the chromosphere, have no predominant longitudes (Glackin, 1973). Such compact regions exist during about 10 rotations and are continuously carried away toward the east with the velocity of about 0.6 deg per day. The coincidence of active longitudes of flares and sunspots, best expressed for important phenomena, was obtained by Vitinsky (1969a) and Maris (1972). Similar results of the coincidence of active longitudes of a great number of characteristics were obtained by Warwick (1965) for sunspots, flares, proton flares, plages, prominences, type-IV bursts, noise storms, important sunspot groups with an area of more than 1000.

The results obtained are consistent when data for large time intervals are utilized. In the case of short intervals, however, active longitude systems can be obtained with a real spectrum of values of the synodical period of solar rotation as it was shown by Wilcox and Schatten (1967) on the basis of Warwick's data. Fung et al. (1971) have discovered an interesting detail in connection with the flare longitude distribution. Active longitudes are obtained, taking into consideration all observed flares, however, if one by one the flares are sampled randomly from each active region, random distribution results. Hence, active longitudes of flares, to a considerable extent, are caused by flare concentration in active regions and by the existence of active longitudes in active regions.

The existence of active longitudes of calcium flocculi was shown by Godoli *et al.* (1966, 967). Ambrož (1973a) has examined the distribution of calcium flocculi in longitude and has tried to determine the periods, with which resulting structures rotate. He has revealed row systems rotating at different velocities due to which they form a net of intersecting straight lines on the 'longitude-time' diagram. The rows themselves exist for about 10 revolutions (active region) and form streams. The regions of crossing streams are characterized by increased activity. The whole structure has a lifetime of 20–25 solar rotations (1.7 yr). The only exception consists in the longitude distribution of hydrogen filaments, active longitudes of which are quite real, but they are situated within intervals between active longitudes of other indices (Vitinsky, 1973b).

Active longitudes are produced by bright formations in the corona (5303) according to the data by Sýkora (1971) and Cimakhovich (1972). According to Nesmyanovich and Khomenko (1969), active longitudes exist also for the S-component of solar radio emission.

On the basis of the above one should expect that the activity center distribution must repeat the longitude distributions of other indices and must contain the same active longitudes. According to the results by Vitinsky (1968), at the 11-year cycle descending branch, the activity centers have both an average lifetime of 3.3 rotations and real maxima in longitude distribution. When the minimum epoch enters, the lifetime decreases 1.5 times, and distribution in longitude becomes random. Further investigations (Vitinsky, 1971b) have shown that in maximum epochs the longitude distribution of activity centers bears a random character too. However, one can reveal active longitudes during a cycle, as a whole, and they remain during two cycles. Similar studies were performed by Trellis (1971a, b). Distribution of activity centers has a relatively stable structure, especially in longitude for zones of sunspot formations. As the mean sunspot area increases, structural stability increases (up to several cycles). Zones of increased activity are outstanding for their typical spatial distribution of the mean area index.

Therefore, for the overwhelming majority of important phenomena of solar activity, irregular longitude distribution is typical, with the formation of active longitudes existing during several cycles and not influenced by the differential rotation. These peculiarities are weaker or even are absent in distributions of indices which either reflect the frequency or describe minor formations.

2.9. SUMMARY

All the above confirms the concept of two classes of indices representing the frequency and importance of phenomena on the Sun. The reality of this concept is illustrated by observational data and facts. The main problem which remains is the building of a solar activity theory and that consists of a number of smaller problems. It is somewhat tiresome to tell in detail about each of these problems but the main problems must be mentioned.

TABLE III

Period	Frequency 11 yr	Importance 80–90 yr
Dependence on latitude	Spoerer's law	Increase of maximal latitude
N–S asymmetry	?	+ ?
Compactness of active zones	Large in latitude small in longitude	Active longtitudes
Type of rotation	Differential	Rigid

(1) The build up of a complete and exhaustive picture of solar activity phenomena in order to obtain a functionally adequate system model with the most important couplings.

(2) The build up of a structurally adequate model of solar activity which is the last step to a physical theory.

(3) The clearing up of the meaning of the correlations between the planet locations and the solar activity variations.

(4) The study and classification of different solar activity phenomena in respect to time and space regularities.

(5) The build up of physically consistent mathematical models of various phenomena changes.

(6) The build up of an evolution model of solar activity considering time and spatial couplings.

(7) The clearing up of sense of difference and similarity of frequency and importance indices, phenomena, concepts.

References

Afanas'ev, A. N.: 1961, *Soln. Dan.*, No. 5, 62.
Alfvén, H.: 1945a, *Monthly Notices Roy. Astron. Soc.* **105**, 3.
Alfvén, H.: 1945b, *Monthly Notices, Roy. Astron. Soc.* **105**, 382.
Ambrož, P.: 1971, *Solar Phys.* **19**, 480.
Ambrož, P.: 1973a, *Bull. Astron. Inst. Czech.* **24**, 80.
Ambrož, P.: 1973b, *Bull. Astron. Inst. Czech.* **24**, 130.
Anderson, C. N.: 1939, *Bell. Syst. Tech. J.* **18**, 292.
Anderson, P. N.: 1954, *J. Geophys. Res.* **59**, 455.
Antalova, A.: 1971, *Bull. Astron. Inst. Czech.* **22**, 352.
Antalova, A. and Gnevyshev, M. N.: 1965, *Astron. Zh.* **42**, 253.
Antonucci, E.: 1974, *Solar Phys.* **34**, 471.
Basu, D. and Das Gupta, M. K.: 1966, *Ind. J. Phys.* **40**, 117.
Bell, B.: 1960, *Smithson. Contrib. Astrophys.* **5**, 17.
Becker, U.: 1959, *Z. Astrophys.* **48**, 88.
Berdichevskaya, V. S.: 1967, *Astron. Zh.* **44**, 358.
Berdichevskaya, V. S.: 1968, *Astron. Zh.* **45**, 459.
Bezrukova, A. Ya.: 1958, *Izv. Gl. Astron. Obs. Pulkovo*, No. 159.
Bezrukova, A. Ya.: 1963, *Izv. Gl. Astron. Obs. Pulkovo* **23**, 57.
Bigg, E. K.: 1967, *Astron. J.* **72**, 463.
Blizard, J. B.: 1965, *Astron. J.* **70**, 667.
Bonov, A. D.: 1957, *Soln. Dan.*, No. 3.
Bonov, A. D.: 1958, *Bull. Vses. Astron. Geodez. Ob.*, No. 21.

Bonov, A. D.: 1961, *Soln. Dan.*, No. 3, 56.
Bonov, A. D.: 1964, *Soln. Dan.*, No. 3, 67.
Bonov, A. D.: 1966, *Godishnik Sof. Univ. Fiz. Fak.* **59**, 75.
Bonov, A. D.: 1972a, *Izv. Sekc. Astron. Bulg. Akad. Nauk* **5**, 33.
Bonov, A. D.: 1972b, *Izv. Sekc. Astron, Bulg. Akad, Nauk* **5**, 41.
Bonov, A. D.: 1973, *Izv. Sekc. Astron. Bulg. Akad. Nauk* **16**, 15.
Cheng Byao and In Chung-lin: 1965, *Acta Astron. Sin.* **13**, 89.
Chertoprud, V. E.: 1966, *Astron. Zh.* **43**, 390.
Chertoprud, V. E. and Kotov, V. A.: 1965, *Astron. Tsirk.*, No. 318, 1.
Chirkov, N. P.: 1971, *Soln. Dan.*, No. 11, 75.
Chistyakov, V. F.: 1959a, *Soln. Dan.*, No. 2.
Chistyakov, V. F.: 1959b, *Bull. Vses. Astron. Geodez. Ob.*, No. 25.
Chistyakov, V. F.: 1961, *Soln. Dan.*, No. 7, 78.
Chistyakov, V. F.: 1963, *Soln. Dan.*, No. 6, 65.
Chistyakov, V. F.: 1965, *Izv. Gl. Astron. Obs. Pulkovo*, **24**, 60.
Chistyakov, V. F.: 1966, *Soln. Dan.*, No. 8, 83.
Chistyakov, V. F.: 1968, *Solnech. Aktivnost'*, Moskva, No. 3, 140.
Chistyakov, V. F.: 1973, *Ciklicheskaya Deyatel'nost' Solnca*, Vladivostok.
Cimakhovich, N. P.: 1972, *Appar. Metody Obrab. Radiosatr. Nabl.*, Riga, p. 105.
Cohen, T. J. and Limtz, P. R.: 1974, *Nature* **250**, 398.
Cole, T. W.: 1973, *Solar Phys.* **30**, 103.
Covington, A. E.: 1974, *J. Roy. Astron. Soc. Can.* **68**, 35.
Currie, R. G.: 1973, *Astrophys. Space Sci.* **20**, 509.
Danvillier, A.: 1970, *Compt. Acad. Sci.* **270**, B1119.
Danvillier, A.: 1973, *Bull. Rend. Sci. Acad. Roy. Belg.* **59**, 917.
Dezsö, L. and Gerlei, O.: 1964a, *Publ. Debrecen Heliophys. Observ.* **1**, 3.
Dezsö, L. and Gerlei, O.: 1964b, *Publ. Debrecen Heliophys. Observ.* **1**, 35.
Dezsö, L. and Gerlei, O.: 1965, *Publ. CSAV, Astron. Ustav*, No. 51, 41.
Dingle, L. A. *et al.*: 1973, *Solar Phys.* **31**, 243.
Djurkovič, P.: 1956, *Soln. Dan.*, No. 6.
Dmitrieva, M. G.: 1965, *Soln. Dan.*, No. 4.
Dmitrieva, M. G. *et. al.*: 1971, *Issled. Geom., Aeron., Fiz. Solnca, Irkutsk*, vyp. 2, 180.
Dodson, H. W. and Hedeman, R. E.: 1972, in E. R. Dyer (General Editor), *Solar-Terrestrial Phys. 1970*, Dordrecht, Part 1, p. 151.
Dolginov, A. Z. *et. al.*: 1972, *Astron. Vestn.* **6**, 195.
Drozdov, O. A.: 1950, *Tr. Gl. Geofiz. Obs.* **19**(81), 102.
Eigenson, M. S.: 1963, *Solnce, pogoda, klimat*, Gidrometeoizdat.
Eigenson, M. S. and Mandykina, T. L.: 1962a, *Cirk. Astron. Observ. Lvov Univ.*, No. 37–38, 80.
Eigenson, M. S. and Mandykina, T. L.: 1962b, *Cirk. Astron. Observ. Lvov Univ.*, No. 37–38, 82.
Eigenson, M. S. and Mandykina, T. L.: 1962c, *Cirk. Astron. Observ. Lvov Univ.*, No. 37–38, 91.
Ferris, G. A. Y.: 1969, *J. Brit. Astron. Assoc.* **79**, 385.
Fokker, A. D.: 1963, *Bull. Astron. Inst. Neth.* **17**, 84.
Fredga, K.: 1965, *Stellar and Solar Magnetic Fields*, Amsterdam, p. 310.
Fung, P. C. W. *et. al.*: 1971, *Solar Phys.* **18**, 90.
Gabrovski, I. *et al.*: 1967, *Izv. Sekc. Astron. Bulg. Akad. Nauk* **2**, 139.
Gabrovski, I. *et. al*: 1968, *Geomagnetizm i Aerononsyia* **8**, 929.
Giovanelli, G.: 1964, *Observatory* **84**, 57.
Glackin, D. L.: 1973, *Publ. Astron. Soc. Pacific* **85**, 241.
Gleissberg, W.: 1955, *Naturwissenschaften* **42**, 410.
Gleissberg, W.: 1958, *Z. Astrophys.* 46, 219.
Gleissberg, W.: 1966, *J. Brit. Astron. Assoc.* **76**, 265.
Gleissberg, W.: 1968, *Solar Phys.* **4**, 93.
Gnevyshev, M. N.: 1963, *Astron. Zh.* **40**, 401.
Gnevyshev, M. N.: 1966, *Usp. Fiz. Nauk* **90**, 291.
Gnevyshev, M. N.: 1967, *Solar Phys.* **1**, 107.
Gnevyshev, M. N. and Antalova, A.: 1965, *Publ. CSAV, Astron. Ustav*, No. 51, 47.
Gnevyshev, M. N. and Krivsky, L.: 1966, *Astron. Zh.* **43**, 385.
Gnevyshev, M. N. and Ol', A. I.: 1948, *Astron. Zh.* **38**, 18.
Godoli, G. and Mazzucconi, F.: 1967, *Astrophys. J.* **147**, 1131.
Godoli, G. *et al.*: 1966, *Ann. Geophys.* **19**, 395.
Godoli, G. *et al.*: 1967, *Ann. Geophys.* **20**, 265.

Golubtsov, V. V.: 1965, *Soln. Dan.*, No. 6, 70.
Granova, V. D. *et al.*: 1972, *Astron. Tsirk.*, No. 703, 1.
Greatrix, G. R.: 1970, *Astron. Astrophys.* **5**, 171.
Gudzenko, L. I. *et al.*: 1965, *Astron. Tsirk.*, No. 342, 1.
Gudzenko, L. I. and Chertoprud, V. E.: 1962, *Astron. Zh.* **39**, 758.
Gudzenko, L. I. and Chertoprud, V. E.: 1966, *Astron. Zh.* **43**, 113.
Hakura, Y.: 1974, *Solar Phys.* **39**, 493.
Howard, R. F.: 1965, *Stellar and Solar Magnetic Fields*, Amsterdam, p. 132.
Henkel, R.: 1972, *Solar Phys.* **25**, 498.
Jakimiec, M.: 1969, *Acta Astron.* (*Polska*) **19**, 59.
Jose, P. D.: 1965, *Astron. J.* **70**, 193.
Kandaurova, K. A.: 1971, *Soln. Dan.*, *No.* 5, 107.
Kasinsky, V. V.: 1973a, *Soln. Dan.*, No. 2, 78.
Kasinsky, V. V.: 1973b, *Soln. Dan.*, No. 7, 77.
Kasinsky, V. V. and Tomozov, V. M.: 1975, *Soln. Dan.*, No. 2, 84.
Kashirin, G. F.: 1962, *Cirk. Astron. Observ. Lvov Univ.*, No. 37–38, 96.
Kolomeets, E. V. *et al.*: 1974, *Geomagnetizm i Aeronomyia* **14**, 728.
Kolomiets, A. R.: 1969, *Astron. Tsirk.*, No. 509, 7.
Kopecký, M.: 1958a, *Publ. CSAV Astron. Ustav*, No. 42.
Kopecký, M.: 1958b, *Bull. Astron. Inst. Czech.* **9**, 34.
Kopecký, M.: 1962, *Bull. Astron. Inst. Czech.* **13**, 63.
Kopecký, M.: 1964, *Bull. Astron. Inst. Czech.* **15**, 178.
Kopecký, M.: 1967, *Adv. Astron. Astrophys.*, vol. 5, Academic Press, N.Y.–London, p. 189.
Kopecký, M.: 1972, *Bull. Astron. Inst. Czech.* **23**, 107.
Kopecký, M.: 1974, *Bull. Astron. Inst. Czech.* **25**, 271.
Kopecký, M.: 1975, *Bull. Astron. Inst. Czech.* (in press).
Kopecký, M. and Kuklin, G. V.: 1969, *Bull. Astron. Inst. Czech.* **20**, 22.
Kopecký, M. and Kuklin, G. V.: 1971, *Issled. Geomagn., Aeron., Fiz. Solnca*, Irkutsk, vyp. 2, 167.
Kopecký, M. and Künzel, H.: 1962, *Astron. Nachr.* **286**, 193.
Kotov, V. A. *et al.*: 1965a, *Astron. Tsirk.*, No. 331.
Kotov, V. A. *et. al*: 1965b, *Astron. Tsirk.*, No. 333, 6.
Kozelov, V. P.: 1972, *Geofiz. Issled. v Zone pol. Siyan.*, Apatity, 128.
Kozelov, V. P.: 1975, *Subburi i vozm. v. magnitosf.*, Leningrad, 274.
Kozelov, V. P. and Mingaleva, G. I.: 1975, *Subburi i vizm. v magnitosf.*, Leningrad, 264.
Kozhevnikov, N. I.: 1970, *Soob. Gos. Astron. Inst. Shternb.*, No. 162, 16.
Kozhevnikov, N. I.: 1973, *Astron. Tsirk.*, No. 802, 3.
Kreplin, R. W.: 1970, *Ann. Geophys.* **26**, 567.
Křivský, L. and Knoška, S.: 1967, *Bull. Astron. Inst. Czech.* **18**, 325.
Křivský, L. and Knoška, S.: 1968, *Bull. Astron. Inst. Czech.* **19**, 365.
Křivský, L. and Krüger, A.: 1966, *Bull. Astron. Inst. Czech.* **17**, 243.
Kuklin, G. V.: 1962, *Soln. Dan.*, *No.* 1, 72.
Kuklin, G. V.: 1971a, *Soln. Dan.*, No. 2, 75.
Kuklin, G. V.: 1971b, *Soln. Dan.*, No. 6, 89.
Kuklin, G. V.: 1973, *Soln. Dan.*, No. 2, 53.
Kuklin, G. V.: 1971c, *IAU Symp.* **43**, 737.
Kuleshova, K. F.: 1962, *Astron. Zh.* **39**, 272.
Levitsky, L. S.: 1967a, *Izv. Krymsk. Astrofiz. Observ.* **37**, 137.
Levitsky, L. S.: 1967b, *Izv. Krymsk. Astrofiz. Observ.* **37**, 158.
Liese, R.: 1971, *Hinweise auf Zusammenhänge zwischen den Planeten und den Sonnenfleckenperioden*, Techn. Univ. Hannover.
Link, F.: 1963, *Bull. Astron. Inst. Czech.* **14**, 226.
Loewe, F.: 1973, *Gerlands Beitr. Geophys.* **82**, 25.
Macris, C. J.: 1960. *Mem. Nat. Obs. Athenes*, ser. I, **7**.
Makarenko, N. G.: 1973, *Tr. Astrofiz. Inst. Akad. Nauk KazSSR* **23**, 92.
Maksimov, I. V.: 1952, *Dokl. Akad. Nauk. SSSR* **42**, 1149.
Maksimov, I. V. and Smirnov, N. P. 1967, *Soln. Dan.*, No. 10, 104.
Maris, G.: 1972, *Stud. Si Cerc. Astron.* **17**, 71.
Minasyants, G. S. and Obashev, S. O.: 1969, *Soln. Dan.*, No. 6, 108.
Molchanov, A. M.: 1966, *Dokl. Akad. Nauk. SSSR* **168**, 284.
Morozov, N. N. and Obashev, S. O.: 1969, *Tr. Astrofiz. Inst. Akad. Nauk KazSSR* **15**, 38.
Morozov, N. N. *et al.*: 1973, *Tr. Astrofiz. Inst. Akad. Nauk KazSSR* **23**, 48.

Mursalimova, G. G.: 1957, *Tr. Tashk. Astron. Obs.*, ser. 2 **5**, 145.
Muzalevsky, Yu. S. and Zhukov, L. V.: 1968a, *Astron. Tsirk.*, No. 488, 7.
Muzalevsky, Yu. S. and Zhukov, L. V.: 1968b, *Soln. Dan.*, No. 12, 77.
Nelson, J. H.: 1963, *The Effect of Disturbances of Solar Origin on Communications*, Pergamon Press, p. 293.
Nemeth, T.: 1966, *Pure Appl. Geophys.* **63**, 205.
Nesmyanovich, A. T. and Khomenko, Yu. A.: 1969, *Soln. Dan.*, No. 12, 98.
Newcomb, T.: 1901, *Astrophys. J.* **13**, 1.
Ol', A. I.: 1960, *Astron. Zh.* **37**, 222.
Ol', A. I.: 1972, *Soln. Dan.*, No. 12, 105.
Pathak, P. N.: 1972, *Solar Phys.* **25**, 489.
Pavelyev, S. V. and Pavelyeva, Z. S.: 1965, *Tr. Gl. Geofiz. Obs.*, vyp. 181, 92.
Phadke, M. S. and Wu, S. M.: 1974, *J. Amer. Statist. Assoc.* **69**, 325.
Predtechensky, P. P. and Gurevich, B. S.: 1948, *Tr. Gl. Geofiz. Obs.* **8(70)**, 33.
Prokudina, V. S.: 1973a, *Astron. Tsirk.*, No. 804, 3.
Prokudina, V. S.: 1973b, *Soob. Gos. Astron. Inst. Shternb.*, No. 181, 11.
Ramanathan, A. S. and Nagarajan, V.: 1965, *Observatory* **85**, 188.
Rao, K. R.: 1973, *Solar Phys.* **29**, 47.
Rima, A.: 1961, *Geophys. Meteorol.* **9**, 39.
Ringnes, T. S.: 1962, *Astrophys. Norv.* **8**, 17.
Ringnes, T. S.: 1964, *Astrophys. Norv.* **8**, 303.
Ringnes, T. S.: 1965, *Astrophys. Norv.* **10**, 27.
Ringnes, T. S.: 1968, *Astrophys. Norv.* **10**, 189.
Rodionov, A. V.: 1962, *Cirk. Astron. Observ. Lvov Univ.*, No. 37–38, 99.
Romanchuk, P. R.: 1963, *Geofiz. i Astron. Inform. Bull.*, No. 5, 248.
Romanchuk, P. R.: 1965a, *Soln. Dan.*, No. 5, 65.
Romanchuk, P. R.: 1965b, *Soln. Dan.*, No. 7, 65.
Romanchuk, P. R.: 1965c, *Soln. Dan.*, No. 8, 74.
Romanhcuk, P. R.: 1974a, *Vestn. Kiev. Univ. Ser. Astron.*, No. 16, 17.
Romanchuk, P. R.: 1974b, *Vestn. Kiev. Univ. Ser. Astron.*, No. 16, 28.
Romanchuk, P. R. and Sergeeva, A. N.: 1974, *Astrometr. i Astrofiz.*, vyp. 24, 107.
Rubashev, B. M.: 1949, *Bull. Komis. Issled. Solnca*, No. 2.
Rubashev, B. M.: 1962, *Soln. Dan.*, No. 10, 52.
Rubashev, B. M.: 1964, *Problemy solnechnoj aktivnosti*, Moskva-Leningrad, Nauka.
Schmidt, L.: 1962, *Bull. Astron. Ist. Czech.* **13**, 246.
Schove, D. J.: 1955, *J. Geophys. Res.* **60**, 127.
Shapiro, R. and Ward, F.: 1962, *J. Astrophys. Sci.* **19**, 506.
Sheeley, N. R. Jr.: 1964, *Astrophys. J.* **140**, 731.
Shnitnikov, A. V.: 1951, *Bull. Komis. Issled. Solnca*, No. 7, 47.
Shodo, E. L.: 1955, *Astron. Tsirk.*, No. 161, 9.
Shodo, E. L.: 1957, *Astron. Tsirk.* No. 178, 15.
Stuvalov, V. M.: 1970, *Astron. Vestn.* **4**, 198.
Stanek, W.: 1972, *Solar Phys.* **27**, 89.
Suda, T.: 1962, *J. Meteorol. Soc. Japan, Ser. II*, **40**, 287.
Sýkora, J.: 1971, *Solar Phys.* **18**, 72.
Szymanski, W.: 1970, *Postepy Astron.* **18**, 305.
Szymanski, W.: 1971, *Postepy Astron.* **19**, 153.
Szymanski, W.: 1974, *Postepy Astron.* **22**, 45.
Takahashi, K.: 1967, *J. Radio Res. Labs.* **14**, 237.
Takahashi, K.: 1968, *Solar Phys.* **4**, 598.
Toman, K.: 1967, *J. Geophys. Res.* **72**, 5570.
Trellis, M.: 1966a, *Compt. Rend. Acad. Sci.* **AB262**, B221.
Trellis, M.: 1966b, *Compt. Rend. Acad. Sci.* **AB262**, B312.
Trellis, M.: 1966c, *Compt. Rend. Acad. Sci.* **AB262**, B376.
Trellis, M.: 1971a, *Compt. Rend. Acad. Sci.* **272**, B549.
Trellis, M.: 1971b, *Compt. Rend. Acad. Sci.* **272**, B1026.
Trellis, M.: 1973, *Compt. Rend. Acad. Sci.* **277**, B183.
Trotter, D. E. and Billings, D. E.: 1962, *Astrophys. J.* **136**, 1140.
Tschegoleva, G. P.: 1965, *Soln. Dan.*, No. 8, 70.
Tuominen, J.: 1942, *Z. Astrophys.* **21**, 96.
Turner, H.: 1925, *Monthly Notices Roy. Astron. Soc.* **85**, 467.

Vasilyev, O. B.: 1970, *Soln. Dan.*, No. 1, 92.
Vasilyev, O. B. and Vitinsky, Yu. I.: 1969, *Soln. Dan.*, No. 9, 105.
Vasilyev, O. B. and Kandaurova, K. A.: 1968, *Soln. Dan.*, No. 12, 85.
Vasilyev, O. B. and Kandaurova, K. A.: 1970a, *Soln. Dan.*, No. 2, 106.
Vasilyev, O. B. and Kandaurova, K. A.: 1970b, *Soln. Dan.*, No. 11, 109.
Vasilyev, O. B. and Rubashev, B. M.: 1972, *Soln. Dan.*, No. 1, 95.
Vasilyeva, G. Ya *et al.*: 1971, *Soln. Dan*, No. 9, 96.
Vasilyeva, G. Ya *et al.*: 1972a, *Soln. Dan.*, No. 2, 99.
Vasilyeva, G. Ya *et al.*: 1972b, *Soln. Dan.*, No. 8, 106.
Vasilyeva, G. Ya *et al.*: 1974, *Soln. Dan.*, No. 6, 99.
Vasilyeva, G. Ya *et al.*: 1975, *Soln. Dan.*, No. 2, 76.
Vertlib, A. B. and Kuklin, G. V.: 1971a, *Soln. Dan.*, No. 5, 76.
Vertlib, A. B. and Kuklin, G. V.: 1971b, *Soln. Dan.*, No. 9, 79.
Vertlib, A. B. and Kuklin, G. V.: 1971c, *Soln. Dan.*, No. 10. 86.
Vitinsky, Yu. I.: 1958, *Soln. Dan.*, No. 3.
Vitinsky, Yu. I.: 1960, *Soln. Dan.*, No. 9, 78.
Vitinsky, Yu. I.: 1961a, *Izv. Gl. Astron. Obs. Pulkovo* **22**, 121.
Vitinsky, Yu. I.: 1961b, *Soln. Dan.*, No. 11, 64.
Vitinsky, Yu. I.: 1962a, *Soln. Dan.*, No. 2, 66.
Vitinsky, Yu. I.: 1962b, *Izv. G. Astron. Obs. Pulkovo* **22**, 111.
Vitinsky, Yu. I.: 1963a, *Soln. Dan.*, No. 3, 64.
Vitinsky, Yu. I.: 1963b, *Soln. Dan.*, No. 5, 71.
Vitinsky, Yu. I.: 1965a, *Soln. Dan.*, No. 11, 62.
Vitinsky, Yu. I.: 1965b, *Soln. Dan.*, No. 12, 53.
Vitinsky, Yu. I.: 1967, *Soln. Dan.*, No. 6, 80.
Vitinsky, Yu. I.: 1968a, *Soln. Dan.*, No. 2, 90.
Vitinsky, Yu. I.: 1968b, *Soln. Dan.*, No. 3, 100.
Vitinsky, Yu. I.: 1968c, *Soln. Dan.*, No. 9, 90.
Vitinsky, Yu. I.: 1968d, *Izv. Gl. Astron. Obs. Pulkovo*, No. 184, 66.
Vitinsky, Yu. I.: 1969a, *Solar Phys.* **7**, 210.
Vitinsky, Yu. I.: 1969b, *Soln. Dan.*, No. 4, 88.
Vitinsky, Yu. I.: 1969c, *Soln. Dan.*, No. 10, 107.
Vitinsky, Yu. I.: 1971a, *Izv. Gl. Astron. Obs. Pulkovo*, No. 186, 20.
Vitinsky, Yu. I.: 1971b, *Izv. Gl. Astron. Obs. Pulkovo*, No. 189–190, 10.
Vitinsky, Yu. I.: 1972, *Soln. Dan.*, No. 5, 84.
Vitinsky, Yu. I.: 1973a, *Soln. Dan.*, No. 6, 82.
Vitinsky, Yu. I.: 1973b, *Soln. Dan.*, No. 9, 80.
Vitinsky, Yu. I.: 1973c, *Ciklichnost' i prognozy solnechnoj aktivnosti*, Leningrad.
Vitinsky, Yu. I. and Ikhsanov, R. N.: 1970, *Soln. Dan.*, No. 8, 103.
Vitinsky, Yu. I. and Kopecký, M.: 1968, *Izv. Gl. Astron. Obs. Pulkovo*, No. 184, 73.
Vitinsky, Yu. I. and Rubashev, B. M.: 1974, *Soln. Dan.*, No. 1, 84.
Vitinsky, Yu. I. and Tschegoleva, G. P.: 1971, *Soln. Dan.*, No. 3, 74.
Vlasov, V. A. *et al.*: 1974, *Kratkie soob. po fiz.*, No. 12, 9.
Waldmeier, M.: 1935, *Astron. Mitt. Zürich*, No. 133.
Waldmeier, M.: 1957, *Z. Astrophys.* **43**, 149.
Waldmeier, M.: 1966, *Astron. Mitt. Zürich*, No. 274.
Waldmeier, M.: 1973, *Solar Phys.* **28**, 389.
Warwick, C. S.: 1965, *Astrophys. J.* **141**, 500.
Wilcox, J. M. and Colburn, D. S.: 1969, *J. Geophys. Res.* **74**, 2388.
Wilcox, J. M. and Schatten, K. H.: 1967, *Astrophys. J.* **147**, 364.
Wood, R. M. and Wood, K. D.: 1965, *Nature*, **208**, 129.
Xanthakis, J.: 1960, *Prakt. tis Akad. Athen.* **35**, 352.
Xanthakis, J.: 1962a, *Mem. Soc. Astron. Ital.* **33**, 291.
Xanthakis, J.: 1962b, *Ann. Astrophys.* **25**, 342.
Xanthakis, J.: 1966, *Bull. Astron. Inst. Czech.* **17**, 215.
Xanthakis, J.: 1969, *Solar Phys.* **10**, 168.
Xanthakis, J.: 1970, *Compt. Rend. Acad. Sci.* **271**, B1009.
Zabza, M.: 1962, *Acta Astron. (Polska)* **12**, 210.
Zabza, M.: 1964, *Prace Wrocl. towarz. nauk, B*, No. 112, 57.
Zhukov, L. V. and Muzalevsky, Yu. S.: 1969, *Soln. Dan.*, No. 8, 88.
Zhukova, L. N. and Mitrofanova, L. A.: 1973, *Soln. Dan.*, No. 6, 65.

DISCUSSION

Krause: Estimations of the energy of the solar magnetic fields are strongly depending on the model you consider. For example, if you interpolate the 1 G surface field as a dipole-like-field which penetrates the whole Sun you will get quite a different value compared with a model where the a.c.-character is taken into account and which penetrates therefore only a surface layer. The same is true for the differential rotation. There are suggestions that the Sun under the convection zone rotates much faster. So my question is, on what models have you based your estimations of the magnetic energy and the differential rotation?

Kuklin: The first estimation is the value of rotation energy, not only of differential rotation energy. The second one is the value of convection motion energy and the differential rotation energy is of the same order if using well known formula for differential rotation.

Stix: How deep, then, do the sunspots extend?

Kuklin: My estimations are limited within a layer of 20 000 km thickness but I took into account the fact that small sunspots have smaller depth and used the sunspot distribution in respect to sunspot area and assumed $H \sim L/3$.

SOME RESULTS OF THE SPACE DISTRIBUTION OF THE ACTIVATION VECTOR OF THE FILAMENTS IN THE 11-YEAR CYCLE

V. V. KASINSKIJ, G. Ya. SMOLKOV, and G. N. ZUBKOVA

Sibizmiran, Irkutsk, U.S.S.R.

In order to obtain new information on the manifestation of basic mechanisms of solar activity with the sunspot cycle development we have studied peculiarities of the latitude-time distribution of activation positions and suddenly disappearing filaments with respect to sunspot groups of cycles 18 to 20.

Interest in such an analysis is caused by the particular role of filaments which is conditioned by their close relationship with the appearance and variation of sunspot groups, by the development of flares, topology and magnetic field disturbances, as well as by the development of coronal transients in the solar atmosphere, detected by 'Skylab'.

Assuming that activation and a sudden disappearance of filaments (later on, simply activation) are triggered by the sunspots, or flares influencing the former along the line between them or along the filament, we have chosen, in terms of coupling parameters, differences in their latitudes ($\Delta\varphi$) and longitudes ($\Delta\lambda$), or displacement vectors ($\bar{R}$) produced in them, after which $\Delta\varphi$-, $\Delta\lambda$-, mean-diagrams were then constructed. The activated filaments were identified with the neighbouring sunspot groups up to a distance R less than 25° in latitude and longitude. Those active regions were chosen as the triggering centres for which R has a local minimum. An analysis of these diagrams yielded the following results.

1. Averaged φt-Diagrams of Activation

The regularity of the filament activation position distribution with respect to sunspot groups with the phase of the sunspot cycle evolution, is indicated by the unaccidental character of this phenomenon:

(a) The distribution looks like a butterfly diagram which presents the mean value filament activation position. This distribution is closely associated with other phenomena of solar activity and reflects the effect of basic laws and mechanisms of the solar activity development (Figures 1 to 3). On all pictures the scale of vectors corresponds to the scale of latitude.

(b) There are two latitude zones of filament activation phenomena with the cycle evolution, marked in both their filament displacement direction and magnitude with respect to sunspot groups: high-latitude and near-equator zones have filament activation displacement (represented by small arrows) toward the poles and toward the equator, respectively. In the high-latitude zone displacements are greater than in the near-equator zone.

Bumba and Kleczek (eds.), Basic Mechanisms of Solar Activity, 191–197. All Rights Reserved.
Copyright © 1976 by the IAU.

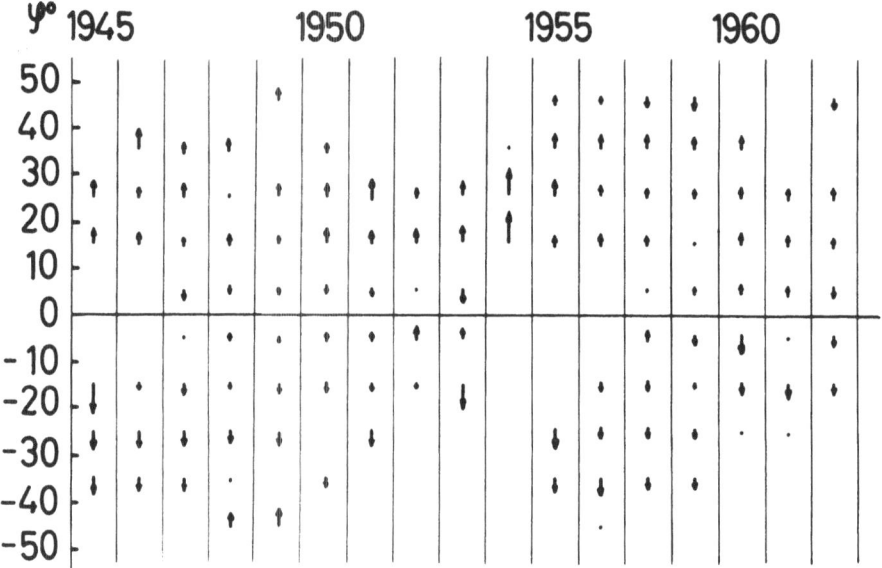

Fig. 1.   The diagram of the latitudinal (meridional) displacements of the sudden disappearances of filaments, 1945–1962, cycles Nos. 18, 19.

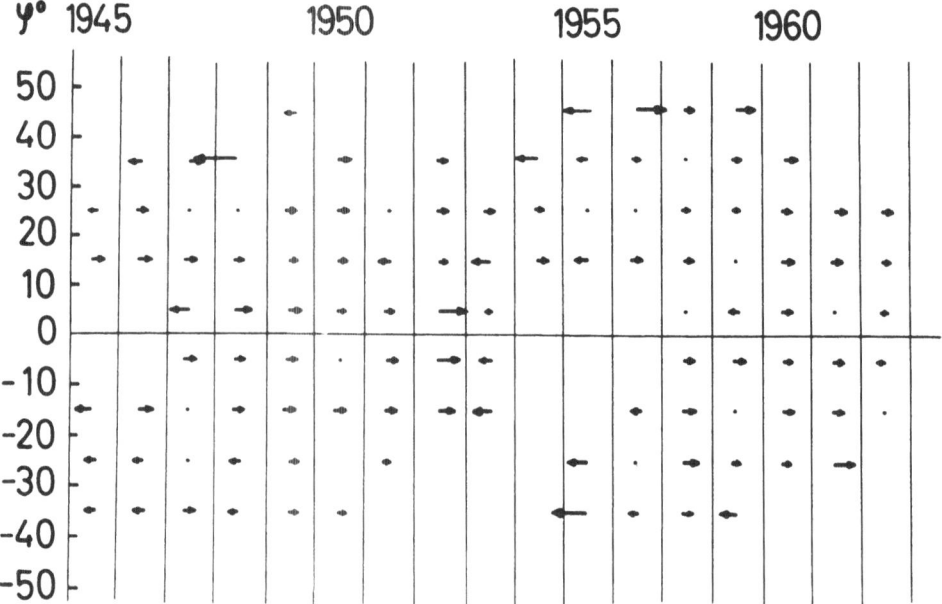

Fig. 2.   The butterfly diagram of the longitudinal displacements of the sudden disappearances of filaments, 1945–1962.

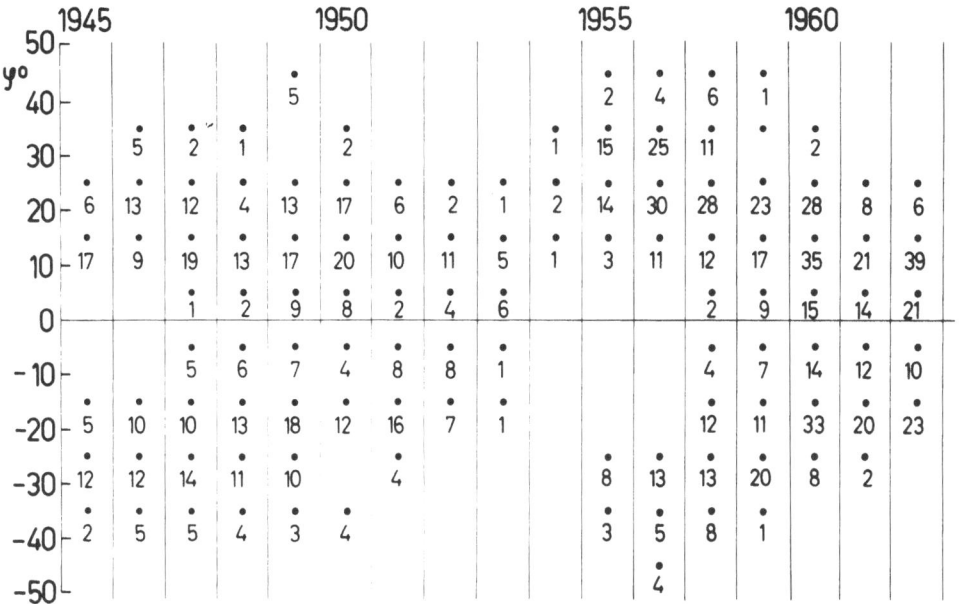

Fig. 3.   The butterfly diagram of the total number of activations.

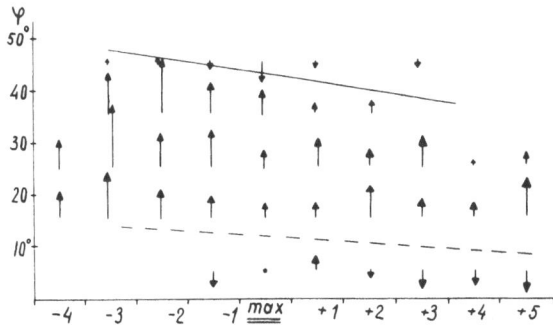

Fig. 4.   The mean diagram of the latitudinal displacements of the filament activations for the single solar hemisphere.

Polar filament activation displacements are the largest at the increasing phase of the solar cycle, then they decrease. The near-equator type of displacements is more pronounced at the decreasing activity phase (Figure 4).

In addition, during the whole cycle, two marked zones of sign change of latitude displacement are retraced. The high-latitude, or conversion zone is at the latitudes near 45°, and the low-latitude or divergation zone is in the latitude range of 8–10°. Outside these zones, the equatorial type of displacement is observed, inside the zone, polar displacements are seen.

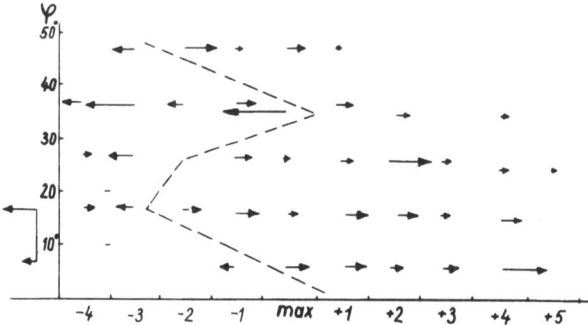

Fig. 5.   The average diagram of the longitudinal displacements of the filament activations in one
hemisphere.

(c) The directions of filament activation displacement in longitude vary with the cycle evolution from an eastward direction at the beginning of the cycle to a westward one at the end. In the cycle maximum such displacements are less noticeable than during minima epochs of solar activity (Figure 2).

The transition from the eastward displacement to the westward one takes place not gradually, but as it is seen from the behavior of the zero line at Figure 5, by fits and starts and resemble displacement or torsional oscillations. Variations in sign change of displacement are found only on the ascending branch of the sunspot cycle when interaction between the fields of the preceding and the present sunspot cycles still takes place. After the maximum, only westward displacements are observed.

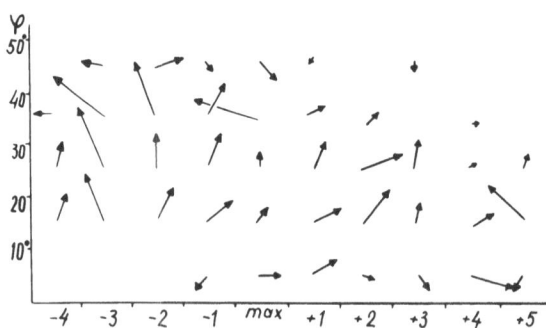

Fig. 6.   The constructed latitudinal-longitudinal diagram of activation displacements for one hemisphere.

(d) The directions of displacement vectors tend, with the cycle evolution, to vary gradually from the polar-eastward to the polar-westward sense in the high-latitude zone, and from equatorial-eastward to equatorial-westward sense in the low-latitude zone (Figure 6). The polar type of displacements is predominant. Therefore, displacement vector staggering is observed instead of simple displacement direction change, found by one of the authors for the flares [1, 2].

2. The Fine Structure of φt-diagrams of Activation

The fine structure of φt-diagrams (Figure 7b) together with the mentioned (average) common features may be shown by means of sum $(N+S)$ and subtract $(N-S)$ two pictures in the N and S-hemispheres thus drawing the zero line separating the region of polar displacements from the equatorial ones. Both zones exist simultaneously through the whole cycle. The equatorial zone, extending up to latitudes of 20–30° at the beginning of a cycle, reveals itself more completely after the activity maximum. During the 20th cycle the zero line has a deviation to the high latitudes by 1966–68. This fact is one more piece of evidence for an anomaly in the 20th cycle, noted earlier from other phenomena of solar activity (sunspots, polar prominences, corona).

The filament activation position distribution in the northern and southern hemispheres is marked by considerable asymmetry (Figure 7a). This asymmetry is stronger at the beginning of a cycle, and at high latitudes during practically the whole cycle. At intermediate and low latitudes, especially after the maximum epoch, nearly symmetric distribution of activation takes place (vectors tend to zero).

The distributions of activation displacements modulus with respect to latitude are appreciably different from year to year (Figure 8), also exhibiting asymmetry of hemispheres. Most activations take place in the northern hemisphere. In it, at the

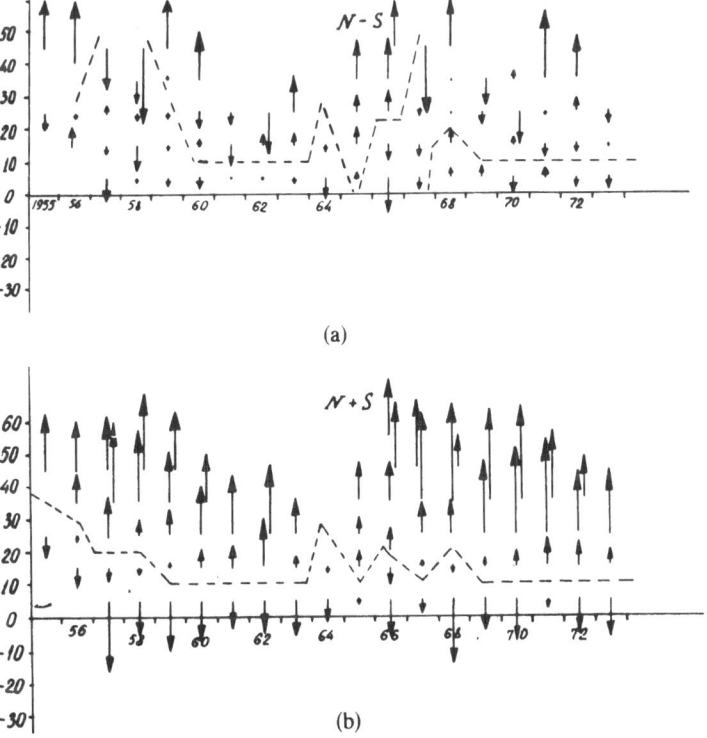

Fig. 7.   The fine structure and the zero line asymmetry drift for the meridional displacements of the filament activations in the 19th and 20th cycles, 1955–1973.

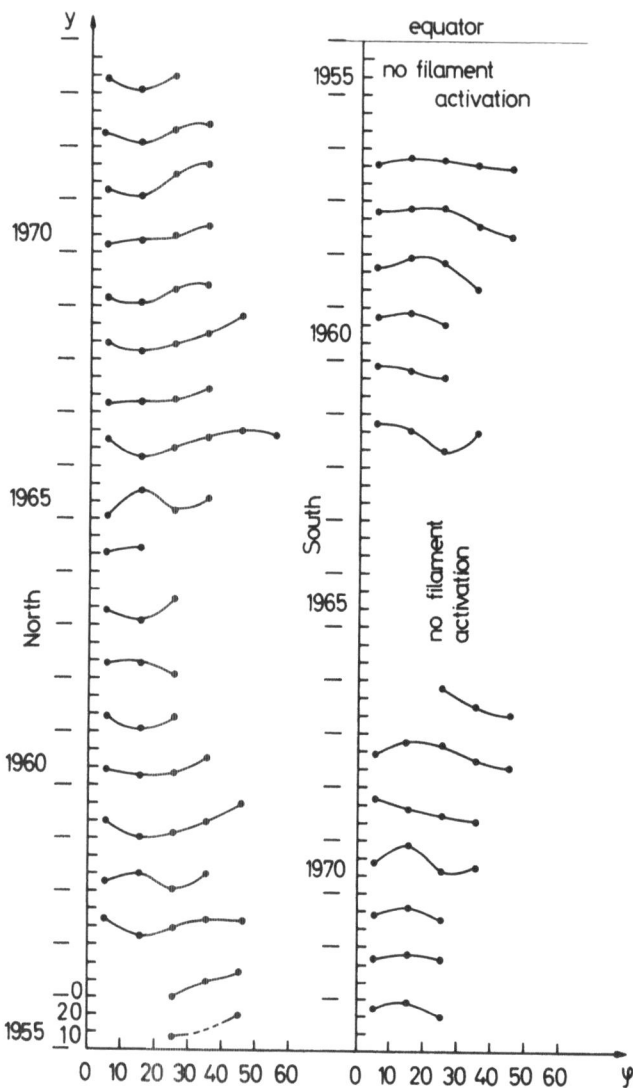

Fig. 8.   The changes of modulus of distance 'sunspot group – activation of filament' within the time of the cycle in both hemispheres of the Sun, 1955–1973.

latitude of 15–20°, a marked minimum of activation displacements is observed. Later on, displacements increase with latitude. In the southern hemisphere (less active) such a minimum is less pronounced. During some years, it is observed at higher latitudes (25–30°) than that in the northern hemisphere, an effect which seems likely to be associated with cycle delay in the southern hemisphere.

When summarizing we have obtained the following results.

(1) On the material of *Cartes Synoptiques de la Chromosphère Solaire* the proper position of the active filaments relative to the adjacent sunspot's group (D.B. phenomena) are considered [3, 4]. The vector-displacement for each activation of

the filament (each D.B.) are summarised inside the space-time region on the time-latitude diagram along the latitudinal intervals $\Delta\varphi = 10°$ and time $\Delta t = 1$ yr of the cycle.

(2) The 'butterfly diagram' constructed has been analysed from different points of view. The two populations of active filaments are picked up. One of them is the equatorial zone filaments and the other is the middle and high latitude ones. The middle-high latitude type shows regular rotation of the proper vector from the polar-eastern direction to the polar-western in the end of cycle. In the middle of the 11-year cycle there is a well-defined polar direction of the vectors. For the different latitudinal zones the vector $\bar{R}$ is constructed (1945–1973).

The results obtained are discussed from the point of view of common regularities of 11-year activity of the sunspots and the filaments on the Sun.

The full text of this report will be published in *Issled. Geomag. Aeron. Fiz. Solnza* **38** (1976), 225.

References

[1] Kasinskij, V. V.: 1973, *Solar Data Bulletin*, NN 2, 7, 1973.
[2] Kasinskij, V. V. and Plusnina, L. A.: 1974, *Issled. Geomag. Aeron. Fiz. Solnza* **31**, 44.
[3] Cartes Synoptiques de la Chromosphère Solaire, Méudon, 1945–1973.
[4] *Quart. Bull. Solar Activity*, Zürich, 1955-1973.

SEQUENCES OF LARGE SUNSPOT GROUPS

P. KOTRČ

Astronomical Institute of the Czechoslovak Academy of Sciences,
Ondřejov, Czechoslovakia

Abstract. The concept of a sequence of large sunspot groups is defined. From distribution functions of large groups from the period of sequences and that outside sequences it has been found that there are relatively more groups with a smaller maximum area in the period of sequences than in the period outside sequences. With an increased frequency of the occurrence of large sunspot groups on the solar disc their size decreases.

1. Introduction

Kopecký and Kotrč (1974, 1976) submitted that after time intervals in which the occurrence of large sunspot groups at a heliographic longitude is more or less fortuitous and in time sporadic, there is a sudden increase in the activity of large groups. Such periods were called by the authors 'outbursts of solar activity'. Among other things, they manifest themselves by an increased frequency of the occurrence of large groups appearing in such periods simultaneously on a large part of the solar surface, practically at all heliographic longitudes.

2. Sequences of Large Sunspot Groups

For large sunspot groups which during their passage across the solar disc have attained a mean area of at least 500 millionths of the solar hemisphere we determined according to *Greenwich Photoheliographic Results* the course of frequency of their occurrence in cycles 14 to 19.

An increase in the frequency of the occurrence of large groups high over a usual level given by the 11-year cycle phase takes place very quickly. As a rule, the state of an increased frequency of occurrence lasts for one to several rotations of the Sun. During such a period we observe an abnormally large number of large sunspot groups. In such periods large groups do not accumulate in short longitudinal intervals but are densely distributed in a long belt of heliographic longitudes. Such a trend in the occurrence of large groups was ascertained in the entire investigated period.

Figure 1 reflects a typical example of the described phenomenon from the first half of 1958. Even at a high level of activity (maximum of the 19th cycle) a dense series of large groups containing objects from both hemispheres is quite distinct. The distribution of groups in this zone exceeding one rotation of the Sun is very conspicuous in the diagram reflecting the heliographic longitude and the time of the first appearance and differs from the form of the occurrence of large groups in adjacent rotations.

For such a phenomenon the working term 'sequence of large sunspot groups' was selected. Figure 2 demonstrates the course of such a sequence in the first half of 1950. It is evident from Figures 1 and 2 that characteristic features of sequences in the period of lower solar activity and of those in the maximum phase are similar. The

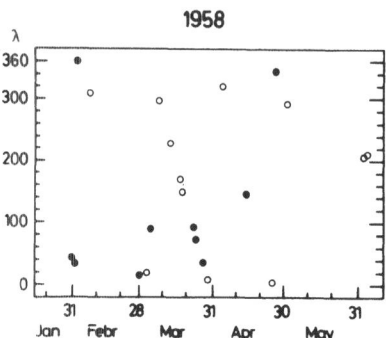

Fig. 1.   A sequence of large sunspot groups from February to April 1958. On the vertical axis the Carrington heliographic longitude, on the horizontal one the time of the first appearance. Groups from the northern hemisphere of the Sun are represented by empty circles, those from the southern one by full circles.

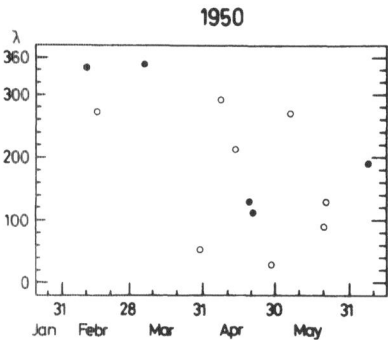

Fig. 2.   A sequence of large sunspot groups from March to May 1950. Marking identical with Figure 1.

conclusion relating to common features verified on data obtained from cycles 18 and 19 was conducive to our efforts to delimit the term 'sequence of large sunspot groups'. We wish to define the above term mainly in order to establish a firm basis for the study of laws of spot-forming activity discovered from the course of the frequency of the occurrence of large sunspot groups.

Under a sequence of large sunspot groups we understand a formation of sunspot groups attaining, during their passage over the solar disc, an area of at least 500 millionths of the solar hemisphere and having the following features:

(1) at least six large sunspot groups appear during 27 subsequent days.*

(2) those large groups are distributed in a zone of heliographic longitudes at least 180 deg long.

(3) as border objects belonging to the respective sequence we consider those groups which from the neighbouring ones belonging to the respective sequence

* A minimum number of groups stipulated under item 1 is in another definition describing the above phenomenon inversely proportional to the magnitude of a mean area of the respective group or another criterion determining the border of the basic statistical set. It is also directly proportional to the total number of large groups of the basic set occurring during the 11-year cycle under consideration.

according to features (1) and (2) are distant not more than 150 deg of heliographic longitude.

In accordance with the above definition altogether six sequences of large groups in cycle 18 and ten sequences in cycle 19 were found and their main characteristics were also ascertained.

3. Some Characteristics of a Statistical Distribution of Large Sunspot Groups in Sequences and Outside Sequences

Characteristic features of the frequency distribution of sets of large sunspot groups from the period of sequences and from the period outside them both according to mean and maximum areas were found. It is evident from them that the relative number of large groups from sequences decreases with the growing mean area quicker than the relative number of groups not belonging to the sequences. See Figures 3a and 4a. This is interrelated also with the fact that the maximum of the distribution function of large groups according to maximum areas appears for groups in sequences already in an interval of areas from 700 to 900 millionths and is relatively narrow. However, for groups not belonging to sequences, the maximum of the distribution function according to maximum areas appears only in the interval of maximum areas around 1000 millionths and in cycle 19 even around 1200 millionths and is relatively wide. See Figures 3b and 4b.

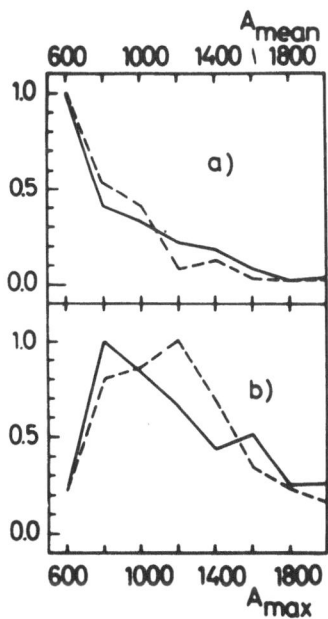

Fig. 3. Frequency distribution of large sunspot groups from cycle 18. In part (a) is the distribution according to area A_{mean}, in part (b) according to Area A_{max}. Full lines represent groups from sequences, dashed ones groups not belonging to sequences.

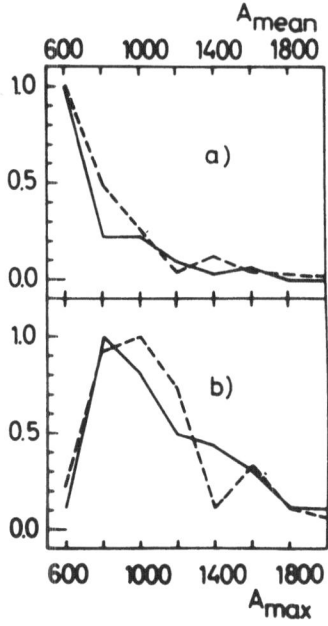

Fig. 4.   Frequency distribution of large sunspot groups in cycle 19. Marking of individual curves the same
as in Figure 3.

We may thus conclude that in the period of a high frequency of the occurrence of
large sunspot groups, i.e. in the period of sequences there are relatively more groups
with a smaller maximum and mean area than in the period outside sequences. With
an increased frequency of the occurrence of large groups their size decreases. The
hypothesis to the effect that in the period of high areal density of large sunspot
groups, on the Sun are such conditions that cause a smaller probability of the
formation of sizeable large groups, may serve for clarification. The search for
interrelations of this kind which might clarify the substance of sequences of large
groups will be the object of further investigations in this field.

The present paper is a brief summary of the article which will be published in *Bull.
Astron. Inst. Czech.*

References

Greenwich Photoheliographic Results 1903–1964.
Kopecký, M. and Kotrč, P.: 1974, *Bull. Astron. Inst. Czech.* **25**, 171.
Kopecký, M. and Kotrč, P.: 1976, Paper presented at the 7th Regional Consultation on Solar Physics,
    Sept. 1973, Starý Smokovec, (To be printed in *Publ. Astron. Obs. Skalnaté Pleso*).

CHARACTERISTIC FEATURES OF
CYCLIC CHANGE OF SOLAR ACTIVITY AFTER 1610

A. D. BONOV

University of Sofia, Bulgaria

Abstract. The most characteristic feature of the solar activity is its cyclic change in the course of time. The basic cycle is the 11-year one.

The law discovered by Hale and Nicholson (1925) concerning the change of sign of magnetic polarity of the bipolar groups of solar spots proved the existence of the 22-year cycle. Gnevyshev and Ol' (1948) point out that this cycle consists of two consistent 11-year cycles with an even and an odd number (according to the Zürich numeration).

Taking into consideration that the growth (t) of the 11-year cycle is a parameter closely connected with the characteristics of the 11-year cycle (Waldmeier, 1935), we calculate for all 22-year cycles after 1610 the relation

$$S = \frac{(t_{2n} + t_{2n+1})'}{(t_{2n} + t_{2n+1})''}.$$

The change of S shows some interesting characteristics of the cyclic recurrence of solar activity after 1610:

(1) A 44-year cycle is clearly delineated;

(2) The consistent 44-year cycles are grouped together two by two in one secular cycle with a mean duration of 88 yr.

(3) After 1610 the activity of the Sun is presented by two supersecular cycles (a supersecular 180-year cycle, as we conditionally call it.

(4) At the end of a 180-year cycle and at the beginning of the next one there is a 'catastrophical' decrease of solar activity. This fact as well as other ones obtained on the basis of some other investigations lead us to the conclusion that up to the end of the 20th century the activity of the Sun will be considerably weaker in comparison with its level in the 22-year cycle (18, 19);

(5) It is pointed out that the 180-year cycle is a part of a quite longer cycle of solar activity.

Bumba and Kleczek (eds.), Basic Mechanisms of Solar Activity, 203. All Rights Reserved.

PART 2

SOLAR CONVECTION AND DIFFERENTIAL ROTATION

THEORY OF CONVECTION IN A
DEEP ROTATING SPHERICAL SHELL,
AND ITS APPLICATION TO THE SUN

PETER A. GILMAN

National Center for Atmospheric Research,* Boulder, Colo. 80303, U.S.A.

Abstract. The theory of convection in a rotating spherical shell, when applied to the Sun, should ultimately satisfy at least three broad constraints. The radial heat flux by the convection must not vary significantly with latitude; an equatorial acceleration must be produced; and the convection must give the right dynamo action to produce the gross features of a solar cycle.

Important quantities to look for in the observations with which a convection theory can be compared include evidence of global velocities of giant convection cells, differences in motion features between low latitudes and high, persistence of motion features over successive rotations, evidence of excess brightness and variations in total solar luminosity, correlations of velocities between northern and southern hemispheres, and between north-south and east-west motions, and time periodic changes in motion fields.

The general theory of convective motions in a rotating spherical shell such as the convection zone of the Sun has developed rapidly over the last several years, but is still a long way from providing quantitative models which agree in a satisfactory way with the main solar observations. Most work has (for mathematical reasons) employed the Boussinesq approximation, and most has been either linear or nearly so. Early work demonstrated the basic ability of global convection to transport momentum toward the equator, but whether an equatorial acceleration results depends on the effects of competing angular momentum transport processes, namely transport in the radial direction by convection, and transport by axisymmetric meridional circulations. The end result can be assessed only by nonlinear calculations.

Some recent calculations directed at this question by the author indicate that, at least at Prandtl number of unity, convection growing from an initial state of solid rotation produces equatorial acceleration only when equatorial modes dominate and the rotational constraint is sufficiently strong, which results in a convective heat flux which is strongly dependent on latitude. When the rotational constraint is broken enough to give convection at all latitudes and therefore more nearly uniform heat flux, equatorial deceleration and angular momentum mixing in latitude results. A few calculations for Prandtl number substantially smaller than one give results which suggest it is possible to produce equatorial acceleration and nearly uniform heat flux, but these solutions appear not to be unique. In particular, the amount of differential rotation present in the initial conditions appears to be important in determining the final state.

1. Introduction

The theory of thermal convection for a deep rotating spherical shell such as the convection zone of the Sun is still in its early stages of development. On the one hand, it is clearly still impractical (and unwise) to include in a computer model all the physics that could be said to be important for this problem. On the other hand, there is much room for progress in developing simple, but still meaningful models. Working with such simpler models, I and others have obtained a number of promising results, but at the same time have revealed some perplexing and difficult problems.

Before going any further, let me indicate I am talking about modeling those scales of motion which are large enough in physical scale, and persistent enough in time, to be significantly influenced by rotation. Thus, for the Sun we are talking about the differential rotation, together with its fluctuations in time as well as global scale

* The National Center for Atmospheric Research is sponsored by the National Science Foundation.

eddies or giant cells imbedded in it. In this context, granules and even supergranules are too small in physical extent and too short in lifetime to be influenced much by rotation. Because of the central importance of differential rotation in maintaining solar activity, this paper will give particular emphasis to the angular momentum transport effects of global scale convection.

2. Broad Constraints Imposed by Observations

Even when developing models with highly simplified physics compared to the real Sun, certain broad observational constraints should be kept in mind for testing model progress. First, the heat flux coming out of the convection zone does not vary substantially with latitude – certainly no more than a few parts in 10^3, and perhaps substantially less. Thus, to be interesting for the Sun, the convection which is produced in a spherical shell model should also not show large differentials in heat flux. As we will see later, this imposes quite severe constraints on convective theory.

Second, even simple models should produce rotation rates which are higher in equatorial regions than in higher latitudes, as occurs on the real Sun. This is a substantially less severe constraint than on the heat flux, but still we find that the convection has a way of *de*celerating the equator when we least want it to.

Third, even if the convective model passes both of the above tests, it must still be shown to give hydromagnetic dynamo action which either produces the gross features of a solar cycle, or which does not interfere with some other mechanism which does. To test how well a convective spherical shell model satisfies this third constraint will clearly require considerable effort.

In addition to the broad constraints given above, there is also other observational evidence we should pay attention to in building and testing a model. For example, from Howard's observations (reviewed earlier at this meeting) it is clear that the kinetic energy at the surface in any giant cells on the sun is not large compared to the kinetic energy present in the shears of the mean differential rotation itself and may be substantially smaller. Furthermore, the fluctuations in the differential rotation (which may be hard to separate from giant cells) are of roughly the same magnitude or less than this mean shear. In addition, any axisymmetric meridional circulation (i.e., north-south motion) is at least one order of magnitude smaller than the differential rotation, and perhaps smaller still. Also, sunspots appear to rotate systematically faster at the surface than does the solar plasma, but some structures of very large latitudinal extent, such as coronal holes, show rather little differential rotation and rotate at nearly the equatorial plasma rate.

3. Important Parameters which are Poorly Known

Despite the fact there are many observational facts (alluded to above) with which we can compare a model, there are also a number of important parameters which we simply do not know well. For example, while we have some knowledge of the energy spectrum of motions at the surface (granules are somewhat more energetic than supergranules, which in turn are more energetic than giant cells or the differential rotation), we have really very little idea how this spectrum varies with depth. We suspect on theoretical grounds the dominant eddy size increases in scale with increasing depth and the velocities decrease in magnitude, but we have little in the way of observational evidence to support this. Consequently, parameterizing the effects of small scale motion, as we must in order to make progress, is a very uncertain matter.

In addition, while the depth of the convection zone can probably be reasonably estimated from solar structure calculations, we have little idea how much the convection may overshoot into the interior. Consequently, proper boundary conditions are difficult to define.

4. What Should be Looked for in Observations

The observations of the global circulation of the Sun have already been reviewed for this symposium by Howard and Yoshimura. I would like to list certain effects which, if found, could provide very useful guidance to theoreticians such as myself in attempting circulation and dynamo models.

(1) Differences in the form of large scale motion when comparing low latitudes with high, on scales substantially larger than supergranules. In this regard, Howard and Yoshimura's evidence of velocity anomalies in low and middle latitudes in each hemisphere is most encouraging.

(2) Evidence for actual giant convective cells – which Howard and Yoshimura's discovery may be. One particularly important parameter to know is the wavelength in longitude.

(3) Persistence of motion features (in addition to the mean differential rotation itself) for several solar rotations.

(4) Any large scale persistent features of excess brightness, no matter how small in amplitude, which cannot be attributed to plages or other obvious active region features. Long-lasting variations in both longitudinal and latitudinal directions are important, because they could be evidence of the thermal properties of giant convective cells. The important recent observational work of Hill and colleagues at the University of Arizona on solar oblateness, excess brightness and oscillations encourages me to believe it may be possible to look for such variations.

(5) Fluctuations in the solar constant of periods of weeks, months or longer, if any are found with new measurements, could also be related to thermal structure in giant convective cells. I note that Schwarzschild (1975) has recently invoked time dependent giant convective cells for red giants and supergiants as a possible explanation of their irregular fluctuations in energy output. Perhaps giant cells on the Sun, of which presumably only a few would be on the observable disk at any time, could produce small fluctuations in the solar flux visible from the Earth as a function of time, both from their own possible time dependence, and from rotation of new convective cells onto the disk. In turn, small modulations of the giant cells by the changing magnetic fields of the solar cycle could give longer period variations in heat flux.

(6) Any tendency in velocity anomalies to show symmetry between the two hemispheres, or correlation in the fluctuations in time of motions in the two

hemispheres. Theoretical models of spherical shell convection with rotation do indicate certain symmetries. With regard to time correlations, I recently did a simple analysis of the published parameters b and c in Howard and Harvey's (1970) differential rotation Ω as a function of latitude ϕ

$$\Omega = a + b \sin^2 \phi + c \sin^4 \phi$$

Howard and Harvey reported separate values for northern and southern hemispheres for a, b and c for each day observations were taken. I correlated $b(n)$ with $b(s)$, and $c(n)$ with $c(s)$, obtaining significantly positive values in both cases:

$$r(b(n), b(s)) = 0.27 \pm 0.11$$

$$r(c(n), c(s)) = 0.34 \pm 0.11$$

(with attached 95% confidence limits determined according to Hoel (1947), p. 122.) Thus, it would appear that mid and high latitudes in the two hemispheres tend to speed up and slow down together. It has already been established by Yoshimura (1971) and later by Wolff (1975) that within a single hemisphere, mid-latitudes tend to speed up while higher latitudes are slowing down, and vice versa. Since all these results are found from the result of least squares fitting done by Howard and Harvey (1970), it is possible that some of these effects, if real, represent not actual changes in the differential rotation but rather correlations (which are linked between the two hemispheres) in the global scale convective cell structure. The effect needs to be looked for in new, larger data samples.

(7) Any evidence from the large scale velocity anomalies which allows us to distinguish between the longitudinal and latitudinal motions in them. In cases where the velocity anomaly persists for a few days on both sides of the central meridian, and its rotation rate can be determined, then by successively adding and subtracting matched data from the eastern and western hemispheres the two components can be separated (provided the radial motion is small) since the Doppler shift of the longitudinal component changes sign with central meridian passage, while that of the latitudinal component remains the same.

(8) If such a separation of longitudinal and latitudinal velocity anomalies can be made (call them u' and v', respectively) then the latitudinal angular momentum transport

$$r = \cos \phi\, u'\, v'$$

can be computed. Theory indicates this ought to indicate transport toward the equator, in order to help maintain the equatorial acceleration there. Sunspot motion statistics computed by Ward (1965) indicate equatorward transport.

(9) Some knowledge of u' and v' separately might also allow us to estimate the radial component of vorticity, as well as the velocity divergence in horizontal surfaces. These quantities could provide comparison points with convective structures generated by the models.

(10) Knowledge of eddy latitudinal motions should also allow the estimation of transport in latitude of radial magnetic flux. This is an important quantity to know for understanding the solar dynamo.

(11) Periods of any long-lived fluctuations in either the differential rotation or cell velocities could indicate the presence of inertial oscillations, as well as helping in determining the nature of nonlinear interactions between the cells and the differential rotation.

5. Recent Developments in the Theory of Convection in a Rotating Spherical Shell

Having commented on various limitations to our knowledge of the dynamical properties of the Sun, let me turn to the theory, and describe the present state of the subject as I see it. Where it is possible to do so, I will try to relate theory back to the observational questions raised earlier.

There have been quite a variety of theories proposed for the solar differential rotation, only some of which involve explicit calculation of the effects of global scale convection. Much of the more general work will be discussed in the invited papers by Durney and Weiss. Here I will concentrate only on the theory of non-axisymmetric convection, and what it has to say about differential rotation. Most of the important contributions to this subject have been made in the last five to seven years, by Busse (1970a, 1973), Durney (1970, 1971), Gilman (1972, 1975), Yoshimura (1971, 1972, 1974), and Yoshimura and Kato (1971). The work of Roberts (1968), Busse (1970b), and Gierasch (1975) is also related and relevant. Simon and Weiss (1968) have considered simple models of global solar convection which include compressibility but leave out rotation.

5.1. MODEL ASSUMPTIONS

There are certain assumptions common to most of the spherical shell convection models referenced above which should be mentioned. Some of these are made to render the problem mathematically more tractable and are not particularly well justified in physical terms. In particular, all except Gierasch (1975) consider a so-called Boussinesq system, that is, one in which all density variations are ignored except where coupled with gravity. This clearly cannot be justified for the solar convection zone, but does provide a relevant analogue, with which to concentrate on rotational effects. In addition, small scale diffusion of heat and momentum is parameterized in linear constant eddy viscosity and thermal diffusivities, clearly a considerable oversimplification. Finally, except for Gilman (1972) and, to a lesser degree Busse (1973), the models presented are at best only slightly nonlinear. Consequently, not much can be said from most of the models about amplitudes of convection and differential rotation. Gilman (1972) demonstrated a number of nonlinear effects. Recent results obtained by me with a much more nonlinear model than presented previously, on which I will report shortly, indicate the nonlinearities are even more important.

Some of the early models (particularly Busse, 1970, 1973) considered only thin shells (depth ≪ radius) but other results (Gilman, 1972, 1975) indicate deep shell effects are important, particularly Coriolis forces in the radial direction affecting the radial momentum flux. Similarly, Yoshimura and Kato (1971) and Yoshimura (1971,

1974) assumed the convective motions were hydrostatic, which also excludes deep shell Coriolis force effects.

The boundary conditions at the top and bottom of the convection zone for the real Sun are clearly quite complex, but not well known. Models to date have restricted themselves to quite simple conditions, usually assuming stress-free fixed temperature top and bottom. In a few cases, a non slip bottom has been considered.

5.2. Fluid dynamical parameters

There are several dimensionless parameters which need to be defined in order to facilitate description of spherical shell convection theory. These are the Rayleigh number R, Taylor number T, Prandtl number P, and a depth parameter β. They are summarized below

$$R = \frac{g_0 \alpha \Delta \Theta d^3}{\kappa \nu}$$

in which g_0 is gravity at the outer edge of the shell, α is the coefficient of volume expansion ($=$ temperature^{-1} in a perfect gas), $\Delta \Theta$ is the imposed radial temperature difference, d is the depth of the convecting layer, κ is the thermal diffusivity, and ν the kinematic viscosity. The Rayleigh number measures the strength of the buoyancy forces. The Taylor number

$$T = 4\Omega^2 d^4 / \nu^2,$$

in which Ω is the rotation rate of the coordinate system in which the relative angular momentum is zero. The Taylor number measures the relative importance of Coriolis and viscous forces. The Prandtl number

$$P = \nu / \kappa$$

is simply a measure of the relative importance of viscous and thermal diffusion processes. Finally $\beta =$ radius of inner spherical shell/depth of convecting layer. The various results which will be referred to are for a variety of values of these parameters. For linear results and perturbation nonlinear results, R must be near R_c, the critical Rayleigh number for onset of convection. Nonlinear calculations have been done for the spherical shell up to $R \approx 10^5$. Mixing length arguments can provide only extremely crude estimates of the effective Rayleigh number for the Sun. One indirect approach is to compare motion amplitudes to what observations are available, but this, too, is very approximate, since compressibility is not included.

The Taylor number T for the Sun, if we use $d = 1.4 \times 10^{10}$ cm for the total depth of the convection zone, a mean rotation rate $\Omega \approx 2.6 \times 10^{-6}$ s^{-1}, and an eddy viscosity of 10^{12} cm^2 s$^{-1} \leq \nu \leq 10^{14}$ cm^2 s^{-1}, falls in the range $10^2 \leq T \leq 10^6$. At $T = 10^2$, rotation is a minor perturbation on the nonrotating case; for $T = 10^6$, the flow is very strongly rotationally influenced. With κ, ν representing eddy diffusivities, we should expect to take the Prandtl number $P \sim 1$. This assumption will be re-examined later. Busse and Yoshimura have concentrated on thin shell theory, i.e., $\beta \gg 1$, while Durney and Gilman have taken a finite depth, typically 25% of the inner radius, or $\beta = 4$, which is a reasonable depth for the actual solar convection zone.

5.3. LINEAR ANALYSES: THE BASIC MECHANISM FOR PRODUCING EQUATORIAL ACCELERATION

The early work of Busse (1970a), Durney (1970, 1971), and Yoshimura and Kato (1971) principally illustrated the fundamental mechanism by which global convection can generate and maintain an equatorial acceleration. Without rotation, convection in a spherical shell is degenerate in that many different modes first become unstable at the same Rayleigh number. They showed that when a small amount of rotation perturbs these modes, i.e., Taylor number small, then one mode is selected as the initial unstable mode. This mode is highly elongated in latitude and has the desirable property of transporting angular momentum towards the equator from higher latitudes. The mechanism of this transport is perhaps best illustrated in Figure 1, taken from my own linear numerical work, which shows the horizontal velocity vectors associated with the most unstable mode for $T = 10^3$ compared to $T = 0$. The Coriolis forces acting in the horizontal plane have turned the horizontal velocity vectors to the right, so that flow in longitude in the direction of rotation has a component toward the equator, flow against the rotation has a component toward the pole. Thus the Reynolds stress $u'v'$ gives momentum flux toward the equator. The effect remains even as T is increased to the point where the original perturbation analysis employed for small T by several authors is no longer valid. This results in horizontal motions with the form of closed swirls as seen, for example, in the bottom figures for $T = 3 \times 10^4$. The profile of momentum flux is shown in Figure 2 indicating it typically peaks for these modes somewhere in the neighborhood of 20° north latitude, and extends in depth largely through the convecting layer.

In the actual nonlinear system, of course, this equatorward transport of momentum by global convection must compete with other mechanisms of transport. In particular, all these modes also transport momentum in the radial direction through the Reynolds stress $u'w'$, or correlation between longitudinal motion u and radial motion w. This is particularly important if the depth of the convecting layer is large enough that Coriolis forces in the radial direction are active. Some profiles of radial momentum flux I have calculated for the most unstable modes at various Taylor number are shown in Figure 3 for increasing Taylor number. We see that at low T, the transport is negative (dashed contours) or radially inward. Only at and above $T = 10^5$ is the flux outward. Thus at low T, even though momentum is transported toward the equator near the surface, it is also carried below. Whether an equatorial acceleration would still result at the surface is not clear. On the other hand, at large T, an equatorial acceleration seems much more likely, since momentum is being brought to the equatorial outer surface both from high latitudes and greater depths. The reason for the switch from inward to outward momentum flux is illustrated in Figure 4. At $T = 0$, rising and sinking fluid particles have no rotational momentum to carry. At low but nonzero T, e.g., $T = 10^3$, fluid particles essentially conserve total angular momentum (except for frictional effects) and thus transport momentum relative to the rotating frame inward. Rising particles acquire a component of longitudinal motion opposite to rotation, sinking particles a component in the direction of rotation. At high $T = 10^5$ organized longitudinal pressure torques develop which accelerate rising particles in the direction of rotation in longitude, sinking particles in

P. A. GILMAN

HORIZONTAL VELOCITY VECTORS

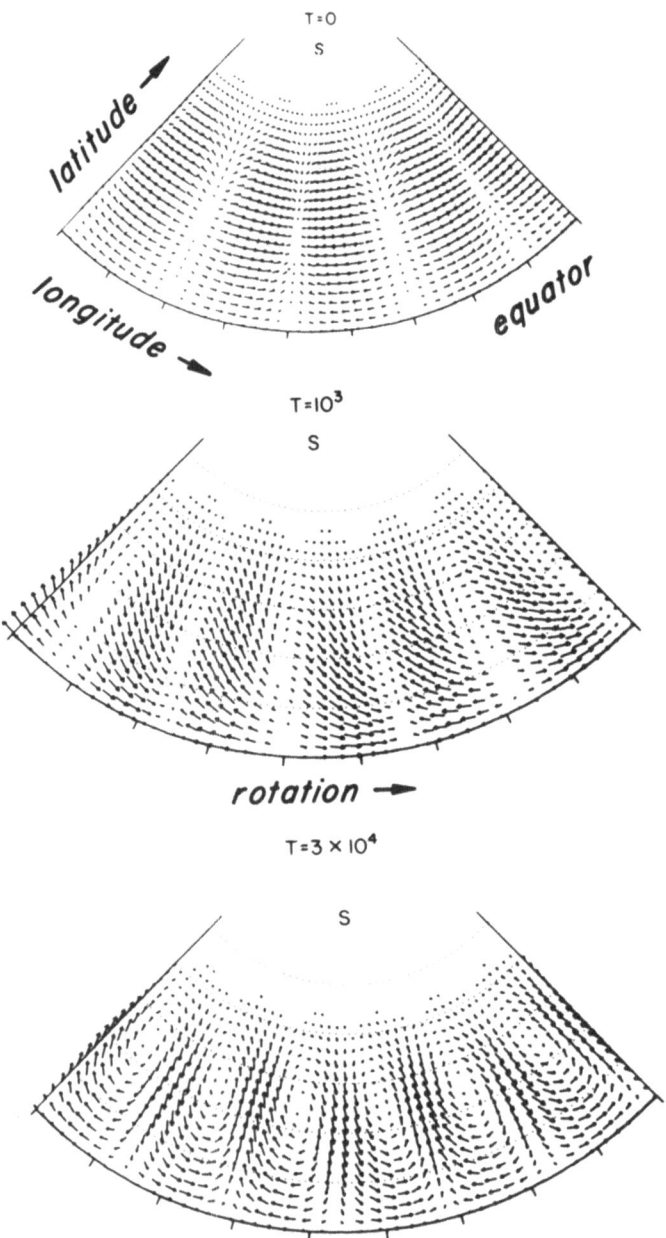

Fig. 1.   Horizontal velocity vectors near outer surface of spherical shell in northern hemisphere for most unstable longitudinal wave number m, as function of Taylor number T (defined in text) (for $T = 0$, $m = 9$, $T = 10^3$, $m = 10$, $T = 3 \times 10^4$, $m = 12$). Light dotted arcs are latitude belts spaced by $10°$. Tick marks are spaced $10°$ in longitude.

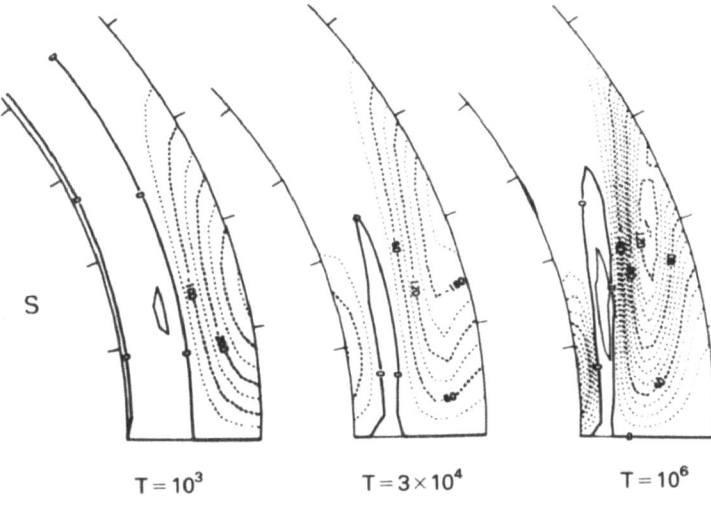

LATITUDE MOMENTUM FLUX

Fig. 2. Latitude-radius sections showing latitudinal angular momentum flux associated with most unstable mode m as function of Taylor number T. Dashed contours denote negative, or equatorward, momentum flux. Equatorial plane is at lower edge of each figure.

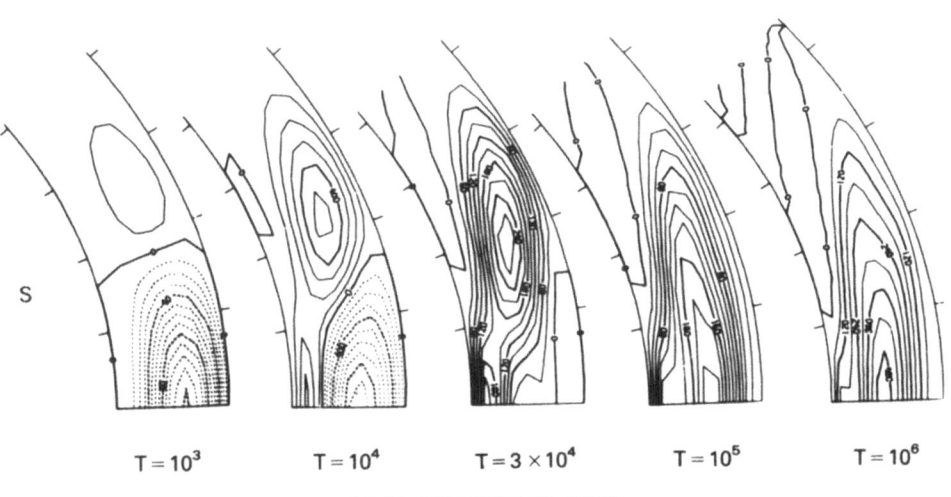

RADIAL MOMENTUM FLUX

Fig. 3. Latitude-radius sections showing radial angular momentum flux associated with most unstable modes as function of Taylor number T. Solid contours denote positive or outward momentum flux; dashed contours negative or inward flux.

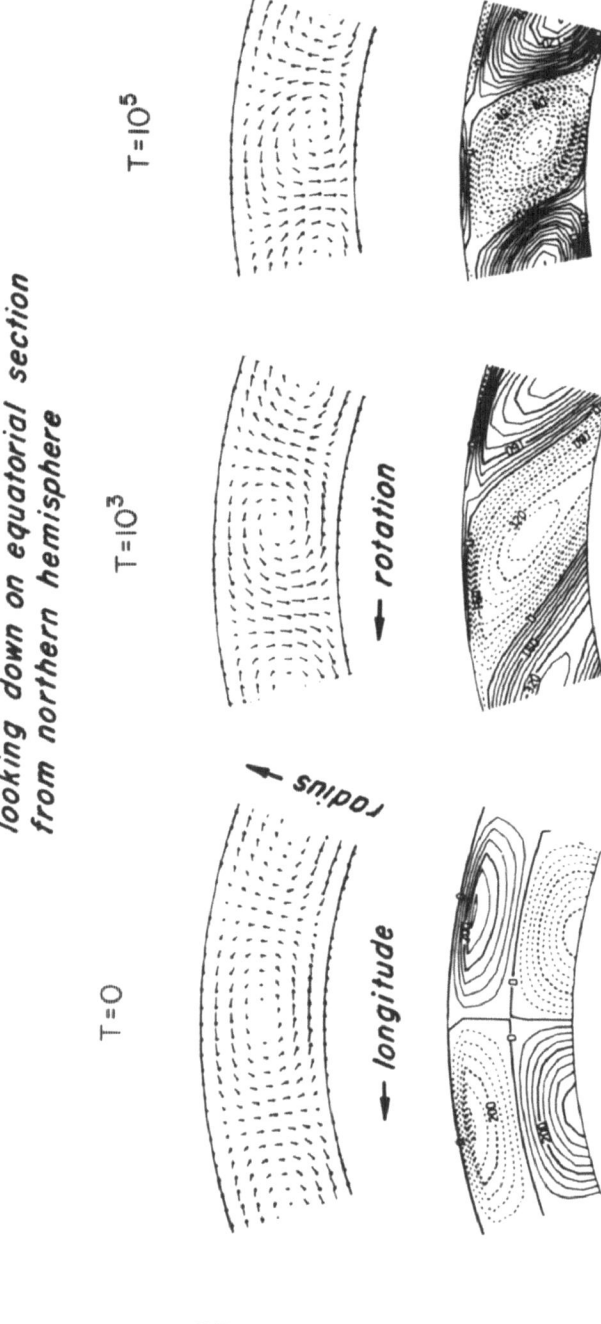

Fig. 4.   Longitude-radius section in equatorial plane showing velocity vectors and perturbation pressure fields associated with modes with same T whose radial momentum flux is shown in Figure 3.

the opposite direction, resulting in a net outward flux of momentum. These pressure forces are slightly larger than needed to keep the flow in heliostrophic balance (pressure forces balancing Coriolis forces). This effect is also described in Gilman (1975).

A third mechanism of angular momentum transport arises from the fact that other Reynolds stresses, namely v^2, w^2, and vw, as well as thermal stresses, force an axisymmetric meridional circulation. As Busse (1975) has pointed out, this circulation occurs even in the limit of small rotation, provided some mode selectivity is present. Consequently, the lower the Taylor number the more important is this meridional circulation relative to the other momentum transport mechanisms. In general, it tends to produce equatorial deceleration because the rings of fluid being advected around in the circulation tend to conserve their angular momentum.

The problem of competing and perhaps conflicting mechanisms of momentum transport becomes even more acute when one realizes that at small Taylor number, the modes with the most desirable properties are only slightly favored over other modes with different properties. The behavior of the system can (and does) change markedly when nonlinear effects come in. We will illustrate this below.

From Gilman (1972, 1975) it is clear that as Taylor number is increased beyond the point where small T effects are dominant, the favoring at instability onset of modes which have the desirable momentum transport properties becomes much stronger, and equatorial acceleration is more nearly assured. These modes also produce a differential rotation which is symmetric about the equator, which the Sun's rotation is, predominantly. Gilman (1975) showed that for Prandtl numbers near unity the most unstable mode structures in latitude for each longitudinal wave number m break at high T into just two classes: a large number of modes above a certain m ($m \leq 4$ for $T \geq 10^3$) all of which peak at or near the equator, and which transport momentum towards it from higher latitudes; and a small number of very differently structured low m modes which peak near the pole. (This result is one reason why it would be very interesting to see if any differences between equatorial and polar regions occur in large scale convection on the real Sun.)

The equatorial modes for Prandtl number $P \sim 1$ are excited at much lower Rayleigh numbers than are the polar modes. This leads to the prospect of the convective heat flux being a strong function of latitude in the nonlinear case, clearly not observed on the Sun. The key question to be answered in nonlinear calculations then is, does a regime of solutions exist in which the heat flux has been more or less equalized in latitude, which at the same time still gives equatorial acceleration? As will be shown below, so far we have not found one for $P \approx 1$, but solutions for $P \ll 1$ look promising.

5.4. NONLINEAR ANALYSES AT PRANDTL NUMBER $P = 1$:
TESTING THE MOMENTUM TRANSPORT MECHANISMS

In the Spring of 1975 I began running a nonlinear spherical shell convection model to try to find out, among other things, what kind of differential rotation is produced. I am able to give some preliminary results. In the model calculations, the nonlinear Boussinesq convection equations are solved numerically as an initial value problem, using a grid of points in the meridian plane (latitude-radius section) with all variables represented by Fourier series in longitude. Typically, up to 7 different longitudinal modes were retained, most often longitudinal wave numbers $m = 0, 4, 8, 12, 16, 20, 24$, in order to span the most unstable convective modes (denser arrays of wave numbers are much more time consuming on the computer, and can be done only for a few cases).

The calculations were done for a fluid layer 20% of the outer radius, and at first we focused on calculations with the Prandtl number $P = 1$, as well as stress free, fixed temperature upper and lower boundary conditions.

The basic regimes of convection which occur are described in Figure 5. The stability boundary below which no convection occurs is given by the lower solid curve. Above this line, but below the next one, a spectrum of convection occurs (that

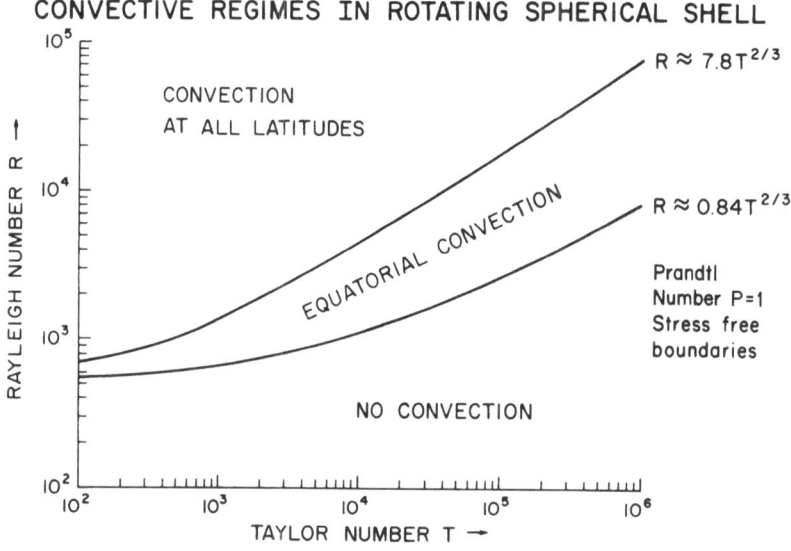

Fig. 5.   Simplified regime diagram showing convection occurrence as function of Rayleigh number R and
Taylor number T for Prandtl number $P = 1$.

is, many modes, each with different m, are needed to adequately describe it) but is confined to equatorial latitudes. Above the second curve, the polar modes come in, and convection occurs at all latitudes, though at high Taylor number, it is definitely concentrated near the poles for Rayleigh numbers near the line. Substantially above the second line, the heat flux is more nearly equal at low and high latitudes, but some variations remain.

What about the differential rotation that is produced? Figure 6 summarizes this. We have found that in general, equatorial regions rotate faster in the nonlinear case only when the convection itself is confined to equatorial regions. Even here, as we increase R for a given T, a deceleration near the equator starts to appear. For R high enough that convection is occurring at all latitudes, there is a strong tendency for high

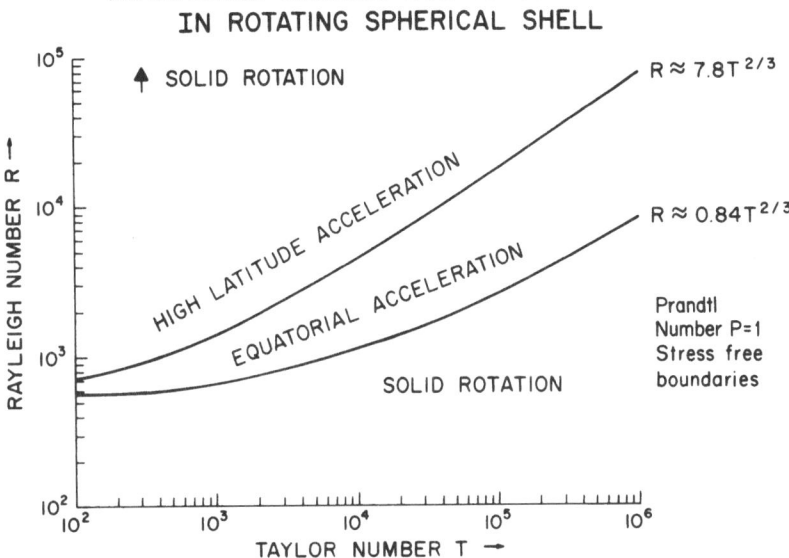

Fig. 6.   Rotational regimes corresponding to convection regimes in Figure 5, as deduced from integrations of the nonlinear spherical shell convection equations.

latitudes to rotate faster than low, exactly the opposite of the result we want for the Sun. So at least at Prandtl number $P = 1$, the price paid for equalizing the heat flux at low and high latitudes is the destruction of equatorial acceleration.

Let me give a few examples of the more detailed results, and explain what causes the effect.

Let us consider a set of integrations at $P = 1$, $T = 10^5$, for R between 4000 and 80 000, for stress free top and bottom, which spans the regime diagram. At the beginning, the fluid is in solid rotation and the temperature field is given by conduction alone. Then essentially arbitrary temperature perturbations are added which grow if the fluid is convectively unstable. At $R = 4000$ the resulting convection is confined to equatorial regions, has a spectrum which peaks rather sharply near the most unstable mode. It drives a very simple differential rotation, essentially constant on cylinders, shown in Figure 7. So at this R, the rotation falls off nicely with latitude and decreases inward. A weak meridional circulation with the profile shown in Figure 8 is also produced, including poleward flow near the surface, equatorward flow underneath. The differential rotation profile is maintained by equatorward and radially outward angular momentum flux in the convection cells, as expected from the linear results quoted earlier. Momentum transport by meridional circulation is too weak to alter this. The total heat flux peaks at the equator, since the convection is confined to low latitudes.

Now, as the Rayleigh number R is increased, so the convection becomes stronger, the differential rotation produced begins to evolve substantially. Figures 9–12 show the differential rotation for $R = 10^4$, 2×10^4, 4×10^4, 8×10^4, respectively. At $R = 10^4$ (Figure 9) we see that the maximum velocity has moved away from the

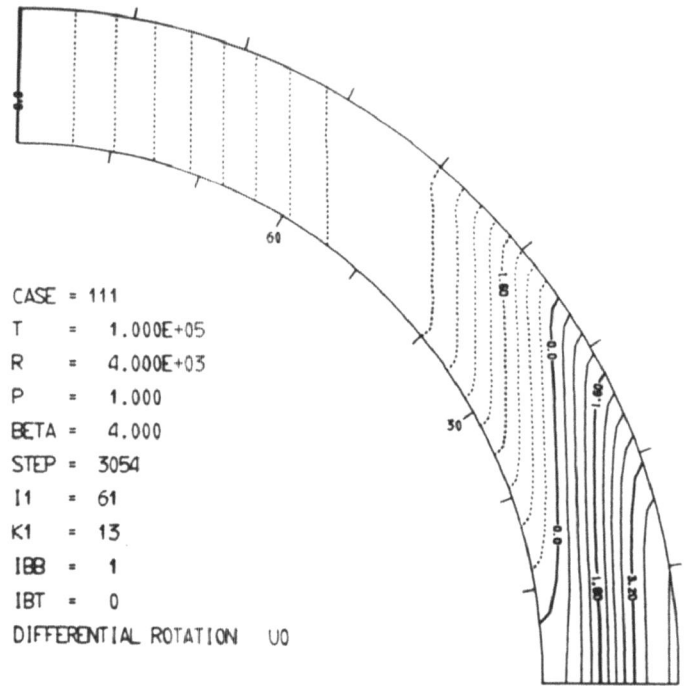

Fig. 7. Computer produced meridional cross section of linear rotation velocity u_0 relative to uniformly rotating frame, for Taylor number $T = 10^5$, Rayleigh number $R = 4 \times 10^3$, Prandtl number $P = 1$. Units dimensionless, with velocity scaled by κ/d, in which κ is thermal diffusivity, d is convection zone depth. Positive u_0 indicated by solid contours; negative u_0 by dashed contours.

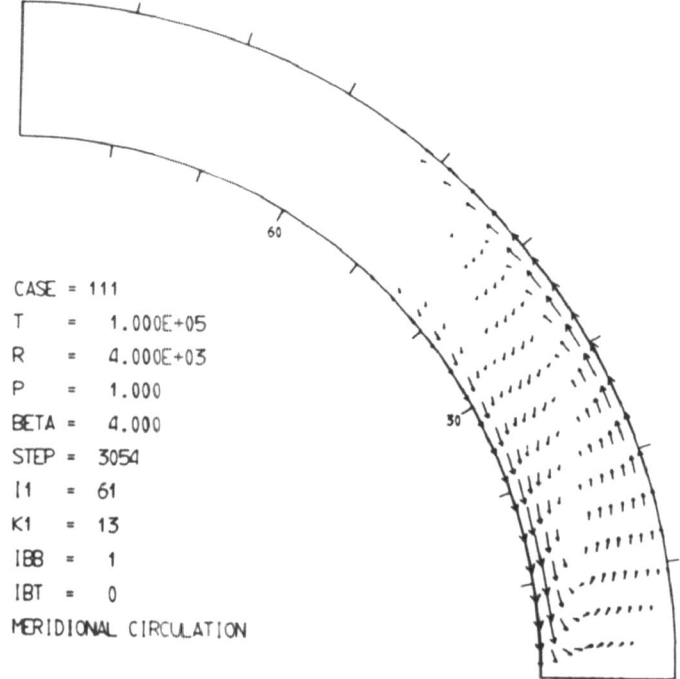

Fig. 8. Meridional circulation vectors for same case as Figure 7.

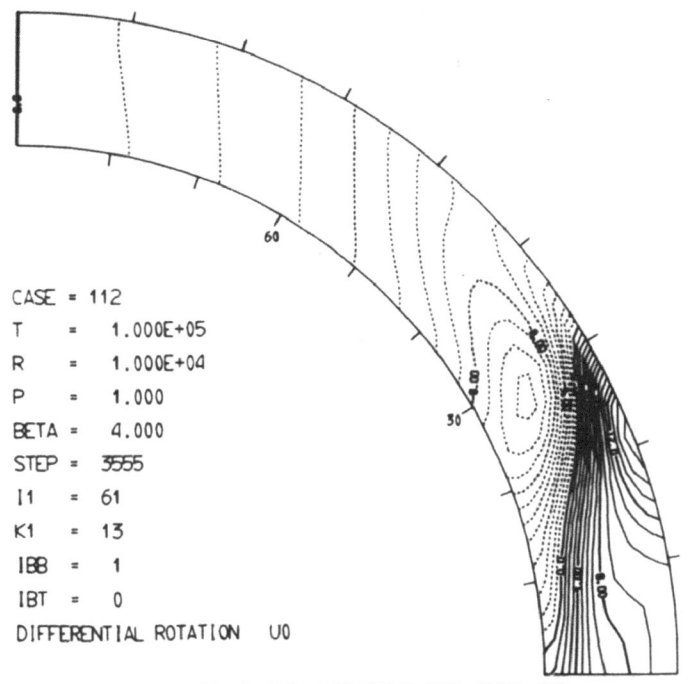

Fig. 9. Typical differential rotation profile for Rayleigh number R increased to 10^4.

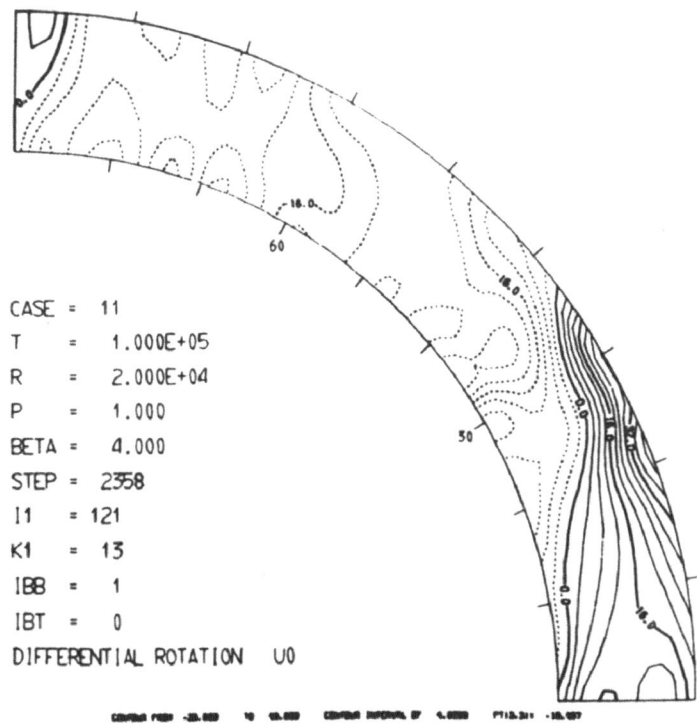

Fig. 10. Typical differential rotation for Rayleigh number R raised to 2×10^4.

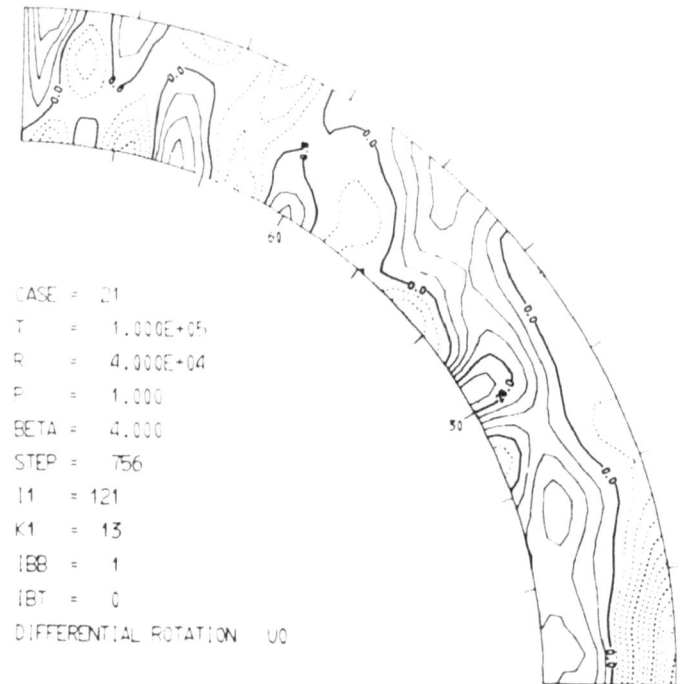

Fig. 11. Typical differential rotation for Rayleigh number R raised to 4×10^4.

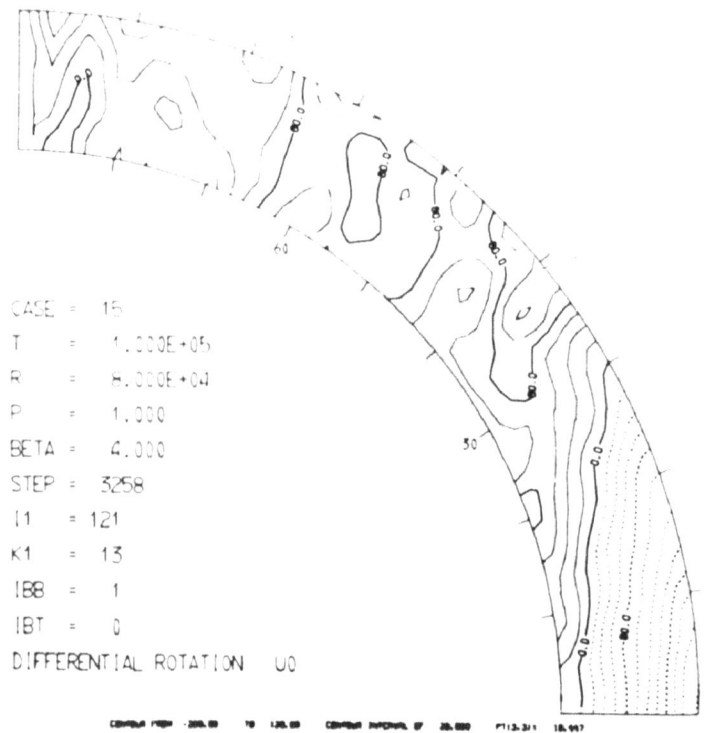

Fig. 12. Typical differential rotation for Rayleigh number R raised to 8×10^4.

equator to about 20° north, so a slight equatorial deceleration has appeared. By $R = 2 \times 10^4$ (Figure 10) this effect is more pronounced and by $R = 4 \times 10^4$ (Figure 11) the relative velocity has actually changed sign. By $R = 8 \times 10^4$, the profile is again nearly cylindrical, but with angular velocity *increasing* with latitude up to about 60°, as well as increasing with depth.

The change in rotational velocity is so pronounced that if one plots the total angular momentum of the system (the coordinate system + flow relative to it) as a function of latitude (Figure 13), it is nearly constant with latitude up to about 30° (the curve u_0 in Figure 13). In other words, the angular momentum, originally that of solid rotation (curve c in Figure 13) has been mixed in low latitudes so much its gradients are wiped out. This is, of course, exactly the opposite of what apparently occurs on the Sun. At the lower values of R for which we do get an equatorial acceleration, its magnitude is significantly smaller than needed for the Sun, although the $R = 2 \times 10^4$ profile comes within a factor of two.

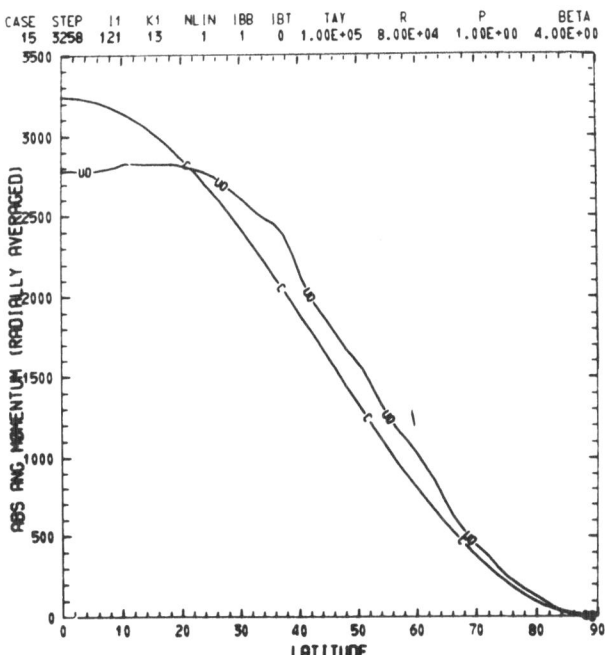

Fig. 13. Radially averaged angular momentum in convecting layer, for $T = 10^5$, $R = 8 \times 10^4$, $P = 1.0$ case. Curve c: angular momentum of rotating reference frame ($= \frac{1}{2}PT^{1/2} r^2 \cos^2 \phi$, radially averaged in which r is dimensionless radius, and ϕ is latitude). Curve u_0: same total angular momentum as curve c but redistributed according to profile of relative velocity u_0 which is built up by convection and meridional circulation ($= (\frac{1}{2}PT^{1/2}r \cos \phi + u_0)r \cos \phi$, radially averaged).

Three processes have conspired to produce this result. First, the meridional circulation has grown very strong and carries angular momentum to higher latitudes. Second, the radial angular momentum transport has changed sign as R increased, so it decelerates the outer layers near the equator by producing net inward flux. Third, the equatorward momentum flux by the cells, so dominant at low R, has been essentially destroyed by the nonlinear interactions. All three changes are due to nonlinear effects breaking the rotational constraints.

In so doing, the total heat flux has become much more nearly equal in latitude, but the differential rotation is now quite unsuitable for the Sun. Furthermore, the meridional circulation, which for $R = 4000$, was very small compared to the differential rotation amplitude, is of equal magnitude by $R = 4 \times 10^4$, which is much too large for the Sun.

Increasing the Taylor number does not help, because it only raises the Rayleigh number required to get convection at all latitudes, at which point the angular momentum again becomes mixed.

At smaller Taylor numbers, a similar effect simply occurs at much smaller Rayleigh number. For example, at $T = 10^3$, angular momentum is well mixed in the lowest 30° by $R = 4800$. The low Taylor number solutions have the additional disadvantage that at all R, forcing cells are very large in amplitude compared to the differential rotation they force, which is apparently not the case on the Sun. Therefore, nonlinear results obtained by Busse (1973) and others as an expansion in small amplitude and small Taylor number do not represent the dominant processes and results as R is increased significantly above what is required for convection to occur.

5.5. THE SEARCH FOR BETTER ANALOGUES TO THE SUN

Given that the solutions described above are not nearly as good an analogue of the Sun as we would like, what conditions or parameters should we change to find better ones? One possibility we have tried is to change the boundary condition on velocity at the bottom of the convecting layer to be nonslip rather than stress free. Gierasch (1975) has argued this may be a more appropriate boundary condition for the Sun. What this does is to impose the angular momentum distribution of the solidly rotating coordinate system on the flow, and in effect makes the interior below the bottom boundary an infinite source or sink of momentum, rather than allowing no momentum to pass through, which is what the stress free condition does. We have run a few cases to compare the differential rotation produced and found, as we should expect, that the nonslip condition does inhibit the angular momentum mixing in latitude somewhat, but the basic tendency is still there.

The basic problem with the $P = 1$ solutions is that in order to get convection occurring at all latitudes in the model, i.e., excite polar as well as equatorial modes, the Rayleigh number must be raised so high that the rotational constraint in low latitudes is effectively broken, and equatorial deceleration results. As indicated in Gilman (1975), however, it appears that for $P \ll 1$, convection should set in at much more nearly the same Rayleigh number at all latitudes, because the polar modes become overstable (growing oscillations) at relatively low R. Because the linear solutions oscillate, finding the linear stability boundary by numerical integrations is not easy, but nonlinear calculations can still be done. We have very recently done a few of these, for P down to 0.01 using as initial conditions results at the same T, R, but higher P. While much more careful analysis is needed, preliminary indications are that it should be possible to find some solutions which give amplitude and structure of differential rotation more nearly like those of the Sun, as well as more nearly uniform heat flux.

For example, as seen in Figure 14, solutions at $P = 0.01$ and $R = 10^4$, $T = 10^6$ (with $m = 0, 4, 8, 12, 16, 20, 24$) show an excess of angular momentum *near* the equator above that of solid rotation rather than mixed as we found for the $P = 1$ solutions. In addition, the solutions have the desirable properties (from the point of view of the Sun) that the meridional circulation remains small compared to the differential rotation, and the convective velocities themselves are also relatively small. In addition, very low longitudinal wave numbers seem to dominate in the convective spectrum, which may relate to very large scale solar features such as sectors (and may make it possible to describe the convection with a relatively small number of modes).

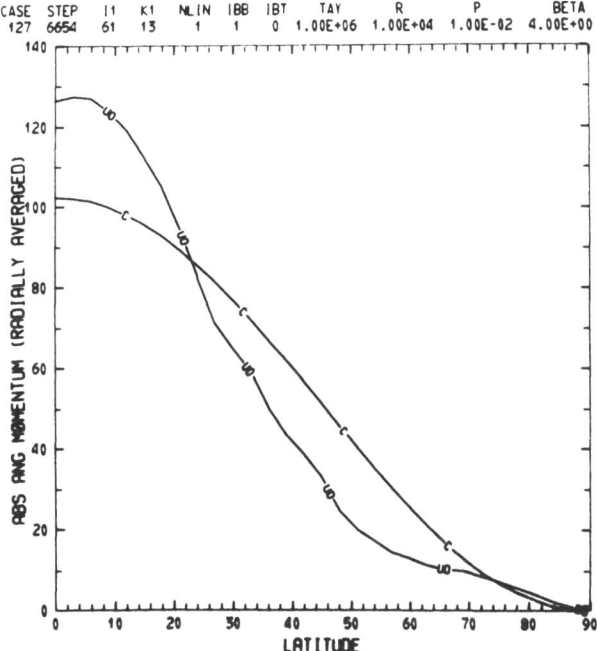

Fig. 14.   Same as Figure 13 except for $P = 0.01$, $T = 10^6$, $R = 10^4$.

One caution is, however, that the uniqueness of the solutions for different initial conditions is in doubt. We did a few calculations at low P starting from a state of solid rotation and more arbitrary initial temperature perturbations. The resulting differential rotation is much smaller in amplitude than found from initial conditions from solutions at higher P; and moreover does not include an equatorial accelera-tion. Clearly, the presence or absence of shear in the initial conditions is important.

Now, the *molecular* Prandtl number for the Sun is very small compared to 1, but it is not clear how to justify a small P for such large-scale convection as we are dealing with, since the small scale turbulence should be taken into account. It is true that with several interacting modes present in the model at once, we are in effect including a nonlinear viscosity, which may suffice to represent this process.

5.6. FURTHER MODEL STUDIES AND IMPROVEMENTS

Clearly, the next step in the context of the Boussinesq model is to look carefully at the small Prandtl number nonlinear solutions and to also look carefully at the effect of the presence or absence of differential rotation in the initial conditions, even for $P = 1$. Assuming we find reasonable analogues for the solar problem, their dynamo properties should then be tested; we hope to generalize our model to do this within six months. Finally, of course, we need to modify the model to include a number of new physical effects, the principal one being some approximation to the large density variation with radius characteristic of the real compressible convection zone of the Sun.

Acknowledgments

I wish to thank Mr Jack Miller for his very careful programming of all computations reported here, and Professor Douglas Gough for offering useful comments on the manuscript.

References

Busse, F.: 1970a, *Astrophys. J.* **159**, 629.
Busse, F.: 1970b, *J. Fluid Mech.* **44**, 441.
Busse, F.: 1973, *Astron. Astrophys.* **28**, 27.
Durney, B.: 1970, *Astrophys. J.* **161**, 1115.
Durney, B.: 1971, *Astrophys. J.* **163**, 353.
Gierasch, P. J.: 1975, *Astrophys. J.* **190**, 199.
Gilman, P. A.: 1972, *Solar Phys.* **27**, 3.
Gilman, P. A.: 1975, *J. Atmospheric Sci.* **32**, 1331.
Hoel, P. G.: 1947, *Introduction to Mathematical Statistics.* John Wiley, New York, 331 pp.
Howard, R. and Harvey, J.: 1970, *Solar Phys.* **12**, 23.
Roberts, P. H.: 1968, *Phil. Trans. Roy. Soc.* (London) Series A, **263**, 93.
Schwarzschild, M.: 1975, *Astrophys. J.* **195**, 137.
Simon, G. W. and Weiss, N. O.: 1968, *Z. Astrophys.* **69**, 435.
Ward, F.: 1965, *Astrophys. J.* **141**, 534.
Wolff, C. L.: 1975, *Solar Phys.* **41**, 297.
Yoshimura, H.: 1971, *Solar Phys.* **18**, 417.
Yoshimura, H. and Kato, S.: 1971. *Publ. Astron. Soc. Japan* **23**, 57.
Yoshimura, H.: 1972, *Solar Phys.* **22**, 20.
Yoshimura, H.: 1974, *Publ. Astron. Soc. Japan* **26**, 9.

DISCUSSION

Weiss: First let me comment on the dual role played by viscosity. As well as providing friction, viscosity governs diffusion of angular momentum. In the Boussinesq approximation, a low Prandtl number is necessary to permit redistribution of angular momentum and allow a finite amplitude instability. So it may be necessary to introduce a low Prandtl number in order to permit physical processes that we might expect in the Sun. Hence, it is not so terrible if you need a turbulent Prandtl number less then unity.

Secondly, have you tried using a fixed flux (rather than a fixed temperature) thermal boundary condition at the base of the convecting layer?

And how large a variation of heat flux with latitude appears in your models?

Gilman: We have not tried constant heat flux boundary conditions but what should occur is that large latitudinal temperature gradients on the boundaries appear. When equatorial modes dominate, the differential in heat flux can be as large as a factor of 3.

Stepanov: Is it possible to get accordance with observations putting convection zone depth a function of latitude?

Gilman: I do not know, but it is an interesting possibility, one that has occurred to us. It seems unlikely the unstable convection layer depth could vary much with latitude, but perhaps the amount convection overshoots into the interior does.

Vandakurov: It is known that the solar magnetic field which is extended over a great distance from the Sun and which in all probability is rooted in the deep interior of the convection zone has rigid rotation. Do you not think that this experimental evidence is in contradiction to your theory? Is the convection determined by the rotation rather than by superdiabaticity of the medium in your case?

Gilman: The model is not accurate enough yet to say under what circumstances we should expect solid rotation at the bottom of the convecting layer. Also compressibility may be very important for this particular question, since the inertia of the bottom layers becomes so high compared to the top, contrary to the incompressible case. Finally, I note that large scale magnetic and velocity disturbances can still rotate rigidly even in the presence of a differential rotation.

Superdiabaticity is required for the convection to occur, but rotation modifies it significantly.

Deinzer: Can you incorporate a core rotating faster than the convective envelope? I am asking because this is perhaps a possibility to get the observed butterfly diagram despite an angular velocity decreasing inward through the convection zone.

Are there other possibilities to get the desired butterfly diagram?

Gilman: If we impose a fast-rotating core and nonslip bottom boundary, the convecting layer will spin up to that rate. No really large radial gradient can be maintained. Weak radial gradients in angular velocity are possible. The dynamo properties of the convection and differential rotation in the model must be explicitly calculated to answer the question of whether the right butterfly diagram is produced.

Newkirk: You suggest that a way out of the dilemma of the reversal of equatorial acceleration might be to introduce a low Prandtl number into the model after the convection has been established. What would be the physical origin of such a temporal change?

Gilman: The point really is that if the initial condition contained significant differential rotation, it may remain indefinitely, and be different from what would be produced starting from initial conditions with solid rotation. Our calculations indicate this happens at low P, but it may also happen at $P = 1$. This needs to be tested.

Stenflo: I think there is strong evidence from observations of the rotation of solar magnetic fields that the angular velocity increases with depth. The magnetic fields should reflect the situation in deeper layers where they are anchored or from which they are expelled, and they systematically rotate faster than the surface layers.

Gilman: This is one reasonable interpretation of the observations but it is also true that hydromagnetic disturbances can and do propagate at different rates from the flow in which they are embedded.

Roxburgh: I would like to question the case of the Boussinesq approximation for modelling the solar convective zone. In the eddy transport approximation it is entropy that is conducted, not temperature as you assume in your calculations. If you use the convection of entropy then the Rayleigh number is not necessarily large, the convective zone may be stable against the large scale convection, or perhaps marginally unstable. Indeed Unno showed some years ago that one can interpret the mixing length theory as the convective layer being marginally unstable.

Gilman: I agree we should include compressibility when possible, but the present model allows us first to gain an understanding of rotational effects in a more simple, but still relevant system. The change to entropy from temperature is not that troublesome, really. Your comment is really an indication of the limits of mixing-length theory, rather than evidence that giant cells should not exist in the solar convection zone.

Durney: I do not think that the solution of the heat flux problem lies in the choice of the Prandtl number. Furthermore it appears reasonable to consider the Boussinesq expression for the energy flux an approximation of the compressible mixing-length expression of convective heat flux. There is then no choice of the Prandtl number.

Schröter: After this interesting discussion between theoreticians now a question from an observer.

In your Figure 8 you show an example of meridional circulation with poleward motion on the surface and equatorward motion in deeper layers. We (Dr Wöhl and myself) observed in 1974 a predominant poleward motion of Ca^+-fine mottles. Now, my question is: What is the velocity of this poleward circulation at the surface, a tenth of a m s^{-1}, several m s^{-1}, or more? I should mention that the modern computer-controlled method in tracing solar fine structures as used by us is capable of measuring systematic motions above a few m s^{-1}.

Gilman: Typically a few meters per second and virtually always poleward flow at the surface when equatorial modes dominate. Such measurements as you mention will be extremely valuable.

Mestel: You remark the axisymmetric meridional circulation driven by the Reynolds' stresses persists in the limit of zero rotation. What determines the axis if $\Omega = 0$? Is this another possible example of different initial conditions determining different asymptotic states?

Gilman: The meridional circulation persists provided a vestige of mode selectivity, present when rotation is small but not zero, remains. Indeed, initial conditions may be very important in determining the final state.

Howard: How will you handle temporal changes in the latitude dependence of the rotation within the content of your model?

Gilman: The calculations are themselves done as a function of time, so time variations, due to interactions among modes, will occur naturally. We intend to study the time dependent behavior more carefully soon.

THE PATTERN OF CONVECTION IN THE SUN

N. O. WEISS

Dept. of Applied Mathematics and Theoretical Physics, University of Cambridge, England

Abstract. The structure of solar magnetic fields is dominated by the effects of convection, which should be incorporated in any model of the solar cycle. Although mixing length theory is adequate for calculating the structure of main sequence stars, a better description of convection is needed for any detailed dynamo model. Recent work on nonlinear convection at low Prandtl numbers is reviewed. There has been some progress towards a theory of compressible convection, though there is still no firm theoretical evidence for cells with scales less than the depth of the convecting layer. However, it remains likely that the pattern of solar convection is dominated by granules, supergranules and giant cells. The effects of rotation on these cells are briefly considered.

1. Introduction

Over the past decade our understanding of the Sun (and particularly of small scale photospheric features) has been transformed by a wealth of detailed observations. Theoreticians lag behind observers but it has become clear in the last few years that progress requires detailed calculations rather than qualitative, order-of-magnitude arguments. Although few of these calculations have yet been carried out it is at least possible to outline a sequence of increasingly complicated problems that must be solved if we are to explain what has been observed. We have attained the stage that geographers had reached by 1500. The age of fantasy is over: now we can map out areas of ignorance and have to determine programmes of systematic exploration.

Solar magnetic fields are intimately related to the pattern of convection in the outer layers of the Sun. A full description of the structure of convection is necessary for any theory of the solar cycle. In particular, we need to supply a detailed velocity field that can be fed into more sophisticated kinematic dynamo models. Astrophysical convection has recently been reviewed by Spiegel (1971b, 1972). Here I shall first outline our limited understanding of the problem and then indicate what progress is being made towards a proper theory. Laboratory convection is still poorly understood and in stars further difficulties are posed by compressibility – the density changes by a factor of 10^5 over the Sun's convective zone – and rotation; since Dr Gilman and Dr Durney have already discussed the latter in some detail I shall concentrate on compressibility and its effect on the characteristic scale of convective motion. Finally, I shall attempt to provide what seems to me the best available description of the pattern of solar convection and its interactions with magnetic fields.

2. Mixing Length Theory

Mixing length theory, as developed by Biermann 40 years ago and elaborated since, still provides the only quantitative method of relating the temperature gradient to the heat flux in a stellar convective zone. Consider a plane horizontal layer, referred to

Bumba and Kleczek (eds.), Basic Mechanisms of Solar Activity, 229–242. All Rights Reserved.

cartesian coordinates with the z-axis pointing downwards in the direction of the gravitational acceleration **g** and a temperature gradient

$$\frac{\mathrm{d}T}{\mathrm{d}z} = \beta_0 > \beta_{\mathrm{ad}},\tag{1}$$

where β_{ad} is the adiabatic gradient. Then the degree of superadiabaticity is measured by the dimensionless Rayleigh number

$$R = \frac{g\alpha\beta l^4}{\kappa\nu},\tag{2}$$

where l is a characteristic length scale for convection, $\beta = \beta_0 - \beta_{\mathrm{ad}}$, α is the coefficient of thermal expansion ($1/T$ for a perfect gas) and κ, ν are the thermal and viscous diffusivities. When $\kappa \gg \nu$ it is convenient to introduce a modified Rayleigh number

$$S = \frac{g\alpha\beta l^4}{\kappa^2} = pR,\tag{3}$$

where the Prandtl number

$$p = \nu/\kappa.\tag{4}$$

The efficiency of convection is given by the Nusselt number

$$N = \frac{(\text{Total heat flux}) - c_p\rho\kappa\beta_{ad}}{c_p\rho\kappa\beta},\tag{5}$$

which is a dimensionless measure of the superadiabatic heat flux, where ρ is the density and c_p the specific heat at constant pressure. The aim of a convection theory is to predict N as a function of R and p.

To derive mixing length theory in its simplest form, let us consider vigorous convection, with eddies of a characteristic length scale l, so that $N \approx F/c_p\rho\kappa\beta$, where F is the convective heat flux. If w is the upward vertical velocity and θ the temperature excess of a rising fluid element then

$$F \sim c_p\rho w\theta.\tag{6}$$

For rapid convection, heat losses through (radiative) diffusion can be ignored and so the potential temperature variation

$$\theta \sim \beta l.\tag{7}$$

The velocity can be estimated by balancing the rate of working of the buoyancy force against the rate of dissipation of energy through the nonlinear inertial term in the equation of motion:

$$\rho g\alpha\theta \sim \rho w^2/l.\tag{8}$$

Then, from (7) and (8), the reduced free fall velocity

$$w \sim (g\alpha\beta)^{1/2}l \qquad\qquad (9)$$

and the flux

$$F \sim c_p\rho(g\alpha\beta^3 l^4)^{1/2} . \qquad\qquad (10)$$

Finally, from (6),

$$N \sim (pR)^{1/2} = S^{1/2} . \qquad\qquad (11)$$

This result holds for any local Boussinesq theory of convection, whether it is expressed in terms of bubbles, cells or eddies. To calculate the heat flux it is only necessary to calibrate various constants, all of which can be absorbed into the mixing length l.

In practice, l is set equal to a multiple of the local pressure (or density) scale height and the arbitrary constant is calibrated by ensuring that a solar model, with a given metal abundance, evolves from the zero age main sequence to the Sun's present radius and luminosity in its known lifetime of 4.7×10^9 yr. This procedure can be followed for Biermann's simple theory (described above) or for the more elaborate local theories of Öpik and Böhm-Vitense, which allow for lateral radiative losses. When this is done, all local theories give the same depth, around 150 000 km, for the solar convection zone and also provide consistent models of lower main-sequence stars (Gough and Weiss, 1976). (Mullan's (1971) application of Öpik's theory employs arbitrary constants with values that reduce the efficiency of convection but are incompatible with evolution of the Sun.)

It is not difficult to devise descriptions of convection that are adequate for calculating the structure of stars like the Sun. (Radiative losses are significant only in a very shallow photospheric zone; of course, these losses are important at levels where the granulation is observed (Travis and Matsushima, 1973; Spruit, 1974) but all available theories of convection are too crude to be valid in this region.) It is only necessary to calculate the jump in entropy across a narrow region, about 1000 km deep, immediately below the photosphere, where the temperature gradient is strongly superadiabatic. Below this region, the stratification is virtually adiabatic throughout the convective zone. Unfortunately, this simple description no longer holds for red giants (Schwarzschild, 1975; Gough and Weiss, 1976), nor can it be used to predict the detailed structure of convection in the Sun.

Mixing length theory depends on the assumption that there is a characteristic local length scale l, related to some local scale height. Provided the viscosity is small, the theory then predicts that the Nusselt number $N \propto S^{1/2}$ and is independent of ν. Both this result and the underlying assumption need to be verified. However, it is difficult to compare the theory with laboratory experiments, which are dominated by the effects of thermal boundary layers. Moreover, few experiments have so far been carried out at low Prandtl numbers ($p = \frac{1}{40}$ for mercury but, owing to radiative diffusion, $p \approx 10^{-9}$ in the Sun) and significant density variations across the layer cannot be reproduced in an experiment. Hence any improvement must rely on theory and, for such a nonlinear problem, on numerical experiments. Various idealized problems have been studied over the last few years and the following sections describe the progress that has so far been made.

3. Low Prandtl Number Convection

It is generally accepted that in a fluid with a low Prandtl number kinetic energy is dissipated through some turbulent process and that the heat transport should not depend explicitly on the viscosity. Thus N should depend not on the Rayleigh number but on the product $S = pR$. If heat transport in the boundary layers is laminar and limited by the thermal diffusivity then $N \propto S^{1/3}$; if the heat flux F is independent of κ also then $N \propto S^{1/2}$ as in Equation (11) (Spiegel, 1971a, b). However, there is as yet no firm theoretical basis for this belief.

Numerical experiments on Boussinesq convection between free boundaries in two-dimensional rolls showed that at high Reynolds numbers N is proportional to $R^{0.36}$ and independent of Prandtl number as $p \to 0$ (Veronis, 1966; Moore and Weiss, 1973). It might appear that three-dimensional geometry, which introduces an asymmetry between hot fluid rising at the centre of a cell and cold fluid sinking at its periphery, would cause a reduction in N as p decreases for a fixed Rayleigh number. Indeed, Gough et al. (1975) studied a simple model of nonlinear convection in which the horizontal ground plan of the cells was specified (the modal approximation) and found that $N \sim (S \ln S)^{1/5}$ for $p \le 1$, $S \gg 1$. However, computations of steady laminar convection in an axisymmetric cylindrical cell showed no such effect: N depended on R only and was independent of p as $p \to 0$, as for two-dimensional rolls, so the heat transport did depend on the viscosity (Jones et al., 1976).

What happens in these numerical experiments is that fluid in the cell turns over many times, gradually picking up speed as it does so. Although viscosity is slight, the frictional force increases with the speed until it is eventually able to balance the buoyancy force and an equilibrium is reached. This flywheel mode of convection has velocities much greater than the free-fall velocity in (9) and can therefore carry far more energy than a cell that only turns over once. Are such flywheels likely to be realized or will their growth be limited by some instability?

One possibility is that cells become unstable to non-axisymmetric perturbations and split up into segments (Jones, 1975). Indeed, inspection of any photograph of the photospheric granulation shows a number of exploding granules (Musman, 1972) that resemble the unstable vortex rings described by Widnall and Sullivan (1973) and Widnall (1975). However, it is also likely that the flywheels suffer from some collective instability that limits the growth of the velocity. Such an instability would allow adjacent vortex rings to merge and disappear, like opposing magnetic fields at a current sheet. Energy would then be dissipated neither by laminar viscous friction nor through an inertial range of eddies (as in homogeneous turbulence) but locally in spasmodic bursts.

To describe such a process requires a fully three-dimensional calculation. For the moment, we must assume that more sophisticated - and considerably more expensive – time-dependent computations would then demonstrate that the lifetime of a cell is of the order of the turnover-time $\tau \sim l/w$ (as seen in the solar granulation), that the velocity would be of the same order as that in (9) and that the heat flux might follow a law of the form predicted by mixing length theory in Equation (11).

4. Effects of Compressibility

The density gradient in the solar convective zone introduces an asymmetry between upward and downward motion. Rising elements of fluid expand and dominate convection; sinking elements contract and can disappear into the interstices between the rising columns (Schwarzchild, 1961). In a Boussinesq fluid, cells with fluid rising or sinking at their centres are equally probable but this degeneracy is removed by non-Boussinesq effects. In air (where the predominant effect is the increase of viscosity with temperature) motion is downward at the centres of convection cells. In the Sun the density gradient is important, favouring upward motion in the core. This asymmetry is of course observed, and is essential for the α-effect in solar dynamos (Steenbeck *et al.*, 1966).

Mathematically, the dominance of rising and expanding gas is expressed through the continuity equation

$$\frac{\partial \rho}{\partial t} = -\nabla \cdot (\rho \mathbf{u}) \tag{12}$$

where $\mathbf{u}$ is the velocity. Provided that convective motion is slow compared with the sound speed c_s (so that the Mach number $M = |\mathbf{u}|/c_s \ll 1$) we may filter out acoustic waves by adopting the anelastic approximation (Gough, 1969), which is a generalization of the familiar Boussinesq approximation. Equation (12) then simplifies to

$$\nabla \cdot (\bar{\rho} \mathbf{u}) = 0, \tag{13}$$

where $\bar{\rho}(z)$ is a horizontally averaged density: the flow is constrained by the mean density variation.

Suppose now that convection occurs in cells with a horizontal dimension L. We expect that L is comparable with the density scale height

$$H = -(\mathrm{d} \ln \bar{\rho}/\mathrm{d}z)^{-1} \tag{14}$$

at some level $z = z_0$ near the base of the cell. But if the cell penetrates to a level $z \ll z_0$, the local scale height H may be small compared with L. From (13) we see that $\bar{\rho} u/L \sim \mathrm{d}(\bar{\rho} w)/\mathrm{d}z$, where u is the horizontal component of $\mathbf{u}$. If, for the moment, we neglect the variation of w with z then

$$u \sim \left(\frac{L}{H}\right) w. \tag{15}$$

Thus $u \gg w$ if $H \ll L$ and the buoyancy force is spent in driving rapid horizontal motions, which do not contribute to the heat transport. This inefficiency favours convection on a smaller scale. More precisely, we might expect local instabilities to develop, deriving their energy either from the sheared velocity or from the superadiabatic stratification itself. At some level these instabilities will grow fast enough to form smaller convection cells that are able to transport energy more effectively than the original, larger cells.

Somewhat more generally, we can consider a polytropic atmosphere, such that the horizontally averaged temperature and density are given by

$$\bar{T} = \beta_0 z, \qquad \bar{\rho} = \rho_0 z^m, \tag{16}$$

where the polytropic index $m = g/(\mathcal{R}\beta_0) - 1$ and $\mathcal{R}$ is the gas constant. If we suppose that $w \propto z^n$ then, from (13), $u \propto w/z \propto z^{n-1}$ and the horizontal component of the velocity increases upwards for $n < 1$. Moreover, if all the energy is carried by convection then, from (6),

$$\theta \sim \frac{F}{c_p \bar{\rho} w} \propto z^{-(m+n)} \tag{17}$$

and so the superadiabatic gradient $\beta \propto z^{-(m+n+1)}$. For the conventional mixing length theory of Section 2, with l everywhere proportional to the local scale height H, $n = -\frac{1}{3}m$, from (8). If the cell extends over many scale heights, so that dissipation of energy is dominated by horizontal motions, Equation (8) must be replaced by

$$\frac{u^3}{L} \sim g\alpha\theta w \sim \left(\frac{gF}{c_p}\right)\frac{1}{\bar{\rho}\bar{T}} \tag{18}$$

and it follows that $n = \frac{1}{3}(2 - m)$ and hence that $u \propto z^{-(m+1)/3}$. (In particular, for $m = \frac{3}{2}$, as in the deep solar convection zone, $w \propto z^{1/6}$ and $u \propto z^{-5/6}$, so that the vertical velocity varies only slightly while the horizontal velocity increases upwards.) The ratio of the superadiabatic gradient for a cell extending over many scale heights to that for a local cell is proportional to $z^{-2/3}$. Eventually, therefore, small scale convection should take over (Simon and Weiss, 1968).

This crude discussion needs to be supported by a proper calculation. Unfortunately, computations have so far provided no evidence for the existence of any vertical scale other than the layer depth, even in a compressible atmosphere. The solution of the linearized marginal stability problem for a polytropic atmosphere (Spiegel, 1965; Gough et al., 1976) shows that, for any cell width, instability occurs first for the fundamental vertical mode, with no internal zeros in the eigenfunctions θ and w. (Vickers' (1971) solutions showing a reversal in θ at the upper boundary are wrong, apparently owing to a numerical error in treating the boundary conditions.) Growth rates have been calculated for small perturbations to models of the convective zone computed using mixing length theory (Böhm, 1963, 1967; Vickers, 1971; Vandakurov, 1975): the highest growth rates are shown by small scale modes with greatest amplitudes near the surface, where the superadiabatic gradient is largest, but all modes extend throughout the region and there is no direct evidence for smaller cells. Of course, linear modes are solutions to a simplified problem. The nonlinear constraint of constant heat flux is not included (for example, both w and θ are relatively small near $z = 0$ in the eigenfunctions for an infinite polytropic atmosphere, so that (17) could not be satisfied). Do non-linear models allow multiple cells to develop in a compressible layer?

The only non-linear solutions are those of Graham (1975) for two-dimensional convection in a fully compressible atmosphere. His numerical experiments, with densities varying by a factor up to 10 (4 pressure scale heights) across the layer and Rayleigh numbers up to 100 times the critical value, all gave cells that filled the entire convecting region. To demonstrate the development of smaller cells it may be necessary to have a much greater density variation, or to proceed to time-dependent three-dimensional models. An alternative possibility is that Graham's results (like the linear solutions of Gough et al.) are affected by the assumption of a constant molecular viscosity $\mu = \rho\nu$. At the top of the layer, where the density is small, the viscous term dominates the equation of motion and (for stress-free boundary conditions) forces a horizontal velocity u that is independent of z. If u is constant, $w \propto 1/z$ and this rather unrealistic constraint may inhibit the growth of instabilities and so stabilize cells extending over many scale heights. Turbulent viscosity in the Sun is better represented by a constant diffusivity ν: calculations with $\mu \propto 1/\bar{\rho}$ might allow greater variation in u and so permit the development of other scales of motion (cf. Parker, 1973).

The pattern of cellular motion is not the only feature of compressible convection that is poorly understood. It is not obvious that the functional dependence of N on R and p will be the same as for a Boussinesq fluid: once the temperature scale height H_T becomes comparable with the layer depth the rate of working against pressure and viscous forces makes a significant contribution to the energy equation and the dissipation rate over a cell $(\bar{\rho}w^3 l^2)$ becomes comparable with the energy flux (Fl^2). From (9) and (10),

$$\bar{\rho}\frac{w^3}{F} \sim \frac{g\alpha l}{c_p} = \frac{l}{H_T} \tag{19}$$

and for a polytrope $H_T = mH$. Graham's (1975) numerical experiments already show many details and further computations are badly needed.

Fully compressible computations, especially in three dimensions, require vast amounts of computer time if it becomes necessary to follow sound waves in regions where the Mach number is small. Hence it seems advisable to use the anelastic approximation, with the simplified form (13) of the continuity equation and corresponding modifications to the momentum and energy equations (Gough, 1969). The modal approximation (with a fixed horizontal ground plan) has been adapted to the anelastic approximation by Latour et al. (1975) and used to compute the extent of overshooting from convective zones in A-type stars. Unfortunately this model does not allow the development of smaller cells. However, a two-dimensional anelastic code is being developed at Cambridge. This, combined with Graham's recent three-dimensional compressible calculations may allow us to carry out a systematic study of compressible convection.

5. Cellular Convection in the Sun

It is clear that there is no firm theoretical basis for assuming that energy is everywhere carried by cells with a scale comparable with the local density scale height, as in normal mixing length theory. The crude argument outlined above suggests that cells might extend over about three density scale heights (Simon and Weiss, 1968) but there are still no reliable calculations to support this estimate. On the other hand, observations show the presence of at least two distinct scales of motion at the surface of the Sun. Photospheric granules have a radius (1000 km) about twice the local density scale height. Supergranules are intimately associated with strong sub-photospheric magnetic fields and must therefore correspond to motion below the surface of the Sun; their radius (15 000 km) corresponds to the density scale height at about 15 000 km depth. Bumba (1967) inferred that there should be a third scale, around 150 000 km, corresponding to giant cells extending throughout the whole convective zone. There is some observational evidence for such giant cells from Doppler measurements of azimuthal velocities (Howard, 1971; Howard and Yoshimura, 1976) and the distribution of magnetic features. Indeed, it seems likely that their presence will be demonstrated from observations before we can succeed in providing a proper theoretical description.

No one has suggested that eddies far from a boundary should be limited by a length-scale smaller than the local scale height and there is general agreement that energy must be carried by large scale cellular motions over the bulk of the convective zone. There is less unanimity over the relation between these giant cells and the observed smaller scales of supergranules and granules. Spiegel (1968) has suggested that they are formed as a result of shear instabilities in thermal boundary layers. In a normal convecting layer, large-scale motions can easily carry energy except near the upper and lower boundaries, where an enhanced temperature gradient develops. In the Sun (or any star with a convection zone produced by ionization of hydrogen) the superadiabatic gradient is high over a narrow region near the surface, whose depth is comparable with the local scale height. At the base of the zone, convection becomes less efficient; Böhm and Stückl (1967), using a non-local mixing length theory, found an enhanced superadiabatic gradient over the bottom 30 000 km of the convective zone. One possibility is that the granules result from small scale turbulence in the upper superadiabatic boundary layer, while the supergranules are similarly generated in the lower boundary layer and somehow penetrate to the upper surface.

The alternative hypothesis, which I myself prefer, is that the giant cells develop smaller scale instabilities (supergranules) which take over the energy transport until they themselves become unstable, allowing granules to carry energy towards the solar surface. This description is an obvious oversimplification. The velocity pattern of supergranules, and apparently that of giant cells too, penetrates into the photosphere, though no corresponding temperature variation has been observed. (Horizontal temperature variations, unless they are associated with energy transport, seem to be eliminated over a height comparable with the local scale height H, presumably by rapid horizontal motions). Moreover, other scales of motion must also be present: at any level we might expect smaller parasitical eddies, carrying little energy but affecting, for example, the diffusion of magnetic fields. Indeed, small scale motions in a region where energy transport is dominated by supergranules appear to be necessary in order to explain the slow decay of sunspots (Meyer *et al.*, 1974). It is not yet possible to determine observationally whether there is a short wavelength cut-off in the photospheric velocity spectrum (Harvey and Schwarzschild, 1975) and the spectrum at the surface may be a distorted form of that at greater depths. Experiments by Busse and Whitehead (1974) on convection at high Rayleigh number in a viscous fluid showed the development of a semi-permanent large scale cellular structure within which irregular motions on a smaller scale (comparable with the layer depth) could be seen. If such a pattern were present on the Sun, and could be observed only indirectly through its effect on the magnetic field, only the large scale structure would be seen.

6. Effects of Rotation

The interaction between convection and rotation has been discussed in several reviews (Spiegel, 1972; Gilman, 1974, 1976) and some recent calculations have already been described by Durney (1976) and by Gilman (1976). The generally acepted recipe for a solar dynamo (Parker, 1955) has two essential ingredients, differential rotation and helicity. Helicity is generated by the Coriolis force, acting on cellular convection, and individual dynamo models prove sensitive to the assumed variation of the angular velocity Ω with position in the convective zone. I shall therefore attempt to summarize the possible effects of rotation on the cellular pattern that I have described above.

The importance of rotation in a convection cell can be estimated from the parameter

$$\sigma = 2\Omega l / w = 4\pi\tau/\tau_{\rm rot}, \tag{20}$$

where τ is the turnover time and $\tau_{\rm rot}$ the period of rotation. This parameter (the reciprocal of the Rossby number) measures the ratio of the Coriolis force to the inertial term in the equation of motion, and so the extent to which a fluid element, conserving its angular momentum, is deflected as it traverses a cell. For granules, with a lifetime of minutes, $\sigma \approx 5 \times 10^{-3}$; for supergranules, lasting for a day, $\sigma \approx 0.4$ and for giant cells, with a turnover time of a month, $\sigma \approx 20$. Hence any effect of rotation on granules must be imperceptible. Supergranules will be significantly affected and Coriolis forces will dominate the motion in any giant cell.

When σ is large, two different effects can be distinguished. The first is a consequence of the Proudman-Taylor theorem. In a uniformly rotating system, the rate of generation of vorticity by the Coriolis force is, in the anelastic approximation,

$$\nabla \wedge (2\bar{\rho}\Omega \wedge \mathbf{u}) = 2\Omega \cdot \nabla(\bar{\rho}\mathbf{u}), \tag{21}$$

which vanishes if $\bar{\rho}\mathbf{u}$ does not vary in the direction parallel to the axis of rotation. For a Boussinesq fluid, with $\bar{\rho}$ constant, the constraint imposed by rotation disappears provided $\mathbf{u}$ itself does not vary in this direction. Consider for the moment the simplified problem of convection in an infinite self-gravitating cylinder, rotating about its axis. If convection is everywhere in rolls parallel to the axis of rotation then the Coriolis force can be balanced by a pressure gradient and the motion is unaffected by rotation (except insofar as the density perturbation is coupled to the pressure through the equation of state) even in a compressible fluid.

Convection in a sphere is more complicated. Let (r, θ, φ) be spherical polar co-ordinates, with Ω along the axis $\vartheta = 0$. In a Boussinesq fluid, instability first appears as a ring of rolls parallel to the rotation axis (Roberts, 1968; Busse, 1970a) and Busse (1975) has demonstrated experimentally that this pattern persists into the non-linear regime. In a spherical shell the eigenfunctions of the marginal stability problem are specified by the spherical harmonics $P_n^m(\cos\theta)\,e^{im\varphi}$. In the absence of rotation, the problem is degenerate with respect to m; Busse (1970b, 1973) used a double perturbation expansion to show that rotation favoured the sectorial harmonics with $m=n$. These sectorial modes (banana cells) show the effect of the Proudman-Taylor constraint even with spherical geometry. Similar results for the non-linear regime were obtained by Durney (1970, 1971), using the mean field approximation, though Gilman's (1976) recent computations show a more complicated pattern of behaviour.

The second effect is the redistribution of angular momentum. The Coriolis force expresses the conservation of angular momentum; with differential rotation such that $r^2\sin^2\vartheta\Omega$ is everywhere constant, this constraint is relaxed. If the viscosity is sufficiently small, convection itself can alter the angular velocity distribution so that the angular momentum is nearly uniform except in narrow boundary layers. This redistribution, which allows the development of subcritical instabilities (Veronis, 1959), was found in two-dimensional computations by Veronis (1968). Weir's (1975, 1976) numerical experiments on axisymmetric Boussinesq convection in a sphere show large regions of constant angular momentum, with a sharp gradient near the axis, where Ω is finite but the angular momentum drops to zero.

The Sun's convective zone, with a thickness $\Delta r\approx0.2\,R_\odot$ can be divided schematically into two regions, separated by the cylindrical surface parallel to the axis of rotation that encloses the radiative zone. The equatorial region, spanning latitudes less than $35°$, resembles the cylindrical model discussed above: **g** and Ω are almost perpendicular, though $\bar\rho$ is no longer constant on cylinders. Convection should be dominated by the Proudman-Taylor constraint, favouring cells elongated parallel to the rotation axis. Motion in the plane perpendicular to this axis is effective at transporting heat and need not violate the constraint. In the polar region Ω and **g** are nearly parallel and $\Omega\cdot\nabla(\bar\rho\mathbf{u})$ cannot be small if convection is effective in radially transporting heat. (For the linear eigensolution with $m=n$ there is no convective heat flux at the poles.) In this region we might expect to find normal cellular convection, redistributing angular momentum in the radial direction, so that $r^2\Omega$ is approximately constant.

This simplified description suggests that giant cells in the equatorial region will be elongated, like a ring of truncated bananas. It is then tempting to identify this region with the sunspot zone (Hide, 1960, private communication) and to relate the elongated cells to the velocity variations described by Howard and Yoshimura (1976). The angular velocity in this region would be approximately uniform and the rotation period of the convection pattern (not necessarily equal to that of the gas itself) would then be the familiar sidereal period of 25 days. This model is consistent with the existence of active longitudes and the sector structure of magnetic fields. In the polar regions, giant cells would form a tesselated pattern and redistribution of angular momentum could reduce angular velocity at the surface by up to one third; this is consistent with the measured rotation period of 37 days at the poles. Of course, the two regions must merge smoothly together. It is clear from this brief discussion that a full description of solar convection requires nonlinear, nonaxisymmetric solutions for anelastic convection in a thick spherical shell, with a uniform heat flux maintained at the inner boundary and, say, a fixed temperature at the outer surface. Such an ambitious calculation must be approached by gradual stages. Gilman's (1976) computations already show the range of complicated behaviour that can result as more sophisticated models are investigated. It is only from systematic numerical experiments that convection, and its interaction with rotation, will ultimately be understood.

The chromospheric network does not vary between the poles and the equator, nor should the pattern of convection in supergranules be affected by rotation. Nevertheless, some redistribution of angular momentum will occur. This would suffice to account for the difference of 5% between the rotation rates of deep-seated magnetic features and those directly measured in the photosphere (Foukal, 1972; Foukal and Jokipii, 1975). Both in supergranules and in giant cells motion is dominated by rising columns at the centres of the cells, which spread outwards over most of their height. The rising and expanding gas will be acted upon by Coriolis forces, so as to produce cyclonic motions and the helicity needed to maintain a dynamo. In supergranules, where this effect is relatively small, we might expect the helicity $\overline{\mathbf{u} \cdot \nabla \wedge \mathbf{u}}$ to vary as $\cos \vartheta$ but the distribution of helicity in giant cells must be obtained from a numerical solution. In the equatorial region the α-effect would be reduced by the formation of elongated cells.

7. Cellular Convection and Magnetic Fields

The kinematic effects of convection on a weak magnetic field can be summarized briefly.

(i) Toroidal fields are drawn out from poloidal fields by differential rotation. In an axisymmetric configuration the local rate of production of the toroidal field is given by $\mathbf{B} \cdot \nabla \Omega$ and there are contributions from both radial and latitudinal components of the gradient of the angular velocity. The relative importance of these components varies with position for any choice of $\Omega(\mathbf{r})$. The discussion above suggest that $|\partial \Omega / \partial \vartheta|$ should be large near the boundary of the sunspot zone and that $-\partial \Omega / \partial r$ should be largest in the polar region.

(ii) Rising fluid expands and rotates, dragging up toroidal flux and generating from it a reversed poloidal field, at a rate depending on the local helicity.

(iii) Horizontal velocities rapidly concentrate the magnetic field between cells. This process is seen in high resolution observations of intergranular magnetic structures (Dunn and Zirker, 1973; Mehltretter, 1974; Stenflo, 1976). Magnetic flux is expelled from most of a convection cell, though the lifetime of an eddy is too short for flux to be eliminated from its centre.

(iv) Magnetic flux ropes, concentrated at the boundaries of convection cells, are brought sufficiently close together for reconnection to occur, with the annihilation of oppposing fields. The reconnection process itself must be dynamically driven, at a velocity comparable with the Alfvén speed (Priest and Soward, 1976).

(v) In three-dimensional convection cells, sinking fluid forms a continuous network while rising fluid is confined to an array of isolated columns. This topological difference allows the motion to pump horizontal field downwards, as demonstrated by Drobyshevski and Yuferev (1974). Topological pumping would not occur for elongated two-dimensional cells but should be effective in the polar region. Differences in horizontal velocities caused by compressibility could also concentrate fields at the base of the convective zone. This geometrical pumping has been investigated by Moore and Proctor (1976). Both these mechanisms act in the opposite direction to the well known buoyancy of magnetic flux ropes (Parker, 1955, 1975).

Strong magnetic fields are no longer passively distorted. The concentration of magnetic flux is limited by forces exerted by the field, though the details of this process are not yet properly understood. The concentrated fields are strong enough locally to hinder convection, and the α-effect is quenched either by excluding flux from regions of strong helicity or by suppressing the cyclonic motion. In addition, magnetic fields may affect the mean flow and, in particular, the extent of differential rotation.

All these processes must be included in a proper treatment of the solar cycle. As simplified models the kinematic dynamos of mean field electrodynamics seem convincing. Now we need to see more elaborate treatments, including large scale cellular motions and discontinuous flux ropes. As our understanding of convection gradually improves, the results should be incorporated into more sophisticated dynamo models, which may ultimately provide an accurate and detailed picture of magnetic fields in the Sun.

Acknowledgments

Some of the work described here was supported by a grant from the Science Research Council. The views expressed here have been influenced by discussion with many colleagues, in Cambridge and in Munich.

References

Böhm, K.-H.: 1963, *Astrophys. J.* **137**, 881.
Böhm, K.-H.: 1967, in R. N. Thomas (ed.), *Aerodynamic Phenomena in Stellar Atmospheres*, Academic Press, London, p. 366.
Böhm, K.-H. and Stückl, E.: 1967, *Z. Astrophys.* **66**, 487.
Bumba, V.: 1967, in P. A. Sturrock (ed.), *Plasma Physics*, Academic Press, London, p. 77.
Busse, F. H.: 1970a, *J. Fluid Mech.* **44**, 441.
Busse, F. H.: 1970b, *Astrophys. J.* **159**, 629.
Busse, F. H.: 1973, *Astron. Astrophys.* **28**, 27.
Busse, F. H.: 1975, *Geophys. J. Roy. Astron. Soc.* **42**, 437.
Busse, F. H. and Whitehead, J. A.: 1974, *J. Fluid Mech.* **66**, 67.
Drobyshevski, E. M. and Yuferev, V. S.: 1974, *J. Fluid Mech.* **65**, 33.
Durney, B. R.: 1970, *Astrophys. J.* **161**, 1115.
Durney, B. R.: 1971, *Astrophys. J.* **163**, 353.
Durney, B. R.: 1976, this volume, p. 243.
Foukal, P.: 1972, *Astrophys. J.* **173**, 439.
Foukal, P. and Jokipii, J. R.: 1975, *Astrophys. J.* **199**, L71.
Gilman, P. A.: 1974, *Ann. Rev. Astron. Astrophys.* **12**, 47.
Gilman, P. A.: 1976, this volume, p. 207.
Gough, D. O.: 1969, *J. Atmospheric Sci.* **26**, 448.

Gough, D. O. and Weiss, N. O.: 1976, *Monthly Notices Roy. Astron. Soc.* (in press).
Gough, D. O., Spiegel, E. A., and Toomre, J.: 1975, *J. Fluid Mech.* **68**, 695.
Gough, D. O., Moore, D. R., Spiegel, E. A., and Weiss, N. O.: 1976, *Astrophys. J.* (in press).
Graham, E.: 1975, *J. Fluid Mech.* **70**, 689.
Harvey, J. and Schwarzschild, M.: 1975, *Astrophys. J.* **196**, 221.
Howard, R.: 1971, *Solar Phys.* **16**, 21.
Howard, R. and Yoshimura, H.: 1976, this volume, p. 19.
Jones, C. A.: 1975, Ph.D. Dissertation, University of Cambridge.
Jones, C. A., Moore, D. R., and Weiss, N. O.: 1976, *J. Fluid Mech.* **73**, 353.
Latour, J., Spiegel, E. A., Toomre, J., and Zahn, J.-P.: 1976, *Astrophys. J.* (in press).
Mehltretter, J. P.: 1974, *Solar Phys.* **38**, 43.
Meyer, F., Schmidt, H. U., Weiss, N. O., and Wilson, P. R.: 1974, *Monthly Notices Roy. Astron. Soc.* **169**, 35.
Moore, D. R. and Proctor, M. R. E.: 1976 (in preparation).
Moore, D. R. and Weiss, N. O.: 1973, *J. Fluid Mech.* **58**, 289.
Mullan, D. J.: 1971, *Monthly Notices Roy. Astron. Soc.* **154**, 467.
Musman, S.: 1972, *Solar Phys.* **26**, 290.
Parker, E. N.: 1955, *Astrophys. J.* **121**, 491.
Parker, E. N.: 1973, *Astrophys. J.* **186**, 643.
Parker, E. N.: 1975, *Astrophys. J.* **198**, 205.
Priest, E. A. and Soward, A. M.: 1976, this volume, p. 353.
Roberts, P. H.: 1968, *Phil. Trans. Roy. Soc.* **A263**, 93.
Schwarzschild, M.: 1961, *Astrophys. J.* **134**, 1.
Schwarzschild, M.: 1975, *Astrophys. J.* **195**, 137.
Simon, G. W. and Weiss, N. O.: 1968, *Z. Astrophys.* **69**, 435.
Spiegel, E. A.: 1965, *Astrophys. J.* **141**, 1068.
Spiegel, E. A.: 1968, in L. Perek (ed.), *Highlights of Astronomy*, Reidel, Dordrecht, p. 261.
Spiegel, E. A.: 1971a, *Comm. Astrophys. Space Sci.* **3**, 53.
Spiegel, E. A.: 1971b, *Ann. Rev. Astron. Astrophys.* **9**, 323.
Spiegel, E. A.: 1972, *Ann. Rev. Astron. Astrophys.* **10**, 261.
Spruit, H. C.: 1974, *Solar Phys.* **34**, 277.
Steenbeck, M., Krause, F., and Rädler, K.-H.: 1966, *Z. Naturforsch.* **21a**, 369.
Stenflo, J. O.: 1976, this volume, p. 69.
Travis, L. D. and Matsushima, S.: 1973, *Astrophys. J.* **180**, 975.
Vandakurov, Y. V.: 1975, *Solar Phys.* **40**, 3.
Veronis, G.: 1959, *J. Fluid Mech.* **5**, 401.
Veronis, G.: 1966, *J. Fluid Mech.* **26**, 49.
Veronis, G.: 1968, *J. Fluid Mech.* **31**, 113.
Vickers, G. T.: 1971, *Astrophys. J.* **163**, 363.
Weir, A. D.: 1975, *Mem. Soc. Roy. Sci. Liège* (6) **8**, 37.
Weir, A. D.: 1976, *J. Fluid Mech.* (in press).
Widnall, S.: 1975, *Ann. Rev. Fluid Mech.* **7**, 141.
Widnall, S. and Sullivan, J.: 1973, *Proc. Roy. Soc.* **A332**, 335.

DISCUSSION

Giovanelli: Since there are differences in the convective behaviour in polar and equatorial regions, could you predict any observable differences to be expected between supergranules in these regions?

Weiss: Supergranules are only slightly affected by rotation, so I do not expect that their shapes or velocities would show any observable variation with latitude. Have any differences been observed?

Schröter: Since observers are asked to comment on Dr Giovanelli's question, I shall try to summarize our experiences during our observations last year and this year (Dr Wöhl and myself). As I reported yesterday, we had every day to preselect 10–15 fine Ca$^+$-mottles in different solar latitudes for our differential rotation program. My experience (and Dr Wöhl reported to me a very similar impression) is that there was no problem in finding well defined tiny Ca$^+$-mottles showing an arrangement similar to supergranules in latitudes near the equator. When searching for well defined bright Ca$^+$-mottles in medium latitudes, we had some problems. In rather high latitudes I easily found tiny bright Ca$^+$-mottles again, but this time they looked like single, not specifically aligned features. In interpreting this, please do not forget two facts:

(a) These observations refer to a time close to the solar activity minimum (e.g. the Ca^+-mottles close to the equator may well reflect the solar activity belt, the polar Ca^+-mottles the polar faculae, as investigated by Waldmeier).

(b) Our observations refer to Ca^+-structures and not to a velocity pattern which defines supergranules. However, we know the close correlation between both phenomena.

Stenflo: Supergranulation is defined by its velocity pattern, which is associated with magnetic fields and brightness enhancements. The network seen in magnetograms and Ca spectroheliograms varies strongly with latitude and with the solar cycle, but there seems to be no observational evidence that the velocity pattern associated with supergranulation varies with heliographic latitude.

Giovanelli: Some years ago Dr Beckers mentioned to me that he was unable to identify supergranule cells in the chromosphere well away from equatorial regions; this is certainly associated with magnetic differences between the two regions. As far as I am aware, there have been no differences observed in the line-of-sight velocities. Therefore, Dr Weiss' expectation seems to be confirmed

Roxburgh: Were you suggesting that there are steady convective cells extending over several scale heights and if so why does motion not become turbulently unstable since the Reynolds number is very high, of the order of 10^{13}.

Weiss: I certainly do not imagine that there are steady convection cells in the Sun. However, cells lasting for about one turnover time may extend over several scale heights without being prevented by shear instabilities, regardless of the Reynolds number.

Durney: I think that you said that Vickers' results were difficult to understand. Heard has obtained results that are somewhat similar; the large-scale convective motions are large in the lower part of the convection zone and very small in the upper part.

Weiss: I said that Vickers' result, that the temperature perturbation could change sign without a corresponding change in the velocity perturbation was incorrect. Indeed, one can show that such an eigenfunction cannot satisfy the temperature equation. However, it seems a fairly general result that large scale convective motions have lower velocities in regions where the scale-height is small compared with the scale of the motion. This is found in our linear solutions as well as in Vickers' and Heard's results.

Gilman: You have pointed out two mechanisms for convection giving net *inward* transport of magnetic flux. Dr Parker has pointed out that magnetic buoyancy should cause a net *outward* movement of flux. Would you care to speculate on the relative importance of these two effects?

Weiss: It would be rash to make a prediction without any proper calculation. However, my guess is that topological pumping may be able to keep flux ropes deep in the convective zone.

Parker: The network of downflow represented by the supergranule boundaries blocks the escape of magnetic lines of force from the interior of the Sun provided that the downflow velocity v exceeds the rate of rise of the field. Magnetic buoyancy causes a horizontal magnetic flux tube in a region of convective instability to rise at a rate of the general order of magnitude of the Alfvén speed. Thus, very roughly, the lines of force can be blocked from rising to the surface if $V_A < v$. Strong fields come up through regardless of the downflow, as we know from observation.

We must not overlook the fact, however, that any field – weak or strong – can come up in the rising currents at the center of a supergranule, forming a local bipolar region there.

Deinzer: Are there theoretical arguments for the occurrence of three distinct linear scales in the convection zone, represented by granulation, supergranulation and giant cells?

Weiss: There is still nothing more substantial than the crude arguments that I have repeated in my paper. The exact computations so far carried out all show a single cell across the entire unstable layer.

ON THEORIES OF SOLAR ROTATION

B. R. DURNEY

National Center for Atmospheric Research*
Boulder, Colo. 80303, U.S.A.

Foreword

About twelve years ago the first attempts to explain the Sun's differential rotation appeared in the literature. In these twelve years our understanding of the processes that could give rise to the large-scale circulation of the solar convection zone has greatly increased.

It therefore seems appropriate at this time to critically review the degree of development of theories of solar rotation. This could perhaps be achieved by referring the reader to the relevant papers. In this type of review the reader must, however, spend a considerable effort to gain a proper understanding of the subject. Consequently, in this paper a different approach has been followed, the aim of which has been to present a unified and critical exposition of theories of solar rotation. There is no doubt that this paper would be very different had it been written by another contributor to this field: a subject can be understood in several distinct ways; furthermore, different authors would weight very differently the diverse theories that are put forward to explain the Sun's differential rotation since there does not exist at present a commonly accepted explanation of this phenomenon.

This review will have achieved its goal if it allows the reader to gain an easy understanding of this subject and if it clearly shows where the main difficulties lie.

Abstract. The main theories of solar rotation are critically reviewed.

The interaction of large-scale convection with rotation gives rise to a transport of angular momentum towards the equator and therefore to differential rotation with equatorial acceleration. (Large-scale convection is defined as follows: in a highly turbulent fluid, the small-scale turbulence acts as a viscosity and organizes fluid motions on a much larger scale.) This transport of angular momentum towards the equator arises because of the highly non-axisymmetric character of the large-scale convective motions in the presence of rotation. These motions tend to be concentrated near the equator. It is not surprising, therefore, that for magnitudes of large-scale convection which are needed to generate the observed solar differential rotation, large *and unobserved* pole-equator differences in flux appear in the Boussinesq approximation.

It is important, therefore, to take the variations in density into account. Studies of large-scale convection in a compressible rotating medium are still in a very early stage; these studies suggest, however, that the surface layers must indeed rotate differentially.

The interaction of rotation with convection appears to be especially efficient in generating a pole-equator difference in flux, $\Delta\mathscr{F}$. Such a $\Delta\mathscr{F}$ drives meridional motions, and the action of Coriolis forces on these motions gives rise to differential rotation. In the 'large-viscosity' approximation the problem separates; the meridional motions can be determined first (from the radial and latitudinal equations of motions, and the energy equation) and the angular velocity can be determined next from the azimuthal equation of motion. Since very little is known about compressible large-scale convection, it has been assumed in the development of this theory that the stabilizing effect of rotation on *turbulent convection* depends on the polar angle θ and on depth. The solution for the angular velocity in the large viscosity approximation gives a differential rotation that varies slowly with depth. As a consequence, the large

* The National Center for Atmospheric Research is sponsored by the National Science Foundation.

Bumba and Kleczek (eds.), Basic Mechanisms of Solar Activity, 243–295. All Rights Reserved.
Copyright © 1976 by the IAU.

viscosity approximation is not valid over most of the convection zone, the Coriolis term being larger than the viscous term; a thin layer at the top excepted. (It appears, however, that if the angular velocity, Ω, is a slowly varying function of depth and the azimuthal stresses vanish at both ends of the convection zone, then the general behavior of Ω will be very much like that predicted by the large viscosity approximation.)

The stabilizing effect of rotation on turbulent convection is neglected; if differential rotation is significant over the entire convection zone, and if the meridional and large-scale convective velocities are not too large, then in the radial and latitudinal equations of motion, the main balance of forces is between pressure gradients, buoyancy and Coriolis forces. If rotation is not constant along cylinders, then the differential rotation gives rise to latitudinal variations in the convective flux which are proportional to $\Omega_0^2 T/g$ (where T is the temperature and g is gravity). In the lower part of the convection zone, $\Omega_0^2 T/g$ is of the order of the superadiabatic gradient itself. Therefore large pole-equator differences in flux, $\Delta\mathcal{F}$, will be present unless the angular velocity is constant along cylinders. The meridional velocities associated with this rotation law are not small, however, and could generate a significant $\Delta\mathcal{F}$. It could well be that large $\Delta\mathcal{F}$'s can be avoided only if rotation is uniform in the lower part of the convection zone. (To be certain of these results, however, it is important to estimate the magnitude of the stabilizing effect of rotation on turbulent convection.)

Turbulent convection is driven by the buoyancy force which thus introduces a preferred direction: gravity. In consequence, the turbulence in the sun should be anisotropic and if this is the case the convection zone cannot rotate uniformly. The degree of anisotropy is not known and must be determined from the observed solar differential rotation. The anisotropy is such that the horizontal exchange of momentum is larger than the vertical.

The normal mode of vibrations and the inner rotation of the Sun are briefly discussed.

1. Observations

The observations that have a bearing on theories of solar rotation are the following.

1.1. ROTATION RATES

Different features of the solar surface rotate at various rates. In Figure 1 (from Stenflo, 1974) the angular velocity is plotted as a function of latitude for (i) the sunspots (Newton and Nunn, 1951), (ii) the photosphere measured from Doppler shifts (Howard and Harvey, 1970), (iii) the photospheric magnetic field measured by an auto correlation technique (Wilcox and Howard, 1970; Wilcox *et al.*, 1970a; Stenflo, 1974). For the curve labeled 'longitudinal magnetic field' Stenflo calculated the autocorrelation curves for the strength (with sign) of the radial component of the field. Only the sign was used in the curve labeled 'sign of the longitudinal magnetic field.' To some extent the dashed, solid and dotted curves of Figure 1 correspond to increasing values of the flux density.

The chromosphere and transition region appear to rotate like the photosphere (Dupree and Henze, 1972; Henze and Dupree, 1973). The results of OSO 6 (Henze and Dupree) indicate a smaller differential rotation than the data of OSO 4 (Dupree and Henze). If only the brightest points in the EUV spectroheliograms are included, then their rotation rate is similar to that of the sunspots (Simon and Noyes, 1972).

Measurements of the chromospheric rotation rate in the Hα line give larger angular velocities than when metallic lines are used (Livingston, 1969). The K-corona rotates as the sunspots (Hansen *et al.*, 1969).

The above data have been interpreted by Stenflo (1974) as indicating that observations which utilize the Doppler and Zeeman effects refer to different regions of the Sun: the Doppler shift measurements refer to non-magnetic regions which do not contribute to the Zeeman-effect observations. Furthermore, Stenflo suggests

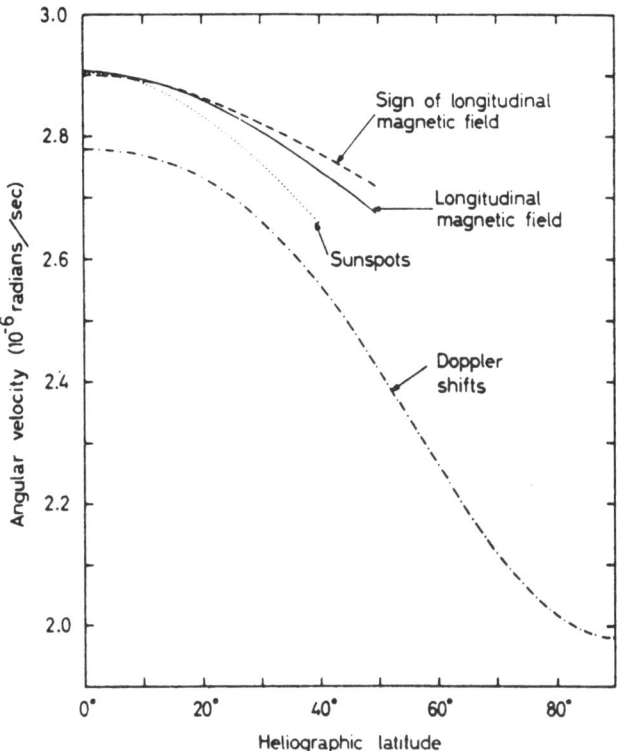

Fig. 1. Angular velocity of rotation as a function of heliographic latitude. The solid and dashed lines are from Stenflo (1974), the curve for recurrent sunspots is from Newton and Nunn (1951), and the Doppler shift curve is from Howard and Harvey (1970) (from Stenflo: 1974, *Solar Phys.* **36**, 495).

that the smaller the differential rotation of the magnetized plasma, the deeper the magnetic field pattern is rooted. It will be seen later that this picture is not without difficulties.

1.2. TIME VARIATIONS AND CORRELATIONS

Howard and Harvey expressed the Sun's angular velocity in the form

$$\Omega = a + b \cos^2 \theta + c \cos^4 \theta \tag{1.1}$$

where θ is the polar angle. The average values of a, b, and c were found to be: $a = 2.78 \times 10^{-6}$ rad s^{-1}, $b = -3.51 \times 10^{-7}$ rad s^{-1}, and $c = -4.43 \times 10^{-7}$ rad s^{-1}. Time variations of a, b, and c are large and correlations exist between these time variations (cf. Howard and Harvey, 1970). In Figure 2 we have plotted the correlation between b, and c as given by Yoshimura (1972a). A weaker correlation exists between a and c (Wolff, 1975).

1.3. RIGID ROTATION, THE SECTOR STRUCTURE AND PREFERRED LONGITUDES

The interplanetary sector structure (Wilcox and Ness, 1965) has been shown to be an extension of the solar sector structure (Ness and Wilcox, 1966; Wilcox and Ness,

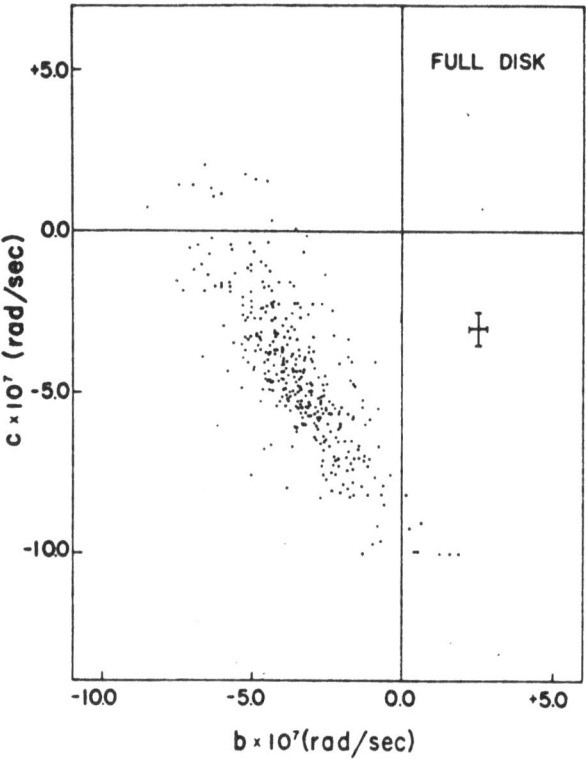

Fig. 2.   A plot of the values of *b* versus *c* for the whole disk. The values of *b* and *c* are from Howard and
Harvey (1970) (from Yoshimura: 1972, *Solar Phys.* **22**, 20).

1967). Boundaries of the solar sector structure do not evolve as expected from the
solar differential rotation, but appear to have a rigid-rotation component (Wilcox
and Howard, 1968). Large-scale photospheric magnetic fields can also display, at
some latitudes, both rigid and differential rotation properties (Wilcox *et al.*, 1970b).
Švestka (1968a, b) has found that the sources of proton flare rotate rigidly around the
Sun in the opposite direction of rotation. Also, the long-lived coronal activity shows
rigid rotation in the latitude interval ±57.5° (Antonucci and Svalgaard, 1974).

Flare activity, especially the proton flare activity, occurs preferentially near the
sector boundaries (Bumba and Obridko, 1969); a marked enhancement is found
within one day of the (− +) solar sector boundaries (Dittmer, 1975).

1.4. Giant magnetic field structures, meridional motions and angular momentum transport

Giant magnetic field structures have been observed by Bumba *et al.* (1964), Bumba
(1967), McIntosh (1975) and others. Their possible origin has been discussed by
Bumba (1970).

From spectroscopic measurements, Howard (1971) finds no evidence of meridional motions in the photosphere with an upper limit to the line-of-sight velocity of
30 m s^{-1}. Sunspots, on the other hand, show a poleward drift of about $0.01 \text{ deg day}^{-1}$

for latitudes higher than ~20° and a drift towards the equator for smaller latitudes (Tuominen, 1955; Richardson and Schwarzschild, 1953).

From a statistical analysis, Ward (1965) found that the longitudinal and latitudinal components of the proper velocities of the sunspots are correlated: $\langle U_\phi U_\theta \rangle > 0$, that is, on the average the sunspots with $U_\phi - \langle U_\phi \rangle > 0$ move towards the equator (and those with $U_\phi - \langle U_\phi \rangle < 0$ towards the pole), giving rise to a transport of angular momentum towards the equator (cf. Starr and Gilman, 1965). Ward's results have been criticized by Leighton (1966) and until further evidence becomes available, the correlation $\langle U_\phi U_\theta \rangle > 0$ cannot be considered as established.

1.5. POLE-EQUATOR DIFFERENCES IN FLUX AND TEMPERATURE

The pole-equator differences in flux ($\Delta \mathscr{F}$) are very small (Dicke and Goldenberg, 1967; Hill *et al.*, 1974); the upper limit of $\Delta \mathscr{F}/\mathscr{F}$ is probably not larger than a few parts in 10^{-4}. Pole-equator differences in temperature (ΔT), if present, are also small (Appenzeller and Schröter, 1967; Caccin *et al.*, 1970; Altrock and Canfield, 1972a, 1972b; Noyes *et al.*, 1973; Rutten, 1973; Falciani *et al.*, 1974). Noyes *et al.* (1973) find that $\Delta T \cdot \tau \leq 0.3$ K if $\tau < 10^{-2}$ (τ is the optical depth).

1.6. NORMAL MODE OF VIBRATIONS

Evidence has been accumulating (Deubner, 1972; Kaufman, 1972; Kobrin and Korshunov, 1972; Fossat and Ricort, 1973) that normal modes of vibration of the entire Sun are present. (Fossat (1975) has, however cautioned that the long-period oscillation of the data in Fossat and Ricort's paper could be contaminated by atmospheric noise.) Recent observations by the SCLERA group (cf. Hill *et al.*, 1976) dispel any doubt about the existence of these normal modes of vibration.

It is unlikely that observations of solar-type stars, following their arrival at the main sequence, could be relevant for theories of the solar differential rotation. These observations are essential, however, for theories of the solar inner rotation (the radiative core).

1.7. *Average Rotational Velocities of Main Sequence Stars*

The average rotational velocity for stars in the main sequence increases from ~180 km s^{-1} for a B0 star to ~225 km s^{-1} for a B5 star and decreases thereafter (~100 km s^{-1} for an F0 star), the decrease becoming very sharp at about F6 (Abt and Hunter, 1962). Figure 3 (from Kraft, 1969) shows the log of the average angular momentum density versus the log of the mass for stars in the main sequence ($\langle J \rangle$ is the average angular momentum, assuming solid body rotation, of main-sequence stars with a given mass, divided by the star's mass M). The extrapolated line, $\langle J \rangle \sim M^{2/3}$, was added by Dicke (1970a, b).

1.8. *Angular Velocity as a Function of Age and Ca II Emission*

Stars later than F6 rotate faster in young than in old clusters (Kraft, 1967). Figure 3 shows that rotation rate becomes very small in the main sequence for stars later than ~F6. It is just at this place that Ca II emission begins (Wilson, 1966a, 1966b).

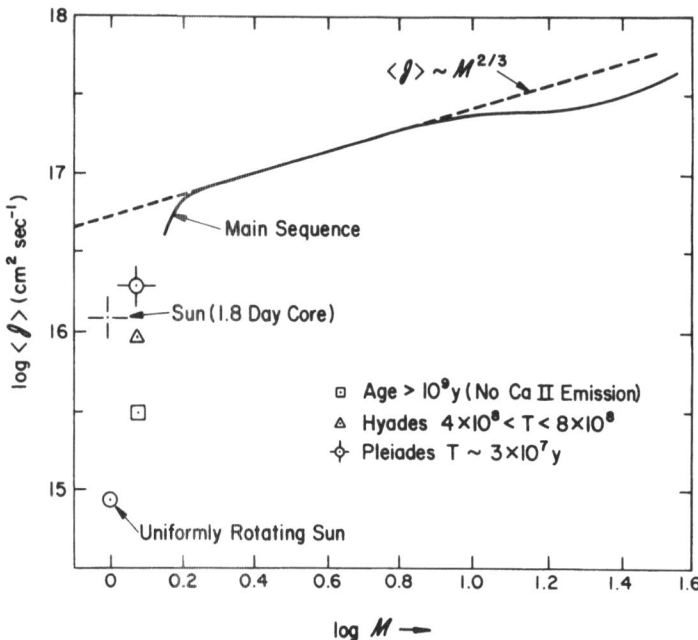

Fig. 3. The log of the angular momentum, per unit mass, of stars of various masses (assumed to be rotating uniformly) versus the log of the mass (from Kraft, 1969, in *Stellar Astronomy*, vol. 2 (ed. by Chiu, Warasila and Remo), Gordon and Breach, New York.

Furthermore, Wilson (1966b) has shown that stars with Ca II emission are nearer the zero-age main sequence than stars without it. Figure 4 (from Kraft, 1967) shows that, for stars less massive than $M/M_\odot = 1.25$, the largest rotational velocities are associated with stars having active chromospheres (Ca II emission). For more details about the above observations the reader is referred to the excellent review articles by Kraft (1969, 1970).

1.9. *Lithium and Beryllium Abundances*

There is a correlation between the Li abundance and the age of the star: the older the star, the lower the Li abundance; that is, there is a progressive loss of Li with time (cf. Wallerstein and Conti, 1969).

Stars with masses less than $1.1\ M_\odot$ have much lower abundance of Li than stars with $M > 1.1\ M_\odot$ (Wallerstein *et al.*, 1965). Stars with appreciable rotation have the largest Li content (Conti, 1968).

The abundances of beryllium show a much lower dispersion than the abundances of Li (cf. Wallerstein and Conti, 1969; Grevesse, 1968; Hauge and Engvold, 1968; Ross and Aller, 1974). Ross and Aller find that, in the photospheric layers of the Sun, beryllium is depleted below the solar system abundance by a factor of about two. It is not possible, therefore, to rule out a small depletion of Be in the Sun. In contrast, solar-type stars show a very strong depletion of lithium.

Lithium and beryllium are destroyed at distances from the center of the Sun equal to $\sim 0.63\ R_\odot$ and $\sim 0.47\ R_\odot$, respectively (cf. Dicke, 1970b, Table 2). In relation to

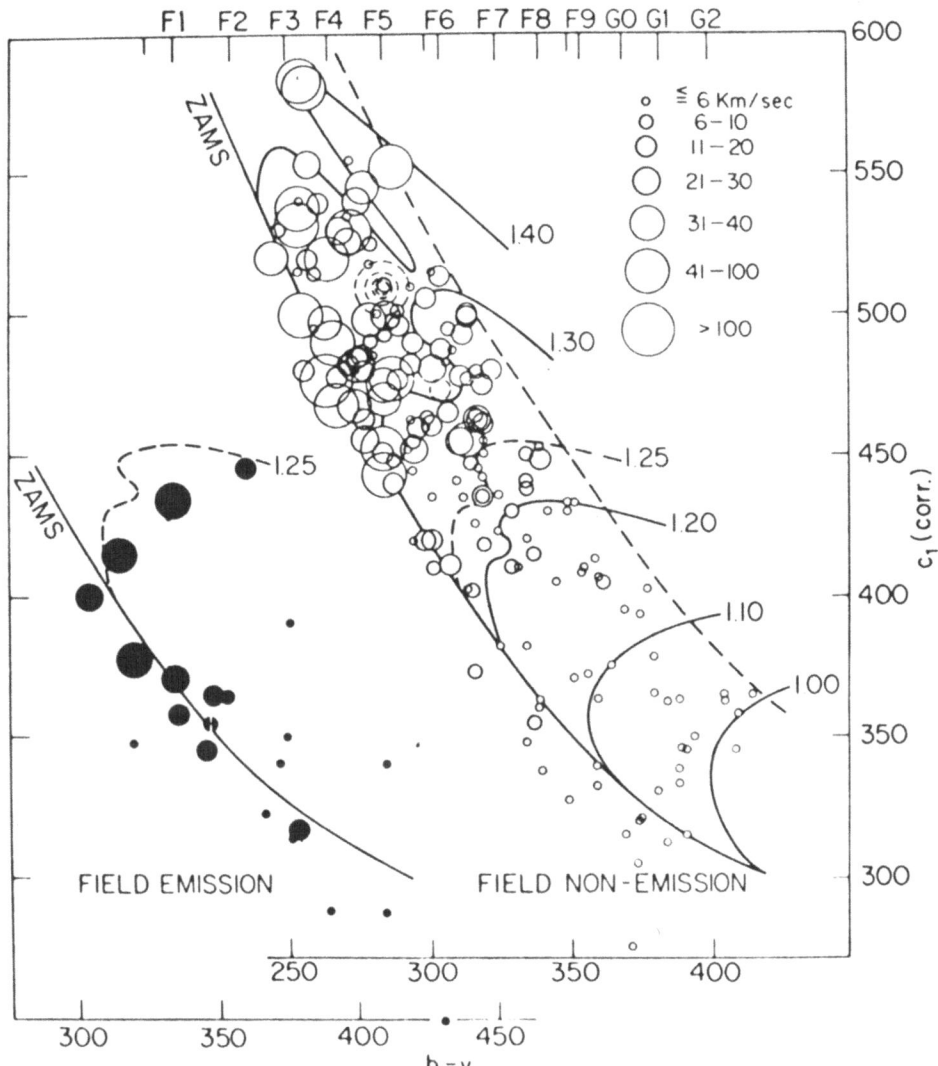

Fig. 4. Strömgren diagram representation of the HR diagram for both field emission stars (dark circles) and field-free emission stars (open circles). The circle size gives an indication of the rotation rate (from Kraft: 1967, *Astrophys. J.* **150**, 551).

theories of the solar rotation, the importance of these observations lies in the fact that they indicate that there is very little mixing of matter from the surface down to $0.47 R_\odot$. To account for the Li depletion, however, a mixed region must exist, extending down to $0.63 R_\odot$. This mixing has been attributed to turbulence induced by gradients in the angular velocity (Goldreich and Schubert, 1967; Spiegel, 1968; Howard *et al.*, 1967).

2. Differential Rotation as a Consequence of the Interaction of Rotation with Convection

2.1. BOUSSINESQ THEORIES

Typical values of the Rayleigh ($\mathscr{R}$) and Prandtl (σ) numbers in the solar convection zone are $\mathscr{R} \sim 10^{12}$ to 10^{20} (depending mainly on whether one takes the length appearing in $\mathscr{R}$ as a scale height or as the depth of the convection zone; cf. Spiegel, 1971) and $\sigma \sim 10^{-9}$ because the thermal diffusion is radiative (Ledoux et al., 1961). Convection in the Sun should therefore be highly turbulent.

An idea that goes back to the beinning of turbulence studies (Boussinesq, 1877, 1897) is that of a turbulent viscosity: the momentum exchange due to the turbulent motions is assumed to act as a viscosity and to organize relatively steady large-scale motions. The effect of rotation on this large-scale convection will be especially important and this leads to an appealing and natural theory of differential rotation: the radial and latitudinal variations of the angular velocity are assumed to be generated by the interaction of this global convection with rotation (Durney, 1968a, 1970; Busse, 1970, 1973; Yoshimura and Kato, 1971). As a starting point, let us ignore the compressible character of the solar convection zone and consider a spherical layer of a rotating convective fluid in the Boussinesq approximation. The usual boundary conditions imposed on this problem are specified temperatures at the top (R_0) and bottom (R_c) of the convection zone, as well as zero stresses and radial velocities at $r = R_0$ and $r = R_c$. These boundary conditions are very far from being ideal. Assuming that the flow problems in the inner radiative region and in the convective zone can be separated (and it could well be that this is *not* possible), it seems more appropriate to specify the energy flux and the temperature at the inner boundary of the convection zone, letting the outer surface choose its own temperature and flux. In particular, a boundary condition of uniform heat flux at $r = R_c$ could significantly alter the nature of the solutions. Furthermore, the boundary conditions of zero stresses at $r = R_c$ could be questioned (Gierasch, 1974; Durney, 1976). We define dimensionless quantities as follows:

$$\mathbf{U} = \kappa U'/R_0 ; \qquad \mathbf{r} = R_0 \mathbf{r}' ; \qquad t = R_0^2 t'/\kappa ;$$
$$T = \Delta T T' ; \qquad \mathbf{G}(r) = -g(r)\hat{r} = -g(R_0)g'(r')\hat{r}' \tag{2.1.1}$$

$\hat{r}$ and $\hat{r}'$ are unit vectors and the primed quantities are the dimensionless variables for the velocity, radial distance, time, temperature, and gravity ($\mathbf{G}(r)$), respectively. (If M_0 is the mass of the Sun and if we neglect the mass of the convection zone then $g(R_0) = G_c M_0/R_0^2$ and $g'(r') = 1/r'^2$; G_c is the gravitational constant). In Equations (2.1.1), κ is the thermometric diffusivity and $\Delta T = T_c - T_0$, that is, the difference in temperature between the inner and outer surfaces of the convection zone. In the rest of this section all quantities will be assumed to be dimensionless and the primes will be dropped.

Differential rotation is an axisymmetric mode of the velocity field $\mathbf{U}$. To understand its origin it is convenient to expand the velocity field in poloidal and toroidal vectors (Chandrasekhar, 1961, Appendix III) and the temperature field in spherical harmonics. We retain only the lowest axisymmetric modes of **U** and *T*, that are furthermore symmetric about the equator.

For the angular velocity we obtain (Chandrasekhar, Appendix III, Equation (15))

$$U_\phi = -\frac{t_1(r)}{r}\frac{\partial Y_1^0}{\partial\theta} - \frac{t_3(r)}{r}\frac{\partial Y_3^0}{\partial\theta}$$

$$= \frac{r\sin\theta}{2}\left(\frac{3}{\pi}\right)^{\frac{1}{2}}\left[T_1(r) + \frac{(21)^{\frac{1}{2}}}{2}T_3(r)(5\cos^2\theta - 1)\right] \tag{2.1.2}$$

Above, $T_1(r)(= t_1(r)/r^2)$ and $T_3(r)(= t_3(r)/r^2)$ are some functions of r; θ is the polar angle; and the spherical harmonics have been normalized according to Condon and Shortley (1951)

$$\left(Y_L^m(CS) = \frac{(-1)^m}{\sqrt{2\pi}}\left[\frac{2L+1}{2}\frac{(L-m)!}{(L+m)!}\right]^{1/2}Y_L^m(\text{Chandrasekhar})\right).$$

The meridional motions (i.e., the poloidal components of the axisymmetric velocity field) are also determined by two scalars, $p_2(r)$ and $p_4(r)$ (stream functions), which define motions with one and two cells, respectively, in each hemisphere ($0 < \theta < \pi/2$ and $\pi/2 < \theta < \pi$ for the northern and southern hemisphere, respectively) of the convection zone.

(Explicitly,

$$U_r = \sum_{L=2,4}\frac{L(L+1)}{r^2}p_L Y_L^0; \qquad U_\theta = \sum_{L=2,4}\frac{1}{r}\frac{\partial p_L}{\partial r}\frac{\partial Y_L^0}{\partial\theta},$$

cf. Equation (A3a) of Appendix 1.)

For the temperature we write

$$T = T_c/\Delta T + (\eta/r - 1)/(1 - \eta) + \psi(r, t) + \Theta(r, \theta, \phi, t) \tag{2.1.3}$$

In Equation (2.1.3), $\eta = R_c/R_0$; the first two terms are the purely conductive temperature profile, the third term is the distortion of the average temperature due to convection and the fourth term is the fluctuating component of the temperature, which averages to zero on any surface of radius r. Note that $\Theta(r, \theta, \phi, t)$ is *not* only the axisymmetric part of the fluctuating temperature. (It will be clear later that non-axisymmetric components of the velocity and temperature field play an essential role in the generation of differential rotation and of pole-equator differences in temperature). The lowest axisymmetric mode of Θ that has symmetry about the equator is

$$\Theta_2(r)Y_2^0 = \frac{1}{4}\left(\frac{5}{\pi}\right)^{\frac{1}{2}}\Theta_2(r)(3\cos^2\theta - 1) \tag{2.1.4}$$

$\Theta_2(r)$ defines, therefore, a pole-equator temperature difference.

The main justification for retaining only the lowest terms in the expansions of the axisymmetric modes of **U** and *T* is, of course, mathematical simplicity. (Note, however, that the next term in the expansion for the angular velocity is

$$\frac{T_5(r)}{16}\left(\frac{11}{\pi}\right)^{\frac{1}{2}} r \sin\theta(315\cos^4\theta - 210\cos^2\theta + 15)$$

which does not have a maximum at the equator.)

In a system of coordinates rotating with an angular velocity Ω_0, the equations for $T_1(r)$, $T_3(r)$, $P_2(r)$ $(= p_2(r)/r)$ and $\Theta_2(r)$ are found to be (Durney, 1971):

(a) Equations for the angular velocity

$$\left(\frac{d^2}{dr^2} + \frac{4}{r}\frac{d}{dr}\right)T_1 + \left(\frac{3}{5}\right)^{\frac{1}{2}}\mathcal{T}_1\left(\frac{1}{r}\frac{dP_2}{dr} + \frac{3P_2}{r^2}\right)$$

$$= \frac{(3\pi)^{\frac{1}{2}}}{4\sigma r^4}\int_0^\pi \sin^2\theta\,\frac{\partial}{\partial r}(r^3\langle U_\phi U_r\rangle)\,d\theta,\tag{2.1.5}$$

$$\left(\frac{d^2}{dr^2} + \frac{4}{r}\frac{d}{dr} - \frac{10}{r^2}\right)T_3 + \frac{\mathcal{T}_1}{r}\left[\frac{2}{(35)^{\frac{1}{2}}}\left(\frac{dP_2}{dr} - \frac{2P_2}{r}\right)\right.$$

$$\left. + \frac{5}{(63)^{\frac{1}{2}}}\left(\frac{dP_4}{dr} + \frac{5P_4}{r}\right)\right]$$

$$= \frac{(7\pi)^{\frac{1}{2}}}{16\sigma r}\int_0^\pi \sin^2\theta\left[\frac{(5\cos^2\theta - 1)}{r^3}\frac{\partial}{\partial r}(r^3\langle U_\phi U_r\rangle)\right.\tag{2.1.6}$$

$$\left. + \frac{10}{r}\cos\theta\sin\theta\langle U_\theta U_\phi\rangle\right]d\theta.$$

(b) Equations for the meridional circulation

$$\left(\frac{d^4}{dr^4} + \frac{4}{r}\frac{d^3}{dr^3} - \frac{12}{r^3}\frac{d^2}{dr^2} + \frac{24}{r^4}\right)P_2$$

$$- \mathcal{T}_1\left[\frac{4}{(35)^{\frac{1}{2}}}\left(5T_3 + r\frac{dT_3}{dr}\right) + \frac{1}{(15)^{\frac{1}{2}}}r\frac{dT_1}{dr}\right] - \mathcal{R}_1 g_1(r)\Theta_2$$

$$= -\frac{(5\pi)^{\frac{1}{2}}}{4\sigma r}\int_0^\pi \sin\theta\left[\cos\theta\sin\theta\frac{\partial}{\partial r}(r\langle \mathbf{U}\cdot\nabla\mathbf{U}\rangle_\theta)\right.$$

$$\left. + \langle\mathbf{U}\cdot\nabla\mathbf{U}\rangle_r(3\cos^2\theta - 1)\right]d\theta.\tag{2.1.7}$$

(c) Equations for the pole-equator temperature difference

$$\left(\frac{d^2}{dr^2} + \frac{2}{r}\frac{d}{dr} - \frac{6}{r^2}\right)\Theta_2 + \frac{6}{r}P_2\left(\frac{\eta}{(1-\eta)r^2} - \frac{\partial\psi}{\partial r}\right)$$

$$= \frac{(5\pi)^{\frac{1}{2}}}{4r}\int_0^\pi \sin\theta\left[(3\cos^2\theta - 1)\frac{1}{r}\frac{\partial}{\partial r}(r^2\langle U_r\Theta\rangle)\right.\tag{2.1.8}$$

$$\left. + 6\langle U_\theta\Theta\rangle\cos\theta\sin\theta\right]d\theta.$$

where $g_1(r) = g(r)/r$; $\mathcal{T}_1 = 2\Omega_0 R_0^2/\nu$ $(= \mathcal{T}^{1/2}/(1-\eta)^2)$; $\mathcal{R}_1 = \alpha\Delta Tg(R_0)R_0^3/\kappa\nu$ $(= \mathcal{R}/(1-\eta)^3)$; $P_4(r) = p_4(r)/r$; σ $(= \nu/\kappa)$ is the turbulent Prandtl number and $\langle AB \rangle$ is defined by $\langle AB \rangle = (1/\pi)\int_0^{2\pi} AB \, d\phi$. We chose R_0 and not the thickness of the convective shell as the unit of distance. This is the reason why $\mathcal{T}_1$ is not the square root of the Taylor number $(\mathcal{T})$; also the usual Rayleigh number is $\mathcal{R}$ and not $\mathcal{R}_1$.

Turbulent convection is assumed to determine ν and κ in Equations (2.1.5)–(2.1.8); $\mathbf{U}$ and Θ are the large-scale velocity and temperature. Following Unno (1961), it will be assumed that $\sigma \sim 1$. The value of the turbulent viscosity, ν, does not change much in the convection zone and at a depth of 1.8×10^{10} cm, $\nu \sim 8 \times 10^{12}$ cm^2 s^{-1} (see Table I in Section 2.2). The value of $\mathcal{T}_1$ is known; the value of $\mathcal{R}_1$ which determines the strength of the large-scale convection, is not known. In these theories a natural way of determining $\mathcal{R}_1$ is by requiring that the value of $T_3(R_0)$ (which determines the latitudinal differential rotation at the surface by Equation (2.1.2) should agree with the observed value. The thickness of the convective layer is not well-known either: one can assume that it is a scale height or the actual depth of the solar convection zone.

The boundary conditions corresponding to specified temperatures and zero radial velocities and stresses at $r = R_c$, R_0 are $\Theta_{2L} = P_{2L} = P_{2L}'' = T_{2L+1}' = 0$ at $r = \eta$, 1 ($L = 1, 2$); $T_1' = 0$ at $r = \eta$; and $\int_\eta^1 r^4 T_1(r) \, dr = 0$. (If $T_1' = 0$ at $r = \eta$, then in the steady state also $T_1' = 0$ at $r = 1$; $\int_\eta^1 r^4 T_1(r) \, dr = 0$ implies that, in the rotating system of coordinates, the total angular momentum of the spherical shell is zero.)

The only approximation used in deriving Equations (2.1.5)–(2.1.8) is the Boussinesq approximation. In other words, suppose that we have a convective, rotating spherical layer of fluid and that $P_4(r)$ is negligible (or known), then if we could measure $\psi(r)$ and the azimuthal averages of the fluctuating quantities appearing in the right-hand side of Equations (2.1.5)–(2.1.8), these equations would allow us to determine $T_1(r)$, $T_3(r)$, $P_2(r)$, and $\Theta_2(r)$. In the right-hand side of Equations (2.1.5)–(2.1.8), Θ has been defined by Equation (2.1.3) and $\mathbf{U}$ is the total velocity field; $\mathbf{U}$ and Θ contain, therefore, non-axisymmetric as well as axisymmetric components, and it is apparent that the solution of Equations (2.1.5)–(2.1.8) poses a formidable problem. An approximate method of solution that clearly shows how T_1, T_3, P_2 and Θ_2 are generated is the following:

(i) The problem of a rotating, convective spherical layer of fluid is first solved in the quasi-linear approximation (Herring, 1963, 1964, 1969; Durney, 1968a). In this approximation, which has been shown to be qualitatively successful, the fluctuating self-interactions (i.e., the terms $\mathbf{U} \cdot \nabla \mathbf{U}$, $\mathbf{U} \cdot \nabla \Theta$) are neglected; the only non-linear term that is retained is the product of the distortion of the mean temperature and the velocity $[U_r(\partial\psi/\partial r)]$ (the equations for the different modes of the temperature and velocity field in this approximation are given in Appendix 1). The solution of this problem shows that convection in the presence of rotation *is highly non-axisymmetric*.

(ii) With the values of $\mathbf{U}$ and Θ evaluated in (i) we can calculate the right-hand side of Equations (2.1.5)–(2.1.8) (RHS (2.1.5–2.1.8)), and solve for T_1, T_3, P_2 and Θ_2 (for ψ, we use the value found in (i)). It is now clear how the large-scale convection in the presence of rotation generates, through the products of the fluctuating quantities appearing in the RHS (2.1.5–2.1.8), the axisymmetric modes of the temperature and velocity field. In Appendix II we give an order-of-magnitude estimate, based on dimensional considerations, of the RHS (2.1.5–2.1.8) and of T_3. In the method of solution described above, we have neglected the effect of the axisymmetric modes on the large-scale non-axisymmetric convection and on ψ. Gilman (1972) has taken these effects into account in the framework of the mean field approximation.

If $U_r = P_2 = T_1' = 0$ at $r = \eta$, the integration of Equation (2.1.5) gives

$$r^4 \frac{dT_1}{dr} + \left(\frac{3}{5}\right)^{\frac{1}{2}} \mathcal{T}_1 r^3 P_2 = \frac{(3\pi)^{\frac{1}{2}}}{4\sigma} \int_0^\pi \sin^3 \theta \langle U_\phi U_r \rangle \, d\theta \qquad (2.1.9)$$

which expresses that the angular momentum transport across a spherical surface due to (a) viscosity and (b) the action of Coriolis forces on the meridional circulation, is equal (in the steady state) to the angular momentum transport due to the Reynold's stresses. Equations (2.1.5) and (2.1.9) are equations for the radial differential rotation.

Equation (2.1.6) shows how the vertical $[(\partial/\partial r)(r^3 \langle U_\phi U_r \rangle)]$ and latitudinal $(\langle U_\theta U_\phi \rangle)$ transports of angular momentum generate the latitudinal differential rotation. The action of Coriolis forces on a meridional circulation can, just by itself, give rise to differential rotation, as is shown by the term in $\mathcal{T}_1$ in Equation (2.1.6).

A pole-equator temperature difference can be generated (a) by a vertical variation in the radial heat flux, the term $[(\partial/\partial r)(r^2 \langle U_r \Theta \rangle)]$ in Equation (2.1.8), (b) by a latitudinal heat flux $(\langle U_\theta \Theta \rangle)$, or (c) by a meridional circulation, if the average temperature varies with r (the term in P_2 in the left-hand side of Equation (2.1.8)).

A serious problem plaguing this theory of differential rotation from the beginning has been the large (and unobserved) pole-equator difference in flux $(\Delta\mathcal{F})$ needed to generate the observed solar differential rotation (cf. Figure 4 and the discussion in Durney, 1970 and Gilman, 1972). The basic reason for this difficulty is the following: to generate the observed differential rotation, a sizeable fraction of the energy flux must be transported by the large-scale convection (Equation (B5) of Appendix II). Transport of angular momentum towards the equator arises when the convective motions are mainly longitudinal (i.e. when the longitudinal velocities are large) since in this case the Coriolis forces transport the fast-rotating fluid particles towards the equator and the slow-rotating ones towards the poles (see Yoshimura, 1972a and Appendix II). If the convective motions have large longitudinal components, then something is known about the shape of the convective cells: m has to be large (see Equation (A3) of Appendix I). If m is large, the large-scale convective motions are peaked at the equator and significant pole-equator differences in flux appear unavoidable (if $m = L$, the θ dependence of U_r is of the form $\sin^L \theta$; if $m \neq L$, the rate of angular momentum transport is not maximum at the equator, Yoshimura, 1972a, Figure 4). It is unlikely that a mixture of modes can solve the heat flux problem. In Figure 5 we give the convective heat flux at the middle of the convective shell as a function of θ, in the case of axisymmetric convection, for a Rayleigh number of 1500, a Taylor number of 500 and $\eta = 0.8$ (see Figures 1 and 6 of Durney, 1968b). The rotation is fast enough to strongly inhibit convection and the quasi-linear approximation can be trusted qualitatively. The axisymmetric mode, plotted in Figure 5, is essentially identical to the polar modes (with $m = 0$) plotted in Figure 6 of Gilman

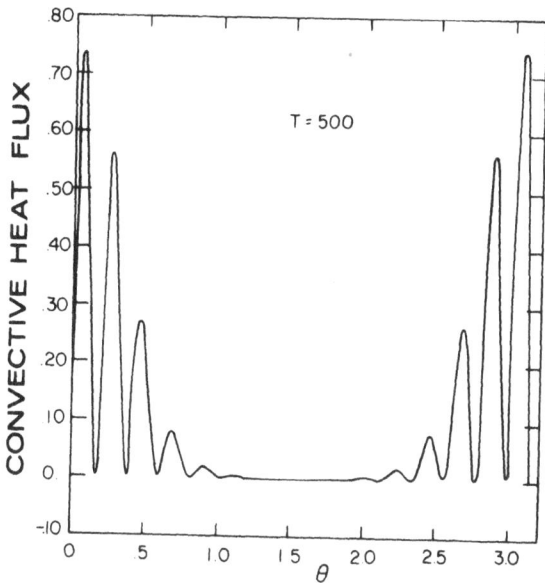

Fig. 5.   Convective heat flux versus polar angle for a Rayleigh number of 1500 at the middle layer of a convective shell. Convection is axisymmetric and the rotation rate is fast enough to inhibit convection almost completely.

(1975). Gilman's calculations (for the case of marginal instability and for a variety of Taylor numbers and values of m) show furthermore that for large Taylor numbers most modes peak at the equator, and that a few modes (with small m's) peak at the poles, with virtually no modes in between.

The right mixture of modes could give a small enough $\Delta\mathscr{F}$; it would, however, in all probability give a wrong profile for the latitudinal differential rotation since the modes that transport angular momentum towards the equator are also those that give rise to a large $\Delta\mathscr{F}$.

It is tempting to assume, therefore, that the main effect of the interaction of rotation with convection is the generation, through the RHS (2.1.8), of a small pole-equator difference in flux; this $\Delta\mathscr{F}$ will drive a meridional circulation (P_2) which, by Equations (2.1.5) and (2.1.6), will generate a radial and latitudinal differential rotation. It is of interest to calculate the magnitude of P_2 that could explain the observed solar differential rotation. From Equation (2.1.6) we obtain, very crudely, $T_3 \sim \mathscr{T}_1 P_2/(35)^{\frac{1}{2}}$. Comparing this value of T_3 with Equation (B3) of Appendix II, we obtain $P_2 = 2/15(\pi/5)^{\frac{1}{2}}\sigma$; if $\sigma = 1$ and $\kappa = 2 \times 10^{12}$ cm^2 s^{-1}, we find $U_r \sim 6$ cm s^{-1}.

To generate a latitudinal differential rotation with equatorial acceleration, the meridional ciculation must, however, rise at the poles and sink at the equator. This point, which is treated in detail in Section 2.3, can also be understood as follows: near the outer surface we approximate Equation (2.1.6) to $10 \, T_3/r^2 = 2\mathscr{T}_1 P_2'/(35)^{\frac{1}{2}}r$. For the equator to rotate faster, T_3 and therefore $P_2' = dP_2/dr$ must be negative. Since $P_2(r = R_0) = 0$, P_2 must be positive below the surface and this implies that the motions must sink at the equator. There is no reason to expect this to happen in a

Boussinesq fluid. In fact, since rotation inhibits convection preferentially at the poles, a circulation rising at the equator (which would increase the equatorial flux) appears more plausible. If differential rotation is generated by a meridional circulation the strong variation of ρ with r must, therefore, be included (Durney, 1972a).

It has been argued here that the angular momentum transport by the Reynold's stresses, in a Boussinesq fluid, cannot explain by itself the observed solar differential rotation. *The strong variation of density with depth in the solar convection zone appears to be essential for the understanding, even in a qualitative way, of the solar differential rotation.*

This result suggests that differential rotation cannot be generated in the lower part of the convection zone by the interaction of rotation with the global convection. It would be premature to conclude, however, that this is certainly the case, since the action of rotation on the turbulent diffusivity and viscosity has not been taken into account; it is not known what effect this action could have on theories of the thermodynamics and dynamics of the lower part of the convection zone.

There can be little doubt that global convection exists in the Sun: the observations (in particular 1.3 and the existence of giant magnetic field structures seem to corroborate it (cf. also Howard and Yoshimura's (1976) paper in these proceedings) *Global convection could, however, be very nearly critical and therefore transport little energy flux.* In this context the existence of periods in the past when the solar cycle appeared to have died off could be of importance; it has been suggested by Yoshimura (1976a) that during these periods global convection was very weak and could even have ceased to exist. It is easy to imagine this happening if global convection were very nearly critical. (In Yoshimura's model of the solar cycle, global convection regenerates the poloidal magnetic field from the toroidal field.)

Rigid rotation appears in this theory of differential rotation in a natural way *because of the existence of waves.* Waves of the velocity field have been discussed in Appendix I. Yoshimura (1971, 1972a, b) has shown that Švestka's observations pertaining to the rigid rotation of complexes of activity (cf. Section 1.3) can be explained, in terms of the time dependence of the magnetic field, as resulting from its interaction with rotation and the global convection.

2.2. NON-BOUSSINESQ THEORIES

The study of convection in a compressible medium (cf. Skumanich, 1955; Böhm, 1963; Spiegel, 1964, 1965; Vickers, 1971; Heard, 1973; Vandakurov, 1975a, b) is a difficult subject even in the absence of rotation; it is not surprising, therefore, that non-Boussinesq theories of differential rotation (Vandakurov, 1975b) are in a more primitive state of development than Boussinesq theories. *This subject could, however, be basic for an understanding of the solar differential rotation and a brief discussion is called for here.*

The convection zone is strongly stratified with depth. In Table I (cf. Table 1 of Cocke, 1967, and Baker and Temesvary, 1966) the values of the following quantities are tabulated as a function of depth: the density (ρ); the temperature (T); the mixing length (l), chosen equal to $1.5 \times$ pressure scale height; the turbulent convective velocity (u_c); the dynamic turbulent viscosity $(\eta = \rho u_c l/3)$; the kinematic turbulent viscosity $(\nu = u_c l/3)$; $|d\eta/dr|/\rho$; the ratio of the convective flux to the total energy flux (F_C/F_T); the superadiabatic gradient $(\nabla \Delta T)$; the square root of the local turbulent Taylor number $(\mathcal{T}^{\frac{1}{2}} = 2\Omega_0 l^2/\nu = 6\Omega_0 l/u_c$; more appropriately, therefore, $\mathcal{T}^{\frac{1}{2}}$ is proportional to the inverse of the Rossby number), and the turbulent Rayleigh number $(\mathcal{R} = \alpha g \nabla \Delta T d^4/\kappa)$. An approximate expression for the superadiabatic gradient was evaluated from the mixing length expression for the convective flux:

$$\mathcal{F} = \frac{l}{2} c_p \rho u_c \nabla \Delta T; \qquad \nabla \Delta T = T\left[\left(1 - \frac{1}{\gamma}\right)\frac{\nabla p}{p} - \frac{\nabla T}{T}\right] \qquad (2.2.1)$$

with $\mathcal{F} = (R_0/r)^2 \mathcal{F}_0 (F_C/F_T)$, $\mathcal{F}_0$ being the energy flux at the surface; $c_p = \gamma R_g/\mu(\gamma - 1)$ with $\gamma = 5/3$, R_g the gas constant, and μ the molecular weight $(= 0.6)$. In the Rayleigh number $\alpha = 1/T$, d $(= 1.9 \times 10^{10}$ cm) is the depth of the convection zone and $\kappa = \nu$ (the turbulent Prandtl number is equal to one).

The problem of global convection in a compressible medium (cf. Vickers, 1971; Heard, 1973; Vandakurov, 1975a, b) can be clarified with the help of the following simple and crude model: since the kinematic turbulent viscosity, ν, does not change much in the convection zone, Equation (2.2.1) gives

$$\nabla \Delta T = \alpha/\rho; \qquad \alpha = 2\mathcal{F}/3c_p\nu \qquad (2.2.2)$$

wherein the first approximation α can be taken as a constant. The equation of hydrostatic equilibrium, the gas equation and Equation (2.2.2) determine the unperturbed state. If motions are present, the relevant equations could be taken as the Navier-Stokes equations with a kinematic viscosity ν, the usual continuity and gas equations and the following energy equation

$$\text{div } \mathcal{F} = c_p \rho \mathbf{U} \cdot \nabla \Delta T \qquad (2.2.3)$$

where $\mathbf{U}$ is the large-scale velocity and $\mathcal{F}$ is given by Equation (2.2.1).

The problem of marginal stability (Vickers, 1971; Heard, 1973) consists, then, in determining the value of $\alpha(\alpha_c)$ such that the system is stable for $\alpha < \alpha_c$ and unstable for $\alpha > \alpha_c$. Vickers' (1971) calculations (performed, however, with a different energy equation, which in the unperturbed state is $\kappa \, dT/dz = $ constant) lend support to the idea that the value of α, as calculated from Baker and Temesvary's tables, for example, is larger than α_c. *In other words, the solar convection zone is unstable against global convection.* Vickers and Heard (who also included rotation) furthermore find that *the large-scale convection is concentrated in the lower part of the convection zone* (i.e., the preferred modes are damped in the upper portions of the layer). This result is somewhat surprising since the Rayleigh number is large in the upper part of the convection zone (cf. Table I); could this result be due to the boundary condition $U_r = 0$ at $r = R_0$? In the Sun the radial velocities do not vanish at the surface but can overshoot into the stable photospheric layers. Whatever the case, it is important to establish firmly (with respect to both the interpretation of the observations and to the generation of the solar differential rotation) whether the global convection is indeed concentrated in the lower part of the convection zone. If this were the case the Boussinesq theories should be qualitatively correct (see, however, the last paragraphs of this section) and, therefore, we would again expect that (i) a relation should

TABLE I

Solar convective parameters

d(cm)	ρ(gm cm^{-3})	T(K)	l(cm)	u_c(cm s^{-1})	$\eta=\frac{1}{3}\rho u_c l$	$\nu=\frac{1}{3}u_c l$	$\rho^{-1}|d\eta/dr|$	F_c/F_T	$\nabla\Delta T$	$\mathcal{T}^1$	$\mathcal{R}$
3.4(6)	4.0(−7)	6.4(3)	2.3(7)	2.4(4)	7.3(4)	1.8(11)	3.1(5)	6.1(−4)		1.5(−2)	1.8(13)
1.3(7)	4.6(−7)	9.1(3)	3.3(7)	2.2(5)	1.1(6)	2.4(12)	5.7(4)	9.3(−1)	1.1(−4)	2.3(−3)	7.4(12)
3.7(7)	8.3(−7)	1.2(4)	4.5(7)	1.5(5)	1.9(6)	2.3(12)	5.4(4)	1.0	6.7(−5)	4.5(−3)	3.9(12)
1.5(8)	7.2(−6)	1.8(4)	8.3(7)	7.6(4)	1.5(7)	2.1(12)	2.9(4)	1.0	8.5(−6)	1.7(−2)	4.0(11)
3.6(8)	6.2(−5)	2.7(4)	1.5(8)	4.2(4)	1.3(8)	2.1(12)	1.6(4)	1.0	9.9(−7)	5.5(−2)	3.0(10)
1.0(9)	8.2(−4)	6.8(4)	4.5(8)	2.2(4)	2.6(9)	3.2(12)	8.4(3)	1.0	5.0(−8)	3.3(−1)	2.7(8)
3.3(9)	8.3(−3)	2.6(5)	1.6(9)	1.1(4)	4.9(10)	5.9(12)	4.0(3)	1.0	2.9(−9)	2.2(0)	1.2(6)
1.1(10)	7.6(−2)	1.0(6)	5.5(9)	5.4(3)	7.5(11)	9.9(12)	1.6(3)	8.8(−1)	2.1(−10)	1.6(1)	1.0(4)
1.8(10)	1.8(−1)	1.9(6)	7.8(9)	2.9(3)	1.4(12)	7.8(12)	1.4(3)	2.7(−1)	4.5(−11)	4.0(1)	2.5(3)
1.9(10)	2.1(−1)	2.1(6)	8.1(9)	2.0(3)	1.1(12)	5.2(12)		9.7(−2)	2.1(−11)	6.4(1)	2.5(3)

exist between the angular momentum transport by the Reynolds stresses and the energy carried by the large-scale convection, and (ii) other processes than this angular momentum transport should be of importance in the generation of the solar differential rotation.

If, on the other hand, the velocities of the global convection are large only in the surface layers, then point (i) above ceases to be valid: *the large-scale convection could carry small amounts of energy and at the same time the large surface azimuthal velocities could give rise to a significant transport of angular momentum towards the equator.* If the Sun's differential rotation is generated in the upper part of the convection zone by angular momentum transport, then differential rotation should be 'large,' i.e., we would expect the Reynolds number of the azimuthal flow to be of the order of the critical Reynolds number. Let $\Delta\Omega$ be the difference in angular velocity between the equator and the poles. On a sphere of radius R_0, differential rotation gives rise to a shear flow with a Reynolds number given by

$$\mathcal{R}_e = \Delta\Omega R_0^2/\nu \tag{2.2.4}$$

where ν is the turbulent viscosity. The turbulent viscosity associated with the granules is easy to evaluate; we take $u \sim 0.4 \text{ km s}^{-1}$ (cf. Beckers and Morrison, 1970; Mehltretter, 1971; Mattig and Nesis, 1974) and $l \sim 10^3$ km. Therefore, $\nu \sim ul \sim 4 \times 10^{12} \text{ cm}^2 \text{ s}^{-1}$, and $\mathcal{R}_e \sim 9 \times 10^2$. The Reynolds number evaluated with the turbulent viscosity due to the supergranules appears to be somewhat smaller. The supergranular velocities decrease sharply with depth (Appenzeller and Schröter, 1968). We take $u \sim 0.14 \text{ km s}^{-1}$ and $l \sim 1.5 \times 10^4$ km, and find $\mathcal{R}_e \sim 1.7 \times 10^2$. These values of the Reynolds number should be compared with the critical Reynolds number, which is about 10^3. An important question that can be raised at this point is whether differential rotation penetrates deeply into the convection zone in the case when it is generated in the surface layers.

If the convection zone is highly unstable against large-scale convection ($\alpha \gg \alpha_c$) then a more reliable method of calculating the dominant modes is to calculate growth rates (Vandakurov, 1975a); a model of the convection zone is assumed to be given (in our example, the value of α) and the growth rates of different large-scale perturbations are calculated. Vandakurov (1975a) takes $\nabla\Delta T$ proportional to T and assumes that $\Omega_0 T^{\frac{1}{2}}/(\nabla\Delta T)^{\frac{1}{2}} g^{\frac{1}{2}} \ll 1$ where g is gravity; this inequality is valid in the upper portions of the convection zone (see Table I). The dominant mode is assumed to be the one with the largest growth rate. Vandakurov (1975b) has generalized his method to include (in a crude way) nonlinear terms and finds that in general the upper parts of stellar convection zones should rotate differentially, with no appreciable variations in the heat flux with latitude. However, large variations of $\mathcal{F}$ with θ are expected only in the lower part of the convection zone (see 2.4) and this problem cannot be considered solved at present. There can be no doubt that in order to fully explain the solar differential rotation, we need a much deeper understanding of the large-scale convection in a compressible model of the solar convection zone.

We have used Equations (2.2.3) and (2.2.1) as the energy equation for a compressible fluid and Equation (A1b) of Appendix 1 as the Boussinesq energy equation. It is important to understand under which approximations Equations (2.2.3) and (2.2.1) reduce to Equation (A1b). Equation (2.2.3) can also be written in the following form

$$\text{div } \mathscr{F} = \mathbf{U} \cdot [\nabla p - c_p \rho \nabla T] . \tag{2.2.5}$$

With the help of the hydrostatic relation the right hand side of Equation (2.2.5) becomes $-\rho c_p \mathbf{U}[\nabla T + g/c_p]$. The expression for the convective flux can be written $\mathscr{F} = -(l/2)\rho u_c c_p [\nabla T + g/c_p]$. Therefore, if the variations of ρc_p can be neglected we obtain

$$\text{div } \mathscr{F} + \mathbf{U}(\nabla T + g/c_p) ; \qquad \mathscr{F} = -\frac{lu_c}{2}(\nabla T + g/c_p) . \tag{2.2.6}$$

Apart from the factor g/c_p (the adiabatic gradient) Equation (2.2.6) is the usual energy equation for an incompressible fluid (Equation (A1b) of Appendix 1). The thermometric diffusivity is equal to $lu_c/2$ (which is approximately constant in the Sun) *and the Prandtl number is close to one,* $(\sigma = \nu/\kappa = 2/3)$.

2.3. MERIDIONAL CIRCULATION

The idea that the Sun's differential rotation could be generated by a meridional circulation driven by a pole-equator temperature difference (due to the interaction of rotation with convection) was first put forward by Weiss (1965) and Veronis (1966) (see also Osaki, 1970; Roxburgh, 1970; Durney and Roxburgh, 1971; Durney, 1974; and Gierasch, 1974).

We will neglect the Reynolds stresses due to the large-scale convection and approximate the action of the turbulent convection by a turbulent viscosity (assumed furthermore to act as a molecular viscosity). We can limit ourselves to axially symmetric flows since it is clear that any pole-equator difference in temperature generated by the interaction of rotation with convection will be axially symmetric.

The time-dependent Navier-Stokes equations can then be written

$$\rho\left[\frac{DU_r}{Dt} - \frac{U_\theta^2 + U_\phi^2}{r}\right] = -\frac{\partial p}{\partial r} - \rho g + R_r \tag{2.3.1}$$

$$\rho\left[\frac{DU_\theta}{Dt} + \frac{U_r U_\theta}{r} - \frac{U_\phi^2 \cot g \, \theta}{r}\right] = -\frac{1}{r}\frac{\partial p}{\partial \theta} + R_\theta \tag{2.3.2}$$

$$\rho\left[\frac{DU_\phi}{Dt} + \frac{U_\phi U_r}{r} + \frac{U_\theta U_\phi \cot g \, \theta}{r}\right] = R_\phi \tag{2.3.3}$$

where $D/Dt = U_r \partial/\partial r + (U_\theta/r)\partial/\partial\theta$ and R_r, R_θ, R_ϕ are the components of the viscous force written down explicitly in Appendix III.

It is readily seen that if γ (the ratio of the specific heats) and c_v are constants, then the time-independent energy equation

$$\rho\mathbf{U} \cdot \nabla c_v T - \frac{p}{\rho}\mathbf{U} \cdot \nabla\rho + \text{div } \mathscr{F} = 0 \tag{2.3.4}$$

can also be written as in Equation (2.2.3); $\mathscr{F}$ and $\nabla\varDelta T$ are defined in Equation (2.2.1). Equation (2.2.1) is the expression for the turbulent convective flux only; no allowance is made for a possible contribution from the large-scale convection. The

reason is simple: not enough is known about the large-scale compressible convection. In the development of these theories, it is therefore necessary to assume that
rotation acts as a perturbation on the turbulent convective flux.

An illustrative example that clarifies the basic ideas behind these theories of
differential rotation is the following (cf. Belvedere and Paterno, 1976): consider a
rotating spherical layer of fluid with a thermal conductivity of the form $\kappa =$
$\kappa_0[1 + \xi(r)P_2(\cos \theta)]$. The factor in the bracket mimics the effect of rotation on
convection, which depends on latitude and depth. Even if the fluid is not convective,
meridional motions will be set up and the action of Coriolis forces on these motions
will generate differential rotation. The replacement of 'thermal conduction' by
'turbulent convection' gives us a clear picture of the theories of differential rotation
discussed in this section. The fluid, however, could also be convective, which would
correspond to the large-scale convection. The action of rotation on these large-scale
convective motions is not taken into account. This effect could be important; in
particular, the dependence on depth of the interaction of rotation with (i) the
large-scale convection and (ii) the turbulent convection could be different.

The action of rotation on turbulent convection can be introduced by replacing
Equation (2.2.1) with Equation (2.3.5), for example:

$$\mathscr{F} = \left[1 + \varepsilon \frac{2\Omega_0 l^2}{\nu} P_2(\cos \theta)\right]\frac{l}{2} u_c c_p \rho \nabla \Delta T \qquad (2.3.5)$$

(a version of this equation was used in Durney and Roxburgh, 1971). The interaction
of rotation with convection depends on latitude (since the angle between the angular
velocity and gravity depends on θ) and on depth (since the scale of the convective
motions is a strong function of r). This interaction is contained in the factor $\varepsilon 2\Omega_0 l^2/\nu$
of Equation (2.3.5) and has, therefore, been assumed to be proportional to the
square root of the Taylor number or the inverse of the Rossby number (see Table I).
This appears to be a reasonable choice since in the Boussinesq theories the effect of
rotation manifests itself through $\mathscr{T}^{\frac{1}{2}}$. The proportionality factor ε can be fixed by
requiring that the calculated and observed values of the latitudinal differential
rotation at the Sun's surface are in agreement.

An understanding of these calculations, and of the relation between the meridional circulation and differential rotation, is particularly simple in the limit of large
viscosity and slow rotation. In this case, Equations (2.3.1) and (2.3.2) simplify to

$$-\frac{\partial p}{\partial r} - \rho g + R_r = 0; \qquad -\frac{1}{r}\frac{\partial p}{\partial \theta} + R_\theta = 0. \qquad (2.3.6)$$

The viscous terms R_r and R_θ depend only on U_r and U_θ (see Appendix III); U_r and
U_θ are defined in terms of the stream function, $\psi_1(r)$, by

$$U_r = \frac{\partial \psi_1}{\partial \theta}\bigg/\rho r^2 \sin \theta ; \qquad U_\theta = -\frac{\partial \psi_1}{\partial r}\bigg/\rho r \sin \theta \qquad (2.3.7)$$

Equations (2.3.6) allow us to express p and ρ (and therefore also T, by the gas
equation) in terms of ψ_1. The energy equation (Equation 2.2.3) is then an equation
for ψ_1. Once ψ_1 is known, the azimuthal equation of motion determines the angular
velocity. An appropriate value of ε would then reproduce the observed differential

rotation at the Sun's surface. A method of solution could be the following: we first
expand all quantities in terms of Legendre polynomials:

$$p = p_u(1 + p_2(r)P_2(\cos \theta)) ; \qquad \rho = \rho_u(1 + \rho_2(r)P_2(\cos \theta))$$

$$T = T_u(1 + T_2(r)P_2(\cos \theta)) \tag{2.3.8}$$

$$U_r = \frac{2\psi(r)}{\rho r^2} P_2(\cos \theta) ; \qquad U_\theta = -\frac{\partial \psi}{\partial r} \sin \theta \cos \theta / \rho r . \tag{2.3.9}$$

The subscript u denotes the unperturbed non-rotating state and we have chosen a
particular form for the stream function $\psi_1(= \sin^2 \theta \cos \theta \psi(r))$. Equations (2.3.9)
describe motions with one latitudinal cell in each hemisphere. Appendix IV contains
the expressions for p_2 and ρ_2 in terms of ψ.

With p, ρ, and T given by Equation (2.3.8), the expression for the superadiabatic
gradient becomes

$$\nabla \Delta T = (\nabla \Delta T)_u \hat{i}_r + \{T_2(\nabla \Delta T)_u + T_u[(1 - 1/\gamma)p_2' - T_2']\}P_2(\cos \theta)\hat{i}_r$$

$$- 3\frac{\sin \theta \cos \theta}{r} T_u[(1 - 1/\gamma)p_2 - T_2]\hat{i}_\theta \tag{2.3.10}$$

where $\hat{i}_r$ and $\hat{i}_\theta$ are unit vectors in the r and θ directions, respectively. We write
Equation (2.3.10) in the condensed form

$$\nabla \Delta T = (\nabla \Delta T)_u \hat{i}_r + \delta(\nabla \Delta T) ,$$

where the definition of $\delta(\nabla \Delta T)$ follows immediately from Equation (2.3.10). To the
first order in the perturbed quantities the energy equation (Equation 2.2.3) can be
written (we use Equations (2.2.1), (2.3.9), and (2.3.10) for $\mathcal{F}$, U, and $\nabla \Delta T$,
respectively)

$$\text{div} \left(\frac{l}{2}u_c c_p \rho_u \delta(\nabla \Delta T)\right) - 2\psi(r)c_p P_2(\cos \theta)(\nabla \Delta T)_u/r^2$$

$$= -\frac{2\varepsilon \Omega_0}{\nu r^2}P_2(\cos \theta)\mathcal{F}_0 R_0^2 \frac{\partial l^2}{\partial r} \tag{2.3.11}$$

where $\mathcal{F}_0$ is the solar energy flux (erg cm^{-2}) at the surface. We have taken, therefore,
a simplified version of the mixing length expression for the convective flux, namely,
$\mathcal{F} = l/2u_c c_p \rho_u \nabla \Delta T$, with $lu_c/2 = \frac{3}{2}\nu$ a constant. It is reasonable to expect that the
scale of variation of T_u, p_2, and T_2 is smaller than r, that is, $|\partial A/\partial r| > A/r$ where A
stands for T_u, p_2, and T_2. We shall therefore neglect the term containing the
θ-derivative in div$[(l/2)u_c c_p \rho_u \delta(\nabla \Delta T)]$. Furthermore, since $(\nabla \Delta T)_u$ is small we
neglect $T_2(\nabla \Delta T)_u$ with respect to $T_u[(1 - 1/\gamma)p_2' - T_2']$ in $\delta \nabla \Delta T$. Equation (2.3.11)
then reduces to

$$\frac{d}{dr}\left[r^2 \frac{l}{2}\frac{u_c}{\gamma - 1}p_u((\gamma - 1)p_2' - \gamma T_2')\right] - 2\psi c_p(\nabla \Delta T)_u = -\frac{2\varepsilon \Omega_0}{\nu}\mathcal{F}_0 R_0^2 \frac{\partial l^2}{\partial r}$$

$$\tag{2.3.12}$$

The ratio (Q) of the second term to the first term in the left-hand side of Equation (2.3.12) is of the order $Q \sim \mathscr{F}_0/\rho(l_\psi/v)^3(l/r)^2$ where $l_\psi^3 = \psi/\psi'''$. In the estimate of Q, Equation (2.2.2) was used for $(\nabla \Delta T)_u$, and $(p_u p_2')'$ was evaluated with the help of Equation (D2) $((p_u p_2')' \sim v\psi'''(\rho'/\rho)(p'/p))$. It is clear that if v is sufficiently large $(> \sim 2 \times 10^{13}$ cm^2 s$^{-1})$, the term $2\psi c_p(\nabla \Delta T)_u$ in Equation (2.3.12) can be neglected. Furthermore, if we set $r = R_0$ in the first term of Equation (2.3.12) we obtain

$$(\gamma - 1)p_2' - \gamma T_2' = -\frac{4\Omega_0(\gamma - 1)\varepsilon}{3v^2} \mathscr{F}_0 l^2/p_u \qquad (2.3.13)$$

The integration constant was chosen equal to zero for the following reasons: by virtue of Equations (2.3.5) and (2.3.10), the pole-equator difference in flux is given by $\mathscr{F}_0[\gamma T_2 + T_u[(\gamma - 1)p_2' - \gamma T_2']/(\nabla \Delta T)_u + 2\varepsilon\Omega_0 l^2\gamma/v]P_2(\cos \theta)/\gamma$, which reduces to $\mathscr{F}_0 T_2 P_2(\cos \theta)$ (a small quantity) if Equation (2.3.13) is satisfied. (Since $v = lu_c/3$ and $\mathscr{F}_0$ is given by Equation (2.2.1) the right-hand side of Equation (2.3.13) can also be written $-2\Omega_0\varepsilon l^2(\nabla \Delta T)_u\gamma/v T_u)$. Equation (2.3.13) will be solved with the boundary condition $(\gamma - 1)p_2 - \gamma T_2 = 0$ at $r = R_0$. Perhaps a better boundary condition would have been $T_2 = 0$ at $r = R_0$. It is unlikely, however, that these solutions would differ much. (It would be much better, of course, to impose that the latitudinal variations of the energy flux and temperature vanish at $r = R_c$, the lower boundary of the convection zone; the calculations determine then $\Delta \mathscr{F}$ and T_2 at $r = R_0$). With the help of Equations (D1) and (D2) of Appendix IV, we can substitute p_2 and T_2 in the integrated version of Equation (2.3.13) and obtain explicitly the equation for ψ. This is Equation (D3) of Appendix IV. Equation (D3) will be solved by assuming that the unperturbed state satisfies a polytropic relation: $p_u = p_{uc}(\rho_u/\rho_{uc})^\gamma$; the pressure, density, and temperature are then given by

$$p_u = p_{uc}(1 - (r - R_c)/L)^{\gamma/(\gamma - 1)}; \qquad \rho_u = \rho_{uc}(1 - (r - R_c)/L)^{1/(\gamma - 1)}$$

$$T_u = T_{uc}(1 - (r - R_c)/L) \qquad (2.3.14)$$

with

$$L = \gamma R_g T_c/(\gamma - 1)\mu g \qquad (2.3.15)$$

The subscript c denotes quantities at the bottom of the convection zone, as given for example by Baker and Temesvary's tables. We take $l = 1.5 p_u/p_u'$; the integral in the right-hand side of Equation (D3) can then be evaluated in closed form (see Equation (D4) of Appendix IV). Appropriate boundary conditions for Equation (D3) (with the right-hand side given by Equation (D4)) are zero radial velocities and stresses at $r = R_c, R_0$:

$$\psi = 0 ; \qquad \psi'' = \psi'(2/r + \rho'/\rho) = 0 ; \qquad r = R_c, R_0. \qquad (2.3.16)$$

In the large viscosity approximation it is therefore possible to determine ψ (apart from a multiplicative factor, since ε is not yet known) independently of the angular velocity. To obtain the equation for the angular velocity, we expand Ω in Legendre polynomials and neglect polynomials of a higher order than two:

$$\Omega = \Omega_0(1 + \omega_0(r) + \omega_2(r)P_2(\cos \theta)). \qquad (2.3.17)$$

This is a good approximation for the observed angular velocity at the surface (cf. Durney, 1976). Howard and Harvey's (1970) observations give

$$\Omega_0 = 2.57 \times 10^{-6} \text{ rad s}^{-1} ; \qquad \omega_2(R_0) = -0.189 \qquad (2.3.18)$$

The equations for $\omega_0(r)$ and $\omega_2(r)$ are Equations (E1) and (E2) of Appendix V. The solution of the problem at hand is particularly simple if we neglect the nonlinear terms in Equations (E1) and (E2): the equations for ψ, $\omega_0(r)$, and $\omega_2(r)$ can be solved first with $\varepsilon = 1$. Since $\omega_2(r)$ is proportional to ψ, the value of $\omega_2(r)$ at the surface will be -0.189 (which is the observed differential rotation) if $\varepsilon = -0.189/\omega_2(R_0, \varepsilon = 1)$. The numerical calculations (with $\gamma = 5/3$, $\nu = 2 \times 10^{12} \text{ cm}^2 \text{ s}^{-1}$, and $s = 1$) give $\varepsilon = -1.4 \times 10^{-6}$. (It should be noted, cf. Equation (D3), that $\psi(r)$ is proportional to ν^{-3} and therefore $\omega_2(r)$ (cf. Equation E2) is proportional to ν^{-4}; had the equations been solved with $\nu = 2 \times 10^{13} \text{ cm}^2 \text{ s}^{-1}$ a much larger value of ε would have been needed to reproduce the observed differential rotation. With $\nu = 2 \times 10^{13} \text{ cm}^2 \text{ s}^{-1}$ we can justify neglecting the term $2\psi c_p (\nabla \Delta T)_u$ in Equation (2.3.12).) If ε is negative then $2\varepsilon \Omega_0 l^2 P_2(\cos \theta)/\nu$ is larger than zero at the equator, and it follows from Equation (2.3.5) that rotation stabilizes the poles more than the equator. Therefore, differential rotation with equatorial acceleration is generated, in this model, if rotation stabilizes turbulent convection preferentially at the poles. This result, of course, should not be taken too seriously and it could well depend on the approximations used in solving Equation (2.3.11) (see, for example, Durney and Roxburgh, 1971). *It is important to note that unlike the case of the large-scale convection, we do not know whether the turbulent convection is preferentially stabilized by rotation at the equator or at the poles.* Furthermore, as stressed by Iroshnikov (1969), the effect of rotation on the turbulent viscous stress tensor could also be important. Table I shows that at a depth of 1.1×10^{10} cm, a typical time associated with the turbulent convection is $l/u_c \sim 10^6$ s, *which is not small* in relation to $1/\Omega_0 \sim 10^6/2.57 \text{ s}^{-1}$. *Therefore, the effect of rotation on the turbulent convection, particularly in the lower half of the convection zone, cannot be ignored.*

Cowling (1951) has studied the local conditions for instability in polytropic rotating stars. He finds that in the case of uniform rotation, instability will occur if $(\alpha = (1 - \gamma/\Gamma)/\gamma H)$

$$\alpha[m^2 + (l \cos \theta - n \sin \theta)^2] > 4\Omega_0^2 n^2 . \qquad (2.3.19)$$

Here Γ is the polytropic index $(p/p_0 = (\rho/\rho_0)^\Gamma)$; H is the scale height, defined by $H = p/\rho g = p/|\nabla p|$; θ is, as usual, the polar angle and m, l, and n are defined by $\partial/\partial R = il$, $\partial/R \partial\phi = im$, $\partial/\partial z = in$ (Cowling takes cylindrical polar coordinates R, ϕ, and z with the star's center as the pole and its axis of rotation as the polar axis). In a spherical system of coordinates, let us define N and L by $\partial/\partial r = iN$, $\partial/r\partial\theta = iL$; $1/N$ and $1/L$ are therefore the radial and latitudinal dimensions of the turbulence. At the equator, Equation (2.3.19) can then be written $\alpha[m^2 + L^2] > 4\Omega_0^2 L^2$; whereas at the poles, $\alpha[m^2 + L^2] > 4\Omega_0^2 N^2$. Therefore, turbulent convection will be preferentially stabilized at the poles if $N > L$ or $1/N < 1/L$, that is, if the convective cells are flattened in the radial direction; if, on the other hand, *the turbulent motions are elongated in the radial direction, then it is the convection at the equator that will be preferentially stabilized by rotation.* The shape of the convective cells is presumably determined by complex phenomena such as nonlinear interactions of the velocity and temperature fields, the variations of density with height, and the action of rotation on the convective motions; it is not known at present whether the convective cells are elongated or flattened in the radial direction. (It should be noted, however, that Simon and Weiss, 1968, and Parker, 1973a, 1973b have argued that the convective motions, in the absence of rotation, extend over several scale heights.) Very little is also known about the effect of nonuniform rotation on the turbulence (cf. Cowling, 1951).

We return now to the results of the numerical calculations pertaining to the solutions of Equations (D3), (D4) and the linearized versions of Equations (E1) and (E2) (as stated above $s = 1$, $\gamma = \frac{5}{3}$, $\nu = 2 \times 10^{12}$ cm^2 s^{-1}). In Figures 6a and 6b, $\omega_0(r)$ and $\omega_2(r)$ are plotted against r; $\omega_0(r)$ and $\omega_2(r)$ are in excellent agreement with previous calculations (Durney, 1974a); it should be noted that $\omega_0(r)$ increases with depth, whereas $\omega_2(r)$ decreases inward. *This appears to be a typical behavior of the solutions of Equations (E1) and (E2) (with $s = 1$ and free stress boundary conditions at $r = R_c, R_0$) when $\omega_0(r)$ and $\omega_2(r)$ are slowly varying functions of r* (cf. Section 3).

The pole–equator differences in flux are small: $\Delta \mathcal{F}/\mathcal{F}_0 \sim 10^{-3}$ in the bulk of the convection zone and $\sim 10^{-6}$ in the surface layers. We have neglected here the energy carried by the meridional motions; if this energy is important we would expect two cells to develop in the radial direction (Durney, 1972a): assume that convection is preferentially stabilized at the poles by rotation. In the lower part of the convection zone where the effect of rotation on convection is large, meridional motions should be generated that will increase the equatorial flux; these motions should therefore

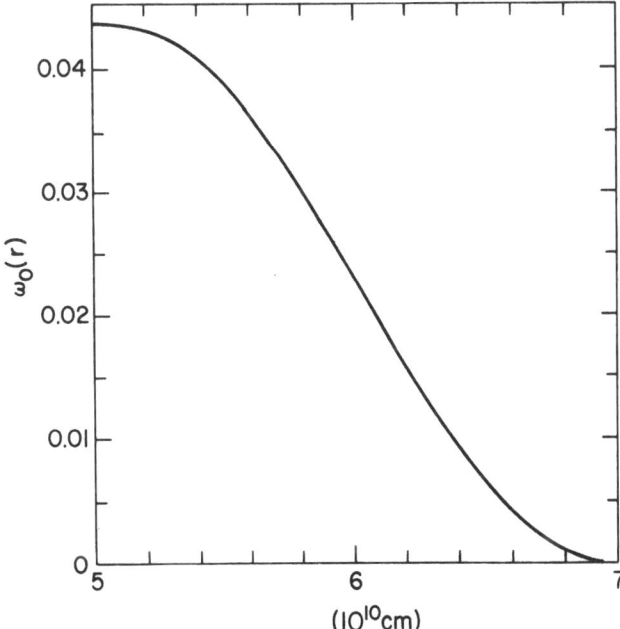

Fig. 6a. $\omega_0(r)$ versus depth. The angular velocity is given by $\Omega = \Omega_0(1 + \omega_0(r) + \omega_2(r)(3 \cos^2 \theta - 1)/2)$.

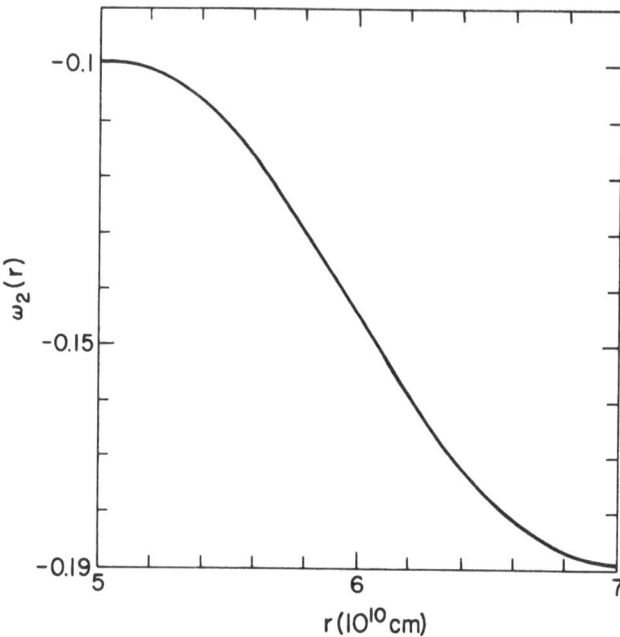

Fig. 6b. $\omega_2(r)$ versus depth.

rise at the equator and sink at the poles. In the upper part of the convection zone the effect of rotation on convection becomes negligible and a counter cell should develop (motions sinking at the equator and rising at the poles) as the convection zone relaxes to a state in which the effect of rotation is unimportant, i.e., to a nonrotating state. This counter cell redistributes the pole–equator differences in flux set up in the lower part of the convection zone, and the direction of these motions is such that they give rise to differential rotation with equatorial acceleration (Kippenhahn, 1963). It should be noted that if only one cell is present and the motions rise at the poles and sink at the equator, then the equatorial flux can be larger than the polar flux at the base of the convection zone $(r = R_c)$ if the meridional motions penetrate (even slightly) into the radiative region. We take ($\mathscr{F}_r$ is the latitudinally dependent component of the radial flux)

$$\mathscr{F}_r = \mathscr{F}(r)P_2(\cos\theta); \qquad U_r = \frac{2\psi(r)}{\rho r^2}P_2(\cos\theta) \tag{2.3.20}$$

and approximate the energy equation by

$$dr^2\mathscr{F}_r/dr = r^2 c_p\rho U_r(\nabla\Delta T)_u. \tag{2.3.21}$$

Since in the stable radiative core $(\nabla\Delta T)_u < 0$, the left-hand side of the above equation will be larger than zero if $U_r < 0$ (as we have assumed to be the case at the equator). Therefore in the radiative region downward motions transport energy upward and, in consequence, the equatorial flux will be larger than the polar flux at the base of the convection zone $(r = R_c)$ if the energy flux is spherically symmetric inside the radiative core. It should also be noted that since $|\nabla\Delta T|$ is 'large' in the radiative region, very small velocities can carry large amounts of energy flux.

The meridional motions in the convection zone which are needed to generate the observed solar differential rotation are of the order of $U_r \sim 10$ cm s^{-1} (Durney, 1974a). The same value is obtained from the solution of Equations (D3) and (D4). It is of interest to calculate the energy flux carried by these motions. We assume again that the radial components of the energy flux and velocity field are given by Equation (2.3.20) and that the energy equation is given by Equation (2.3.21). Since $v = lu_c/3$, an integration of Equation (2.3.21) gives

$$\delta \mathscr{F}_r = \mathscr{F}_r(r = R_0) - \frac{R_c^2}{R_0^2} \mathscr{F}_r(r = R_c) = \tfrac{2}{3} \mathscr{F}_0 \int_{R_c}^{R_0} (U_r/v) \, dr \,.$$

If we take $v = 2 \times 10^{12}$ cm^2 s^{-1} and $U_r(r) = 10$ cm s^{-1}, we obtain

$$\delta \mathscr{F}_r / \mathscr{F}_0 \sim \int_{R_c}^{R_0} (U_r/v) \, dr \sim P_2(\cos \theta)/10 \,. \tag{2.3.22}$$

If $\mathscr{F}_r(r = R_0) = 0$ (no pole–equator difference in flux at the surface) then Equation (2.3.22) shows that the pole–equator difference in flux at $r = R_c$ is the order of $\Delta \mathscr{F}/\mathscr{F} \sim \frac{1}{10}$. The values of $\Delta \mathscr{F}/\mathscr{F}$ found from the solution of Equations (D3), (D4), (E1), and (E2) are smaller than $\frac{1}{10}$: the value used for the turbulent viscosity $(2 \times 10^{12}$ cm^2 s$^{-1})$ is too small to justify neglecting the energy flux carried by the meridional motions. There can be little doubt that the terms of the energy equation that were neglected in Equation (2.3.21) would lower the values of $\Delta \mathscr{F}/\mathscr{F}$. Equation (2.3.22) *suggests, however, that in this theory of differential rotation, the value of $\Delta \mathscr{F}/\mathscr{F}$ has to be significant at the base of the convection zone.*

It is apparent from Figure 6b that $|\omega_2(r)|$ decreases slowly inward. As a consequence, the large viscosity approximation is not valid in the lower part of the convection zone. We assume conservatively large values for $U_r' \sim \rho' U_r/\rho$; $U_\theta \sim r\rho' U_r/\rho$; $U_\theta' \sim (r\rho'/\rho)^2 U_r/r$. It is readily seen from Equations (2.3.1) and (2.3.2) that the largest viscous terms of the r and θ equations of motions are $V_r = (r\rho'/\rho)^2 v U_r\rho/r^2$ and $V_\theta = (r\rho'/\rho)^3 v U_r\rho/r^2$, respectively; the largest terms containing the meridional velocities are $M_r = (r\rho'/\rho)^2 U_r^2 \rho/r$ and an identical expression for the θ-equation $(M_\theta = M_r)$. In Table II the ratios $2\Omega_0^2 \rho |\omega_2| r/V_r$ and $2\Omega_0^2 r |\omega_2/V_\theta|$ are tabulated as a

TABLE II

Ratio of the Coriolis force to the viscous terms as a function of depth in the radial and latitudinal equations of motion

d (cm)	3.4(6)	1.3(7)	3.7(7)	1.5(8)	3.6(8)
r-Eq	2.6($-$1)	2.9($-$1)	2.5	4.4(1)	2.5(2)
θ-Eq	8.4($-$6)	3.6($-$5)	8.7($-$4)	6.3($-$2)	8.7($-$1)
$|r\rho'/\rho|$	3.1(4)	8.1(3)	2.8(3)	7.0(2)	2.9(2)
d (cm)	1.0(9)	3.3(9)	1.1(10)	1.8(10)	1.9(10)
r-Eq	1.3(3)	7.2(3)	4.2(4)	1.3(5)	2.1(5)
θ-Eq	1.2(1)	2.4(2)	5.3(3)	2.9(4)	5.1(4)
$|r\rho'/\rho|$	1.0(2)	3.0(1)	8	4.3	4

function of r for $U_r = 10$ cm s^{-1} and $|\omega_2| = 0.1$; the value of ν was taken from Table I and $r\rho'/\rho$ was calculated by assuming that ρ is given by a polytropic state (cf. Equations (2.3.14)) with $\gamma = \frac{5}{3}$. The values of p_{uc}, ρ_{uc}, T_{uc} (the pressure, density, and temperature at the bottom of the convection zone) are from Baker and Temesvary's tables, and the value of gravity in Equation (2.3.15) was chosen so that $L = 1.9 \times 10^{10}$ cm ($g = 3.7 \times 10^4$ cm^2 s^{-1}). The ratios of the Coriolis force to M_r and M_θ behave very much in the same way as do the values listed in Table II for the r-equation since U_r and v/r are of the same order. It is clear from Table II that the large viscosity approximation is valid only in the surface layers of the convection zone. It is important, however, to realize that the results plotted in Figures (6a) and (6b) and the corresponding values of the meridional motions are more general than the large viscosity approximation: if the angular velocity is a slowly varying function of r and the azimuthal stresses vanish at $r = R_c$, R_0, then $\omega_0(r)$ and $\omega_2(r)$ will behave very much as in Figure 6 for motions rising at the poles and sinking at the equator (see Section 3.1.).

2.4. Rotation in Cylinders

If differential rotation penetrates deeply into the convection zone, then the large viscosity approximation considered in the previous section is not valid. In fact, the *inviscid* radial and latitudinal equations of motion appear to be better approximations (Gierasch, 1974). It is not possible at present to rule out the possibility that differential rotation is a surface effect. The opposite appears more likely, however. (i) According to Parker (1975), the solar dynamo must be driven in the lower part of the convection zone; in the upper part magnetic buoyancy is so efficient that the magnetic field rapidly floats to the surface. (ii) according to Yoshimura (1975a), a latitudinal differential rotation must be present for the solar cycle models to reproduce the observed butterfly diagram.

We will assume, therefore, that differential rotation is not a surface phenomenon and that $\omega_2 \sim = -0.1$ is a typical value of $\omega_2(r)$ in the lower part of the convection zone.

It was seen in the last section that for all reasonable values of an axisymmetric velocity field, the viscous terms and the terms in U_r and U_θ in the r- and θ-equations of motion can be neglected in the lower part of the convection zone (Gierasch, 1974; the reader is referred to this paper for an order-of-magnitude estimate of all terms in these equations).

Global convection is essentially non-axisymmetric. We estimate now the largest values of the large-scale convective velocities (U_r^c, U_θ^c, U_ϕ^c) such that the terms in U_r^c, U_θ^c, U_ϕ^c of the r- and θ-equations of motion are nevertheless smaller than the Coriolis force. As a guide to the relative magnitude of U_r^c, U_θ^c, and U_ϕ^c, we use Equations (B1) of Appendix II: $U_r^c \sim U_\phi^c$ and U_θ^c somewhat smaller than U_r^c or U_ϕ^c. For the derivatives with respect to r, θ, and ϕ we take $\partial/\partial r \sim \rho'/\rho$; $\partial/r\partial\theta \sim L/r$; $\partial/r\partial\phi \sim L/r$; an appropriate value of L is ~ 10 (Appendix I). In consequence, $r\rho'/\rho \gtrsim L$ since in the lower part of the convection zone $r\rho'/\rho \sim 5$ and $r\rho'/\rho$ increases with r. Therefore, $\partial/\partial r \gtrsim (\partial/r\partial\theta, \partial/r\partial\phi)$. The Navier-Stokes equations for the case of nonaxial symmetry have been written down, for example, by Pai (1956). It is readily seen that for the r-equation: the largest viscous term is $V_r = (r\rho'/\rho)^2 \nu U_r^c \rho/r^2$ and that the largest inertial terms containing the global convective velocities are $I_{1r} = \rho'(U_r^c)^2$ and $I_2 = 2\rho\Omega_0 U_\phi^c$. For the θ-equation the largest viscous term is $V_\theta = (r\rho'/\rho)^2 \nu U_{\theta\rho}^c/r^2$ and the largest intertial terms (apart from the Coriolis term, $2\Omega_0^2\rho\omega_2 r$) are $I_{1\theta} = \rho' U_r^c U_\theta^c$ and $I_2 = 2\rho\Omega_0 U_\phi^c$.

In Table II (r-equation), the ratio of the Coriolis force ($2\Omega_0^2\rho|\omega_2|r$) to V_r and V_θ are tabulated for $U_r^c = U_\theta^c = 10$ cm s^{-1}. It is clear that even if U_r^c and U_θ^c are as large as 10^4 cm s^{-1} (larger, therefore, than the turbulent convective velocities; cf. Table I), the viscous terms due to the global convection can be neglected in the r- and θ-equations (in the lower part of the convection zone).

Consider now the inertial terms. The values of U_r^c and U_ϕ^c (such that $I_{1r} \sim I_{1\theta}$ and I_2 are of the order of the Coriolis force, $2\rho\Omega_0^2\omega_2 r$) are $U_r^c \sim 2\times 10^4$ cm s^{-1} and $U_\phi^c \sim 10^4$ cm s^{-1} (we have used $r\rho'/\rho = 10$; $|\omega_2| = 0.1$, and $r = 6\times 10^{10}$ cm). *Therefore if the global convective velocities are not larger than the turbulent velocities, then the viscous and inertial terms due to the large-scale convection can be neglected, in the r- and θ-equations, in the lower part of the convection zone.* Combining this result with the previous one pertaining an axisymmetric circulation, we conclude that in the lower part of the convection zone the r- and θ-equations of motion can be written

$$\rho U_\phi^2/r - \rho g = \partial p/\partial r \tag{2.4.1}$$

$$\rho U_\phi^2 \cot g\, \theta/r = \partial p/r\,\partial\theta. \tag{2.4.2}$$

Equations (2.4.1) and (2.4.2) were used by Gierasch (1974) in his work on differential rotation.

We evaluate now the perturbations in the convective flux due to differential rotation. The convective flux and the superadiabatic gradient are given by Equation (2.2.1). Since $u_c = l/2(g \cdot \nabla\Delta T/T)^{1/2}$ we can write

$$\mathscr{F}_r = \frac{l^2}{4} g^{1/2}\rho c_p A_r^{3/2}/T^{1/2}; \qquad \mathscr{F}_\theta = \frac{l^2}{4} g^{1/2}\rho c_p A_r^{1/2} A_\theta/T^{1/2} \tag{2.4.3}$$

with

$$A_r = (1 - 1/\gamma)\frac{T}{p}\frac{\partial p}{\partial r} - \frac{\partial T}{\partial r}; \qquad A_\theta = (1 - 1/\gamma)\frac{T}{pr}\frac{\partial p}{\partial\theta} - \frac{1}{r}\frac{\partial T}{\partial\theta}. \tag{2.4.4}$$

The absence of rotation defines the unperturbed state designated by a sub-index, u. We neglect second-order terms in the perturbation (rotation). The superadiabatic gradient is 'small'. Therefore, we expect the perturbations in the superadiabatic gradient to be the ones that are the most important, i.e. $\delta\nabla\Delta T/(\nabla\Delta T)_u \gg \delta p/p_u$, $\delta\rho/\rho_u$, $\delta T/T_u$. We take, therefore,

$$\mathscr{F}_r = \frac{l_u^2}{4} g^{1/2}\rho_u c_p A_r^{3/2}/T_u^{1/2}; \qquad \mathscr{F}_\theta = \frac{l_u^2}{4} g^{1/2}\rho_u c_p A_{ru}^{1/2} A_\theta/T_u^{1/2}. \tag{2.4.5}$$

It is important to keep in mind that we neglected the effect of rotation on l. With the help of Equations (2.4.1) and (2.4.2) it is readily found that A_r and A_θ can be written

$$A_r = (-g + U_\phi^2/r)/c_p - \partial T/\partial r \tag{2.4.6}$$

$$A_\theta = U_\phi^2 \cot g\, \theta/rc_p - (1/r)\partial T/\partial\theta. \tag{2.4.7}$$

To evaluate A_r and A_θ we need to calculate $\partial T/\partial r$ and $\partial T/\partial \theta$ as given by Equations (2.4.1) and (2.4.2), and the gas equation. Equations (2.4.1) and (2.4.2) will be solved with the help of expansions of the form

$$U_\phi = \Omega_0 r \sin \theta (1 + \omega_0(r) + \omega_2(r)P_2(\cos \theta)) ;$$

$$p = p_u + p_0 + p_2 P_2(\cos \theta) + p_4 P_4(\cos \theta) ; \qquad \rho = \rho_u + \cdots ; \qquad (2.4.8)$$

$$T = T_u + \cdots .$$

Differential rotation will be assumed to be small, that is ω_0^2, ω_2^2, and $\omega_0\omega_2$ will be neglected (as well, of course, as the products of p_i, ρ_i, T_i ($i = 0, 2, 4$)). With the help of these approximations we can evaluate T in terms of the unperturbed state and of $\omega_0(r)$ and $\omega_2(r)$. The mathematical details are given in Appendix VI.

The perturbed spherically symmetric part of $A_r[-\frac{2}{3}(\Omega_0^2 r/c_p)(1+2\omega_0-2\omega_2/5)-\partial T_0/\partial r]$ must be zero since rotation cannot alter the emerging flux. This is just an equation for $\partial T_0/\partial r$. To evaluate the remaining terms of A_r and A_θ, we approximate the unperturbed state by a polytrope (Equations (2.3.14) and (2.3.15)); in this case the following equality holds:

$$\mu/R_g + \rho'_u T_u/g\rho_u = 1/c_p . \qquad (2.4.9)$$

Substituting Equations (F1), (F2), and (F4) of Appendix VI into Equations (2.4.6) and (2.4.7), we obtain, with the help of Equation (2.4.9):

$$A_r = \frac{2}{7} \frac{\Omega_0^2 T_u}{g} \left[\frac{1}{3} \frac{d}{dr}(rB_1)P_2 + \frac{3}{5} \frac{d}{dr}(rB_2)P_4 \right] \qquad (2.4.10)$$

$$A_\theta = \frac{2}{7} \frac{\Omega_0^2 T_u}{g} \left[\frac{1}{3} B_1 \frac{\partial P_2}{\partial \theta} + \frac{3}{5} B_2 \frac{\partial P_4}{\partial \theta} \right] \qquad (2.4.11)$$

where

$$B_1 = 12\omega_2 + r\omega'_2 + 7r\omega'_0 , \qquad B_2 = r\omega'_2 - 2\omega_2 . \qquad (2.4.12)$$

The radial and latitudinal components of the convective flux are given by Equation (2.4.5). In Table III we list the values of $\Delta\mathcal{F} = (l_u^2/4)g^{1/2}\rho_u c_p A_{ru}^{1/2}\frac{3}{2}[\frac{2}{7}(\Omega_0^2 T_u/g)]/T_u^{1/2} = \frac{3}{2}\mathcal{F}[\frac{2}{7}(\Omega_0^2 T_u/g)/(\nabla\Delta T)_u]$ as a function of depth; $\Delta\mathcal{F}$ gives an indication of the pole–equator differences in flux if the terms inside the bracket of Equation (2.4.10) do not vanish. Furthermore, $\mathcal{F}_\theta \sim \Delta\mathcal{F}$ if the terms inside the bracket of Equation (2.4.11) do not vanish. It is impossible to believe that such large values of $\Delta\mathcal{F}$ and $\mathcal{F}_\theta$

TABLE III

Values of $\Delta\mathcal{F} = \frac{3}{2}\mathcal{F}[2\Omega_0^2 T_u/7g(\nabla\Delta T)_u]$ as a function of depth d

d(cm)	3.4(6)	1.3(7)	3.7(7)	1.5(8)	3.6(8)
$\Delta\mathcal{F}$(erg cm^{-2} s)	2.4(1)	5.3(2)	1.2(3)	1.4(4)	1.8(5)

d(cm)	1.0(9)	3.3(9)	1.1(10)	1.8(10)	1.9(10)
$\Delta\mathcal{F}$(erg cm^{-2} s)	8.8(6)	6.0(8)	2.8(10)	7.8(10)	6.3(10)

could exist inside the convection zone. We conclude, therefore that B_1 and B_2, given by Equation (2.4.12), must be small. If this is the case then

$$\omega_2(r) = \omega_2(R_c)(r/R_c)^2; \qquad \omega_0(r) + \omega_2(r) = \text{constant} \qquad (2.4.13)$$

and (we take the constant appearing in Equation (2.4.13) as zero).

$$\Omega = \Omega_0(1 - \tfrac{3}{2}\omega_2(R_c)(r/R_c)^2 \sin^2 \theta). \qquad (2.4.14)$$

Therefore the angular velocity is constant along cylinders. What has been shown is the following: *If in the absence of rotation the convection zone is adiabatic, and if Equations (2.4.1) and (2.4.2) are good approximations for the radial and latitudinal equations of motion, then in the lower half of the convection zone the perturbations in the convective flux (produced by differential rotation) are unacceptably large unless the angular velocity is constant along cylinders.* (Of course, this result holds only if the value of the latitudinal differential rotation is not too small in the lower half of the convection zone; we have assumed also that l is not a function of θ.)

Intuitively this result can be seen as follows: assuming that $A_r = A_\theta = 0$, by cross differentiation of Equations (2.4.6) and (2.4.7) we obtain (g and c_p are constants)

$$\frac{1}{r}\frac{\partial}{\partial\theta} U_\phi^2 - \frac{\partial}{\partial r}\text{cotg } \theta U_\phi^2 = 0. \qquad (2.4.15)$$

The general solution of this equation is $U_\phi = f(r \sin \theta)$. This, of course, is not surprising: *if $A_r = A_\theta = 0$ then the structure of Equations (2.4.6) and (2.4.7) is the same as that of the radial and latitudinal equations of motion of an incompressible rotating fluid.* It is well known that in this case Ω must be constant along cylinder (Taylor-Proudam theorem). If Ω is not constant along cylinders then we expect A_r and A_θ to be large because in the unperturbed, nonrotating state, the superadiabatic gradient is 'small': *Therefore any small perturbation can nevertheless give rise to large perturbations in the superadiabatic gradient.*

It should be stressed that it has not been shown that rotation in cylinders is a consequence of the energy and momentum equation. Whether this is indeed the case is an important question that remains to be settled. If the meridional motions are given by Equations (2.3.7), the problem is then to solve Equations (2.4.1) (2.4.2), the azimuthal equation (2.3.3), and the energy Equation (2.2.3). If $\Omega(r, \theta)$ is given by Equation (2.4.14), then the azimuthal equation of motion (2.3.3) determines the meridional velocities. An estimate of these velocities can be obtained from Equation (F6) of Appendix VI. If $\omega_2(R_c) = -0.1$ and $\xi(r \sin \theta) = 0$, then $U_r \sim 40$ cm s^{-1} at $r = 6.2 \times 10^{10}$ cm. Let $\delta\mathcal{F}$ be the perturbation in the energy flux ($\delta\mathcal{F} = \tfrac{3}{2}\mathcal{F}\mathbf{A}/(\nabla\Delta T)_u$ with A_r and A_θ given by Equations (2.4.10) and (2.4.11)). It is readily seen that

$$\delta\mathcal{F}_r = \frac{45}{28}\frac{\nu\Omega_0^2 \rho_u}{g}B_r \qquad (2.4.16)$$

$$\delta\mathcal{F}_\theta = \frac{45}{28}\frac{\nu\Omega_0^2 \rho_u}{g}B_\theta \qquad (2.4.17)$$

where B_r and B_θ are the expressions in brackets in Equations (2.4.10) and (2.4.11). If B_r and B_θ are not 'small' ($B_r \sim B_\theta \sim 1$) then, since the pressure is a strongly varying

B. R. DURNEY

function of r,

$$\operatorname{div} \delta \mathscr{F} = -\tfrac{45}{28}\nu\Omega_0^2\rho_u B_r \tag{2.4.18}$$

and the energy equation (Equation 2.2.3) becomes

$$-\tfrac{45}{28}\nu\Omega_0^2\rho_u B_r \sim c_p\rho_u U_r(\nabla\Delta T)_u = \mathscr{F}U_r/\tfrac{3}{2}\nu. \tag{2.4.19}$$

At $r = 6\times10^{10}$ cm, $\nu \sim 10^{13}$ cm^2 s^{-1} and $\rho_u \sim 7.6\times10^{-2}$ g cm^{-3}. For the left- and right-hand sides of Equation (2.4.19) we obtain 10 B_r and 0.23 cgs units, respectively (we used $U_r = 40$ cm s^{-1}). Therefore, even for values of U_r as large as 40 cm s^{-1}, the energy equation appears to favor small values of B_r. If B_r is small, then rB_1 and rB_2 must be slowly varying functions of r; we take

$$rB_1 = \beta R_c; \qquad rB_2 = \alpha R_c \tag{2.4.20}$$

with α and β, two constants. It is readily seen that the solutions of Equations (2.4.20) are given by

$$\omega_2(r) = -\frac{\alpha}{3}\left(\frac{R_c}{r}\right) + \xi(r/R_c)^2; \qquad \omega_0(r) = \frac{-(11\alpha + 3\beta)}{21}\frac{R_c}{r} - \xi\left(\frac{r}{R_c}\right)^2 \tag{2.4.21}$$

where ξ is constant.

If $\omega_0(r)$ and $\omega_2(r)$ are given by Equations (2.4.21) then the energy equation reduces to

$$\delta\mathscr{F}_\theta/r \sim \mathscr{F}U_r/\tfrac{3}{2}\nu. \tag{2.4.22}$$

At $r = 6\times10^{10}$ cm, the left- and right-hand side of Equation (2.4.22) are equal to 0.6 B_θ and 0.23 cgs units, respectively ($U_r = 40$ cm s^{-1}). These quantities are of the same order if $B_\theta \sim 1$. In the present very crude analysis we have assumed that $\mathbf{U}\cdot\nabla\Delta T \sim U_r(\nabla\Delta T)_u$ with $U_r \sim 40$ cm s^{-1}. It is clear that a much more careful study of this problem is needed before we can resolve whether the momentum and energy equations indeed favor a constant angular velocity along cylinders.

For the mixing length l we have taken its unperturbed value; *we have neglected therefore the latitudinal and radial dependence of the stabilizing effect of rotation on convection* (cf. Equation 2.3.19). An estimate of the magnitude of this effect is an important problem that remains to be solved and it is not known at present if therein lies the solution to the heat flux problem. It is appealing, of course, to think that *differential rotation arises as a way for the convection zone to minimize the constraints imposed by rotation on convection and so to maintain a uniform heat flux.*

It is possible, perhaps to give a non-rigorous argument against the existence of large $\Delta\mathscr{F}$'s inside the convection zone: in the surface layers ($d < \sim 3\times10^9$ km) the effect of rotation on turbulent convection becomes negligible and the convective flux should be well described by Equation (2.2.1), with l the unperturbed mixing length. The values of $\Delta\mathscr{F}$ listed in Table III are small for $d < \sim 10^9$ km. The existence of larger values of $\Delta\mathscr{F}$ would imply that Equations (2.4.1) and (2.4.2) are not good approximations at these depths; therefore, large velocities would have to be present in the Sun's surface layers.

It follows from Table III that in the upper part of the convection zone the perturbations in the convective flux are very small even if Ω is not constant along cylinders. In consequence, if $\Delta\mathcal{F}$ is negligible at a certain depth ($d \sim 2 \times 10^4$ km, for example), the energy equation will be approximately satisfied if $\psi = 0$. It is of interest, therefore, to consider the solution of the equations for $\omega_0(r)$ and $\omega_2(r)$ (Equations (E1) and (E2) of Appendix V) in the case when $\psi = 0$. In Table IV we list the values of $\omega_0(r)$ and $\omega_2(r)$ as a function of r in the upper part of the convection zone for $\psi = 0$, zero stresses at $r = R_0$ and $\omega_2(R_0) = -0.189$. (For the density a polytropic relation with $\gamma = 5/3$ was used.)

Assume that differential rotation is generated in the surface layers of the Sun (by transport of angular momentum by the Reynolds stresses, for example) and that $\Delta\mathcal{F}$ and ΔT are very small at $r = R_0$; it would be of interest to know the dependence of Ω with depth if the momentum and energy equations are solved with boundary conditions expressing that $\Delta\mathcal{F}$ and ΔT are small at the surface and that $\omega_2(r) = -0.189$ at $r = R_0$. If differential rotation is indeed a surface phenomenon, we expect that Ω will behave as follows: from Table IV ($s = 1$), $|\omega_2(r)|$ should first increase inward; with increasing depth, meridional motions will be driven by the pole-equator differences in flux that will appear (cf. Table III). These meridional motions should be such as to decrease $\omega_2(r)$ and make it eventually vanish.

In any theory of the Sun's differential rotation, the value of Ω at the surface should, of course, agree with the observed differential rotation of the Sun. Furthermore, it is well known that theories of the solar cycle (Parker, 1955; Steenbeck and Krause, 1969; Leighton, 1969; Parker, 1971; Deinzer and Stix, 1971; Roberts and Stix, 1972; Lerche and Parker, 1972; Yoshimura, 1972, 1975a, b; Köhler, 1973) are sensitive to the values of $d\Omega/d\theta$ and to $d\Omega/dr$ in particular. The dependence of Ω on depth and latitude must therefore satisfy another important constraint: this dependence must be such that theories of the solar cycle reproduce the observed properties of the solar activity cycle. It could be thought that an angular velocity which decreased inward (as in the case if Ω is constant along cylinders) is necessarily in contradiction to theories of the solar cycle. However this is not so. To ensure that the solar dynamo displays the correct butterfly diagram (i.e., the mean toroidal field migrates towards the equator) the product $\alpha \partial\Omega/\partial r$ must be negative (Stix, 1974). According to Yoshimura (1975a), the α-term (due to the global convection) changes sign in the lower part of the convection zone and is negative. *Therefore rotation in cylinders could be compatible with Yoshimura's model of the solar cycle (Yoshimura, 1975b) if the observed toroidal magnetic field is generated in the lower part of the convection zone, as has recently been argued by Parker* (1975). As discussed more fully by Stix (1976a, 1976b), however, the observations show that there is a phase relation between the toroidal and poloidal magnetic fields which could not be satisfied if Ω decreases inward.

3. Anisotropic Viscosity and Normal Modes of Vibration

3.1 ANISOTROPIC VISCOSITY

Since gravity in the solar convection zone is a preferred direction, there is no reason to expect that the turbulent viscosity in the directions parallel and perpendicular to

TABLE IV

Values of $\omega_0(r)$, $\omega_2(r)$ and $\Omega(r, \theta = \pi/2)$ (angular velocity at the equator in 10^{-6} rad s^{-1}) as a function of r for $\psi = 0$

r(cm)		7(10)	6.9(10)	6.8(10)	6.7(10)	6.6(10)	6.4(10)	6.2(10)	6.0(10)
$s = 0.8$	$\omega_0(r)$	0	5.5(−3)	1.1(−2)	1.7(−2)	2.3(−2)	3.5(−2)	4.7(−2)	5.9(−2)
	$\omega_2(r)$	−0.189	−0.190	−0.192	−0.194	−0.196	−0.203	−0.211	−0.224
	$\Omega_{eq}(r)$	2.81	2.83	2.85	2.86	2.88	2.92	2.96	3.01
$s = 1$	$\omega_0(r)$	0	−3.6(−5)	−1.6(−4)	−3.8(−4)	−6.8(−4)	−1.7(−3)	−3.2(−3)	−5.3(−3)
	$\omega_2(r)$	−0.189	−0.189	−0.190	−0.191	−0.192	−0.197	−0.205	−0.216
	$\Omega_{eq}(r)$	2.81	2.81	2.81	2.81	2.82	2.82	2.82	2.83
$s = 1.2$	$\omega_0(r)$	0	−5.6(−3)	−1.2(−2)	−1.8(−2)	−2.4(−2)	−3.7(−2)	−5.1(−2)	−6.6(−2)
	$\omega_2(r)$	−0.189	−0.188	−0.188	−0.188	−0.189	−0.192	−0.199	−0.209
	$\Omega_{eq}(r)$	2.81	2.80	2.78	2.77	2.75	2.72	2.69	2.67

gravity will be the same. If the viscosity is anisotropic, the convection zone cannot rotate uniformly (Bierman, 1951); this theory of differential rotation has been developed by Kippenhahn (1963), Cocke (1967), Köhler (1970), Sakurai (1966), Iroshnikov (1969), Rüdiger (1974), (1976), Roxburgh (1974), and Durney (1974a). There can be little doubt about the soundness of the basic idea of this theory of differential rotation: the turbulent viscosity should indeed be anisotropic. The question that remains to be answered is how important this anisotropy is in generating the Sun's differential rotation.

The ratio of the kinematic turbulent viscosity coefficients in the directions perpendicular and parallel to gravity $(\nu_\perp/\nu_\parallel)$ will be assumed to be a constant, independent of depth and latitude: therefore (Kippenhahn, 1963) $\nu_{\theta\theta} = \nu_{\phi\phi} = \nu_\perp = s\nu_\parallel = s\nu_{rr} = s\nu$.

We take the meridional motions and the angular velocity to be given by Equations (2.3.9) and (2.3.17), respectively; the equations for $\omega_0(r)$ and $\omega_2(r)$ are then Equations (E1), (E2), and (E3) of Appendix V. In the large viscosity limit, Equation (E2) is satisfied if $\omega_2 = 0$, and Equation (E1) gives then $(\rho r^4 \omega_0')' + 2(1-s)(r^3\rho(\omega_0+1))' = 0$, or $\rho r^4(\omega_0' + 2(1-s)(1+\omega_0)/r) = $ constant. If the azimuthal stress, $T_{r\phi}$, vanishes at $r = R_0$, for example, then it is readily seen from Equation (C2c) of Appendix III that $\omega_0' + 2(1-s)(1+\omega_0)/r$ must also vanish. Therefore,

$$\omega_0 = \omega_0(R_0)(R_0/r)^{2(1-s)} .$$ (3.1.1)

This rotation law replaces solid rotation when $s \neq 1$.

To understand some general properties concerning the behavior of $\omega_0(r)$ and $\omega_2(r)$, we write Equations (E1) and (E3) of Appendix V in the following form:

$$\omega_0' + 2(1-s)(1+\omega_0)/r = \frac{2}{r^4\rho} \int_{R_0}^{r} \left[\omega_2 s\rho r^2 + \frac{\psi}{3\nu}\frac{d}{dr}(r^2(1+\omega_0+2\omega_2/5)) \right] dr$$

$$-\frac{2\psi}{3\nu r^2\rho}(1+\omega_0-\omega_2/5) ,$$ (3.1.2)

$$2\psi/\nu = \frac{r^2\rho[\omega_2' + 2(1-s)\omega_2/r - 5(\omega_0' + 2(1-s)(1+\omega_0)/r)]}{1+\omega_0-5\omega_2/7}$$ (3.1.3)

where we have assumed that the azimuthal stress, $T_{\phi r}$, and the radial velocities vanish at $r = R_0$. If $\omega_0 + 2\omega_2/5$ is a slowly varying function of r, that is, if

$$\left| \frac{d}{dr}r^2(\omega_0+2\omega_2/5) \right| < \frac{2}{r}$$ (3.1.4)

then the term in ψ in the integral of Equation (3.1.2) reduces to $2r\psi/3\nu$. Neglecting ω_0 and $\omega_2/5$ with respect to one, Equation (3.1.2) becomes

$$\omega_0' + 2(1-s)(1+\omega_0)/r = \frac{2}{r^4\rho} \int_{R_0}^{r} \rho r^2[\omega_2 s + 2r\psi/3\nu\rho r^2] dr - 2\psi/3\nu r^2\rho$$

(3.1.5)

If the azimuthal stresses and radial velocities vanish at $r = R_c$, then the integral in

Equation (3.1.5) must be zero for $r = R_c$. Equation (3.1.3) shows that if ω_0 and ω_2 are slowly varying functions of r, the same is true of $2\psi/\nu\rho r^2$. The bracketed quantity in the integral of Equation (3.1.5) is then also a slowly varying function of r, and must therefore approximately vanish if the integral vanishes. This gives an estimate of ω_2.

$$\omega_2 \sim -\frac{2\psi}{3\nu s\rho r} . \tag{3.1.6}$$

For $\omega_2 = -0.2$ and $s = 1$, we obtain $\psi/r^2\rho \sim 10$ cm s^{-1}, in excellent agreement with previous calculations (cf. Figure 2a of Durney, 1974a). Equation (3.1.6) clearly shows that if $\omega_2 < 0$, then ψ has to be larger than zero: the meridional motions must rise at the poles and sink at the equator.

To proceed further we need an estimate of the integral $(1/\rho)\int_{R_0}^r \rho\, dr$ which appears in Equation (3.1.5). For a polytrope, and with the exception of a thin layer at the top, we have $(1/\rho)\int_{R_0}^r \rho\, dr = (\gamma - 1)/\gamma(r - R_0)$. Therefore, the term $(2/r^4\rho)\int_{R_0}^r \rho r^2\omega_2 s\, dr$ in Equation (3.1.5) is of the order of $4s|\omega_2|(R_0 - r)/5r^2 < 4s|\omega_2|(R_0 - R_c)/5r^2 \sim 4s|\omega_2|/25r$ (where we have taken $\gamma = 5/3$). Since the integral in Equation (3.1.5) must vanish for $r = R_c$, we expect its value to be smaller than $4s|\omega_2|/25r$. The last term of Equation (3.1.5) on the other hand, is of the order of $|\omega_2|s/r$. This suggests neglecting the integral in Equation (3.1.5) which then becomes

$$\omega_0' + 2(1 - s)(1 + \omega_0)/r = -2\psi/3\nu r^2\rho . \tag{3.1.7}$$

The integrated version of Equation (E3) in Appendix V then gives

$$\omega_2' + 2(1 - s)\omega_2/r = -4\psi/3\nu r^2\rho . \tag{3.1.8}$$

for $s = 1$ and for a typical value of ψ given by Equation (3.1.6), $\omega_0(r)$ and $\omega_2(r)$ behave very much as in Figures 6a and 6b; in particular $\omega_2(r)$ must increase inward since $\psi > 0$, that is, the latitudinal variations of the angular velocity decrease with depth.

If differential rotation is generated by an anisotropy in the turbulence, it is well known that $s > 1$ ($s < 1$) results in a fast (slow) rotating equator (Kippenhahn, 1963). If $s > 1$ the basic solution, in the limit of large viscosity, *decreases* inward (cf. Equation (3.1.1)). If $s = 1$, Figure 6a or Equation (3.1.7) shows that $\omega_0' < 1$ (since ψ has to be larger than zero); therefore ω_0 *increases* inward. The behavior of $\omega_0(r)$ is important for theories of the solar dynamo.

It was stated in Section (1.1) that Stenflo's interpretation of the angular velocities plotted in Figure 1 was not without difficulties. Following Stenflo, we accept that the three curves labeled 'sign of longitudinal magnetic field,' 'longitudinal magnetic field' and 'sunspots' in Figure 1 give an indication of the Sun's angular velocity at different depth in the convection zone.

In relation to differential rotation, we have come upon the fact in previous sections (2.3 and 2.4) that in the solar convection zone there appears to be two regions where different approximations hold: the surface layers and the lower part of the convection zone. This provides a natural interpretation of the two sets of curves in Figure 1: the curve labeled 'Doppler shifts' (set 1) and the other three curves (set 2) referring to the magnetized plasma could give an indication of the Sun's angular velocity in these two regions. If this is the case, a striking feature of the Sun's angular velocity in the deep layers of the convection zone *is the constant angular velocity at the equator.* It is of interest to note that this behavior is predicted by the approximate Equations (3.1.7) and (3.1.8) if $s = 1$: the angular velocity at the equator is given by $\Omega = \Omega_0(1 + \omega_0(r) - \omega_2(r)/2)$; therefore, $d\Omega/dr = \Omega_0(\omega_0'(r) - \omega_2'(r)/2) = 0$ in virtue of Equations (3.1.7) and (3.1.8) $(s = 1)$. Furthermore Table IV shows that if $s = 1$ and $\psi = 0$ the angular velocity at the equator is also very approximately constant. (In Table IV, the full equations for $\omega_0(r)$ and $\omega_2(r)$ were solved, i.e. Equations (E1) and (E2) of Appendix V.)

According to Parker (1975), the toroidal magnetic field associated with the solar activity cycle is generated in the lower part of the convection zone. It appears natural, therefore, to assume that in the three curves of 'set 2' in Figure 1, the larger is the magnetic field, the deeper in the convection zone is this field rooted. This would imply, however, that differential rotation ($|\omega_2(r)|$) increases with depth and such behavior of $\omega_2(r)$ is very difficult to understand. It is certainly in disagreement with Ω being constant along cylinders. Figure 1 shows also that the angular velocity of the magnetized plasma is larger than the photospheric angular velocity. If $s = 0.8$ and $\psi = 0$, the angular velocity at the equator and $|\omega_2(r)|$ both increase inward (cf. Table IV). This solution is, however, unlikely to be valid outside the surface layers. *Therefore, there could indeed be a region beneath the solar surface with larger equatorial angular velocity and differential rotation than at $r = R_0$.* That this region could extend down to the lower part of the convection zone appears very difficult to understand.

Köhler (1970) has solved the momentum equations with anisotropic viscosity and constant density. For $\nu = 4 \times 10^{12}$ cm^2 s^{-1} the angular velocity is constant along cylinders *as a consequence of the Taylor-Proudman theorem.* The energy equation (and it is clear from previous sections that this equation is the source of all difficulties) has, however, not been included in theories of differential rotation based on an anisotropic turbulent viscosity.

3.2. NORMAL MODES OF VIBRATION

As stated in Section 1.6, evidence has been accumulating for the existence of normal mode of vibrations of the entire Sun. It is at present too early to judge the impact that their existence will have on theories of the large-scale circulation of the solar convection zone. This impact could be of great importance.

Radiative and viscous damping of these modes occurs rather close to the surface; they provide, therefore, a mechanism for a rapid transmission of energy directly from the Sun's core to the surface. Recently Hill *et al.* (1975) have suggested that a non-negligible fraction of the solar energy flux could be carried by these normal modes.

Theoretical studies of the Sun's normal mode of vibrations in view of explaining the rigid rotation of magnetic features and the Sun's differential rotation have been carried out by Wolff (1974a, b). According to Wolff the Sun's differential rotation could be generated as follows: under the influence of nonlinear coupling mechanisms, modes with the most similar rotation rates should lock together and rotate as a single entity (designated by L hereafter). The power contained in the set of normal modes (L) is not distributed uniformly over the solar surface but concentrated mainly into regions running diagonally across the surface from pole to pole in a V-shaped pattern (cf. Figure 1 of Wolff, 1974b). The dissipation of this power in the surface layers of the sun drives large-scale flows which could then generate the observed solar differential rotation.

Even if the explanation of the Sun's differential rotation is not the one suggested by Wolff, observational and theoretical work on the Sun's normal mode of vibrations could be of importance for the understanding of the solar differential rotation. It is obvious that a mechanism coupling the surface to inner regions of the Sun (which could rotate with different angular velocities) must not be overlooked.

4. The Inner Rotation of the Sun

This paper is concerned mainly with theories of the Sun's differential rotation. The very important problem of the solar inner rotation will be discussed only briefly; in fact, it may not be possible to study both problems independently.

The observational evidence which has been summarized in Sections (1.7)–(1.9), indicates that all stars arrive at the main sequence ($t = 0$) with large angular velocities (an angular velocity given by the line $\langle \mathfrak{M} \rangle^{2/3}$ of Figure 3 would perhaps be the most reasonable assumption). Whereas stars more massive than ~F6 do not appear to lose much angular momentum during their stay in the main sequence, the evolution of stars less massive than ~F6 is quite different: at $t = 0$ they are rapid rotators and they must have rather large magnetic fields since they show strong Ca II emission (in the Sun Ca II emission is an indication of strong magnetic fields). Both the angular velocity and the magnetic field (Ca II emission) decrease with time. It is well known that stars less massive than ~F6 have appreciable surface convection zones; it appears reasonable, therefore, to assume that the magnetic field is generated by a dynamo action (as is the case for the Sun). All these observations can then be understood by invoking the angular momentum loss (due to the existence of stellar winds) experienced by stars with convection zones (Schatzman, 1962). This angular momentum loss slows down the star; as a consequence the dynamo action becomes less efficient and the magnetic field decreases. Skumanich (1972) has found that the time dependence of both Ω and B are well described by $B \sim t^{-1/2}$, $\Omega \sim t^{-1/2}$ (for values of t that are not too small; $t = 0$ corresponds to the arrival of the star at the main sequence). Estimates of the angular momentum loss (Dicke, 1964; Modisette, 1967; Weber and Davis, 1967; Alfonso-Faus, 1967) show that

$$dJ/dt = \tfrac{2}{3}\Omega_s r_A^2 dM/dt \qquad (4.1)$$

where dJ/dt and dM/dt are the angular momentum and mass loss, r_A is Alfvenic distance, and Ω_s is the surface angular velocity. From the definition of $r_A(B_A^2/4\pi\rho_A U_A^2 = 1)$ we obtain $B_A^2 r_A^2/U_A 4\pi\rho_A U_A r_A^2 = 1$; since the mass loss is given by $dM/dt = -4\pi\rho_A U_A r_A^2$, it is readily seen that Equation (4.1) can be written

$$dJ/dt = -\tfrac{2}{3}\Omega_s (B_0 R_0^2)^2/U_A \qquad (4.2)$$

where B_0 is the surface magnetic field and R_0 is the radius of the star ($B_0 R_0^2 = B_A r_A^2$). *Equation* (4.2) *shows that in the first approximation the angular momentum loss is proportional to* $\Omega_s B_0^2$: even for stars with such small magnetic fields as the Sun, r_A is much farther than the critical point and we can replace U_A by U_∞. In other words, in Equation (4.2), U_A is not a sensitive function of the magnetic field. (The stellar wind velocities could change with the age of the star, but again we expect Ω_s and B_0^2 in Equation (4.2) to be much stronger functions of t than U_A.) The magnetic field is generated by a dynamo action and is therefore itself a function of Ω_s (cf. Spiegel, 1968). There is both observational (Skumanich, 1972) and theoretical (Roberts, 1974) evidence to suggest that $B_0 \propto \Omega_s$. Equation (4.2) can then be written

$$dJ/dt = -\alpha \Omega_s^3 \tag{4.3}$$

where α can be determined from observations. A convenient way of expressing the present-day torque is in terms of $\tau_0 = -\Omega_0/(d\Omega_0/dt)$, the e-folding time for slowing down the Sun *rotating rigidly* (Dicke, 1972; Ω_0 is the present surface angular velocity of the Sun). Observations show that present-day torque is such that $\tau_0 \sim 10^{10}$ yr. If we assume that the sun rotates rigidly, Equation (4.3) becomes

$$\frac{d}{dt}\left(\frac{\Omega_s}{\Omega_0}\right) + \frac{1}{\tau_0}\left(\frac{\Omega_s}{\Omega_0}\right)^3 = 0: \qquad \tau_0 = 10^{10} \text{ yr.} \tag{4.4}$$

(If the Sun rotates rigidly, J is proportional to Ω_s, Equation (4.3) must therefore be of the type $d\Omega_s/dt = -\beta\Omega_s^3$ with β a constant; for the present-day Sun, Equation (4.4) gives $\tau_0 = -\Omega_0/(d\Omega_0/dt)$ and this determines β). The solution of Equation (4.4) is

$$\Omega_s = \Omega_s(t=0)/(1+2t(\Omega_s(t=0)/\Omega_0)^2/\tau_0)^{1/2}. \tag{4.5}$$

It is reasonable to assume that the Sun arrived at the main sequence rotating with an angular velocity given by the $\langle\mathfrak{M}\rangle^{2/3}$ line of Figure 3, that is (Durney, 1972), $\Omega_s(t=0) = 65\,\Omega_0$. In a time larger than $\tau_0/(2\times65^2) \sim 10^6$ yr, Ω_s as given by Equation (4.5) reduces to

$$\Omega_s = \frac{\Omega_0\tau_0^{1/2}}{\sqrt{2}} t^{-1/2} \tag{4.6}$$

which agrees well with the observational law found by Skumanich (1972). Furthermore, for the present-day Sun, $t = 5\times10^9$ yr $= \tau_0/2$, Equation (4.6) gives $\Omega_s = \Omega_0$. For $t = 3\times10^7$ yr and $t = 4\times10^8$ yr (the age of the Pleiades and Hyades) Equation (4.6) gives $\Omega_s = 13\,\Omega_0$ and $\Omega_s = 3.5\,\Omega_0$, respectively. The predicted angular velocity for the Hyades is in good agreement with observations (Kraft, 1967), but the predicted angular velocity for the Pleiades is somewhat high. *In all, the predictions of such a simple equation as Equation* (4.4) *are surprising.* The angular velocity at $t=0$ is not too important, since the angular momentum loss is initially very fast. In fact Equation (4.6) does not contain any information about the angular velocity at $t=0$; according to this equation, the e-folding time for the slowing down of the Sun is given by

$$\tau_s = \Omega_s/(d\Omega_s/dt) = 2\,t. \tag{4.7}$$

Could the Sun have slowed down uniformly (due to the existence of a magnetic field, for example)? The observations of the SCLERA group (Hill *et al.*, 1974; Hill and Stebbins, 1975a, b) indicate that the Sun has no appreciable quadrupole moment; they lend some support, therefore, for a uniform solar angular velocity. (The quadrupole moment is nevertheless a sensitive function of Ω (cf. Roxburgh, 1964) and the Sun's core could be rotating quite fast (~ 1 week) with the Sun still showing a small quadrupole moment.) However, there is one observation, namely the Li depletion at the surface of solar type stars (cf. Section 1.9), that could be in contradiction with a uniform angular velocity; Li depletion allows for an appealing explanation if the Sun's interior is rotating fast: the angular velocity increases inward and becomes unstable (Goldreich and Schubert, 1967; Fricke, 1968). This instability gives rise to a mild turbulence that mixes the Sun's matter down to $\sim 0.63\, R_0$, where Li is destroyed (cf. Dicke, 1971; 1972).

The simplest explanation of Li depletion would certainly be in terms of the overshooting of the convective motions into the radiative region. If this were the case the observations concerning the Li depletion would be in agreement with a uniform angular velocity. Overshooting, which is larger than earlier estimates, appears to be a complex phenomenon requiring more sophisticated theories of the turbulent energy transport than those generally used in calculating stellar convection zones (cf. Shaviv and Salpeter, 1973). It must be stressed that at present no theory exists which explains the Li depletion in terms of convective overshooting.

However, even if the Sun is rotating uniformly, turbulence could be generated in the upper part of the radiative region by the latitudinal variations of the angular velocity. In the convection zone the variations with r and θ of Ω are determined by processes that have no parallel in the radiative region. Assume, for example, that Ω is constant along cylinders in the convection zone. If Equations (2.4.1) and (2.4.2) are satisfied, the $\partial p / \partial \theta$ term is balanced mainly by Coriolis forces. As stressed by Gierasch (1974), rotation cannot be uniform for $r < R_c$ since then, as a consequence of $\partial p / \partial \theta$, an unbalanced radial gradient in the pressure would appear. *Differential rotation must therefore penetrate into the radiative region.* Because of the low viscosity of this region, instabilities of the differential rotation are expected to play an important role in determining the angular velocity law. The study of these instabilities is a complex subject (cf. Strittmatter, 1969; Spiegel and Zahn, 1970; Fricke and Kippenhahn, 1970; Zahn, 1974) and here only one will be considered, namely a shear instability (cf. Zahn, 1975) which arises for large values of the Reynolds number. If $\Delta\Omega$ is the latitudinal differential rotation ($\Delta\Omega = \Omega_{eq} - \Omega_{pole}$ for a given r), then the Reynolds number corresponding to this differential rotation is defined by $\mathcal{R}_e = r^2 \Delta\Omega / \nu$. For $\Delta\Omega \sim 0.1\, \Omega_0$ and for typical values of ν in the radiative region, $\mathcal{R}_e$ is very large, and even a weak differential rotation should be unstable. It is instructive to compare the growth rates of these instabilities (cf. Zahn, 1975) with the slowing down time of the Sun as given by Equation (4.7); $\tau_s = 2\, t$. Let L be the smallest scale which is unstable in a differential rotation $\Delta\Omega$, then $L^2 \Delta\Omega / \nu = \mathcal{R}_e^{\text{crit}} \sim 10^3$. The instability will grow with an e-folding time, L^2 / ν, ~ 500 yr (much smaller than $2\, t$) if $\Delta\Omega = 10^{-7}$ rad s^{-1}.

The penetration of the differential rotation into the radiative region can be visualized as follows: shear instabilities tend to smooth out $d\Omega/d\theta$, giving rise to unbalanced pressure forces which drive meridional motions; $\Delta\Omega$ will decrease inward, and at a certain r (r_1), $\Delta\Omega$ will become negligible. It is unknown at present if this r_1 is small enough to explain the Li burning.

If, instead of a shear instability, the Golreich-Schubert and Fricke instability is important, then the radiative region would tend to rotate in cylinders but with the angular velocity *increasing inwards*. Again instabilities will occur in the upper part of the radiative region because of the different rotation laws in this region and in the convection zone where rotation could be constant along cylinders but with the angular velocity *decreasing inwards*.

If the solar angular velocity is not uniform (and in particular if no magnetic fields are present) the solar spin-down problem is of great complexity. The most sophisticated treatment at present is that of Sakurai (1975). The reader is referred to this paper and to the review article by Benton and Clark (1974) for further references on this subject.

5. Discussion

It is clear that we are not yet certain of the real origin of the Sun's differential rotation; in fact we do not even know whether it can be completely explained within the framework of the equations commonly used in this subject: assume, for example, that differential rotation is generated as a way for the convection zone to minimize the constraints imposed by rotation on convection; this could be achieved by variations in depth and latitude of certain physical parameters (e.g., the turbulent thermal diffusivity and viscosity) that are taken as constants. In fact, it is difficult to avoid the feeling that some of the theories of differential rotation considered in this paper suffer from this limitation and that the action of rotation on the turbulence cannot be ignored. It would be very surprising, however, if the basic mechanism giving rise to differential rotation were to be found entirely outside the theories discussed in this paper. Considerable importance has been given here to theories of differential rotation based on the interaction of rotation with convection. These theories were discussed on the assumption that the solar convection consists mainly of a turbulent convection and a large-scale convection. The dynamics of the convection zone could, of course, be far more complex than this.

On observation in particular, namely, the smallness of the pole–equator difference in flux, $\Delta\mathcal{F}$, has placed very severe restrictions on theories of differential rotation. It is readily seen that this serious difficulty is not necessarily associated with the use of the Boussinesq approximation: even assuming that the perturbations of the pressure, density, and temperature introduced by differential rotation or a meridional circulation are small in relation to the unperturbed values of p, ρ and T, the perturbation of the superadiabatic gradient, $\nabla\Delta T$, can be large since $\nabla\Delta T$ is very small itself, in the lower part of the convection zone. It appears, therefore, that a deep penetration of the differential rotation inside the convection zone would imply a major perturbation in the thermodynamics of the convection zone.

Three problems emerge as being of special importance for a further understanding of the Sun's differential rotation:

(1) In the lower part of the convection zone turbulent convection interacts with rotation. An estimate of the magnitude of the latitudinal and radial dependence of this interaction would be of great interest.

(2) The strength and structure of the large-scale convection in a compressible medium are unknown. In other words, how much heat flux does the large-scale convection carry and how do the magnitudes of the velocities change with depth? Because of their large scale, the action of rotation on these convective motions will be especially important.

(3) Dynamo theories of the solar cycle suggest that differential rotation is not a surface phenomenon. At present it is not understood how large pole-equator differences in the angular velocity ($\Delta\Omega$) can coexist in the lower part of the convection zone with small pole-equator differences in flux ($\Delta\mathscr{F}$). Of great interest would be an estimate of the typical value of $\Delta\mathscr{F}$ (at $r \sim R_c$) that can be 'wiped out' in the convection zone ($\Delta\mathscr{F}$ very small at $r = R_0$). However, even if significant values of $\Delta\mathscr{F}$ inside the convection zone could disappear towards the surface, this would not necessarily mean that these $\Delta\mathscr{F}$'s are present in the Sun. It is here that the lack of observations of what happens inside the convection zone is particularly felt: it is possible that theories of differential rotation could be developed which agree with the present observations at the solar surface, but which predict significant $\Delta\mathscr{F}$'s inside the convection zone. It is also conceivable that the convection zone is able to minimize the constraints imposed by rotation on convection in a way that we do not understand at present and that the main effect of this constraint is the generation of a differential rotation with small $\Delta\mathscr{F}$'s everywhere.

Other points where understanding is lacking and particularly needed have already been discussed in the text; some of them will be summarized here also.

(a) It appears that a deep penetration of differential rotation below the solar surface would entail a major perturbation of the convection zone. Can differential rotation be large only in the upper part and negligible in the lower part of the convection zone? This would have important consequences for theories of the solar dynamo. In this context it would be of interest to know the dependence of Ω and $\Delta\mathscr{F}$ with depth if the momentum and energy equations are solved with a very small $\Delta\mathscr{F}$ and ΔT at $r = R_0$ and the observed value of $w_2(r)$ ($= -0.189$) at the surface.

(b) If differential rotation penetrates deeply into the convection zone, then it must also penetrate into the radiative region; it is now known whether this could be of importance in theories of Lithium depletion in solar-type stars. The influence that a rapidly rotating radiative core (with a period of a week, for example) could have on the conditions at the base of the convection zone is not known either.

Appendix I

In a system of coordinates rotating with an angular velocity Ω_0, the basic equations can be written

$$\frac{1}{\sigma}\frac{\partial}{\partial t}\nabla\times\mathbf{U}-\nabla\times\nabla^2\mathbf{U}-\mathscr{R}_1\nabla\times g(r)T\hat{r}-\mathfrak{T}_1(\hat{\omega}\cdot\nabla)\mathbf{U}=-\frac{1}{\sigma}\nabla\times(\mathbf{U}\cdot\nabla)\mathbf{U},$$

(A1a)

$$\left(\frac{\partial}{\partial t}-\nabla^2\right)T=-\mathbf{U}\cdot\nabla T,$$

(A1b)

$$\text{div }\mathbf{U}=0$$

(A1c)

where $\mathbf{U}$, T, and $g(r)$ are dimensionless variables defined by Equation (2.1.1), $\hat{\omega}$ is a unit vector in the direction of the axis of rotation, and $\mathcal{R}_1$, $\mathfrak{T}_1$, and σ have been defined following Equations (2.1.5)–(2.1.8).

We expand the velocity field in basic poloidal and toroidal vectors:

$$\mathbf{U} = \sum_{L,m} \{\mathbf{P}[p_L^m(r, t) Y_L^m(\theta, \phi)] + \mathfrak{T}[t_L^m(r, t) Y_L^m(\theta, \phi)]\}. \tag{A2}$$

In spherical coordinates, the components of $\mathbf{P}(p_L^m Y_L^m)$ and $\mathfrak{T}(t_L^m Y_L^m)$ are given by

$$P^{(r)} = \frac{(L+1)L}{r^2} p_L^m Y_L^m, \qquad P^{(\theta)} = \frac{1}{r} \frac{\partial p_L^m}{\partial r} \frac{\partial Y_L^m}{\partial \theta},$$

$$P^{(\phi)} = \frac{1}{r \sin \theta} \frac{\partial p_L^m}{\partial r} \frac{\partial Y_L^m}{\partial \phi} \tag{A3a}$$

and

$$\mathfrak{T}^{(r)} = 0, \qquad \mathfrak{T}^{(\theta)} = \frac{t_L^m}{r \sin \theta} \frac{\partial Y_L^m}{\partial \phi}, \qquad \mathfrak{T}^{(\phi)} = -\frac{t_L^m}{r} \frac{\partial Y_L^m}{\partial \theta}. \tag{A3b}$$

The poloidal and toroidal vectors defined by Equation (A3) form a complete orthogonal set for solenoidal vector fields; $p_L^m(r, t)$ and $t_L^m(r, t)$ are the scalars which, together with the spherical harmonic $Y_L^m(\theta, \phi)$, define a basic poloidal and toroidal vector (Chandrasekhar, 1961, Appendix III).

The dimensionless temperature has been defined in Equation (2.1.3); $\Theta(r, \theta, \phi, t)$ is expanded in spherical harmonics.

$$\Theta(r, \theta, \phi, t) = \sum_{L,m} \Theta_L^m(r, t) Y_L^m(\theta, \phi). \tag{A4}$$

The spherical harmonics $Y_L^m(\theta, \phi)$ appearing in Equations (A3) and (A4) are assumed to be normalized according to Condon and Shortley (1951). To obtain the equations for p_L^m, t_L^m, Θ_L^m, and ψ we substitute expression (A2) for $\mathbf{U}$, and expressions (2.1.3) and (A4) for T into Equations (A1), multiply these equations by unit poloidal and toroidal vectors and integrate over the angle coordinates. Neglecting the fluctuating self-interactions (i.e. the right-hand side of Equations (A1)), we obtain (Durney, 1970)

$$\left(\frac{\partial}{\partial t} - \frac{d^2}{dr^2} - \frac{2}{r}\frac{d}{dr}\right)\psi = -\frac{1}{4\pi r^2} \sum_{L,m}^{|m| \le L} (L+1)L\frac{\partial}{\partial r}(rP_L^m \Theta_L^{*m}), \tag{A5a}$$

$$\left(\frac{\partial}{\partial t} - \mathcal{D}_L\right)\Theta_L^m = \frac{(L+1)L}{r} P_L^m \left[\frac{\eta}{(1-\eta)r^2} - \frac{\partial \psi}{\partial r}\right], \tag{A5b}$$

$$\frac{1}{\sigma}\frac{\partial}{\partial t}\mathcal{D}_L P_L^m - \frac{im}{(L+1)L}\mathfrak{T}_1 \mathcal{D}_L P_L^m - \mathcal{D}_L^2 P_L^m$$

$$= -\mathcal{R}_1 g_1(r)\Theta_L^m + \mathfrak{T}_1[L^{-1}A(L, m)(L-1)(L-2)T_{L-1}^m$$

$$-(L+1)^{-1}A(L+1, m)(L+2)(L+3)T_{L+1}^m$$

$$-L^{-1}A(L, m)(L-1)rT_{L-1}'^m - (L+1)^{-1}A(L+1, m)$$

$$\cdot (L+2)rT_{L+1}'^m], \tag{A5c}$$

$$\frac{1}{\sigma}\frac{\partial}{\partial t}T_L^m - \frac{im}{(L+1)L}\mathfrak{T}_1 T_L^m - D_L T_L^m$$

$$= -\mathfrak{T}_1[L^{-1}r^{-2}A(L, m)(L-1)^2 P_{L-1}^m - (L+1)^{-1}r^{-2}A(L+1, m)$$
$$\cdot(L+2)^2 P_{L+1}^m$$

$$-(rL)^{-1}A(L, m)(L-1)P_{L-1}'^m - [r(L+1)]^{-1}A(L+1, m)$$
$$\cdot(L+2)P_{L+1}'^m], \tag{A5d}$$

where

$$\mathcal{D}_L = \frac{d^2}{dr^2} + \frac{2}{r}\frac{d}{dr} - \frac{(L+1)L}{r^2}; \qquad D_L = \frac{d^2}{dr^2} + \frac{4}{r}\frac{d}{dr} + \frac{2-(L+1)L}{r^2};$$

$$P_L^m = p_L^m/r; \qquad T_L^m = t_L^m/r^2; \qquad T_L'^m = \frac{dT_L^m}{dr}; \qquad P_L'^m = \frac{dP_L^m}{dr}; \tag{A6}$$

$$A(L, m) = [(L+m)(L-m)/(2L+1)(2L-1)]^{1/2}.$$

The asterisk (cf. Equation (A5a)) defines the complex conjugate of Θ_L^m, $\eta(=R_c/R_0)$ is the ratio between the inner and outer radius of the spherical shell and $g_1(r) = g(r)/r$. Free-surface boundary conditions at $r = 1$, for example, imply $P_L'^m = T_L'^m = 0$ ($r = 1$). The spherical harmonics, as defined by Condon and Shortley (1951), satisfy the following relation

$$Y_L^{-m}(\theta, \phi) = (-1)^m Y_L^{m*}(\theta, \phi) \tag{A7}$$

From the expansions (A2) and (A4) it follows that the velocity and temperature fields will be real if

$$P_L^{-m} = (-1)^m P_L^{m*}; \qquad T_L^{-m} = (-1)^m T_L^{m*}; \qquad \Theta_L^{-m} = (-1)^m \Theta_L^{m*} \tag{A8}$$

It is readily seen that these relations are a consequence of Equations (A5).

If Equations (A5) are integrated in time, the quantities P_L^m, T_L^m, Θ_L^m, and ψ which determine the velocity and temperature field do not grow indefinitely. This is due to the non-linear term, $-[(L+1)L/r]P_L^m(\partial\psi/\partial r)$, of Equation (A5b): the interaction of the velocity field with $\partial\psi/\partial r$ (the distortion in the mean temperature gradient produced by convection) stabilizes the fluctuating component of the temperature field, Θ_L^m. In relation to its great simplicity, the mean-field approximation can be considered to be very successful (cf. Herring, 1963, 1964, 1969; Durney, 1968a).

In Equations (A5) we have neglected the fluctuating self-interactions. The expressions of $\mathbf{U}\cdot\nabla\mathbf{U}$ and $\mathbf{U}\cdot\nabla\Theta$ in terms of the defining scalars of the velocity and temperature fields are complex, but they have now been worked out by Young (1974).

For $m \neq 0$, the solutions of Equations (A5) are time-dependent (cf. Busse, 1970, 1973; Durney, 1970; Yoshimura and Kato, 1971). This is a general property of non-radial motions of rotating spheres or spherical shells (cf., for example, Durney and Skumarich, 1968, for the case of marginally unstable, non-radial oscillations of a polytrope). If we multiply Equation (A5c) by $\mathcal{D}_L$, use Equation (A5b) to express $-\mathcal{R}_1 g_1 \mathcal{D}_L \Theta_L^m$ in terms of $\mathcal{R}_1 g_1 \times \text{RHS(A5b)}$ and $-i\omega\mathcal{R}_1 g_1 \Theta_L^m(\partial/\partial t = i\omega)$, and finally replace $-\mathcal{R}_1 g_1 \Theta_L^m$, in this last expression, by $-\mathcal{D}_L^2 P_L^m$ (note that Equation (A5c) reduces to $\mathcal{R}_1 g_1 \Theta_L^m = \mathcal{D}_L^2 P_L^m$ in the case of no rotation), we obtain

$$\frac{i}{\sigma}\left(\omega(\sigma+1)-\frac{m\sigma}{(L+1)L}\mathfrak{T}_1\right)\mathcal{D}_L^2 p_L^m - \mathcal{D}_L^3 P_L^m$$
$$= \mathcal{R}_1 g(r) \times \text{RHS(A5b)} + 0(\mathfrak{T}_1^2) .$$

The first term in this equation vanishes if $\omega_d = 2m\Omega_0/(L+1)L(1+\sigma)$, which is Equation (3.8) of Busse (1970); ω_d in this last expression is the dimensional value of ω. (The reader is referred to Busse, 1973 and Heard and Veronis, 1973 for consistent solutions of the Boussinesq equations for convection, in the case of small rotation, and small convective amplitudes.) For $m = L = 10$, corresponding to the most unstable mode (see below) of a rotating shell of thickness $d = 0.2 R_0$, the value of ω_d equals $\Omega_0/11$ (for $\sigma = 1$). These convective waves, which propagate in the opposite sense of rotation (i.e., are retrograde) for small Taylor numbers, become prograde for large values of $\mathfrak{T}_1$ (Gilman, 1975, Figures 4 and 5).

For values of $\mathcal{R}_1$ and $\mathfrak{T}_1$ that are not very large, the values of L and m corresponding to the most unstable modes are mainly determined, respectively, by the thickness of the spherical shell ($L = 10$ for $\eta = 0.8$; Durney, 1968a) and by rotation ($m = L$) (Busse, 1970; Durney, 1970; Yoshimura and Kato, 1971). The reader is referred to Figures 1 and 2 of Gilman (1975) for a plot (for different values of $\mathfrak{T}_1$) of the critical Rayleigh number versus m.

According to Equations (A5), rotation couples the different values of L in the form $\mathbf{P_L^L} - T_{L+1}^L - P_{L+2}^L - T_{L+3}^L - \cdots = M_L^L$. In the presence of rotation this coupling scheme defines the mode M_L^L (P_L^L has been written in bold face to indicate that P_L^L is the largest poloidal mode). The above statement that $m = L$ is the most unstable mode means, therefore, that the mode $M_{L+2}^L = P_L^L - T_{L+1}^L - \mathbf{P_{L+2}^L} - T_{L+3}^L \cdots$, for example, has a larger critical Rayleigh number than the mode M_L^L. The fluctuating self-interactions couple modes with different values of m, as well as L (cf. Busse, 1973; Heard and Veronis, 1973; Gilman, 1975).

Appendix II

We intend here to estimate, in an approximate way, the right-hand side of Equations (2.1.6) and (2.1.8) to gain a very crude understanding of the relative magnitude of the pole-equator difference in flux and latitudinal differential rotation.

The dimensionless values of r will be assumed to be of the order unity, numerical factors will generally be ignored, and one will be neglected with respect to L. We define $K_L^m(\theta)$ by $Y_L^m = (1/\sqrt{2\pi}) e^{im\phi} K_L^m(\theta)$, and perform the calculations for the mode M_L^m. Equation (A3a) then gives, for the convective velocities:

$$U_r^c \sim L^2 P_L^m(r) K_L^m(\theta) \cos m\phi \qquad (\text{B1a})$$

$$U_\theta^c \sim P_L^m(r) \frac{\partial K_L^m(\theta)}{\partial \theta} \cos m\phi/\xi \qquad (\text{B1b})$$

$$U_\phi^c \sim P_L^m(r) m K_L^m(\theta) \sin m\phi/\xi \sin \theta \qquad (\text{B1c})$$

where $\xi = 1 - \eta$ is the dimensionless thickness of the spherical shell, and dP_L^m/dr has been set equal to P_L^m/ξ. The thinner the spherical shell is, the larger becomes the value of L of the most unstable mode, i.e., $\xi L \sim 1$. In what follows, we assume that the values of ξ and L are related in this way.

An estimate of T_{L+1}^m can be obtained from Equation (A5d), which we simplify to

$$D_{L+1}T_{L+1}^m \sim (d^2/dr^2 - L^2/r^2)T_{L+1}^m \sim \mathfrak{T}_1 LA(L+1, m)P_L^m .$$

Therefore, $T_{L+1}^m \sim \mathfrak{T}_1 P_L^m A(L+1, m)/L$. (The equation $(d^2/dr^2 - L^2/r^2)T_{L+1}^m = \sin(\pi r/\xi)$ can be solved by Green's functions and it is readily seen that $T_{L+1}^m \sim 1/L^2$). With the help of Equation (A3b) we obtain, for the toroidal velocities,

$$U_\theta^\Omega \sim T_{L+1}^m m \sin m\phi K_{L+1}^m(\theta)/\sin \theta$$

$$\sim m P_L^m \mathfrak{T}_1 A(L+1, m) \sin m\phi K_{L+1}^m(\theta)/L \sin \theta$$

$$U_\phi^\Omega \sim P_L^m \mathfrak{T}_1 A(L+1, m) \cos m\phi \, (\partial K_{L+1}^m/\partial\theta)/L .$$

The integrals appearing in the right-hand side of Equation (2.1.6) are

$$I_1 = \frac{\partial}{\partial r} \int \sin^2 \theta r^3 \langle U_\phi^\Omega U_r^c \rangle \, d\theta ;$$

$$I_2 = \int \cos \theta \sin \theta \langle U_\theta^c U_\phi^\Omega \rangle \, d\theta ;$$

$$I_3 = \int \cos \theta \sin \theta \langle U_\phi^c U_\theta^\Omega \rangle \, d\theta .$$

(a) $\langle U_r^c U_\phi^\Omega \rangle = L \mathfrak{T}_1 (P_L^m)^2 A(L+1, m) K_L^m(\theta) \, \partial K_{L+1}^m/\partial\theta$. With the help of Equation (19) of Condon and Shortley (1951) (CS), we can express $\partial K_{L+1}^m/\partial\theta$ in terms of $K_{L+1}^{m+1}(\theta)$; and with the help of their formula (21) we can express $\sin \theta K_L^m(\theta)$ in terms of $K_{L+1}^{m+1}(\theta)$. We obtain for the integral I_1,

$$I_1 \sim L^2 \mathfrak{T}_1 (P_L^m A(L+1, m))^2/\xi \tag{B2a}$$

(b) $\langle U_\theta^c U_\phi^\Omega \rangle \sim \mathfrak{T}_1 A(L+1, m)(P_L^m)^2 (\partial K_L^m/\partial\theta)(\partial K_{L+1}^m/\partial\theta)/\xi L$. with the help of Equations (19) and (21) of CS, we express $\partial K_L^m/\partial\theta$ and $\partial K_{L+1}^m/\partial\theta$ in terms of K_L^{m-1} and K_{L+1}^{m-1}, respectively, and $\cos \theta K_L^{m-1}$ in terms of K_{L+1}^{m-1}. We obtain for I_2

$$I_2 \sim \mathfrak{T}_1 (L-m+2)(P_L^m A(L+1, m))^2/\xi \tag{B2b}$$

(c) $\langle U_\theta^\Omega U_\phi^c \rangle \sim m^2 \mathfrak{T}_1 (P_L^m)^2 A(L+1, m) K_{L+1}^m(\theta) \quad K_L^m(\theta)/\xi L \sin^2 \theta$; we express $\cos \theta K_{L+1}^m(\theta)$ in terms of $K_L^m(\theta)$ (cf. Equation (21) of CS) and obtain

$$I_3 \sim m^2 \mathfrak{T}_1 (P_L^m A(L+1, m))^2/\xi L \tag{B2c}$$

We discuss now whether the angular momentum transport given by the integrals I_1, I_2, and I_3 accelerates or decelerates the equatorial regions.

(a') If $m = L$, then $U_r^c \sim L^2 P_L \sin^L \theta \cos L\phi$ and $U_\theta^c \sim L(dP_L/dr) \sin^{L-1} \theta \cos \theta \cos L\phi$. Near the outer surface $|U_\theta^c| \gg |U_r^c|$, since $U_r^c = 0$ for $r = R_0$. This inequality fails to be satisfied only for values of θ very close to $\pi/2$. Also if, for example, $U_r^c > 0$, then $U_\theta^c < 0 \, (> 0)$ for $\theta < \pi/2 \, (> \pi/2)$. It is readily seen that in both cases the action of Coriolis forces on the convective velocities (which, with the exception of a small region around the equator, are mainly latitudinal near the outer surface) generates a U_ϕ^Ω which is positive. Therefore, with the exception of the above region, the term $\langle U_r^c U_\phi^\Omega \rangle$ transports angular momentum outwards (for $r \sim R_0$), since an upgoing particle carries a $U_\phi^\Omega > 0$ $(\langle U_r^c U_\phi^\Omega \rangle > 0)$. At the equator and just below the surface, $\langle U_r^c U_\phi^\Omega \rangle$ is small in absolute value, and negative (since for $\theta = \pi/2$, $U_\theta^c = 0$, and a radially upgoing particle acquires a negative U_ϕ^Ω). At a small latitude, $\langle U_r^c U_\phi^\Omega \rangle$ becomes positive, reaches a maximum at, say, $\theta = \theta_m$ and vanishes at the poles. Therefore, if the vertical transport of angular momentum is important, one would expect the surface angular velocity to be maximum at $\theta \sim \theta_m$ and not at the equator (cf. Gilman, 1972, Figure 10).

The contribution of $\langle U_\phi^\Omega U_r^c \rangle$ to $T_3(r)$ can be understood as follows: we approximate Equation (2.1.6) by $-10\, T_3(r)/r^2 = \text{RHS}(2.1.6)\ (\mathrm{d}T_3(r)/\mathrm{d}r = 0$ at $r = R_0)$. The equator rotates faster than the poles if $T_3(r) < 0$ (cf. Equation (2.1.2)). Therefore, positive contributions to RHS (2.1.6) give rise to equatorial acceleration. Equation (2.1.6) shows that if

$$\int (5 \cos^2 \theta - 1) \frac{\partial}{\partial r} (r^3 \langle U_\phi^\Omega U_r^c \rangle)\, \mathrm{d}\theta > 0,$$

then the vertical transport of angular momentum contributes negatively to $T_3(R_0)$.

(b') If $U_r^c > 0$, then in the outer layers $U_\theta^c < 0$ and, as seen above, $U_\phi^\Omega > 0$ $(\theta \sim \pi/2)$ excepted). Therefore, the term $\langle U_\theta^c U_\phi^\Omega \rangle$ is negative, and this term is expected to give rise to equatorial deceleration $T_3(R_0) > 0$. From Equation (B2b) it follows that I_2 is smallest (compared with I_1 and I_3) when $m = L$.

(c') If $U_\phi^c > 0$, then the Coriolis forces generate a U_θ^Ω that is also positive. The term $\langle U_\theta^\Omega U_\phi^c \rangle$ transports angular momentum towards the equator (I_3 contributes to a negative $T_3(R_0)$). From Equation (B2c) this transport of angular momentum is largest if $m = L$. (In the above discussion when we say, for example, that if $U_\phi^c > 0$ then U_θ^Ω is also larger than zero, we neglect viscosity and consider that the particle is acted upon only by the Coriolis force. Nevertheless, we expect the conclusions to give us a correct description.)

It is readily seen (cf. Equations (2.1.1) and (2.1.2) that to obtain the observed differential rotation ($\Omega_\text{equator} - \Omega_\text{pole} \sim \Omega_0/5$) the value of $T_3(R_0)$ must be given by

$$T_3(R_0) = -\frac{4\Omega_0}{75} \left(\frac{\pi}{7}\right)^{1/2} \frac{R_0^2}{\kappa}. \tag{B3}$$

If we assume that the main contribution to $T_3(R_0)$ comes from the latitudinal angular momentum transport we obtain, with the help of Equations (2.1.6) (with $\xi L \sim 1$) and (B2c):

$$T_3 = -\frac{(7\pi)^{1/2}}{16\sigma} L^2 \mathfrak{T}_1 (P_L^L)^2 / L.$$

Comparing this equation with Equation (B3), we can estimate the magnitude of the convective velocities needed to generate the observed latitudinal differential rotation:

$$(LP_L^L)^2 \sim L\sigma^2/16. \tag{B4}$$

Since these large-scale convective motions are concentrated near the equator, they give rise to a pole-equator difference in flux. The ratio of the convective flux (at $r = 0.9$, for example) to the purely conductive flux (evaluated in the absence of convective motions) is given by

$$Q = r^2 \int \int U_r^c \, \theta \sin \Theta \, d\theta \, d\phi / (4\pi\eta/(1-\eta))$$

$$\sim \frac{1-\eta}{4\pi} \int \int U_r^c \, \Theta \sin \theta \, d\theta \, d\phi .$$

An estimate of Θ for the (L, L) mode $(\Theta \sim \Theta_L^L)$ can be obtained from Equation (A5b): $\Theta_L^L \sim P_L^L$. This order-of-magnitude relation holds only if $\partial\psi/\partial r \ll \eta/(1-\eta)r^2$, that is, if the distortion of the mean temperature gradient, produced by convection, is small (cf. Durney, 1968a, Figures 2 and 3). With the above value of Θ_L^L, it is readily found that

$$Q \sim (1-\eta)(LP_L^L)^2 \sim (1-\eta)L\sigma^2/16 . \tag{B5}$$

For $L = 10$ and $\sigma = 1$, $Q \sim 0.12$. The value of Q given by Equation (B5) should not be taken too literally (it is probably an underestimate; see Gilman, 1972, Figure 10). Equation (B5) shows, however, a basic difficulty associated with this approach to differential rotation: to generate the observed solar differential rotation, *an appreciable fraction of the energy flux must be carried by the large-scale convection.* To give rise to transport of angular momentum towards the equator, this large-scale convection must be highly non-axisymmetric ($m \sim L$) and must be concentrated, therefore, near the equator. Large pole-equator differences in flux appear unavoidable. It is of interest to note that our estimate of the pole-equator difference in flux (Q) needed to generate the observed differential rotation shows that Q is proportional to *the square of the Prandtl number.*

Appendix III

We choose normalized spherical coordinates; it is thus unnecessary to distinguish between covariant and contravariant components of vectors and tensors. In the case of axial symmetry, the components of the viscous force and the viscous stress tensor are given by

$$R_r = \frac{1}{r^2} \frac{\partial}{\partial r}(r^2 T_{rr}) + \frac{1}{r \sin\theta} \frac{\partial}{\partial\theta}(\sin\theta T_{\theta r}) - \frac{T_{\theta\theta} + T_{\phi\phi}}{r} \tag{C1a}$$

$$R_\theta = \frac{1}{r^3} \frac{\partial}{\partial r}(r^3 T_{r\theta}) + \frac{1}{r \sin\theta} \frac{\partial}{\partial\theta}(\sin\theta T_{\theta\theta}) - \frac{\cot g\,\theta}{r} T_{\phi\phi} \tag{C1b}$$

$$R_\phi = \frac{1}{r^3} \frac{\partial}{\partial r}(r^3 T_{\phi r}) + \frac{1}{r \sin^2\theta} \frac{\partial}{\partial\theta}(\sin^2\theta T_{\theta\phi}) \tag{C1c}$$

with

$$T_{rr} = 2s\eta \frac{\partial U_r}{\partial r} + 2\eta(1-s)\frac{\partial U_r}{\partial r} \tag{C2a}$$

$$T_{r\theta} = \frac{s\eta}{r}\left[\frac{\partial U_r}{\partial\theta} + r\frac{\partial U_\theta}{\partial r} - U_\theta\right] + \frac{\eta(1-s)}{r}\frac{\partial(rU_\theta)}{\partial r} \tag{C2b}$$

$$T_{r\phi} = \sin\theta\left[s\eta r\frac{\partial\Omega}{\partial r} + \frac{\eta(1-s)}{r}\frac{\partial\Omega r^2}{\partial r}\right] \tag{C2c}$$

$$T_{\theta\theta} = \frac{2\eta s}{r}\left(U_r + \frac{\partial U_\theta}{\partial\theta}\right) \tag{C2d}$$

$$T_{\theta\phi} = s\eta\sin\theta\frac{\partial\Omega}{\partial\theta} \tag{C2e}$$

$$T_{\phi\phi} = \frac{2s\eta}{r}[U_r + U_\theta\cot\theta]. \tag{C2f}$$

An anisotropic turbulent viscosity of the following form has been assumed $\eta_{\theta\theta} = \eta_{\phi\phi} = s\eta_{rr} = s\eta$.

Appendix IV

With p, ρ, T, U_r and U_θ defined by Equations (2.3.8), (2.3.9) we obtain from Equations (2.3.6):

$$\rho_u\rho_2 g = -(p_u p_2)' + \nu[2\psi''/r^2 - 2\psi'\rho'/3r^2\rho \\ -\psi(12/r^4 + 8\rho'/3r^3\rho + 8(\rho'/\rho)'/3r^2)] \tag{D1}$$

$$p_u p_2 = \frac{\nu}{3}[\psi''' - \psi''\rho'/\rho - \psi'(6/r^2 + 2\rho'/r\rho \\ + (\rho'/\rho)') + \psi(12/r^3 + 4\rho'/r^2\rho)]. \tag{D2}$$

In these equations the primes denote derivatives with respect to r; the viscosity has been assumed to be isotropic and of the form $\eta = \rho\nu$ with ν (a slowly varying quantity in the Sun) constant.

$$-\gamma\psi''' + \left[\gamma\frac{\rho'}{\rho} + \frac{p'}{p}\right]\psi''' + \left[\frac{12\gamma}{r^2} + \frac{2\gamma\rho'}{r\,\rho} + 2\gamma\left(\frac{\rho'}{\rho}\right)' - \frac{p'}{p}\frac{\rho'}{\rho}\right]\psi''$$

$$+\left[-\frac{24\gamma}{r^3} - \frac{8\gamma\rho'}{r^2\,\rho} - \frac{6p'}{r^2 p} + \frac{2\gamma}{r}\left(\frac{\rho'}{\rho}\right)' - \frac{2}{r}\frac{\rho'}{\rho}\frac{p'}{p} + \gamma\left(\frac{\rho'}{\rho}\right)'' - \left(\frac{\rho'}{\rho}\right)'\frac{p'}{p}\right]\psi'$$

$$+\left[\frac{12}{r^3}\frac{p'}{p} - \frac{12\gamma}{r^2}\left(\frac{\rho'}{\rho}\right)' + \frac{4}{r^2}\frac{p'}{p}\frac{\rho'}{\rho}\right]\psi = -\frac{4(\gamma-1)\Omega_0\mathscr{F}_0\varepsilon g\rho_u}{\nu^3}\int_{R_0}^{r}\frac{l^2}{p_u}\,dr. \tag{D3}$$

For a polytrope and for $l = -1.5 p_u/p_u'$ the right-hand side of Equation (D3) becomes

$$\mathrm{RHS} = \frac{9(\gamma-1)^3 L^2\Omega_0\mathscr{F}_0\varepsilon}{\gamma\nu^3(2\gamma-3)}(1 - (r - R_c)/L)^2\Big|_{R_0}^{r}. \tag{D4}$$

Appendix V

The viscosity coefficient is taken to be proportional to the density: $\eta = \rho\nu$, and the anisotropy factor s is defined by $\nu_{\theta\theta} = \nu_{\phi\phi} = s\nu_{rr} = s\nu$. The meridional motions are assumed to be given by Equation (2.3.9), and the angular velocity Ω by Equation (2.3.17). The equations for $\omega_0(r)$ and $\omega_2(r)$ are then obtained from Equation (2.3.3) (Durney, 1974b):

$$\omega_0'' + \left(\frac{4}{r} + \frac{\rho'}{\rho}\right)\omega_0' - \frac{2\omega_2 s}{r^2} + \frac{2(1-s)}{r}\left(\omega_0' + \frac{\rho'}{\rho}\omega_0 + \frac{3\omega_0}{r}\right)$$

$$- \frac{2\psi}{5r^3\rho\nu}(2\omega_2 + \omega_2' r) + \frac{2}{3}\frac{\psi' r}{r^3\rho\nu}[\omega_0 - \tfrac{1}{5}\omega_2]$$

$$= -\frac{2}{3}\frac{\psi' r}{r^3\rho\nu} - \frac{2(1-s)}{r}\left(\frac{\rho'}{\rho} + \frac{3}{r}\right), \tag{E1}$$

$$\omega_2'' + \left(\frac{4}{r} + \frac{\rho'}{\rho}\right)\omega_2' - \frac{10\omega_2 s}{r^2} + \frac{2(1-s)}{r}\left(\omega_2' + \frac{\rho'}{\rho}\omega_2 + \frac{3\omega_2}{r}\right)$$

$$- \frac{2\psi}{r^3\rho\nu}\left[2\omega_0 + r\omega_0' + \frac{2(2\omega_2 + r\omega_2')}{7}\right] + \frac{4\psi' r}{3r^3\rho\nu}\left(\omega_0 + \frac{4\omega_2}{7}\right)$$

$$= \frac{4\psi}{r^3\rho\nu} - \frac{4\psi' r}{3r^3\rho\nu}. \tag{E2}$$

From the difference $5\rho r^4(E1) - \rho r^4(E2)$ it is easily shown that

$$\frac{d}{dr}\Bigg[5r^4\rho\omega_0' + 10(1-s)r^3\rho(1+\omega_0) - r^4\rho\omega_2' - 2(1-s)r^3\rho\omega_2$$

$$- \frac{10}{7\nu}\psi r^2\omega_2 + \frac{2}{\nu}\psi r^2\omega_0 + \frac{2}{\nu}\psi r^2\Bigg] = 0 \tag{E3}$$

Equation (E3) is a particular form of Equation (E4):

$$\frac{d}{dr}\iint r^3 \sin^2\theta(T_{\phi r} - \rho U_r U_\phi)\, d\theta\, d\phi = 0 . \tag{E4}$$

Equation (E4) can be derived from the steady state azimuthal equation of motion (see Equation (2.3.3)) with only one assumption: the dependence of ρ on θ is small and can be neglected; in particular, the assumption of axial symmetry is unnecessary.

The boundary conditions that are commonly imposed on Equations (E1, E2) are zero stresses and vertical velocities $(T_{\phi r} = U_r = 0)$ at both ends of the convection zone $(r = R_c, R_0)$. In this case the integrated version of Equation (E4) is the compressible version of Equation (2.1.9). It should be noted that in the compressible case we have expanded the angular velocity in Legendre polynomials whereas in the Boussinesq approximation the expansion of the toroidal velocities is given by Equation (A3b) of Appendix I: it is an expansion in terms of $\partial P_L(\cos\theta)/\partial\theta$.

Appendix VI

The perturbed pressure, density, and temperature are expanded in Legendre polynomials according to Equation (2.4.8). We neglect quadratic terms in the differential rotation (i.e., ω_0^2, ω_2^2, $\omega_0\omega_2$) and in the perturbation (i.e., $\chi_i\,\xi_j$; where χ, ξ stand for p, ρ, T and $i, j = 0, 2, 4$). Then

$$U_\phi^2/r = \tfrac{2}{3}\Omega_0^2 r[1 + 2\omega_0 - 2\omega_2/5 - P_2(1 + 2\omega_0 - 10\omega_2/7) - 36\omega_2 P_4/35]$$

(F1)

$$U_\phi^2 \cot g\,\theta/r = \Omega_0^2 r(1 + 2\omega_0 + 2\omega_2 P_2)\sin\theta\cos\theta.$$

(F2)

Substituting Equations (2.4.8) and (F1, F2) into Equations (2.4.1) and (2.4.2), we obtain

$$dp_u/dr = -\rho_u g; \qquad dp_0/dr = -\rho_0 g + \tfrac{2}{3}\Omega_0^2 r\rho_u(1 + 2\omega_0 - 2\omega_2/5);$$

$$dp_2/dr = -\rho_2 g - \tfrac{2}{3}\Omega_0^2 r\rho_u(1 + 2\omega_0 - 10\omega_2/7);$$

(F3a)

$$dp_4/dr = -\rho_4 g - \tfrac{24}{35}\Omega_0^2 r\rho_u\omega_2.$$

$$p_2 = -\tfrac{1}{3}\rho_u\Omega_0^2 r^2(1 + 2\omega_0 + 2\omega_2/7);$$

$$p_4 = -\tfrac{6}{35}\Omega_0^2 r^2\rho_u\omega_2.$$

(F3b)

With the help of the gas equation $[p_i = (R/\mu)(\rho_i T_u + \rho_u T_i), i = 2, 4]$, Equations (F3) allow us to calculate T_2 and T_4; we obtain

$$T_2 = -\tfrac{1}{3}\Omega_0^2 r^2(\mu/R + \rho_u' T_u/g\rho_u)(1 + 2\omega_0 + 2\omega_2/7)$$

$$-\frac{2}{21}\Omega_0^2\frac{rT_u}{g}(7r\omega_0' + r\omega_2' + 12\omega_2)$$

(F4)

$$T_4 = -\tfrac{6}{35}\Omega_0^2 r^2\omega_2(\mu/R + \rho_u' T_u/g\rho_u) - \frac{6}{35}\frac{\Omega_0^2 rT_u}{g}(r\omega_2' - 2\omega_2).$$

We assume that the angular velocity and meridional motions are given by Equations (2.4.14) and (2.3.7), respectively. Keeping only the largest term of R_ϕ ($\nu\rho_u' r\sin\theta\,\partial\Omega/\partial r$) but all inertial terms of Equation (2.3.3), it is readily found that the azimuthal equation of motion can be written

$$\frac{\partial\psi_1}{\partial\theta}\sin\theta - \frac{\partial\psi_1}{\partial r}r\cos\theta$$

$$= \frac{3\omega_2(R_c)\nu\rho_u'(r\sin\theta)^4}{2R_c^2(1 - \tfrac{3}{2}\omega_2(R_c)(r/R_c)^2\sin^2\theta)}.$$

(F5)

Equation (F5) can be solved analytically if the pressure and density are related by a polytropic relation ($p_u = p_{uc}(\rho/\rho_{uc})^\gamma$) with $\gamma = \tfrac{3}{2}$, which is very close to the adiabatic ($\gamma = \tfrac{5}{3}$) value (we neglect perturbations in pressure and density due to the motions). The general solution of Equation (F5) is in this case given by

$$\psi_1 = -\frac{3\omega_2(R_c)\nu\rho_c(r\sin\theta)^4}{(R_c L)^2(1 - \tfrac{3}{2}\omega_2(R_c)(r\sin\theta/R_c)^2)}$$

$$\times\left[(L + R_c)\log\frac{1 + \cos\theta}{\sin\theta} - r\cos\theta\right] + \xi(r\sin\theta)$$

(F6)

where L has been defined in Equation (2.3.15) and $\xi(r\sin\theta)$ is an arbitrary function of $r\sin\theta$.

References

Abt, H. and Hunter, J.: 1962, *Astrophys. J.* **136**, 381.
Alfonso-Faus, A.: 1967, *J. Geophys. Res.* **72**, 5576.
Altrock, R. C. and Canfield, R. C.: 1972a, *Astrophys. J.* **171**, L71.
Altrock, R. C. and Canfield, R. C.: 1972b, *Solar Phys.* **23**, 257.
Antonucci, E. and Svalgaard, L.: 1974, *Solar Phys.* **34**, 3.
Appenzeller, I. and Schröter, E. H.: 1967, *Astrophys. J.* **147**, 1100.
Appenzeller, I. and Schröter, E. H.: 1968, *Solar Phys.* **4**, 131.
Baker, N. and Temesvary, S.: 1966, *Tables of Convective Stellar Envelope Models*, 2nd edn., Goddard Institute for Space Studies, New York.
Beckers, J. M. and Morrison, R. A.: 1970, *Solar Phys.* **14**, 280.
Belvedere, G. and Paterno, L.: 1976, *Solar Phys.* (in press).
Benton, E. R. and Clark, A.: 1974, *Ann. Rev. Fluid Mech.* **6**, 257.
Biermann, L.: 1951, *Z. Astrophys.* **28**, 304.
Böhm, K. H.: 1963, *Astrophys. J.* **137**, 881.
Boussinesq, J.: 1877, 'Essai sur la theorie des eaux courantes', *Mém. prés. par div. savants á l'Acad. Sci., Paris* **23**, No. 1, p. 1.
Boussinesq, J.: 1897, *Théorie de l'écoulement tourbillonnant et tumultueux des liquides dans les lits rectilignes á grande section, I–II*, Gauthier-Villars, Paris.
Bumba, V., Howard, R., and Smith, S. F.: 1964, *Carnegie Inst. of Washington Year Book* **63**, 6.
Bumba, V.: 1967, *Rendiconti della Scuola Internazionale di Fisica "E. Fermi," 39 Corso*, 77.
Bumba, V. and Obridko, V. N.: 1969, *Solar Phys.* **6**, 104.
Bumba, V. and Obridko, V. N.: 1970, *Solar Phys.* **14**, 80.
Busse, F. H.: 1970, *Astrophys. J.* **159**, 629.
Busse, F. H.: 1973, *Astron. Astrophys.* **28**, 27.
Caccin, B., Falciani, R., Moschi, G., and Rigutti, M.: 1970, *Solar Phys.* **13**, 33.
Chandrasekhar, S.: 1961, *Hydrodynamic and Hydromagnetic Stability*, Clarendon Press, Oxford.
Cocke, W. J.: 1967, *Astrophys. J.* **150**, 1041.
Condon, E. U. and Shortley, G. H.: 1951, *The Theory of Atomic Spectra*, Cambridge University Press, Cambridge.
Conti, P. S.: 1968, *Astrophys. J.* **152**, 657.
Cowling, T. G.: 1951, *Astrophys. J.* **144**, 272.
Deinzer, W. and Stix, M.: 1971, *Astron. Astrophys.* **12**, 111.
Deubner, F. L.: 1972, *Solar Phys.* **22**, 263.
Dicke, R. H.: 1964, *Nature* **202**, 432.
Dicke, R. H.: 1970a, *IAU Colloq.* **4**, 289.
Dicke, R. H.: 1970b, *Ann. Rev. Astron. Astrophys.* (ed. by L. Goldenberg), Annual Reviews, California, p. 297.
Dicke, R. H.: 1971, *Phys. Rev. Letters*, **27**, 210.
Dicke, R. H.: 1972, *Astrophys. J.* **171**, 331.
Dicke, R. H. and Goldenberg, H. M.: 1967, *Phys. Rev. Letters* **18**, 313.
Dittmer, P. H.: 1975, *Solar Phys.* **41**, 1975.
Dupree, A. K. and Henze, W.: 1972, *Solar Phys.* **27**, 271.
Durney, B. R.: 1968a, *J. Atmospheric Sci.* **25**, 372.
Durney, B. R.: 1968b, *J. Atmospheric Sci.* **25**, 771.
Durney, B. R.: 1970, *Astrophys. J.* **161**, 1115.
Durney, B. R.: 1971, *Astrophys. J.* **163**, 353.
Durney, B. R.: 1972a, *Solar Phys.* **26**, 3.
Durney, B. R.: 1972b, *Solar Wind*, Proc. of the Asilomar Conf., p. 282 (ed. by C. P. Sonnet, P. J. Coleman, Jr., and J. M. Wilcox), NASA, Washington.
Durney, B. R.: 1974a, *Astrophys. J.* **190**, 211.
Durney, B. R.: 1974b, *Solar Phys.* **38**, 301.
Durney, B. R.: 1976, *Astrophys. J.* **204**, 589.
Durney, B. R. and Roxburgh, I. W.: 1971, *Solar Phys.* **16**, 3.
Durney, B. R. and Skumanich, A.: 1968, *Astrophys. J.* **152**, 255.
Falciani, R., Rigatti, M., and Roberti, G.: 1974, *Solar Phys.* **35**, 277.
Fossat, E.: 1975, Thesis, Univ. of Nice (unpublished).
Fossat, E. and Ricort, G.: 1973, *Solar Phys.* **28**, 311.
Fricke, K.: 1968, *Z. Astrophys.* **68**, 317.
Fricke, K. and Kippenhahn, R.: 1970, *Ann. Rev. Astron. Astrophys.* **10**, 45.

Gierasch, P.: 1974, *Astrophys. J.* **190**, 199.
Gilman, P. A.: 1972, *Solar Phys.* **27**, 3.
Gilman, P. A.: 1975, *J. Atmospheric Sci.* **32**, 1331.
Goldreich, P. and Schubert, G.: 1967, *Astrophys. J.* **150**, 571.
Grevesse, N.: 1968, *Solar Phys.* **5**, 159.
Hansen, R. T., Hansen, S. F., and Loomis, H. G.: 1969, *Solar Phys.* **10**, 135.
Hauge, Ö. and Engvold, O.: 1968, *Astrophys. Letters*, **2**, 235.
Heard, W. B.: 1973, *Astrophys. J.* **186**, 1065.
Heard, W. B. and Veronis, G.: 1973 (unpublished).
Henze, W. and Dupree, A. K.: 1973, *Solar Phys.* **33**, 425.
Herring, J. R.: 1963, *J. Atmospheric Sci.* **20**, 325.
Herring, J. R.: 1964, *J. Atmospheric Sci.* **21**, 277.
Herring, J. R.: 1969, *Phys. Fluids* **12**, 39.
Hill, H. A. and Stebbins, R. T.: 1975a, *Ann. N.Y. Acad. Sci.* **262**, 472.
Hill, H. A. and Stebbins, R. T.: 1975b, *Astrophys. J.* **200**, 484.
Hill, H. A., Clayton, P. D., Patz, D. L., Healy, A. W., Stebbins, R. T., Oleson, J. R., and Zanoni, C. A.:
    1974, *Phys. Rev. Letters*, **33**, 1497.
Hill, H. A., McCullen, J. D., Brown, T. M., and Stebbins, R. T.: 1975 (unpublished).
Hill, H. A., Stebbins, R. T., and Brown, T. M.: 1976, Proc. of the Fifth Int. Conf. on Atomic Masses and
    Fundamental Constr., Paris, France (in press).
Howard, L. N., Moore, D. W., and Spiegel, E. A.: 1967, *Nature* **214**, 1297.
Howard, R.: 1971, *Solar Phys.* **16**, 21.
Howard, R. and Harvey, J.: 1970, *Solar Phys.* **12**, 23.
Howard, R. and Yoshimura, H.: 1976, this volume, p.19.
Iroshnikov, R. S.: 1969, *Astron. Zh.* **46**, 97 (translated in *Soviet Astron.* **13** (1969), 73).
Kaufman, P.: 1972, *Solar Phys.* **23**, 178.
Kippenhahn, R.: 1963, *Astrophys. J.* **137**, 664.
Köhler, H.: 1970, *Solar Phys.* **13**, 3.
Köhler, H.: 1973, *Astron. Astrophys.* **25**, 467.
Kobrin, M. M. and Korshunov, A. I.: 1972, *Solar Phys.* **25**, 1972.
Kraft, R.: 1967, *Astrophys. J.* **150**, 551.
Kraft, R.: 1969, in *Stellar Astronomy*, vol. 2 (ed. by H. Y. Chiu, R. Warasila and J. Remo), Gordon and
    Breach, New York.
Kraft, R.: 1970, in *Otto Struve Memorial Volume* (ed by G. Herbig), p. 385.
Ledoux, P., Schwarzschild, M., and Spiegel, E. A.: *Astrophys. J.* **133**, 184.
Leighton, R. B.: 1966 (unpublished).
Leighton, R. B.: 1969, *Astrophys. J.* **156**, 1.
Lerche, I. and Parker, E. N.: 1972, *Astrophys. J.* **176**, 213.
Livingston, W. C.: 1969, *Solar Phys.* **9**, 448.
McIntosh, P. S.: 1975, *Report UAG* **40** (H_α Synoptic Charts of Solar Activity for the Period of Skylab
    Observations. March 1973–March 1974).
Mattig, W. and Nesis, A.: 1974, *Solar Phys.* **36**, 3.
Mehltretter, J. P.: 1971, *Solar Phys.* **16**, 253.
Modisette, J. L.: 1967, *J. Geophys. Res.* **72**, 1521.
Ness, N. F. and Wilcox, J. M.: 1966, *Astrophys. J.* **143**, 23.
Newton, H. W. and Nunn, M. L.: 1951, *Monthly Notices Roy. Astron. Soc.* **111**, 413.
Noyes, R. W., Ayres, T. R., and Hall, D. N. B.: 1973, *Solar Phys.* **28**, 343.
Osaki, Y.: 1970, *Monthly Notices Roy. Astron. Soc.* **131**, 407.
Pai, S. 1956, *Viscous Flow Theory*, Vol. 1, *Laminar Flow*, Van Nostrand, New York, pp. 39–40.
Parker, E. N.: 1955, *Astrophys. J.* **122**, 293.
Parker, E. N.: 1971, *Astrophys. J.* **164**, 491.
Parker, E. N.: 1973a, *Astrophys. J.* **186**, 643.
Parker, E. N.: 1973b, *Astrophys. J.* **186**, 665.
Parker, E. N.: 1975, *Astrophys. J.* **198**, 205.
Richardson, R. S. and Schwarzschild, M.: 1953, *Academia Lincei, Conv.* **11**, p. 228.
Roberts, P. H.: 1974, *Solar Wind Three*, p. 231 (ed. by C. T. Russell), published by Inst. of Geophys. and
    Planetary Phys., UCLA.
Roberts, P. H. and Stix, M.: 1972, *Astron Astrophys.* **18**, 453.
Ross, J. E. and Aller, L. H.: 1974, *Solar Phys.* **36**, 11.
Roxburgh, I. W.: 1964, *Icarus* **3**, 92.
Roxburgh, I. W.: 1970, *IAU Colloq.* **4**.

Roxburgh, I. W.: 1974, *Astrophys. Space Sci.* **27**, 419.
Rüdiger, G.: 1974, *Astron. Nachr.* **295**, 229.
Rüdiger, G.: 1976, *Solar Phys.* (in press).
Rutten, R. J.: 1973, *Solar Phys.* **28**, 347.
Sakurai, T.: 1966, *Publ. Astron. Soc. Japan* **18**, 174.
Sakurai, T.: 1975, *Monthly Notices Roy. Astron. Soc.* **171**, 35.
Schatzman, E.: 1962, *Ann. Astrophys.* **25**, 18.
Shaviv, G. and Salpeter, E. E.: 1973, *Astrophys. J.* **184**, 191.
Simon, G. W. and Noyes, R. W.: 1972, *Solar Phys.* **26**, 8.
Simon, G. W. and Weiss, N. O.: 1968, *Z. Astrophys.* **69**, 435.
Skumanich, A.: 1955, *Astrophys. J.* **121**, 408.
Skumanich, A.: 1972, *Astrophys. J.* **171**, 565.
Spiegel, E. A.: 1964, *Astrophys. J.* **139**, 959.
Spiegel, E. A.: 1965, *Astrophys. J.* **141**, 1068.
Spiegel, E. A.: 1968, in *Highlights of Astronomy* (ed. by L. Perek), D. Reidel Publishing Co., Dordrecht.
Spiegel, E. A.: 1971, *Ann. Rev. Astron. Astrophys.* **9**, 330 (ed. by L. Goldberg).
Spiegel, E. A. and Zahn, J. P.: 1970, *Comments Astrophys. Space Phys.* **2**, 178.
Starr, V. P. and Gilman, P.: 1965, *Astrophys. J.* **141**, 1119.
Steenbeck, M. and Krause, F.: 1969, *Astron. Nachr.* **291**, 49.
Stenflo, J. O.: *Solar Phys.* **36**, 495.
Stix, M.: 1974, *Astron. Astrophys.* **37**, 121.
Stix, M.: 1976a, this volume, p. 367.
Stix, M.: 1976b (submitted to *Astron. Astrophys.*).
Strittmatter, P. A.: 1969, *Ann. Rev. Astron. Astrophys.* **7**, 665.
Švestka, Z.: 1968a, *Solar Phys.* **4**, 18.
Švestka, Z.: *IAU Symp.* **35**, 287.
Tuominen, J.: 1955, *Z. Astrophys.* **27**, 145.
Unno, W.: 1961, *Publ. Astron. Soc. Japan* **13**, 276.
Vandakurov, Yu. V.: 1975a, *Solar Phys.* **40**, 3.
Vandakurov, Yu. V.: 1975b, *Solar Phys.* (in press).
Veronis, G., 1966, *Tellus* **18**, 67.
Vickers, G. T.: 1971, *Astrophys. J.* **163**, 363.
Wallerstein, G. and Conti, P. S.: 1969, *Ann. Rev. Astron. Astrophys.* **7**, 99.
Wallerstein, G., Herbig, G. H., and Conti, P. S.: 1965, *Astrophys. J.* **141**, 610.
Ward, F.: 1965, *Astrophys. J.* **141**, 534.
Weber, E. J. and Davis, L., Jr.: 1967, *Astrophys. J.* **148**, 217.
Weiss, N. O.: 1964, *Monthly Notices Roy. Astron. Soc.* **128**, 225.
Weiss, N. O.: 1965, *Observatory* **85**, 37.
Wilcox, J. M. and Howard, R.: 1968, *Solar Phys.* **5**, 564.
Wilcox, J. M. and Howard, R.: 1970, *Solar Phys.* **13**, 251.
Wilcox, J. M. and Ness, N. F.: 1965, *J. Geophys. Res.* **70**, 5793.
Wilcox, J. M. and Ness, N. F.: 1967, *Solar Phys.* **1**, 437.
Wilcox, J. M., Schatten, K. H., Tanenbaum, A. S., and Howard, R.: 1970a, *Solar Phys.* **14**, 225.
Wilcox, J. M., Schatten, K. H., Tanenbaum, A. S., and Howard, R.: 1970b, *Solar Phys.* **14**, 255.
Wilson, O. C.: 1966a, *Astrophys. J.* **144**, 695.
Wilson, O. C.: 1966b, *Science* **151**, 1487.
Wolff, C. L.: 1974a, *Astrophys. J.* **193**, 721.
Wolff, C. L.: 1974b, *Astrophys. J.* **194**, 489.
Wolff, C. L.: 1975, *Solar Phys.* **41**, 297.
Yoshimura, H.: 1971, *Solar Phys.* **18**, 417.
Yoshimura, H.: 1972a, *Solar Phys.* **22**, 20.
Yoshimura, H.: 1972b, *Astrophys. J.* **178**, 863.
Yoshimura, H.: 1975a, *Astrophys. J. Suppl.* **29**, 467.
Yoshimura, H.: 1975b, *Astrophys. J.* **201**, 740.
Yoshimura, H.: 1976a, this volume (Discussion).
Yoshimura, H.: 1976b (unpublished).
Yoshimura, H. and Kato, S.: 1971, *Publ. Astron. Soc. Japan* **23**, 57.
Young, R.: 1974, *J. Fluid Mech.* **63**, 695.
Zahn, J. P. 1974, *IAU* **59**, 185.
Zahn, J. P.: 1975, *Coll. Internat. d'Astr. de Liege* (in press).

DISCUSSION

Roxburgh: (1) Is it not possible that ΔF is large in the bulk of the convective zone but becomes small in the surface layers where $\Delta \bar{\nabla} T$ becomes large since with a large $\Delta \nabla T$ a small circulation could possibly destroy the ΔF?

(2) In your calculation you assume that the turbulent model can be applied all the way to the surface of the convective zone, but in the surface layers radiative transport becomes dominant and even if $\Delta F/F$ is large for the convective flux, the convective flux itself goes to zero at the surface so ΔF could also go to zero. In the top 1000 km where radiative transfer is dominant the ΔF can be wiped out.

Durney: (1) It is possible, However, one cannot help thinking that somehow the main effect of the constraint imposed by rotation on convection is only differential rotation and not large pole-equator differences in flux inside the convection zone. We do not know at present how a large $\Delta \Omega$ can be achieved with a small ΔF.

(2) I think that it is unlikely that a large ΔF can be wiped out in the top 1000 km with no manifestations at the surface. Differential rotation could be a surface phenomenon but then no large ΔF's are needed, but this appears to be in contradiction with dynamo models.

Schröter: One of your alternative solutions out of the dilemma is the suggestion of the existence of a large pole-equator flux difference within the convective zone. Would you expect from such a fact an observable difference in the appearance (size, contrast, life-time, etc.) of the solar granulation pattern between pole and equator?

Durney: I think it is unlikely. If a large pole-equator difference does indeed exist inside the solar convection zone, it completely disappears in the surface layers and would not affect, I think, the solar granulation.

Weiss: We can place some constraints on the variation of energy flux with latitude. In the radiative zone ΔF must be very small, any significant variation of flux on an equipotential surface will rapidly be eliminated by mass motions. At the photosphere we observe no appreciable variation. On the other hand, a significant variation in the deep convective zone might be eliminated near the surface just as the supergranular temperature variations are below the limit observation.

Durney: It is possible, and to answer this question with certainty is an important problem that remains to be solved. However, the Sun could satisfy the constraints imposed by rotation on convection by processes of which we do not know yet any idea.

Chvojková: During the maximum of a solar cycle high magnetic fields of about 10^3 G should be expected just below the photosphere. Thus at the top of the convection zone there should exist a layer in which the magnetic pressure prevailed the kinetic one, the motion would become nonisotropic, the field would act against most of the movements of the described mechanism. Hence, the differential rotation should be most probably somewhat smaller during cycle maxima. Is it so or is the layer $H^2/8\pi > \frac{1}{2}\rho V^2$ too thin and insignificant for affecting the result?

Durney: If we accept Parker's point of view, the solar cycle is generated in the lower part of the convection zone and the magnetic flux tube rises at about the Alfvénic speed, thus very rapidly in the surface layers. Large toroidal fields waiting to be dissipated or carried away by the solar wind should be present below the photosphere and could indeed influence the Sun's differential rotation. I think that there are observations confirming this.

Mestel: I seem to recall that Biermann postulated that there should not be any large latitude variations in flux, and then derived the necessity for circulation from momentum balance. Kippenhahn then used this circulation to construct the departure from uniformity of rotation.

Durney: I think that Biermann postulated a different rate of momentum exchange between the direction parallel and perpendicular to gravity. He then proved that conservation of angular momentum precludes solid rotation.

EQUILIBRIUM PROBLEM IN A
ROTATING CONVECTION ZONE

YU. V. VANDAKUROV

Physical and Technical Institute A. F. Ioffe, Leningrad, U.S.S.R.

Abstract. Taking into account effects produced by the convective motions, the equilibrium problem for a rotating star becomes greatly complicated. We consider this problem for the case of slow rotation in the following approximations.

We treat the convection zone as a medium with turbulent viscosity and turbulent thermal conductivity. However, we take into account the nonlinear effects produced by the most rapidly growing perturbations. The corresponding nonlinear terms are calculated by using the solution of linear perturbed equations. Each independent convective mode is supposed to have initially the same amount of kinetic energy.

In the limit of small turbulent viscosity, we show that unstable convective perturbations produce a mean azimuthal force due to which rigid rotation appears not to be in the equilibrium. For the case of small-scale perturbations and latitudinal differential rotation, this force is analogous to the viscous force, but the coefficient of viscosity is negative.

We suggest that such a force maintains the differential rotation of the solar convection zone. Note that in the case under consideration the latitudinal dependence of the solar heat flux is small. However, difficulties arise due to different conditions at different depths in the convection zone. In this connection, a hypothesis is put forward that magnetic fields are also necessary to get balance in full. A model of the solar cycle is discussed which is similar in some respects to the well-known Babcock model. We propose, however, that the field reversal takes place in the lower layers of that zone where fields are intensified.

DISCUSSION

Durney: You consider the modes with the largest growth rate which I will designate by M. The influence of rotation on a mode depends on its strength and dimensions. Could not this influence be larger for modes with smaller growth rates but larger dimensions than M?

Vandakurov: The horizontal mode number L in the final formula for the azimuthal force refers to some dominant mode. It seems likely that this mode corresponds to the mode having the faster growth rate (of course, this rate should be found from nonlinear perturbed equations). If some other value of L is more suitable, it can be inserted in the above formula. To study the problem more exactly an extension of the calculations to cover different values of L is needed.

Gilman: Your theory requires a magnetic field, which, in turn would have cyclic variations with time. By what fraction would your differential rotation change with time to feed back from the magnetic field?

Vandakurov: The problem of time variations of differential rotation has not been considered in detail. We think that our model can be put in accordance with the well-known models by Babcock and Leighton (with the modifications concerning the field-reversal mechanism and the basic justification of the existence of the solar activity cycle). Thus, the explanation of the main features of the cycle seems to be similar to that in the above mentioned models. But indeed, the time variations of rotation velocity should be present in our model. In some layers field amplification is smaller or larger than that required by the equilibrium condition, thus, there, redistribution of angular velocity occurs. The rotation velocity is kept constant only in some averaged value.

Bumba and Kleczek (eds.), Basic Mechanisms of Solar Activity, 297. *All Rights Reserved.*

A BOUSSINESQ MODEL

FOR THE CONVECTION ZONE

AND THE SOLAR ANGULAR VELOCITY

G. BELVEDERE and L. PATERNÒ

Astronomical Institute of the Catania University, Catania Astrophysical Observatory, Italy

Abstract. In this paper we study the dependence of the solar angular velocity on depth and latitude as produced by a meridional circulation in the convection zone.

We assume that the main mechanism responsible for setting up and driving the circulation is the interaction of rotation with convection. We solve the first-order equations (perturbation of the spherically symmetric state) for the motion and energy (diffusion equation) in the Boussinesq approximation and in the steady state for the axisymmetric case. The interaction of convection with rotation is considered through the convective transport coefficient $k(r) = k_0 + \varepsilon k_2(r) P_2(\cos \theta)$, where $P_2(\cos \theta)$ is the 2nd Legendre Polynomial which allows for the latitudinal perturbation and $k_2(r) \sim T_a(r)(F_c(r)/F_T)$, where T_a is the local Taylor number, F_c, F_T are convective and total fluxes, while ε is an expansion parameter. Here T_a takes into account the interaction of convection with rotation and F_c/F_T the energy transport.

The equations are numerically solved in order to determine the stream function ψ, with the boundary conditions that the fluid must be confined in a spherical shell, that there are no stresses at the boundaries and that the temperature and flux are spherically symmetric on the inner boundary. Once determined ψ we solve for the angular velocity and then determine ε by a fit with the observed angular velocity.

We obtained the following results for a Rayleigh number $R_a = 10^3$:

(1) A single cell circulation extending from poles to the equator and with the circulation directed toward the equator at the surface. Radial velocities are of the order of 10 cm s^{-1} and meridional ones of the order of 150 cm s^{-1}.

(2) A flux difference between poles and equator $\Delta F \approx 5 \times 10^{-2}$ at the surface, the poles being hotter.

(3) A negligible temperature difference between poles and equator at the surface.

(4) An angular velocity increasing inward.

(5) Angular velocity constant surfaces of spheroidal shape.

The model is consistent with the fact that the interaction of convection with rotation sets up the circulation (driven by the temperature gradient) which carries angular momentum toward the equator against the viscous friction. Unfortunately also a large ΔF is obtained. Nevertheless it seems that the model has the basic requisites for a correct dynamo action.

DISCUSSION

Paternò: I would like to add one more point to the model of Belvedere and myself. When we made the calculations for a Prandtl number equal to unity we obtained a flux difference between poles and equator $\Delta F = 5 \times 10^{-2}$. But using a Prandtl number $\sigma = 10^{-2}$ we obtained ΔF's as small as a few parts in 10^{-4}. In

Bumba and Kleczek (eds.), Basic Mechanisms of Solar Activity, 299–300. All Rights Reserved.
Copyright © 1976 by the IAU.

this case the same circulation carries a smaller amount of flux towards the surface. However a more appealing idea suggested by Prof. Roxburgh looks to be promising to solve the ΔF dilemma. We also made calculations for a convection zone twice as deep. The preliminary results seem to indicate that in this case the model generates larger differential rotation than flux differences and ΔF's of the order of 10^{-3} can be obtained. On the other hand, the depth of the convection zone, and the extent of convective overshooting into the radiative zone, is still a matter of discussion.

TWO-DIMENSIONAL STOCHASTIC MOTIONS
AND THE PROBLEM OF DIFFERENTIAL ROTATION
FOR UNRESTRICTED ROTATIONAL RATES

G. RÜDIGER

Zentralinstitut für Astrophysik, Potsdam, G.D.R.

Abstract. For dealing analytically with the problem of differential rotation we investigate the spatial dependence of the angular velocity in a rotating turbulent fluid. The original turbulence unaffected by the rotation is assumed to be two-dimensional, where the stochastic motions completely lie in the horizontal planes. From the expression describing the relation between the correlations of rotating and nonrotating turbulent fields the meridional flux of momentum is derived. The resulting rotational law is determined by using Bochner's theorem for homogeneous turbulence as well as the characteristic scales of the turbulence field considered. The conclusions are:

(a) The angular velocity Ω is increasing toward the outer layers.

(b) For $2\,\Omega \ll \omega_c$ (ω_c frequency of turbulent mode) the Biermann-Kippenhahn-theory of anisotropic viscosity is deduced. An equatorial acceleration is only caused by a meridional circulation.

(c) For $2\,\Omega \lesssim \omega_c$ a latitudinal dependence of Ω is possible without any meridional circulation. If the two-dimensional eddy viscosity is negative the equatorial regions are accelerated. The expression for the two-dimensional eddy viscosity which has been derived earlier allows negativity in contrast to that for three-dimensional eddy viscosity. The scale length and the scale time of supergranulation as well as of giant cells lead to negative two-dimensional eddy viscosity.

Ward's observations including a negative sign of the two-dimensional eddy viscosity might represent an independent argument supporting the theory.

Our results agree with Gilman's (1972) numerical approach.

The paper is submitted to *Solar Physics*. It was presented by Dr F. Krause.

DISCUSSION

Stix: The anisotropic turbulent viscosity used by Biermann and Kippenhahn was based on *gravity*, whereas here *rotation* introduces the preferred direction. I wonder whether both effects have similar consequences.

Krause: There is a misunderstanding: Here as with Biermann (1951) the anisotropic behaviour is caused by *both* the gravity *and* the angular velocity.

Durney: You seem to favour Biermann-Kippenhahn's theory of differential rotation based on an anisotropic viscosity. This theory predicts an angular velocity decreasing inward. How do you conciliate this with your dynamo theory?

Krause: This question tends in the direction of one of the dilemmas noted by Dr Parker, which have given rise to many discussions during this meeting. The theories of differential rotation only take into account the convective shell. It may be possible that the core under the convection zone may rotate a bit faster, since the surface of the Sun is (and has been all the time) decelerated by the solar wind.

Bumba and Kleczek (eds.), Basic Mechanisms of Solar Activity, 301–302. All Rights Reserved.
Copyright © 1976 by the IAU.

Durney: If I remember correctly Iroshnikov has considered a similar problem. Biermann's and Kippenhahn's anisotropy is between the turbulent viscosity parallel and perpendicular to gravity. It appears to me the inclusion of vertical motions is very important.

Krause: This is true. But the discussion of pure two-dimensional motions provides for very clear statements. The situation is not completely changed, if to a certain degree vertical motions are taken into account.

Vandakurov: I should like to add that the averaged force produced by the convective motions I have spoken about is the result of the correlation between the radial and azimuthal velocities. It is therefore important to consider radial fluctuations.

PART 3

DYNAMO THEORY
AND MAGNETIC DISSIPATION

MEAN-FIELD MAGNETOHYDRODYNAMICS
OF THE SOLAR CONVECTION ZONE

F. KRAUSE

Zentralinstitut für Astrophysik, Potsdam, G.D.R.

1. Introduction

1.1. Observations of the solar surface show that some of the physical quantities, especially the velocity field and the magnetic field, show random character.

Their time and space variations are irregular and neither can be predicted with certainty. However, at least the existence of the 22-year solar cycle indicates that these quantities show in the average a regular, periodic behavior.

The idea of constructing a theory of the mean quantities was conceived by M. Steenbeck *et al.* (1963) in connection with an explanation of the solar cycle. They suggested to develop a theory of the mean electromagnetic fields where the action of the turbulent motion is regarded by constitutive equations.

1.2. There are previous suggestions to explain some large-scale effects out of the action of small scale effects. In this connection may be noted a paper of Gurewitch and Lebedinskii (1945) where the convective motions of the plasma in the convection zone of the Sun are considered responsible for the build-up of the magnetic fields of the sunspots.

Also of interest is a paper of Biermann (1951) where the influence of an anisotropic convection on the rotation law of a rotating body is discussed. His finding that the large-scale action of the convection makes a rigid rotation impossible, can be regarded as the discovery of the basic phenomena responsible for the differential rotation.

As a next step we note the papers of Sweet (1950) and Csada (1951), who suggested that turbulence may destroy large-scale magnetic fields. The problem originates mainly in the observation of sunspots: A sunspot magnetic field of a diameter $\bar{L} \approx 10^4$ km should have a decay time $\bar{T} = \mu \sigma \bar{L}^2$ of about 3000 yr ($\mu \sim$ permeability, $\sigma \sim$ conductivity). But it is observed to decay in about 3 or 4 months. From this discrepancy it was concluded that turbulence may provide for an enhancement of the magnetic diffusivity or, what comes out to be the same, for a diminution of the conductivity with respect to the large-scale field. The same was expected to be the case for the permeability (Zeldovich, 1956).

Without doubt the most important result in this regard is due to Parker (1955). He discovered that small-scale cyclonic motions on a rotating body can produce a poloidal magnetic field out of a toroidal field. By combining this effect with a constant shear Parker was able to prove the existence of migrating dynamo waves for homogeneous models.

As a last example we mention the random walk migration of magnetic regions as described by Leighton (1964). This is a large-scale behaviour which is caused by

Bumba and Kleczek (eds.), Basic Mechanisms of Solar Activity, 305–321. *All Rights Reserved.*
Copyright © 1976 by the IAU.

turbulent diffusion under the presence of the large-scale shear due to differential rotation.

2. Basic Ideas of Mean-Field Magnetohydrodynamics

2.1. Mean-field magnetohydrodynamics, or – as the first step – mean-field electrodynamics intends to explain those different phenomena in the frame of one theory. It starts from the assumption that the basic laws describing those phenomena are well known, namely the Maxwell equations

$$\text{rot } \mathbf{E} = -\dot{\mathbf{B}}, \qquad \text{rot } \mathbf{H} = \mathbf{j}, \qquad \text{div } \mathbf{B} = 0, \tag{1}$$

combined with the constitutive equations

$$\mathbf{B} = \mu \mathbf{H}, \qquad \mathbf{j} = \sigma(\mathbf{E} + \mathbf{u} \times \mathbf{B}), \tag{2}$$

and the Navier–Stokes equation

$$\rho\left(\frac{\partial \mathbf{u}}{\partial t} + (\mathbf{u} \cdot \text{grad})\mathbf{u}\right) = -\text{grad } p + \mathbf{j} \times \mathbf{B} + \cdots, \tag{3}$$

the equation of continuity

$$\frac{\partial \rho}{\partial t} + \text{div } \rho \mathbf{u} = 0, \tag{4}$$

and equations of state.

Here $\mathbf{E}$ is the electric field strength, $\mathbf{H}$ is the magnetic field strength, $\mathbf{B}$ the magnetic induction, $\mathbf{j}$ the current density, $\mathbf{u}$ the velocity field, p the pressure and ρ the density. By the suspension points in the Navier–Stokes equation we indicate that there are additional quantities which need not be specified in our consideration.

By taking the average we can derive equations for the mean quantities, in this way following the proposal of O. Reynolds (1883) made for the Navier–Stokes equation in connection with his investigations of hydrodynamic turbulence.

2.2. It is very convenient to use statistical means for the theoretical (mathematical) investigations. We adopt the point of view of Gibbs that there are one million or more Suns and we take the average over this set of Suns (Figure 1). This average can be exchanged with all linear operations, especially with integration and differentiation operation, thus simplifying the deductions.

However, from the observational data of the Sun we can only derive averages over the space of the time coordinates. The question arises whether we can apply the deduced results to the mean quantities derived from the observational data.

For a quantity F we denote the average by $\bar{F}$, and the fluctuation $F - \bar{F}$ by F'. For the statistical average we have the rule

$$\bar{\bar{F}} = \bar{F} \tag{5}$$

or, equivalently,

$$\overline{F'} = 0. \tag{6}$$

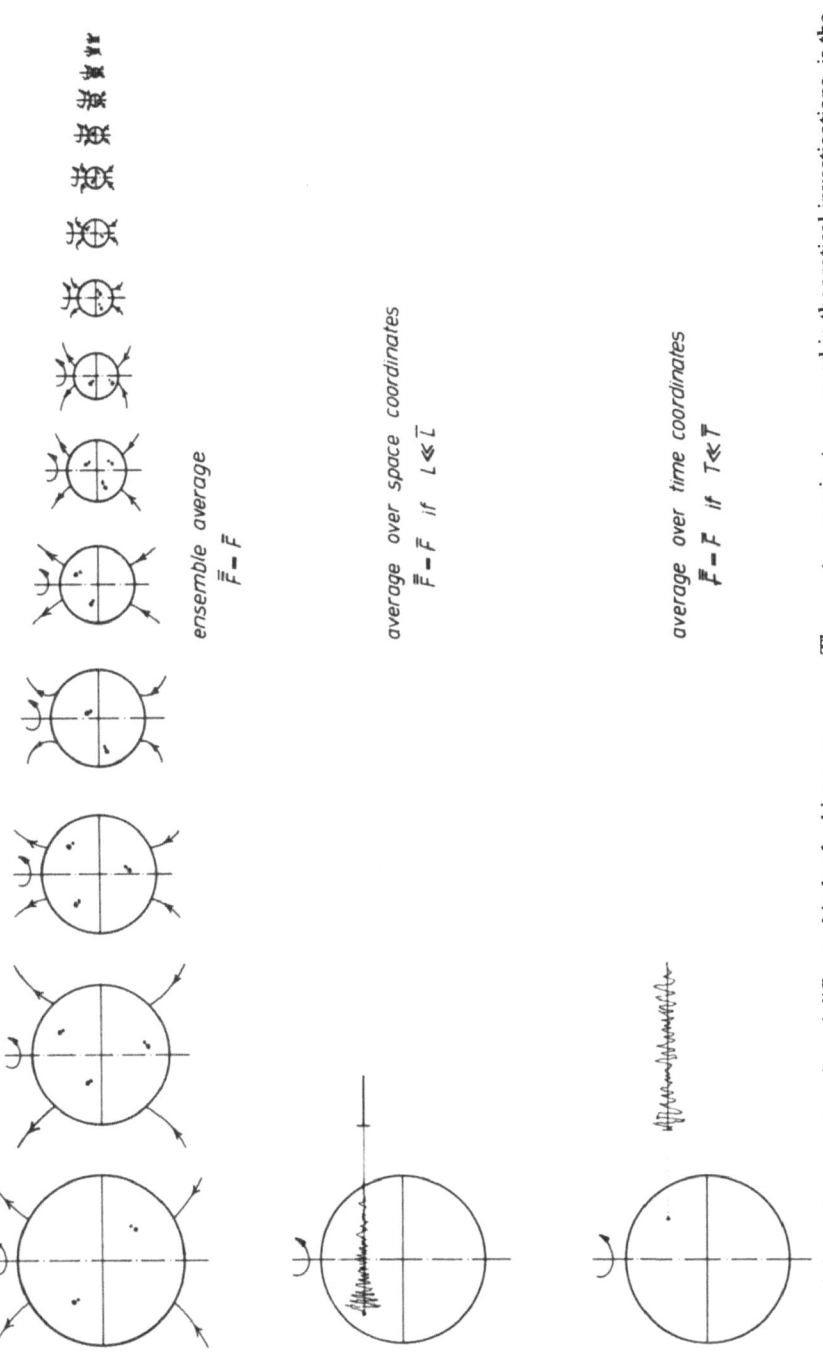

Fig. 1. Schematic representation of different kinds of taking an average. The most convenient one, used in theoretical investigations, is the ensemble average. The idea is that there is a set of a great number of Suns and the average is taken over this set of Suns. By taking each cycle as a representation of a Sun and averaging over all cycles, a possibility for an ensemble average is given. From observational data one can in most cases only take the average with respect to space or time coordinates. Then especially the rule $\bar{\bar{F}} = \bar{F}$ is only approximately valid. (L, T scales of the turbulent field; $\bar{L}$, $\bar{T}$ scales of the mean field.)

For an average with respect to the space or time coordinates the relations (5) or (6) are only approximately valid, and that if the mean quantities behave nearly constant over the scales used for the average. More precisely, let $\bar{L}$ and $\bar{T}$ be the scales of the mean field, and L and T those of the turbulent field, i.e. L the correlation length and T the correlation time. Then (5) and (6) are valid only in the sense

$$\bar{\bar{F}} = \bar{F} + 0\left(\frac{L}{\bar{L}}, \frac{T}{\bar{T}}\right), \qquad \overline{F'} = 0\left(\frac{L}{\bar{L}}, \frac{T}{\bar{T}}\right). \tag{7}$$

Thus the application of the theory is indicated only if

$$L \ll \bar{L}, \qquad T \ll \bar{T}, \tag{8}$$

and is questionable whenever (8) is violated.

For the convection we have $L \approx 10^3$ km, $T \approx 3 \times 10^2$ s. Compared with the scales of the mean magnetic field, $\bar{L} \approx R_0 = 7 \times 10^5$ km, $\bar{T} = 6 \times 10^8$ s (period of the solar cycle), the relation (8) is, indeed, rather well fulfilled. With respect to the space coordinates it is less satisfactory for the supergranulation.

2.3. Under averaging the Maxwell-equations remain unchanged since they are linear; the same holds for the constitutive equation connecting **B** and **H**. However, Ohm's law includes a non-linear term, $\mathbf{u} \times \mathbf{B}$, which gives rise to an additional term in Ohm's law for the mean fields, the turbulent emf

$$\mathscr{E} = \overline{\mathbf{u'} \times \mathbf{B'}}. \tag{9}$$

Thus we find as the basic equations for mean-field electrodynamics

$$\operatorname{curl} \bar{\mathbf{E}} = -\overset{\circ}{\bar{\mathbf{B}}}, \qquad \operatorname{curl} \bar{\mathbf{H}} = \bar{\mathbf{j}}, \qquad \operatorname{div} \bar{\mathbf{B}} = 0, \tag{10}$$

with the constitutive equations

$$\bar{\mathbf{B}} = \mu \bar{\mathbf{H}}, \qquad \bar{\mathbf{j}} = \sigma(\bar{\mathbf{E}} + \bar{\mathbf{u}} \times \bar{\mathbf{B}} + \mathscr{E}). \tag{11}$$

The occurrence of the turbulent emf $\mathscr{E}$ in Ohm's law for the mean electromagnetic fields is the crucial point of the theory. It is essential for providing for an additional equation expressing $\mathscr{E}$ by the mean fields. We start from the well-known induction equation

$$\frac{\partial \mathbf{B}}{\partial t} - \operatorname{curl}(\mathbf{u} \times \mathbf{B}) - \frac{1}{\mu\sigma} \Delta \mathbf{B} = 0, \tag{12}$$

which can be easily derived from the Equations (1) and (2). Taking the average we find

$$\frac{\partial \bar{\mathbf{B}}}{\partial t} - \operatorname{curl}(\bar{\mathbf{u}} \times \bar{\mathbf{B}}) - \frac{1}{\mu\sigma} \Delta \bar{\mathbf{B}} = \operatorname{curl}\overline{(\mathbf{u'} \times \mathbf{B'})}. \tag{13}$$

Subtracting (13) from (12) we have the equation for the fluctuations

$$\frac{\partial \mathbf{B'}}{\partial t} - \operatorname{curl}(\bar{\mathbf{u}} \times \mathbf{B'}) - \frac{1}{\mu\sigma} \Delta \mathbf{B'} - \operatorname{curl}(\overline{\mathbf{u'} \times \mathbf{B'}} - \overline{\mathbf{u'} \times \mathbf{B'}})$$

$$= \operatorname{curl}(\mathbf{u'} \times \bar{\mathbf{B}}). \tag{14}$$

While an exact solution of Equation (14) is hardly possible, we can draw the general conclusion that $\mathbf{B}'$ as well as $\mathscr{E}$ is a linear functional of $\bar{\mathbf{B}}$. Thus we have

$$\mathscr{E} = \mathscr{L}(\bar{\mathbf{B}}) , \qquad (15)$$

with $\mathscr{L}$ denoting the linear functional dependence. This conclusion can be made without any restrictive assumption. If we now assume the relation (8) to be fulfilled, the field $\bar{\mathbf{B}}$ in Equation (14) can be taken constant or a linear function of the space coordinates. Thus $\mathscr{E}$ depends only on the local field, and (15) reduces to

$$\mathscr{E}_i = a_{ik}\bar{B}_k + b_{ikl}\frac{\partial \bar{B}_k}{\partial x_l} , \qquad (16)$$

where the error is of the order $0((L/\bar{L})^2, T/\bar{T})$. a_{ik}, b_{ikl} are certain pseudo-tensors depending on the mean properties of the turbulence. Relation (16) is the needed additional constitutive equation. It must be noted that (16) is valid under the assumption (8), which is a guarantee for the applicability of our theory. (16) is valid independent of what turbulence is regarded. A model is characterized by the pseudo-tensors a_{ik}, b_{ikl}.

3. Ohm's Law for the Mean Electromagnetic Quantities in the Solar Convection Zone

3.1. It is furthermore important to know that more explicit expressions for the tensors a_{ik}, b_{ikl} can be obtained if it is taken into account that these quantities are derivable out of the tensorial mean quantities of the turbulent velocity field by tensorial operations.

The simplest example is given by homogeneous isotropic turbulence, where the only available tensorial quantities are the isotropic tensors, i.e. the Kronecker Tensor δ_{ik} and the Levi-Cività Tensor ε_{ikl}. Hence we have

$$a_{ik} = \alpha\delta_{ik} , \qquad b_{ikl} = \beta\varepsilon_{ikl} , \qquad (17)$$

where α is a pseudo-scalar and β a scalar. Introducing (17) into (16) and then in (11) we find Ohm's law for the main quantities as

$$\mathbf{j} = \sigma_T(\mathbf{E} + \alpha\mathbf{B}) , \qquad (18)$$

where the conductivity with respect to the mean field is given by

$$\sigma_T = \frac{\sigma}{1 + \mu\sigma\beta} . \qquad (19)$$

We find that the action of homogeneous isotropic turbulence changes conductivity and produces a new effect – named α-effect. We will come back to this point later.

Obviously, in the convection zone of the Sun there is some structure caused mainly by two vectorial quantities, the rotational motion, represented by the angular velocity $\boldsymbol{\omega}$, and the gradient of density $\mathbf{g}$ in the radial direction. Restricting to the linear expressions in the vectors $\boldsymbol{\omega}$ and $\mathbf{g}$ we have the general ansatz for the tensors a_{ik}, b_{ikl}

$$a_{ik} = \alpha_0(\mathbf{g} \cdot \boldsymbol{\omega})\delta_{ik} + \alpha_1(\omega_i g_k + \omega_k g_i)$$

$$+ \alpha_2(\omega_i g_k - \omega_k g_i) + \gamma\varepsilon_{ikl}g_l$$

$$= \alpha_0(\mathbf{g} \cdot \boldsymbol{\omega})\delta_{ik} + \alpha_1(\omega_i g_k + \omega_k g_i)$$

$$+ \varepsilon_{ikl}(\gamma g_1 + \alpha_2\varepsilon_{lpq}\omega_p g_q); \tag{20}$$

$$b_{ikl} = \beta\varepsilon_{ikl} + \beta_1\omega_i\delta_{kl} + \beta_2\omega_k\delta_{il} + \beta_3\omega_l\,\delta_{ik}. \tag{21}$$

In the construction we have taken into account that both tensors are skew. The expressions (20) and (21) lead to an Ohm's law for the solar convection zone given by

$$\bar{\mathbf{j}} = \sigma_T(\bar{\mathbf{E}} + \bar{\mathbf{u}} \times \bar{\mathbf{B}} + \alpha_0(\mathbf{g} \cdot \boldsymbol{\omega})\bar{\mathbf{B}} + \alpha_1((\mathbf{g} \cdot \bar{\mathbf{B}})\boldsymbol{\omega} + (\boldsymbol{\omega} \cdot \bar{\mathbf{B}})\mathbf{g})$$

$$+ \bar{\mathbf{B}} \times (\gamma\mathbf{g} + \alpha_2\boldsymbol{\omega} \times \mathbf{g})$$

$$+ (\beta_2 + \beta_3)\,\mathrm{grad}\,(\boldsymbol{\omega} \cdot \bar{\mathbf{B}}) - \mu\beta_3\boldsymbol{\omega} \times \bar{\mathbf{j}}); \tag{22}$$

σ_T is given by (19).

The only unknown quantities are the scalars α_0, α_1, α_2, β, β_2, β_3, γ.

3.2. A theoretical determination of the quantities is rather complicated and not fully solved, since one is here confronted with the difficult closure problem of the theory of turbulence. There are a lot of papers which present approximative calculations of those constants. We will not go into details here and refer the interested reader to review papers (Krause and Rädler (1971), Roberts (1971), Roberts and Stix (1971), Vainshtein and Zeldovich (1972), Roberts and Soward (1975), Krause (1975)). The most often used approximation neglects the term curl $(\mathbf{u}' \times \mathbf{B}' - \overline{\mathbf{u}' \times \mathbf{B}'})$ in Equation (14). The resulting equation for $\mathbf{B}'$ is linear and can be solved by a Green's function method. We will speak of the second order correlations approximation (SOCA)*, since all correlations of higher than second order are neglected. This approximation is valid if

$$\min(R_m, S) \ll 1 \tag{23}$$

where R_m is the magnetic Reynolds number,

$$R_m = \mu\sigma u' L, \tag{24}$$

and S the Strouhal Number

$$S = u'\frac{T}{L}; \tag{25}$$

u' denotes the turbulent rms velocity.

For convection data $\mu = 4\pi \times 10^{-7}\ \Omega S\ \mathrm{m}^{-1}$, $\sigma = 10^3\ \Omega^{-1}\ \mathrm{m}^{-1}$, $u' \approx 300\ \mathrm{m\ s}^{-1}$ we find

$$R_m \approx 3 \times 10^5, \qquad S \approx 10^{-1}. \tag{26}$$

---

* This approximation is sometimes named 'first-order smoothing' or 'Herrings approximation' in hydrodynamic turbulence theory.

The value for S encourages to apply the results derived on the basis of SOCA to the convection zone. If $R_m \gg 1$ one finds in the homogeneous isotropic case

$$\alpha = -\frac{T}{3}\overline{\mathbf{u}' \cdot \operatorname{curl} \mathbf{u}'},\tag{27}$$

$$\beta = \frac{\overline{\mathbf{u}'^2 T}}{3}.\tag{28}$$

(see, e.g., Steenbeck and Krause, 1969a).

From (27) it becomes obvious that $\alpha \neq 0$ is connected with the fact that one kind of helical motions is more probable than the other. For example, α is positive if left-handed helical motions are more probable than right-handed.

3.3. The value of β depends on quantities that can be taken from observations. With the data belonging to (26) we have

$$\beta_{cz} \approx 10^7 \, \mathrm{m^2 \, s^{-1}}.\tag{29}$$

If we derive from Ohm's law (18), with $\alpha = 0$, the induction equation for the mean magnetic field, we find

$$\frac{\partial \bar{\mathbf{B}}}{\partial t} - \left(\frac{1}{\mu\sigma} + \beta\right)\Delta\bar{\mathbf{B}} = 0,\tag{30}$$

and we see that β proves to be the turbulent magnetic diffusivity. For the ratio $\beta/(1/\mu\sigma)$ we obtain from (29)

$$\frac{\beta}{1/\mu\sigma} = \mu\sigma\beta \approx 10^4.\tag{31}$$

This implies especially for the decay time of a large-scale magnetic field that it is shortened by a factor 10^4. One is tempted to apply this result to the decay of sunspots, since the numerical agreement with the observations is very good. Furthermore, more detailed discussions of the decay of sunspot areas by a diffusion model based on Equation (30) by Meyer and Schmidt (1973) and us (Krause and Rüdiger (1975)) show not only qualitatively but also quantitatively a nice agreement with the results derived from observations by Bumba (1963).

However, this discussion ignores that there is a rather strong magnetic field in sunspots, which, perhaps, suppresses the turbulence or at least influences the turbulence, and one has to ask why the magnetic field should be unable to brake its own decay. We will revert to this point later.

A further point of interest in this connection is the period of the solar cycle. Observations show the Sun's global magnetic field alternating with a period of about 22 years; its characteristic length scale $\bar{L}$ must be expected to be of the order of the thickness of the convection zone, that means $\bar{L} \approx 10^5$ km and hence

$$\frac{\bar{L}^2}{\beta} \approx 10^9 \, \mathrm{s} \approx 30 \, \mathrm{yr}.$$

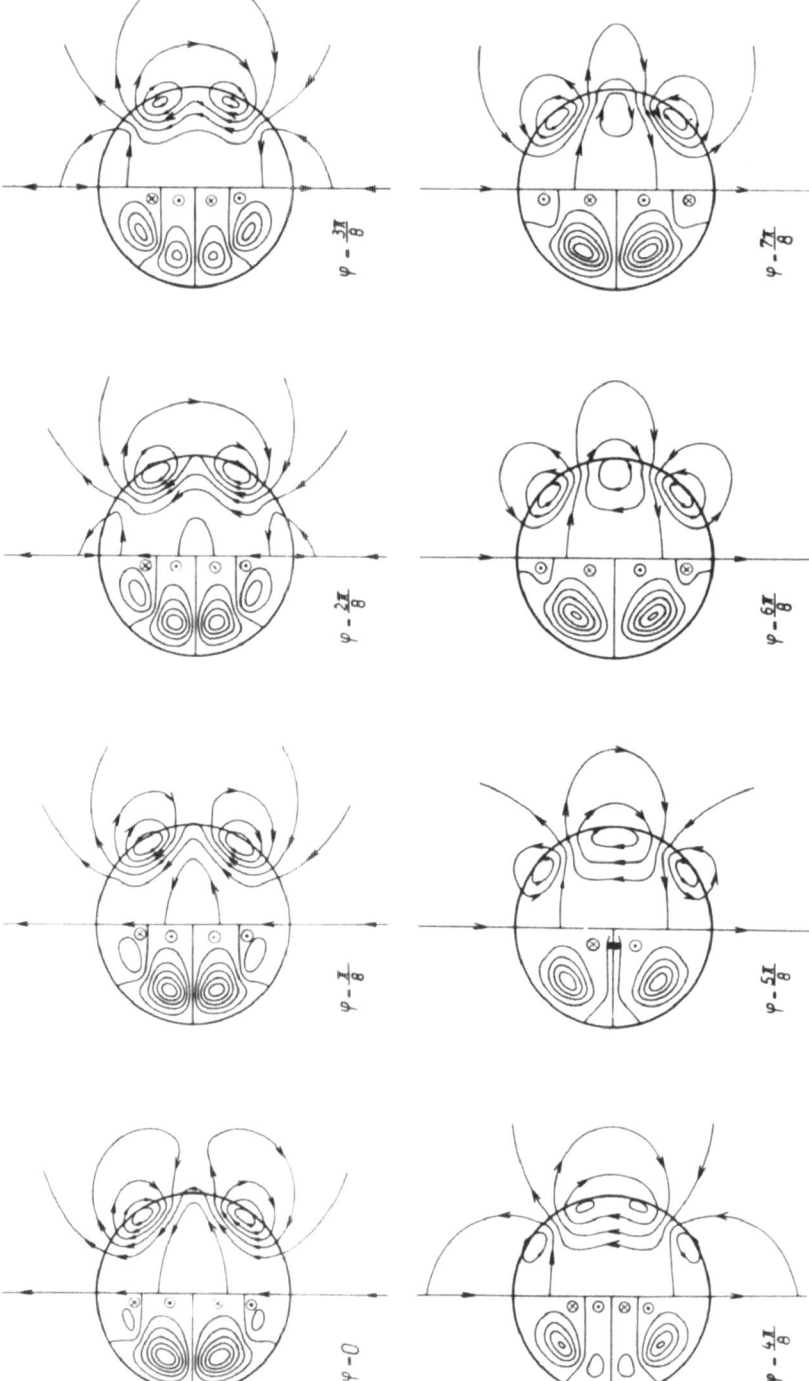

Fig. 2.   A model of a self-excited dynamo for the solar cycle. Eight phases of a half cycle of the magnetic field are represented. The phase $\varphi = 0$ corresponds to the phase of maximum activity (maximum toroidal field). At the left half of each phase picture the curves of constant field strength of the toroidal field are drawn, the field lines of the poloidal field at the right. One realizes the migration of the field towards the equator. (After Steenbeck and Krause, 1969a.)

The use of the turbulent magnetic diffusivity leads already to an agreement in the time scales.

3.4. Of greatest interest in dynamo theory is the quantity $\alpha = \alpha_0(\mathbf{g} \cdot \boldsymbol{\omega})$. An estimation from observations is not possible as can be taken from (27), since there is no direct information about the quantity $\mathbf{u}' \cdot \text{curl } \mathbf{u}'$. Steenbeck *et al.* (1966) derived an expression in the low-conductivity limit, $R_m \ll 1$. For the high-conductivity limit Krause (1968) derived the relation

$$\alpha = -T^2 \overline{\mathbf{u}'^2} \boldsymbol{\omega} \cdot \text{grad } (\ln u'\rho), \tag{32}$$

which for the solar convection zone (Steenbeck and Krause, 1969a) gives

$$\alpha = \frac{T^2 \overline{u'^2} \omega}{H} \cos \vartheta \approx 0.26 \cos \vartheta \text{ m s}^{-1} \tag{33}$$

where H is the scale height. Further calculations of α have been given by Moffatt (1970a, b, 1974) and by Stix (unpublished).

Calculations of spherical dynamo models based on the combined action of α-effect and differential rotation have shown that a simulation of details of the solar cycle is possible to a considerable degree (Figure 2 and Figure 3). I am not going to discuss here the coefficients α_0, α_1, α_2 in detail; it will be done in subsequent papers. The term in (22) $(\beta_2 + \beta_3) \text{ grad } (\boldsymbol{\omega} \cdot \bar{\mathbf{B}})$ is of no further interest since it is compensated by

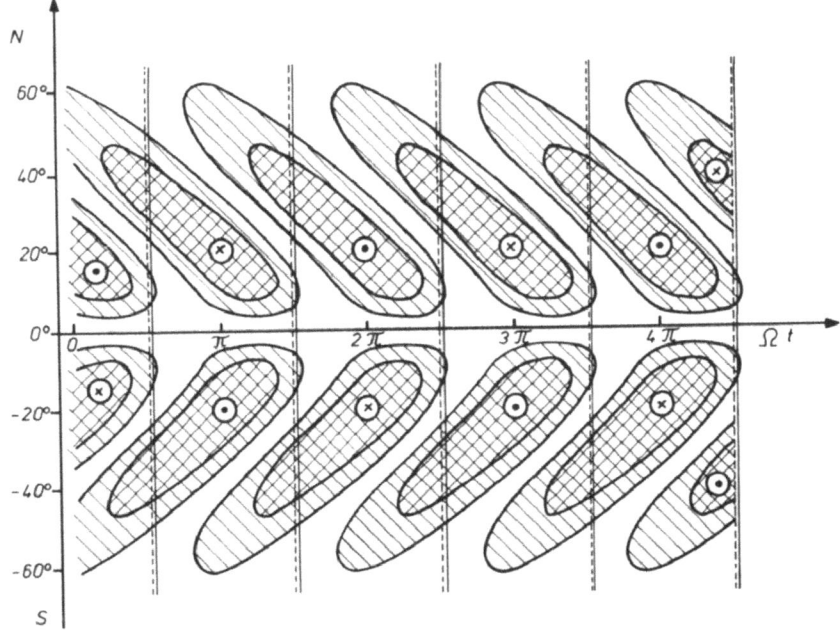

Fig. 3.   Butterfly diagram derived from the model described in Figure 2. In the hatched areas the toroidal field strength is larger than $\frac{1}{3}$, in the cross-hatched larger than $\frac{2}{3}$ of its maximum value. Again one realizes the migration towards the equator just as it is observed at the Sun.

space charges. The last term, $-\mu\beta_3\boldsymbol{\omega}\times\bar{\mathbf{j}}$, can also provide for dynamo action if combined with differential rotation (Rädler, 1969).

A remark on differential rotation may be appropriate here. Ohm's law as given by (22) for the mean electromagnetic fields in the solar convection zone was derived under the assumption of rigid rotation. In the presence of differential rotation further terms will appear describing the turbulent diffusion under the influence of the shear of the differential rotation. In this way Leighton's (1964) heuristic discussion of the random walk of bipolar magnetic regions has found its place in the frame of this deductive theory.

4. The rms of the Magnetic Field Fluctuations

4.1. The decay of a magnetic field in a conductor is governed by a diffusion equation

$$\frac{\partial \mathbf{B}}{\partial t} - \eta\,\Delta\mathbf{B} = 0 \; ; \tag{34}$$

with the magnetic diffusivity $\eta = 1/\mu\sigma$.

The magnetic energy is transformed into heat. The decay of the mean magnetic field also obeys a diffusion equation, namely

$$\frac{\partial \bar{\mathbf{B}}}{\partial t} - (\eta + \eta_T)\,\Delta\bar{\mathbf{B}} = 0 \,, \tag{35}$$

$\eta = 1/\mu\sigma$ and $\eta_T = \beta$, as shown in Section 3.3. But the physical background is quite different: The decay of $\bar{\mathbf{B}}$ is both an Ohmic decay *and* a decay of the mean field in the turbulent field. Especially in a situation as it is met with in the convection zone the decay in the small-scale field is much more efficient than the Ohmic decay.

If we assume $\mathbf{B}'$ to show only Ohmic decay, we have a time scale of

$$\mu\sigma L^2 \approx 10^9 \text{ s} \approx 30 \text{ yr} \,. \tag{36}$$

On the other hand for a large-scale field we have the decay times

$$\bar{T} = \frac{\bar{L}^2}{\eta_T} = \begin{cases} 10^7 \text{ s} \approx 4 \text{ months, if } \bar{L} = 10^4 \text{ km} \\ 10^9 \text{ s} \approx 30 \text{ years, if } \bar{L} = 10^5 \text{ km} \end{cases} \tag{37}$$

Comparing these values we realize that the decay in the small-scale field is more rapid than the Ohmic decay of the small-scale field. Consequently we have to expect in the stationary case that the magnetic energy stored in the fluctuations is larger, or even much larger, than in the mean field.

4.2. According to the general arguments leading to (16) we have the relation

$$\overline{B'^2} = q_{ik}\bar{B}_i\bar{B}_k \,, \tag{38}$$

with a certain tensor q_{ik} depending on the properties of the turbulent motion. The inequalities (8) were a sufficient condition for the validity of (16), and so are they for the validity of (38). An error of the order $0(L/\bar{L}, T/\bar{T})$ must be expected.

In the isotropic case (38) reduces to

$$\overline{B'^2} = q\bar{B}^2 , \tag{39}$$

where $3q$ is the trace of the tensor q_{ik}. In the framework of SOCA one finds in the high-conductivity limit $(R_m \gg 1)$

$$q = \tfrac{1}{3}\mu\sigma u'^2 L \tag{40}$$

(Bräuer and Krause, 1973), which with the data for the convection zone implies

$$\overline{B'^2} \approx 10^4 \cdot \bar{B}^2 . \tag{41}$$

The relations (40) and (41) had been derived already by Steenbeck and Krause (1969a) using physical arguments.

There have been given other estimations of q by less deductive methods than that leading to (40). Parker, in a paper dating from 1963 expected a proportionality of q to $(R_m^{1/4}/\ln R)^2$, which he retraced in 1969 and in 1973 replaced by the relation

$$q \approx R_m . \tag{42}$$

A numerical investigation of Moss (1970) of a two-dimensional model indicates the behaviour

$$q \approx R_m^{0.7} \tag{43}$$

The result (42) of Parker is in agreement with (40) because it rests on the assumption $S \approx 1$, which is often used in investigations of turbulence.

For the convection zone of the Sun we find from (40) for the magnetic field rms

$$B' \approx 100 \, \bar{B} , \tag{44}$$

which is nearly the same value as derived from the result of Moss $(R_m^{0.35} \approx 83)$; Parker's formula gives about five times this value.

4.3. Because of its simplicity we will present here the derivation of (40) given by Steenbeck and Krause (1969a): We start from the induction equation

$$\frac{\partial \mathbf{B}'}{\partial t} = \text{curl} \, (\mathbf{u}' \times \bar{\mathbf{B}}) , \tag{45}$$

where we have omitted the diffusion term since we integrate over a time interval of the length of the correlation time, which is small compared with $\mu\sigma L^2$. For the correlated part we find from (45)

$$B'_{\text{cor}} \approx \frac{u'T}{L} \bar{B} . \tag{46}$$

This field part is now assumed to exist over the time $\mu\sigma L^2$. We assume the turbulence to produce in every time interval of length T one field part (46), so that the complete field is the statistical sum of such field parts. The number N of the field parts existing at the same point of time is equal to

$$N = \frac{\mu\sigma L^2}{T} ; \tag{47}$$

hence we get for B' itself from (46) and (47)

$$B' \approx \sqrt{N} B'_{\text{cor}} \approx \sqrt{\mu \sigma u'^2 T} \bar{B}, \tag{48}$$

in agreement with (40).

4.4. From these results a very interesting feature becomes visible: According to (44) we have to expect the mean field to be of the order of about 1% of the rms of the fluctuating field. This is, indeed, a theoretical prediction, but if we adopt this result, we must conclude that it is hardly possible to detect the mean field by observation. And this may, perhaps, explain why some observers do not believe in the mean field. However, the mean field must exist, since otherwise there would be no explanation for the existence of the small-scale field. The large-scale field is the exciter field in a separate excited dynamo and the small-scale field is the excited field, which in our case is much larger than the exciter field.

In this light one must see the investigations by Howard (1974a, b) who intended to derive the mean magnetic field on the solar surface by averaging the observational data. I asked Dr Howard for his opinion on how strong the magnetic field rms at the solar surface in a quiet region might be. His estimate lies between 1 G and 10 G.

Hence, according to (44), the mean magnetic field will be between 0.01 G and 0.1 G, which leads to the conclusion, that the averages derived by Howard give no information about the mean field. Answering the same question Dr Stenflo argued that Dr Howard probably underestimates the rms by a factor of about 20. His reasons Dr Stenflo already explained in his talk yesterday when he demonstrated that magnetographs record an average over some area. Hence, according to Stenflo, the rms would be between 20 G and 200 G. Therefore the mean field must be expected in the order of 1 G and in this picture the averages derived by Howard give some correct impression of the mean field. A great accuracy cannot be expected because the mean field is only 1% of the rms.

Against a dynamo theory of the solar cycle there is sometimes the objection raised that the rate of field diffusion and reconnection might be too small. Relation (39) with q from (40) can possibly give an answer, especially if written in the form

$$\overline{B'^2} = \frac{\eta_T}{\eta} \bar{B}^2. \tag{49}$$

Now it becomes obvious that a small molecular magnetic diffusivity provides for an enrichment of magnetic energy in the small scales.

Our discussion shows that the fluctuating magnetic field can be, and in the case of the Sun in fact is, much larger than the mean field. Since in the frame of SOCA the equation for the fluctuating field had been linearized one might expect that the results (41) and (49) could be out of the range of parameters where the theory can be applied. This is, however, not the case: The derivation of (40) and subsequently of (41) and (49) is restricted to the assumptions (8) and (23) only, wherefrom no restriction of B' follows. This question is discussed in more detail by Krause and Roberts (1976).

5. The Back-Reaction of the Magnetic Field to the Motion

5.1. Up to this point we have discussed situations where the magnetic energy is assumed to be small compared with the kinetic energy,

$$\frac{1}{2\mu}(\bar{B}^2 + \overline{B'^2}) \ll \frac{\rho}{2}(\bar{u}^2 + \overline{u'^2}) \,. \tag{50}$$

If we also take into account the back-reaction of the magnetic field on the motion, we have to regard the Lorentz-force in the Navier–Stokes equation. Since we have

$$\overline{\mathbf{j} \times \mathbf{B}} = \bar{\mathbf{j}} \times \bar{\mathbf{B}} + \overline{\mathbf{j}' \times \mathbf{B}'} \,, \tag{51}$$

an additional force in the Navier–Stokes equation for the mean fields, the Reynolds equation, will appear. As a consequence the dependence of the electromotive force, $\mathscr{E}$, on the mean magnetic field is no longer linear, the tensors a_{ij} and b_{ijk} in Equation (16) will now depend on the mean magnetic field itself:

$$a_{ij} = a_{ij}(\bar{\mathbf{B}}) \,, \qquad b_{ijk} = b_{ijk}(\bar{\mathbf{B}}) \,; \tag{52}$$

and the same is the case for the quantities α, β, etc.

In the incompressible case investigations have been carried out by Rädler (1974) and Rüdiger (1974) assuming a constant mean magnetic field. Roberts and Soward (1975) take into account a gradient of the mean magnetic field. Putting the tensors a_{ij} and b_{ijk} into the induction equation for the mean field (13), one is led to a non-linear equation. First attempts at solving the corresponding non-linear dynamo problem were carried out by Stix (1972) and by Rüdiger (1973).

5.2. I wish to end my talk with some remarks concerning the influence of strong magnetic fields on turbulent motions.

It is often said that strong magnetic fields suppress turbulence. However, experiments with liquid sodium show not a suppression of turbulence but a transformation of its structure, which appears to become two-dimensional: The motion takes place in planes orthogonal to the magnetic field and does not vary along it (Figure 4). Interesting results were published by Kit and Tsinober (1971). Perhaps the rotational motions in prominences as described by Rompolt (1975) indicate a two-dimensional motion occurring in a certain solar structure.

The conception of two-dimensional turbulence occurring in sunspots may open a possibility for explaining the rapid decay of sunspots by turbulent diffusion, although the strong magnetic field influences the turbulent motion, as shown by Krause and Rüdiger (1975). It is of interest that an estimation of the turbulent magnetic diffusivity derived from a comparison of a model proposed by Meyer and Schmidt (1973) with the observational data for long-lived spot groups as published by Bumba (1963), gives just the value (29) derived on the basis of mean-field magneto-hydrodynamics. Furthermore, the models show also a relation like Waldmeier's rule of proportionality between the maximum area of a sunspot group and its lifetime. A quantitative comparison yields a turbulent magnetic diffusivity about 8 times the value of (29). This is so obviously because the derivation of Waldmeier includes the short-lived sunspot groups.

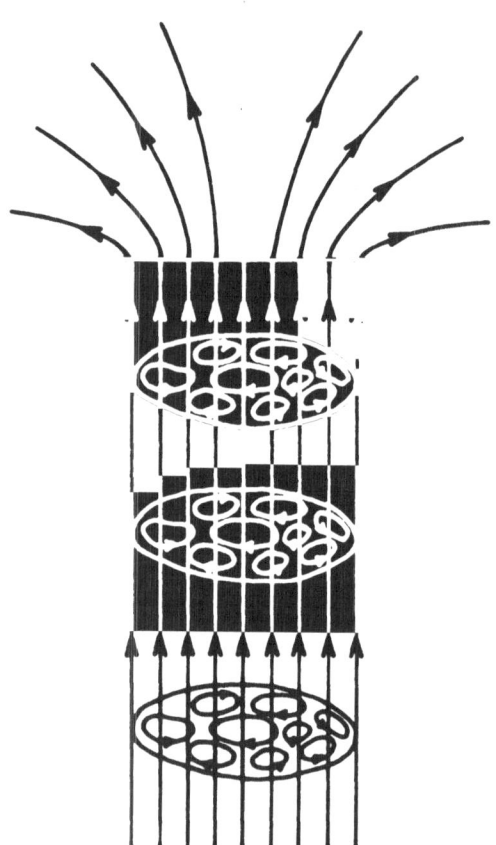

Fig. 4.   Schematic drawing of the two-dimensional motion in a sunspot: The magnetic field is assumed to be perpendicular to the Sun's surface. The turbulent motion under the influence of the strong magnetic field assumes a two-dimensional structure. The velocity vectors lie completely in planes orthogonal to the magnetic field and the pattern does not change in the direction of the field.

It is also possible to fit curves describing the development of the areas of sunspot models based on turbulent diffusion to those observed at short-lived sunspot groups. The turbulent magnetic diffusivities derived in these cases are larger than the value of (29) by a factor up to 100. This probably indicates, compared with normal convection, an enhanced turbulence at moments where sunspots emerge until a few days later.

Acknowledgments

The author is grateful to Dr Rädler and Dr Rüdiger for stimulating discussions.

References

Biermann, L.: 1951, *Z. Astrophys.* **28**, 304.
Bräuer, H. and Krause, F.: 1973, *Astron. Nachr.* **294**, 179.

Bumba, V.: 1963, *Bull. Astron. Inst. Czech.* **14**, 91.

Csada, I. K.: 1951, *Acta Phys. Hungarica* **1**, 235.

Gurewitch, L. E. and Lebedinskii, A. I.: 1945, *Dokl. Akad. Nauk U.S.S.R.* **49**, 92.

Howard, R.: 1974a, *Solar Phys.* **38**, 283.

Howard, R.: 1974b, *Solar Phys.* **38**, 59.

Kit, L. G. and Tsinober, A. B.: 1971, *Magnitnaja gidrodinamika*, **1971**, 27.

Krause, F.: 1968, Habilitationsschrift, Univ. Jena.

Krause, F.: 1975, *Ann. N.Y. Acad. Sci.* **257**, 156.

Krause, F. and Rädler, K.-H.: 1971, in *Handbuch der Plasmaphysik und Gaselektronik* **2** (ed. by R. Rompe and M. Steenbeck), Akademie-Verlag, Berlin.

Krause, F. and Roberts, P. H.: 1976, *J. Math. Phys.* (in press).

Krause, F. and Rüdiger, G.: 1975, *Solar Phys.* **41**, 286.

Leighton, R. B.: 1964, *Astrophys. J.* **140**, 1547.

Meyer, F. and Schmidt, H.-U.: 1973, *Mitt. Astron. Ges.* **32**, 173.

Moffatt, H. K.: 1970a, *J. Fluid Mech.* **41**, 435.

Moffatt, H. K.: 1970b, *J. Fluid Mech.* **44**, 705.

Moffatt, H. K.: 1974, *J. Fluid Mech.* **65**, 1.

Moss, D. L.: 1970, *Monthly Notices Astron. Soc.* **148**, 173.

Parker, E. N.: 1955, *Astrophys. J.* **122**, 293.

Parker, E. N.: 1963, *Astrophys. J.* **138**, 226.

Parker, E. N.: 1969, *Astrophys. J.* **158**, 815.

Parker, E. N.: 1973, *Astrophys. Space Sci.* **22**, 279.

Rädler, K.-H.: 1968a, *Z. Naturforsch.* **23a**, 1841.

Rädler, K.-H.: 1968b, *Z. Naturforsch.* **23a**, 1851.

Rädler, K.-H.: 1969, *Monatsber. Dtsch. Akad. Wissensch. Berlin* **11**, 272.

Rädler, K.-H.: 1974, *Astron. Nachr.* **295**, 265.

Reynolds, O.: 1883, *Trans. Roy. Soc. (London)* **A174**, 935.

Roberts, P. H.: 1971, *Lectures on Applied Math.* **14**, (ed. by W. H. Reid), Am. Math. Soc. Providence, R.I., U.S.A.

Roberts, P. H. and Stix, M.: 1971, The Turbulent Dynamo, a translation of papers by F. Krause, K.-H. Rädler and M. Steenbeck, Techn. Note TN/IA-60 from the National Center for Atmospheric Research, Boulder.

Roberts, P. H. and Soward, A. M.: 1975, *Astron. Nachr.* **296**, 48.

Rompolt, B.: 1975, *Acta Universitatis Wratislaviensis* **252**, 1.

Rüdiger, J.: 1973, *Astron. Nachr.* **294**, 183.

Rüdiger, G.: 1974, *Astron. Nachr.* **295**, 275.

Steenbeck, M. and Krause, F.: 1967, *Magnitnaja gidrodinamika*, 1967/3, 19.

Steenbeck, M. and Krause, F.: 1969a, *Astron. Nachr.* **291**, 49.

Steenbeck, M. and Krause, F.: 1969b, *Astron. Nachr.* **291**, 271.

Steenbeck, M., Krause, F., and Rädler, K.-H.: 1963, 'Elektromagnetische Eigenschaften turbulenter Plasmen', *Sitzungsber. Dtsch. Akad. Wiss. Berlin, Klasse Math.-Phys.-Techn.*, Heft 1.

Steenbeck, M., Krause, F., and Rädler, K.-H.: 1966, *Z. Naturforsch.* **21a**, 1285.

Stix, M.: 1972, *Astron. Astrophys.* **20**, 9.

Sweet, P. A.: 1950, *Monthly Notices Roy. Astron. Soc.* **110**, 69.

Vainshtein, S. I. and Zeldovich, Ja. B.: 1972, *Usp. Fiz. Nauk* **106**, 431 (*Soviet Phys.-Uspechi* **15**, 159 (1972)).

Zeldovich, Ja. B.: 1956, *Zh. Eksp. Teor. Fiz.* **31**, 154 (*Soviet Phys. JETP* **4**, 460 (1957)).

DISCUSSION

Giovanelli: It worries me that observations show sunspots to decay by the transport of small packets of magnetic flux of field-strength approaching that of the sunspot, whereas theories of this type attempt to explain decay by a continuous spectrum of eddy ridges.

Krause: The turbulent decay describes a decay of the large-scale field in small-scale fields. I think your description of a transport of small packets of magnetic flux fits in this picture.

Zwaan: It should be kept in mind that the lowest decay rates discovered by Bumba apply to a *fraction* of the spots. In fact there is a *range* of decay times, from very high rates down to the lowest decay rate as a lower limit. Therefore I am surprised that your dissipation model predicts at the same time Bumba's *lowest* decay rates, and Waldmeier's *average* relation between sunspot areas and lifetimes.

Krause: This is an interesting point. The turbulent magnetic diffusivity derived from Bumba's long-living sunspot groups is $10^7 \ \mathrm{m^2 \ s^{-1}}$, whereas the same quantity taken from Waldmeier's average relation is $8 \times 10^7 \ \mathrm{m^2 \ s^{-1}}$.

Weiss: Two time scales are involved in the growth and decay of sunspots. A few long-lived spots decay slowly at a steady rate. The majority are formed and decay on a time scale of one to six days. Meyer, Schmidt, Wilson and I interpreted slow decay as a result of turbulent diffusion owing to small-scale convection but the rapid growth and break-up were related to supergranular motion, with the same scales as the sunspot itself. So it seems unlikely that the initial growth and fast decay of sunspot groups can be described by an eddy diffusivity.

Krause: The diffusion model fits also the short-living sunspots groups, but with larger magnetic diffusivity.

Bumba: I have to agree with Dr Zwaan's and Dr Weiss' remarks. There are really two types or two velocities of sunspots area diminution, but they are closely related to the supergranular cells. If the cell is completely filled in by the spot, then the spot is more stable, its area diminishes slower than in the case of small spots which are too small in size to be comparable with supergranules. The number of regular stable spots is of course much smaller than the number of all spots.

Krause: In our paper (Krause and Rüdiger, 1975) we showed that also the more rapidly decaying sunspots groups can be described by a diffusion model, but with a larger magnetic diffusivity.

Deubner: I should like to have classified the type or topology of field you have in mind when you are discussing the observability of the *mean* field. Is it *mainly* poloidal or toroidal, what is its radial component, and finally what is its location with reference to the *observed* field? After all, aren't both quantities derived from identical physical structures, using different observational techniques?

Krause: If there is outside an insulator (vacuum) the toroidal field on the surface is zero. In this sense any observable magnetic field is a poloidal field. Observed with the magnetographs is in the most cases the component in the direction of line of sight only.

Parker: I do not see that the rapid decay of some sunspots presents a problem. As I pointed out on Monday, the magnetic field configuration of a sunspot is subject to the hydromagnetic exchange instability, unless that instability is blocked by some other force. Dr Krause has made the point in his lecture that a sunspot (once it ceases to accumulate more flux) must decay at least as fast as required by turbulent diffusion. He has also shown how well this idea fits the decay of the more long-lived spots. That some spots decay more rapidly, merely reminds us that there are disruptive forces, in addition to turbulent mixing. Presumably the hydromagnetic exchange instability assists in the decay of many spots.

Newkirk: I'd like to ask a question and also make a comment. The question is what observable parameters, other than the relation between mean and rms field and the decay rate of sunspots which you have mentioned, can be used to test the validity of the dynamo theory. The comment concerns the value of the observed topology of the large scale photospheric fields. Dynamo theories predict a very regular poloidal and toroidal field. Yet the observed field is much more complete with such patterns as a 'dipole' with a large angle to the rotation axis being quite common.

Krause: Both things you mentioned in your question have not really something to do with dynamo theory, however, they play some role in this theory. The most striking result supporting dynamo theory is that it was possible to construct models simulating the solar cycle by a strict integration of the basic equations. As concerns your comment: Dynamo theory predicts a regular *mean* poloidal and regular *mean* toroidal field and mean-field electrodynamics predicts at the surface a very *irregular* field, since the rms field B' is larger by a factor of 100 than the mean field.

Stenflo: In reply to Dr Newkirk let me note that the rms field strength is not 1–2 kG, the value of the field in the flux tubes that carry most of the net magnetic flux through the solar surface. It is much smaller, since the strong-field flux tubes occupy only a small area in the network. When determining the rms field we have to average over the large non-network regions of weaker fields as well.

Weiss: Perhaps I can allay some of Professor Parker's worries about the stability of sunspots. F. Meyer, H. U. Schmidt and I have been studying this problem. The interchange instability occurs when the curved lines of force are concave towards the surrounding plasma, but the configuration can be stabilized by the vertical pressure gradient outside the sunspot. Energy is released by straightening out a curved flux tube but the tube will be restored by buoyancy forces to its original position. Thus the hydrostatic pressure gradient which causes the curvature of field lines can stabilize the magnetic field. For a simple model we find that a spot is stable if the horizontal component of the field decreases upwards at its boundary. Preliminary results suggest that pores and sunspots will indeed be stable.

Howard: I believe that the Kitt Peak observers have placed an upper limit in the neighbourhood of 50 G on the turbulent magnetic field of the quiet Sun.

Stenflo: My estimate of the ratio between the rms and the mean field strengths that you mentioned do not really refer to a turbulent field. It is based on the observation that most of the magnetic flux recorded by solar magnetographs is due to kG fields which occupy only a very small fraction of the solar surface. The rms of such a field is much higher, by approximately a factor of 20, than if the same magnetic flux were uniformly spread out. The turbulent field you are talking about is a very tangled field, with opposite polarities being mixed on a small scale. Such a field, is very difficult to detect by polarization measurements unless the spatial resolution is very high, since opposite polarity fluxes cancel out over a small area. A turbulent field causes however some extra broadening of spectral lines, depending on the Laudé factor. Attempts have been made to detect this minute effect, starting with Unno in 1959, but so far it has only been possible to set an upper limit of a few hundred Gauss.

Giovanelli: We can scarcely expect any fine-scale 'unresolved' field of random polarity distribution to contribute appreciably to the rms field strength, since the total flux contained in the observed strong-field magnetic points is approximately equal to that of the sunspot from which the active-region-field originated.

Krause: My argumentation mainly counts with the average poloidal field in quiet regions (e.g. polar caps) and not in active regions, where the magnetic field mainly has its origin in the toroidal field. Thus I do not expect a relation between the two fluxes.

Vainshtein: (a) From our study of a nonlinear dynamo-theory it is clear that the field growth cannot be without limit. The conclusion of the nonlinear theory is as follows: The electromagnetic forces influence on the first the α-effect and than the diffusivity.

(b) I am studying now the mean field electrodynamics in the presence of plasma oscillations: ionic sound and plasma oscillations Langmuire oscillations). The theory can be applied to solar flare: to explain the fast magnetic damping one must use the turbulent conductivity caused by ionic sound. And if the ionic sound is excited there is generation and reconstruction of magnetic field.

Kuklin: I should like to recall the results of Severny, Grigoriev and others. Practically any background magnetic field observed with large aperture is composed of totality of fine structure magnetic field elements with different areas, field intensities and polarities. All changes of background fields are really variations of number, of size, of field intensity, of polarity of those elements. The principle fact is that background fields consist of changing magnetic field elements of different polarities.

Gilman: Is the mean field theory, particularly estimates of α, β affected by difference in shape between kinetic and magnetic energy spectra with space, in other words differences in spatial scale of v' and B'.?

Krause: As far as we discuss the mean-field electromagnetics the kinetic energy is assumed to be large compared with the magnetic field. α, β, etc. are integral values involving the (uninfluenced) spectrum of the turbulent velocity by field. If the back-reaction of the magnetic field is taken into account such affections may be indicated by a vanishing or, taking into account diffusion, nearly vanishing denominator involved in the integral expressions.

Stenflo: I would just like to add that Livingston and Harvey have recently recorded with high spatial resolution some kind of tangled field with mixed polarities inside supergranulation cells. Since the visibility of this field is quite seeing dependent, it is not clear how large fluxes and field strengths are involved, but still this observation shows the existence of some kind of tangled or 'turbulent' magnetic field away from the network.

MEAN-FIELD MAGNETOHYDRODYNAMICS
AS A BASIS OF SOLAR DYNAMO THEORY

K.-H. RÄDLER

Zentralinstitut für Astrophysik, Potsdam, G.D.R.

1. Introduction

One of the most striking features of both the magnetic field and the motions observed at the Sun is their highly irregular or random character which indicates the presence of rather complicated magnetohydrodynamic processes. Of great importance in this context is a comprehension of the behaviour of the large scale components of the magnetic field; large scales are understood here as length scales in the order of the solar radius and time scales of a few years. Since there is a strong relationship between these components and the solar 22-years cycle, an insight into the mechanism controlling these components also provides for an insight into the mechanism of the cycle. The large scale components of the magnetic field are determined not only by their interaction with the large scale components of motion. On the contrary, a very important part is played also by an interaction between the large and the small scale components of magnetic field and motion so that a very complicated situation has to be considered.

There have been a number of investigations on this within the framework of mean-field magnetohydrodynamics. This theory developed for magnetohydrodynamic phenomena with irregular character supposes that each of the electromagnetic and hydrodynamic fields is understood as a superposition of a mean and a fluctuating field. The mean fields are defined by taking suitable averages over the original fields and correspond, except for some restrictions, to the large scale fields explained above. From the usual equations of magnetohydrodynamics, equations are deduced governing the behaviour of mean fields in the presence of fluctuations. On this basis a dynamo theory of the solar cycle has been developed which reflects and explains many of the observed features and, in addition, allows conclusions on parameters relating to processes in layers inaccessible for observations.

In this paper a survey will be given on the fundamentals and some special results of mean-field magnetohydrodynamics as far as they are of interest for the dynamo theory of the solar cycle. The representation of mean-field magnetohydrodynamics is closely related to that in the foregoing papers by Krause (see pp. 305–321). As for the dynamo theory itself, only some more general statements will be made. Results obtained for special dynamo models of the solar cycle will be summarized and discussed in the following paper by Stix (see pp. 367–388).

The central and furthest developed part of mean-field magnetohydrodynamics is the mean-field electrodynamics dealing with the question of how mean electromagnetic fields are influenced by the motion of the matter but considering this motion as given. To begin with, we shall be concerned with mean-field electrodynamics and only then include the influence of the electromagnetic fields on the motion, thus arriving at mean-field magnetohydrodynamics per se.

Bumba and Kleczek (eds.), Basic Mechanisms of Solar Activity, 323–344. All Rights Reserved.

2. Mean Field Electrodynamics

2.1. BASIC FEATURES

The mean field electrodynamics, as a deductive theory, has been initiated by
Steenbeck *et al.* (1963, 1966; Rädler, 1966; Krause, 1967) and elaborated in
manifold respect by several authors; for summarizing representations see Krause
and Rädler (1971), Roberts (1971), Roberts and Stix (1971), Roberts and Soward
(1975). To represent the basic features of this theory we turn our attention to the
electromagnetic fields in electrically conducting matter carrying out motions which
are considered to be given. The behaviour of these electromagnetic fields is governed
by Maxwell's and the corresponding constitutive equations. Accepting the suppositions usual in magnetohydrodynamics we have

$$\text{curl } \mathbf{E} = -\frac{\partial \mathbf{B}}{\partial t} \qquad \text{curl } \mathbf{H} = \mathbf{j} \qquad \text{div } \mathbf{B} = 0 \qquad (2.1\text{a, b, c})$$

$$\mathbf{B} = \mu \mathbf{H} \qquad \mathbf{j} = \sigma(\mathbf{E} + \mathbf{u} \times \mathbf{B}) . \qquad (2.1\text{d, e})$$

Here $\mathbf{B}$ and $\mathbf{H}$ are magnetic flux density and field strength, $\mathbf{E}$ and $\mathbf{j}$ electric field
strength and current density, μ and σ the vacuum permeability and the electric
conductivity of the matter, and $\mathbf{u}$ the velocity of the matter. The equations (2.1) allow
the determination of $\mathbf{B}$, $\mathbf{H}$, $\mathbf{E}$, and $\mathbf{j}$ if, apart from initial or boundary conditions, $\mathbf{u}$ is
given. As is well known Equations (2.1) can be reduced to

$$\frac{1}{\mu\sigma} \Delta \mathbf{B} + \text{curl } (\mathbf{u} \times \mathbf{B}) - \frac{\partial \mathbf{B}}{\partial t} = 0 \qquad \text{div } \mathbf{B} = 0 . \qquad (2.2\text{a, b})$$

If then $\mathbf{B}$ and $\mathbf{u}$ are known, from (2.1) we get immediately $\mathbf{H}$, $\mathbf{E}$, and $\mathbf{j}$.

With respect to situations in which the electromagnetic fields and the velocity field
show an irregular character we split each of them into a mean and fluctuating part.
The mean fields are understood as averages of the original ones. The average of a
quantity, say F, is denoted by $\bar{F}$, and $F - \bar{F}$ by F', so that $F = \bar{F} + F'$. The special
definition of the averaging procedure is unimportant. It has only to ensure that
Reynolds' averaging rules as well as the commutation rule for averaging and
derivations or integrations hold. From the theoretical point of view ensemble
averages are to be preferred for which these rules are clearly justified. In order to
avoid difficulties in comparing theoretical results with observations, however, averages taken over certain ranges in space or time should be considered. In this case
Reynolds' rules are solely approximations for situations in which the averaged
quantities vary only weakly within each of these ranges. As before the commutation
rule can easily be justified here.

We now particularize

$$\mathbf{B} = \bar{\mathbf{B}} + \mathbf{B}' \qquad \mathbf{u} = \bar{\mathbf{u}} + \mathbf{u}' \qquad (2.3)$$

with the mean magnetic flux density and the mean velocity, $\bar{\mathbf{B}}$ and $\bar{\mathbf{u}}$, as well as the
fluctuations of magnetic flux density and velocity, $\mathbf{B}'$ and $\mathbf{u}'$. In most of the cases of
interest fluctuations as described by $\mathbf{B}'$ and $\mathbf{u}'$ are specified to represent a turbulence.

But the following general considerations apply also in cases where fluctuations with certain regular features, e.g. periodicity in space or time, occur.

The objective of the mean field electrodynamics is the determination of the mean electromagnetic fields, $\bar{\mathbf{B}}$, $\bar{\mathbf{H}}$, $\bar{\mathbf{E}}$, and $\bar{\mathbf{j}}$, if the mean velocity field, $\bar{\mathbf{u}}$, and some properties of the fluctuating part of the velocity field, $\mathbf{u}'$, are given. Taking the average of Equations (2.1), with regard to (2.3) we get

$$\text{curl } \bar{\mathbf{E}} = -\frac{\partial \bar{\mathbf{B}}}{\partial t} \qquad \text{curl } \bar{\mathbf{H}} = \bar{\mathbf{j}} \qquad \text{div } \bar{\mathbf{B}} = 0 \qquad (2.4\text{a, b, c})$$

$$\bar{\mathbf{B}} = \mu \bar{\mathbf{H}} \qquad \bar{\mathbf{j}} = \sigma(\mathbf{E} + \bar{\mathbf{u}} \times \bar{\mathbf{B}} + \overline{\mathbf{u}' \times \mathbf{B}'}) . \qquad (2.4\text{d, e})$$

The formal correspondence between the basic Equations (2.1) and (2.4) for the original and the mean fields is disturbed only by Ohm's law, i.e. (2.1e) and (2.4e), where in the case of mean fields an additional electromotive force, $\overline{\mathbf{u}' \times \mathbf{B}'}$, appears. In order to determine the mean electromagnetic fields, $\bar{\mathbf{B}}$, $\bar{\mathbf{H}}$, $\bar{\mathbf{E}}$, and $\bar{\mathbf{j}}$, in addition to the mean velocity, $\bar{\mathbf{u}}$, this electromotive force $\overline{\mathbf{u}' \times \mathbf{B}'}$, has to be known.

Basically, by means of Equations (2.2) and (2.3) the quantity $\mathbf{B}'$ can be represented as a functional of $\bar{\mathbf{u}}$, $\mathbf{u}'$, and $\bar{\mathbf{B}}$. Consequently, the quantity $\overline{\mathbf{u}' \times \mathbf{B}'}$ is also to be considered as a functional of $\bar{\mathbf{u}}$, $\mathbf{u}'$, and $\bar{\mathbf{B}}$. After replacing $\overline{\mathbf{u}' \times \mathbf{B}'}$ by an expression of this kind, Equations (2.4) allow the determination of $\bar{\mathbf{B}}$, $\bar{\mathbf{H}}$, $\bar{\mathbf{E}}$, and $\bar{\mathbf{j}}$ from $\bar{\mathbf{u}}$ and quantities depending on $\mathbf{u}'$.

Hence, the crucial point of mean-field electrodynamics is the investigation of the electromotive force $\overline{\mathbf{u}' \times \mathbf{B}'}$. Fortunately some interesting conclusions on $\overline{\mathbf{u}' \times \mathbf{B}'}$ may be drawn even without detailed calculations. From Equations (2.2) and (2.3) follows that $\overline{\mathbf{u}' \times \mathbf{B}'}$, regarded as functional of $\bar{\mathbf{u}}$, $\mathbf{u}'$, and $\bar{\mathbf{B}}$, is linear with respect to $\bar{\mathbf{B}}$. Following the usual notation we therefore write

$$\mathscr{E} = \overline{\mathbf{u}' \times \mathbf{B}'} = \mathscr{E}^B + \mathscr{E}^0 \qquad (2.5)$$

where $\mathscr{E}^B$ means a linear homogeneous functional of $\bar{\mathbf{B}}$, and $\mathscr{E}^0$ a quantity independent of $\bar{\mathbf{B}}$. Obviously, $\mathscr{E}^B$ can be represented by

$$\mathscr{E}_i^B(\mathbf{x}, t) = \int\int_0^\infty K_{ij}(\mathbf{x}, t; \boldsymbol{\xi}, \tau) \bar{B}_j(\mathbf{x} - \boldsymbol{\xi}, t - \tau) \, d\boldsymbol{\xi} d\tau \qquad (2.6)$$

with K_{ij} depending on $\bar{\mathbf{u}}$ and $\mathbf{u}'$ but not on $\bar{\mathbf{B}}$. Here as well as in the following the $\boldsymbol{\xi}$-integration is over all space. Since $\mathscr{E}^B$ for a given time may not depend on $\bar{\mathbf{B}}$ at subsequent times the τ-integration is restricted to $\tau \geq 0$. As far as $\mathscr{E}^0$ is concerned we shall just note that it is zero at least as long as the assumption is justified that $\mathbf{B}'$ vanishes if $\bar{\mathbf{B}}$ is zero.

It is to be expected that fluctuating quantities as $\mathbf{u}'$ and $\mathbf{B}'$ at a given point in space and time show no correlation with any quantities at any points far away from the considered one. Consequently, $\mathscr{E}^B$ for a given point depends only on the behaviour of $\bar{\mathbf{u}}$, $\mathbf{u}'$, and $\bar{\mathbf{B}}$ in confined surroundings of it defined by certain characteristic lengths and times. For the determination of K_{ij} for given $\mathbf{x}$ and t, therefore, $\bar{\mathbf{u}}$ and $\mathbf{u}'$ are needed for such surroundings only, and K_{ij} can be regarded as zero if ξ and τ considerably exceed the characteristic lengths and times.

We now assume that $\bar{\mathbf{B}}$ varies only weakly in space and time so that its behaviour in the respective surroundings of a given point is already determined by $\bar{\mathbf{B}}$ itself and a few of its derivatives in this point. Then $\boldsymbol{\mathscr{E}}^B$ can be represented as a function of $\bar{\mathbf{B}}$ and its derivatives in the point considered. In the simple case in which no other than the first spatial derivatives are included we have

$$\mathscr{E}_i^B = a_{ij}\bar{B}_j + b_{ijk}\frac{\partial \bar{B}_j}{\partial x_k} \tag{2.7}$$

where the coefficients a_{ij} and b_{ijk} are averaged quantities depending on $\bar{\mathbf{u}}$ and $\mathbf{u}'$. Clearly it holds

$$a_{ij} = \int\int_0^\infty K_{ij}(\mathbf{x}, t; \boldsymbol{\xi}, \tau)\, d\boldsymbol{\xi}\, d\tau \qquad b_{ijk} = -\int\int_0^\infty K_{ij}(\mathbf{x}, t; \boldsymbol{\xi}, \tau)\xi_k\, d\boldsymbol{\xi}\, d\tau. \tag{2.8a, b}$$

The complete determination of $\boldsymbol{\mathscr{E}}^B$ requires the knowledge of K_{ij}. Fortunately some far-reaching conclusions on $\boldsymbol{\mathscr{E}}^B$ can be drawn, even without using K_{ij}, from special suppositions on $\bar{\mathbf{u}}$ and the statistical properties of $\mathbf{u}'$. In this way expressions for $\boldsymbol{\mathscr{E}}^B$ can be given for special situations of interest, and only some factors remain to be determined. This will be illustrated below.

Departing from Equations (2.2) and (2.3) a method was developed for the complete determination of $\boldsymbol{\mathscr{E}}^B$ which especially provides for explicit expressions of K_{ij}. In an approximation in which only second order correlations of $\mathbf{u}'$ are considered and all higher ones are neglected it is found

$$K_{ij}(\mathbf{x}, t; \boldsymbol{\xi}, \tau) = \varepsilon_{ikl}\varepsilon_{mnp}\varepsilon_{pqj}\frac{\partial G_{lm}(\mathbf{x}, t; \mathbf{x}-\boldsymbol{\xi}, t-\tau)}{\partial \xi_n}Q_{kq}(\mathbf{x}, t; -\boldsymbol{\xi}, -\tau). \tag{2.9}$$

Here G_{ij} denotes Green's functions of Equation (2.2a) with $\mathbf{u}=\bar{\mathbf{u}}$. In the simplest case, $\bar{\mathbf{u}}=0$, one has

$$G_{ij}(\mathbf{x}, t; \boldsymbol{\xi}, \tau) = \delta_{ij}G(|\mathbf{x}-\boldsymbol{\xi}|, t-\tau) \tag{2.10a}$$

$$G(x, t) = \left(\frac{\mu\sigma}{4\pi t}\right)^{3/2}\exp\left\{-\frac{\mu\sigma x^2}{4t}\right\}. \tag{2.10b}$$

Furthermore, Q_{ij} is the second order correlation tensor of $\mathbf{u}'$ defined by

$$Q_{ij}(\mathbf{x}, t; \boldsymbol{\xi}, \tau) = \overline{u_i'(\mathbf{x}, t)u_j'(\mathbf{x}+\boldsymbol{\xi}, t+\tau)}. \tag{2.11}$$

The approximation can be improved by adding expressions with higher order correlation tensors to the right-hand side of Equation (2.9).

2.2. ILLUSTRATIVE EXAMPLE

As a first example mostly the case is discussed in which the motion is supposed to have no mean part, $\bar{\mathbf{u}}=0$, but to consist of fluctuations, $\mathbf{u}'$, representing a homogeneous isotropic turbulence. Homogeneity and isotropy mean that all averaged quantities depending on the $\mathbf{u}'$ field are invariant against arbitrary translations of this field and against arbitrary rotations of it around arbitrary axes. Returning to Equation (2.7) for $\mathscr{E}^B$ and bearing in mind that this especially holds for the coefficients a_{ij} and b_{ijk} we conclude that they are constant in space and have the structures $\alpha\delta_{ij}$ and $\beta\varepsilon_{ijk}$ so that

$$\mathscr{E}^B = \alpha\bar{\mathbf{B}} - \beta\,\mathrm{curl}\,\bar{\mathbf{B}}. \tag{2.12}$$

Obviously, α is a pseudoscalar but β a scalar, both depending on $\mathbf{u}'$. With respect to $\mathscr{E}^0$ we furthermore suppose that, if another excitation of $\mathbf{B}'$ than that due to $\bar{\mathbf{B}}$ exists at all, the seeds for $\mathbf{B}'$ are homogeneously and isotropically distributed. Then we have to conclude that $\mathscr{E}^0 = 0$. In this way Ohm's law (2.4e) becomes

$$\bar{\mathbf{j}} = \sigma_T(\bar{\mathbf{E}} + \alpha\bar{\mathbf{B}}) \tag{2.13}$$

with

$$\sigma_T = \frac{\sigma}{1+\mu\sigma\beta}. \tag{2.14}$$

A homogeneous isotropic turbulence will always be mirror-symmetric too. In this context, mirror-symmetry is defined by the invariance of all averaged quantities depending on the $\mathbf{u}'$ field against reflexions of it on arbitrary planes. Since the reflexion of a right-handed screw in the flow pattern produces a left-handed one and vice versa, mirror-symmetry means equipartition between right- and left-handed helical motions.

For a homogeneous isotropic and mirror-symmetric turbulence the pseudoscalar α turns out to be zero. Then the total influence of turbulence on the mean fields can be comprehended as a 'turbulent' conductivity, σ_T, different from the original one, σ. This statement is related to conceptions like that of eddy viscosity or turbulent diffusion in hydrodynamics. The introduction of a turbulent conductivity has already been proposed by Sweet (1950) and Csada (1951).

For illustration let us consider the electrical current between two assumed electrodes with a given potential difference. At first assuming the medium to be at rest we envisage a current tube, for which an electrical resistance can be defined. If the medium moves, this tube is generally stretched and narrowed so that the resistance increases. The pattern of the mean current, however, coincides with that of the current occurring in the case of the medium at rest. This picture shows clearly that there is a difference between σ_T and σ, and we can expect that $\sigma_T < \sigma$; see also Steenbeck et al. (1963). As long as the latter relation holds the turbulence enhances the decay of the mean magnetic field.

The value of the turbulent conductivity, σ_T, depends on the coefficient β. In the second-order correlation approximation we get

$$\beta = -\frac{1}{3}\int\int_0^\infty \frac{\partial G(\xi, \tau)}{\partial\xi} f(\xi, -\tau)\xi\,\mathrm{d}\xi\mathrm{d}\tau. \tag{2.15}$$

Here f is a longitudinal correlation function defined by

$$f(\boldsymbol{\xi}, \tau) = \overline{(\mathbf{u}'(\mathbf{x}, t) \cdot \boldsymbol{\xi})(\mathbf{u}'(\mathbf{x}+\boldsymbol{\xi}, t+\tau) \cdot \boldsymbol{\xi})}/\xi^2 \ . \tag{2.16}$$

Incidentally, since isotropy is supposed here, f does not depend on the direction of $\boldsymbol{\xi}$, but on ξ and τ only. Special results for σ_T, all restricted to second-order correlation approximation, has been given by Rädler (1966, 1968b). Departing from (2.15) and (2.16) the relation $\sigma_T < \sigma$ can be confirmed for a wide range of suppositions; see Krause and Roberts (1973a, b), Roberts and Soward (1975). A comparison between theoretical results on σ_T and observational data of the Sun is given by Krause (1975).

Although the case of a homogeneous isotropic turbulence lacking mirror-symmetry has to be considered as unrealistic it is instructive to be studied. In this way a point can be easily seen which is important with respect to more complicated kinds of turbulence as to be observed at rotating bodies. There really is a lack of mirror-symmetry, i.e. a predominance of either right-handed or left-handed helical motions.

For a homogeneous isotropic non-mirror-symmetric turbulence the coefficient α is no longer zero. Then an additional electromotive force, $\alpha\bar{\mathbf{B}}$, has to be taken into account which is parallel or antiparallel to the mean magnetic flux. The occurrence of this electromotive force is called 'α-effect'. Within the framework of a deductive theory, it has been first discussed by Steenbeck *et al.* (1966); it should be noted, however, that it was already involved in ideas expressed by Parker (1955).

The α-effect becomes plausible if we consider the deformation of originally straight magnetic flux tubes by helical motions. Due to the magnetic field being at least partially frozen in the medium it is clear that situations as those in Figure 1 occur. Roughly speaking, in addition to the originally straight flux tubes, annular flux

Fig. 1.   Deformation of originally straight magnetic flux tubes by helical motions. The magnetic flux tubes are represented by the heavy lines, the electric currents are indicated by arrows, and the motions by the thin lines.

tubes result, which are accompanied by currents crossing the latters' planes. Since these planes do not coincide with the plane of the drawn figure there are components of these currents parallel or antiparallel to the original magnetic flux. If there is an equipartition between right- and left-handed helical motions we have also an equipartition between components parallel and antiparallel to the original flux and, therefore, on average no current. However, if this equipartition is disturbed we have on average a current parallel or antiparallel to the magnetic flux depending on whether left- or right-handed helical motions predominate.

In the second-order correlation approximation we have

$$
\alpha = -\frac{1}{3} \int \int\limits_{0}^{\infty} \frac{\partial G(\xi, \tau)}{\partial \xi} \cdot h(\xi, -\tau) \, d\xi \, d\tau
$$

$$
= -\frac{1}{3} \int \int\limits_{0}^{\infty} G(\xi, \tau) h^*(\xi, -\tau) \, d\xi \, d\tau
\tag{2.17}
$$

with correlation functions, h and h^*, defined by

$$
h(\xi, \tau) = \overline{(\mathbf{u}'(\mathbf{x}, t) \times \mathbf{u}'(\mathbf{x} + \xi, t + \tau)) \cdot \xi / \xi}
\tag{2.18a}
$$

$$
h^*(\xi, \tau) = \overline{\mathbf{u}'(\mathbf{x}, t) \cdot \operatorname{curl} \mathbf{u}'(\mathbf{x} + \xi, t + \tau)}
\tag{2.18b}
$$

Obviously, h and h^* are measures for the occurrence of helical motions. As proposed by Moffatt (1970a), $h^*(0, 0)$ is called 'helicity'. Further, h^* is positive or negative if right or left-handed helical motions predominate. Accordingly, α is positive or negative and, consequently, $\alpha \bar{\mathbf{B}}$ parallel or antiparallel to $\bar{\mathbf{B}}$ if we have preferably left- or right-handed motions, respectively.

The most important feature of the α-effect is that it allows the regeneration and the growth of mean magnetic fields. In order to demonstrate this let us consider the situation in an infinitely extended medium where the mean fields satisfy the Equations (2.4) with Ohm's law as specified by (2.13). Then we have the equations

$$
\frac{1}{\mu \sigma_T} \Delta \bar{\mathbf{B}} + \alpha \operatorname{curl} \bar{\mathbf{B}} - \frac{\partial \bar{\mathbf{B}}}{\partial t} = 0 \qquad \operatorname{div} \bar{\mathbf{B}} = 0
\tag{2.19a, b}
$$

allowing solutions

$$
\bar{\mathbf{B}} = (\mathbf{C} \cos (\mathbf{k} \cdot \mathbf{x}) \pm (\mathbf{C} \times \frac{\mathbf{k}}{k}) \sin (\mathbf{k} \cdot \mathbf{x})) \exp \left\{ -\left(\frac{k^2}{\mu \sigma_T} \mp \alpha k \right) t \right\}
\tag{2.20}
$$

with arbitrary vectors $\mathbf{C}$ and $\mathbf{k}$ satisfying $\mathbf{C} \cdot \mathbf{k} = 0$. Obviously, such $\bar{\mathbf{B}}$ modes are stationary if $\mu \sigma_T \alpha / k = \pm 1$, and there are always $\bar{\mathbf{B}}$ modes which grow with time if $\mu \sigma_T |\alpha| / k > 1$.

2.3. REMARKS ON THE APPLICABILITY TO THE SUN

If we want to apply this general conception to conceive electromagnetic phenomena at the Sun, especially in its convection zone, some points have to be reconsidered.

The first point concerns the split-up of fields into mean fields and fluctuations defined on the basis of an averaging procedure. By reasons already mentioned, averages over space or time are to be considered. Furthermore, the averaging procedure should be compatible with the conception used in all previous work too that, e.g., differential rotation is considered as mean motion whereas all convection occurs as fluctuation. Finally, the validity of Reynolds rules supposed above has to be ensured. As mentioned also by Krause (1975) and Stix (1975), averages taken over a spatial volume hardly show the properties required. Such averages cannot even separate differential rotation and convection as far as large convection cells are considered; the reason is that the orders of length scales of the two do not clearly differ. A similar difficulty arises with the magnetic fields. A way out could be to replace the average over a volume by an average over longitude as used by Braginski (1964). Then all requirements mentioned are fulfilled automatically. But we have to put up with the restriction to axisymmetric mean fields; all non-axisymmetric parts of the original fields occur as fluctuations. Following Stix (1975) again we consider it best to use averages over a time span in the order of about one or two years. In this case the requirements formulated above are satisfied at least in a certain approximation. This is due to the fact that the spectrum of characteristic times observed at the Sun shows a gap for times in the order of one or two years.

For the next point to be discussed the magnetic Reynolds number, R_m, and the Strouhal number, S, both in relation to fluctuations, are of interest. They are defined by

$$R_m = \mu \sigma u'_c \lambda_c \qquad S = u'_c \tau_c / \lambda_c \qquad (2.21a, b)$$

where u'_c represents a characteristic value of the fluctuating velocity, e.g. $\sqrt{u'^2}$, and λ_c and τ_c are characteristic length and time, e.g., correlation length and time. It holds $\mu = 4\pi \times 10^{-7}$ Vs A^{-1} m^{-1}, and we accept $\sigma = 10^3 \, \Omega^{-1}$ m^{-1}. With respect to granules we take $u'_c = 2 \times 10^3$ m s^{-1}, $\lambda_c = 2 \times 10^6$ m, and $\tau_c = 3 \times 10^2$ s. Then we have $R_m \approx 5 \times 10^6$ and $S \approx 3 \times 10^{-1}$. If we consider supergranules and choose $u'_c = 4 \times 10^2$ m s^{-1}, $\lambda_c = 3 \times 10^7$ m, and $\tau_c = 5 \times 10^4$ s it results $R_m \approx 2 \times 10^7$ and $S \approx 7 \times 10^{-1}$.

We reconsider now the transition from the general to the special expression of $\mathscr{E}^B$ given by (2.6) and (2.7) and ask whether other derivatives of $\bar{B}$ have to be included. As mentioned above, K_{ij} can be regarded to be zero if ξ and τ considerably exceed certain characteristic values, say λ_K and τ_K. From the expressions of K_{ij} it can easily be read that

$$\lambda_K = \lambda_c \qquad \tau_K = \mu \sigma \lambda_c^2 \qquad \text{for } R_m/S \ll 1 \qquad (2.22a)$$

$$\lambda_K = \sqrt{\tau_c/\mu\sigma} \qquad \tau_K = \tau_c \qquad \text{for } R_m/S \gg 1 . \qquad (2.22b)$$

The contributions to $\mathscr{E}^B$ due to higher derivatives of $\bar{B}$ are, e.g., in the order of $\tau_K(\partial\bar{B}/\partial t)$ or $\lambda_K^2(\partial^2\bar{B}/\partial x_i \partial x_j)$; see also Rädler (1968b). Since always $R_m/S \gg 1$, with the data of granules we have $\lambda_K \approx 5 \times 10^2$ m and $\tau_K \approx 3 \times 10^2$ s, with the data of supergranules $\lambda_K \approx 6 \times 10^3$ m and $\tau_K \approx 5 \times 10^4$ s. Therefore, those contributions seem to be negligible.

Finally we deal with the validity of the above-mentioned second-order correlation approximation for the coefficients occurring in special expressions of $\mathscr{E}^B$. This is of interest because higher order approximations require very lengthy calculations which have been done up to now only in a few exceptional cases. A simple sufficient condition for the validity, which can easily be read from the deductions, is

$$\min (R_m, S) \ll 1 . \qquad (2.23)$$

With the data used above this is scarcely fulfilled. It must be noted, however, that (2.23) is not a necessary condition. The values of the turbulent conductivity which have been obtained within the frame of the second-order correlation approximation fit very well to observational data; see Krause (1967, 1975), Rädler (1968b), Steenbeck and Krause (1969), Krause and Rüdiger (1975). In this way, there is at least some reason to believe that the results obtained from the approximation are reliable for the solar convection zone.

2.4. OHM'S LAW FOR THE SOLAR CONVECTION ZONE

2.4.1. Elaborating the mean-field electrodynamics for the Sun we have to consider that, in contrast to the simple example discussed above, there are firstly a non-vanishing mean motion, at least the differential rotation, and secondly fluctuating motions, like the convection, which can no longer be treated as a homogeneous isotropic turbulence. Already the intensity of these fluctuating motions shows a radial dependence, and there are significant differences between the velocity components in radial and horizontal directions. In addition, these motions are subjected to Coriolis forces. Taking all this into account, the homogeneity is disturbed, and instead of isotropy we have to take into consideration that in each point at least two preferred directions exist, namely a radial and an axial one. Finally, there must be deviations from mirror-symmetry because of the preference of these directions.

Studying now Ohm's law for the mean fields we first consider the cases in which only the radial or only the axial direction is preferred and then pass over to the case where both are included. Furthermore, we first deal only with effects described by the electromotive force $\mathscr{E}^B$ in the form given in (2.7), and we neglect $\mathscr{E}^0$ which will be discussed later.

2.4.2. Let us now suppose that isotropy and mirror-symmetry of the fluctuating motions are disturbed only by the preference of the radial direction. More precisely, all averaged quantities depending on the fluctuating velocity field are considered to be invariant under rotation of this field around a given radius and under reflexion of it at planes containing this radius. By similar arguments as used above $\mathscr{E}^B$ can readily be specified, and Ohm's law for the mean fields becomes

$$\mathbf{j} = \sigma_T(\bar{\mathbf{E}} + \bar{\mathbf{u}} \times \bar{\mathbf{B}}$$
$$- \gamma \hat{\mathbf{g}} \times \bar{\mathbf{B}}$$
$$- \gamma_1 \hat{\mathbf{g}}(\hat{\mathbf{g}} \cdot \operatorname{curl} \bar{\mathbf{B}}) - \gamma_2 \hat{\mathbf{g}} \times (\hat{\mathbf{g}} \cdot \operatorname{grad})\bar{\mathbf{B}} - \gamma_3 \hat{\mathbf{g}} \times (\hat{\mathbf{g}} \cdot \operatorname{grad} \bar{\mathbf{B}}) \qquad (2.24)$$

where $\hat{\mathbf{g}}$ is the unit vector in radial direction and $(\hat{\mathbf{g}} \cdot \operatorname{grad} \bar{\mathbf{B}})_i$ means $\hat{g}_j \partial \bar{B}_j / \partial x_i$. Incidentally, the last three terms within the brackets can be rewritten to be

$$- \mu(\gamma_1 - \gamma_2)\hat{\mathbf{g}}(\hat{\mathbf{g}}\bar{\mathbf{j}}) - \mu\gamma_2\bar{\mathbf{j}} - (\gamma_2 + \gamma_3)\hat{\mathbf{g}} \times (\hat{\mathbf{g}} \cdot \operatorname{grad} \bar{\mathbf{B}}). \qquad (2.25)$$

Again, a turbulent conductivity, σ_T, was introduced for which (2.15) and also (2.16) hold.

Beside the Lorentz force, $\bar{\mathbf{u}} \times \bar{\mathbf{B}}$, an electromotive force, $- \gamma \hat{\mathbf{g}} \times \bar{\mathbf{B}}$, appears where the part of mean velocity, $\bar{\mathbf{u}}$, is played by a vector quantity, $- \gamma \hat{\mathbf{g}}$, antiparallel or parallel to the radial direction. Just as the Lorentz force, this electromotive force corresponds to a transport of magnetic flux. Such transport different from that due to a mean motion is sometimes called 'pumping of magnetic flux.'

Within the frame of second-order correlation approximation we have

$$\gamma = -\frac{1}{2} \int \int_0^\infty \frac{\partial G(\xi, \tau)}{\partial \xi} (k_1(\xi, -\tau) + k_2(\xi, -\tau)) \, d\xi \, d\tau \qquad (2.26)$$

with correlation functions, k_1 and k_2, defined by

$$k_1(\xi, \tau) = \overline{(\mathbf{u}'(\mathbf{x}, t) \cdot \xi)(\mathbf{u}'(\mathbf{x}+\xi, t+\tau) \cdot \hat{\mathbf{g}})} / \xi \qquad (2.27a)$$

$$k_2(\xi, \tau) = \overline{(\mathbf{u}'(\mathbf{x}, t) \cdot \hat{\mathbf{g}})(\mathbf{u}(\mathbf{x}+\xi, t+\tau) \cdot \xi)} / \xi . \qquad (2.27b)$$

For a turbulence deviating only by a gradient of the mean intensity from a homogeneous isotropic and mirror-symmetric turbulence, more detailed results on the electromotive force under consideration and especially on the coefficient γ have been given by Rädler (1966, 1968b, 1969a). It turns out that there is a tendency to push out the mean magnetic flux from the regions of high turbulence intensity. Therefore, this effect is sometimes denoted as 'turbulent diamagnetism'. A similar situation occurs if the gradient of intensity is replaced by, e.g., a gradient of correlation length; see Rädler (1969a), Krause and Rädler (1971). In a turbulent layer like the solar convection zone a gradient in intensity, the direction of which must change within the layer, pushes the magnetic flux from the inner regions to both boundaries whereas a gradient in correlation length may push it from one boundary to the other. If the fluctuating motions are specified to represent convective cell motions the electromotive force under consideration describes the effect of 'topological pumping of magnetic flux' discussed by Drobyshevski and Yuferev (1974). In this case, a pumping from one boundary to another only occurs for higher than second-order correlation approximations.

In addition to the electromotive force discussed there are others connected with spatial derivatives of $\bar{\mathbf{B}}$. A part of them may be described by means of an anisotropic turbulent conductivity too. It shows that (2.27) is equivalent to

$$\bar{\mathbf{j}} = \sigma_T(\bar{\mathbf{E}} + \bar{\mathbf{u}} \times \bar{\mathbf{B}} - \gamma \hat{\mathbf{g}} \times \bar{\mathbf{B}} - (\gamma_2 + \gamma_3)\hat{\mathbf{g}} \times (\hat{\mathbf{g}} \cdot \operatorname{grad} \bar{\mathbf{B}}) \qquad (2.28)$$

with a conductivity tensor, σ_T, given by

$$\sigma_{Tij} = \frac{\sigma_T}{(1 + \mu\sigma_T\gamma_1)(1 + \mu\sigma_T\gamma_2)} ((1 + \mu\sigma_T\gamma_1)\delta_{ij} - \mu\sigma_T(\gamma_1 - \gamma_2)\hat{g}_i\hat{g}_j). \qquad (2.29)$$

In the second-order correlation approximation we have

$$\gamma_1 = -\frac{1}{6} \int \int_0^\infty \frac{\partial G(\xi, \tau)}{\partial \xi} (f(\xi, -\tau)\xi + 3k_2(\xi, -\tau)(\hat{\mathbf{g}} \cdot \xi)) \, d\xi \, d\tau \qquad (2.30a)$$

$$\gamma_2 = \frac{1}{6} \int \int_0^\infty \frac{\partial G(\xi, \tau)}{\partial \xi} (2f(\xi, -\tau)\xi + 3(k_1(\xi, -\tau) + k_2(\xi, -\tau))(\hat{\mathbf{g}} \cdot \xi)) \, d\xi \, d\tau$$

$$(2.30b)$$

$$\gamma_3 = \frac{1}{6} \int\limits_0^\infty \int \frac{\partial G(\xi, \tau)}{\partial \xi} (f(\xi, -\tau)\xi + 3k_1(\xi, -\tau)(\hat{\mathbf{g}} \cdot \xi)) \, d\xi \, d\tau \qquad (2.30c)$$

with f, k_1, and k_2 as given above. In this context we refer to the results obtained for the extreme case of a homogeneous two-dimensional turbulence with no motion in the preferred direction; see Krause and Rüdiger (1975). They can easily be formulated in terms of an anisotropic turbulent conductivity; in this case k_1 and k_2 turn out to be zero. For the above-mentioned turbulence with a gradient of mean intensity, the second-order correlation tensor is linear in this gradient; see Rädler (1966, 1974). Therefore, in the applied approximation γ_1, γ_2 and γ_3 vanish; see Rädler (1966, 1968b).

2.4.3. Now we pass over to the case in which isotropy and mirror-symmetry of the fluctuating motions are violated only by the influence of Coriolis forces, which occur as a consequence of the rotation. Then the averaged quantities dependent from the fluctuating velocity field have to be invariant under rotation of this field around axes parallel to the rotational axis and under reflexions at planes perpendicular to it. For simplicity we suppose the influence of Coriolis forces as weak enough in order to neglect in the following all quantities of higher than first order with regard to the angular velocity. In the same way as above we get

$$\bar{\mathbf{j}} = \sigma_T(\bar{\mathbf{E}} + \bar{\mathbf{u}} \times \bar{\mathbf{B}} - \beta_1(\hat{\boldsymbol{\omega}} \cdot \mathrm{grad})\bar{\mathbf{B}} - \beta_2 \, \mathrm{grad}\,(\hat{\boldsymbol{\omega}} \cdot \bar{\mathbf{B}})) \qquad (2.31)$$

where $\hat{\boldsymbol{\omega}}$ is the unit vector parallel to the rotational axis. If we relinquish linearity in the angular velocity, four other terms have to be added at the right-hand side of this equation. The last two terms within the brackets in (2.31) can be rewritten to be

$$\mu\beta_1\hat{\boldsymbol{\omega}} \times \bar{\mathbf{j}} - (\beta_1 + \beta_2)\,\mathrm{grad}\,(\hat{\boldsymbol{\omega}} \cdot \bar{\mathbf{B}}) . \qquad (2.32)$$

As before, a turbulent conductivity, σ_T, appears for which (2.15) and (2.16) hold.

The most interesting feature of this result is, however, the occurrence of an electromotive force, $\mu\beta_1\hat{\boldsymbol{\omega}} \times \bar{\mathbf{j}}$, often called '$\boldsymbol{\omega} \times \mathbf{j}$-effect'. It reminds one of the Hall effect. This $\boldsymbol{\omega} \times \mathbf{j}$-effect can be described in terms of an anisotropic turbulent conductivity too. From (2.31) and (2.32) immediately follows

$$\bar{\mathbf{j}} = \boldsymbol{\sigma}_T(\bar{\mathbf{E}} + \bar{\mathbf{u}} \times \bar{\mathbf{B}} - (\beta_1 + \beta_2)\,\mathrm{grad}\,(\hat{\boldsymbol{\omega}} \cdot \bar{\mathbf{B}})) \qquad (2.33)$$

with a turbulent conductivity tensor, $\boldsymbol{\sigma}_T$, given by

$$\sigma_{Tij} = \frac{\sigma_T}{1 + (\mu\sigma_T\beta_1)^2}(\delta_{ij} - \mu\sigma_T\beta_1\varepsilon_{ijk}\hat{\omega}_k + (\mu\sigma_T\beta_1)^2\hat{\omega}_i\hat{\omega}_k). \qquad (2.34)$$

Contrary to (2.29) it contains an antisymmetric part. The electromotive force $-(\beta_1 + \beta_2)\,\mathrm{grad}\,(\hat{\boldsymbol{\omega}} \cdot \bar{\mathbf{B}})$ plays a minor part. As far as $\beta_1 + \beta_2$ does not depend on space this force may always be compensated by that part of $\bar{\mathbf{E}}$ which results from space charges.

Just as the α-effect occurring with homogeneous isotropic non-mirror-symmetric turbulence, the $\boldsymbol{\omega} \times \mathbf{j}$-effect is also caused by helical motions, but it does not require a predominance of right-or left-handed helical motions. A similar illustration as given for the α-effect by means of Figure 1 is possible for the $\boldsymbol{\omega} \times \mathbf{j}$-effect too. In this case, the anisotropy of the motions has to be taken into account. Furthermore, due to the gradient in the mean flux density two flux tubes as in Figure 1 may differ in flux density. Therefore the helical motions, despite the equipartition of both types, can produce a mean current.

Within the second-order correlation approximation we have

$$\beta_1 = \frac{1}{10} \int_0^\infty \int \frac{\partial G(\xi, \tau)}{\partial \xi} (3h(\xi, -\tau)(\hat{\boldsymbol{\omega}} \cdot \boldsymbol{\xi}) - l_1(\xi, -\tau)\xi) \, d\xi d\tau \qquad (2.35a)$$

$$\beta_2 = -\frac{1}{10} \int_0^\infty \int \frac{\partial G(\xi, \tau)}{\partial \xi} (2h(\xi, -\tau)(\hat{\boldsymbol{\omega}} \cdot \boldsymbol{\xi}) - 4l_1(\xi, -\tau)\xi$$

$$+ 5l_2(\xi, -\tau)\xi) \, d\xi \, d\tau \quad (2.35b)$$

where h is defined as above and l_1 and l_2 are given by

$$l_1(\boldsymbol{\xi}, \tau) = \overline{((\mathbf{u}'(\mathbf{x}, t) \times \mathbf{u}'(\mathbf{x}+\boldsymbol{\xi}, t+\tau)) \cdot \hat{\boldsymbol{\omega}})} \qquad (2.36a)$$

$$l_2(\boldsymbol{\xi}, \tau) = \overline{(\mathbf{u}'(\mathbf{x}, t) \cdot \boldsymbol{\xi})((\mathbf{u}'(\mathbf{x}+\boldsymbol{\xi}, t+\tau) \times \hat{\boldsymbol{\omega}}) \, \boldsymbol{\xi})}/\xi^2. \qquad (2.36b)$$

For an originally homogeneous isotropic and mirror-symmetric turbulence in an incompressible medium subject to Coriolis forces explicit expressions of β_1 and β_2 have been given by Rädler (1969a, b), and by Roberts and Soward (1975).

2.4.4. Finally, we deal with the situation in which the fluctuating motions are characterized by the simultaneous occurrence of the two preferred directions considered above. Extending the hitherto used conception to this more complicated situation, a rather complex result occurs. Restricting ourselves again to linearity in the angular velocity which is responsible to the Coriolis forces we have

$$\bar{\mathbf{j}} = \sigma_T(\bar{\mathbf{E}} + \bar{\mathbf{u}} \times \bar{\mathbf{B}}$$

$$- \gamma \hat{\mathbf{g}} \times \bar{\mathbf{B}}$$

$$- \gamma_1 \hat{\mathbf{g}}(\hat{\mathbf{g}} \cdot \text{curl } \bar{\mathbf{B}}) - \gamma_2 \hat{\mathbf{g}} \times (\hat{\mathbf{g}} \cdot \text{grad})\bar{\mathbf{B}} - \gamma_3 \hat{\mathbf{g}} \times (\hat{\mathbf{g}} \cdot \text{grad } \bar{\mathbf{B}})$$

$$- \beta_1(\hat{\boldsymbol{\omega}} \cdot \text{grad})\bar{\mathbf{B}} - \beta_2 \text{ grad } (\hat{\boldsymbol{\omega}} \cdot \bar{\mathbf{B}})$$

$$- \alpha_1(\hat{\mathbf{g}} \cdot \hat{\boldsymbol{\omega}})\bar{\mathbf{B}} - \alpha_2(\hat{\mathbf{g}} \cdot \bar{\mathbf{B}})\hat{\boldsymbol{\omega}} - \alpha_3(\hat{\boldsymbol{\omega}} \cdot \bar{\mathbf{B}})\hat{\mathbf{g}} - \alpha_4(\hat{\mathbf{g}} \cdot \hat{\boldsymbol{\omega}})(\hat{\mathbf{g}} \cdot \bar{\mathbf{B}})\hat{\mathbf{g}}$$

$$- \alpha_5(\hat{\mathbf{g}} \cdot \text{grad } (\hat{\boldsymbol{\omega}} \cdot \bar{\mathbf{B}}))\hat{\mathbf{g}} - \alpha_6(\hat{\boldsymbol{\omega}} \cdot (\hat{\mathbf{g}} \cdot \text{grad } \bar{\mathbf{B}}))\hat{\mathbf{g}}$$

$$- \alpha_7(\hat{\mathbf{g}} \cdot \hat{\boldsymbol{\omega}})(\hat{\mathbf{g}} \cdot \text{grad } (\hat{\mathbf{g}} \cdot \bar{\mathbf{B}}))\hat{\mathbf{g}} - \alpha_8(\hat{\mathbf{g}} \cdot \text{grad } (\hat{\mathbf{g}} \cdot \bar{\mathbf{B}}))\hat{\boldsymbol{\omega}}$$

$$- \alpha_9(\hat{\mathbf{g}} \cdot \hat{\boldsymbol{\omega}})(\hat{\mathbf{g}} \cdot \text{grad})\bar{\mathbf{B}} - \alpha_{10}(\hat{\mathbf{g}} \cdot \hat{\boldsymbol{\omega}})(\hat{\mathbf{g}} \cdot \text{grad } \bar{\mathbf{B}})). \qquad (2.37)$$

Again, other representations are possible too. We especially refer to (2.25) and (2.32), and we add that the terms in the last line can also be written in the form

$$\mu \alpha_9(\hat{\mathbf{g}} \cdot \hat{\boldsymbol{\omega}})\hat{\mathbf{g}} \times \bar{\mathbf{j}} - (\alpha_9 + \alpha_{10})(\hat{\mathbf{g}} \cdot \hat{\boldsymbol{\omega}})(\hat{\mathbf{g}} \cdot \text{grad } \bar{\mathbf{B}}). \qquad (2.38)$$

The electromotive forces in the first four lines of (2.37) are to be expected from the foregoing discussion, and all relations given in this connection apply here too. But there are further terms in which both preferred directions, i.e. $\hat{\mathbf{g}}$ and $\hat{\boldsymbol{\omega}}$, appear simultaneously. In all previous representations of this matter the simplifying assumption of linearity in both $\hat{\mathbf{g}}$ and $\hat{\boldsymbol{\omega}}$ was introduced. In this way, the last term of the fifth line and all following terms did not occur. For a number of reasons this assumption seems rather problematic.

Just as in the case of a homogeneous isotropic and non-mirror-symmetric turbulence we have an electromotive force proportional to the mean magnetic flux density, namely $-\alpha_1(\hat{\mathbf{g}} \cdot \hat{\boldsymbol{\omega}})\bar{\mathbf{B}}$, i.e. an α-effect. In contrast to that case, however, this electromotive force is accompanied by the other ones given in the fifth line of (2.37). We shall speak of 'idealized α-effect' as long as only $-\alpha_1(\hat{\mathbf{g}} \cdot \hat{\boldsymbol{\omega}})\bar{\mathbf{B}}$ is considered, and of 'real α-effect' if the other contributions are included.

In accordance with the α-effect occurring with homogeneous isotropic non-mirror-symmetric turbulence, and in contrast to the $\hat{\boldsymbol{\omega}} \times \mathbf{j}$-effect, the α-effect considered here is due to the predominance of right- or left-handed helical motions. The illustration given in the case of homogeneous isotropic non-mirror-symmetric turbulence can easily be modified to fit to the situation under discussion.

In the second-order correlation we have

$$\alpha_1(\hat{\mathbf{g}} \cdot \hat{\boldsymbol{\omega}}) = \frac{1}{2} \int\limits_0^\infty \int \frac{\partial G(\xi, \tau)}{\partial \xi} \left(h(\boldsymbol{\xi}, -\tau) - m_1(\boldsymbol{\xi}, -\tau) \frac{(\hat{\mathbf{g}} \cdot \boldsymbol{\xi})}{\xi} \right) d\boldsymbol{\xi} \, d\tau \qquad (2.39)$$

with h as defined above and m_1 given by

$$m_1(\boldsymbol{\xi}, \tau) = \overline{((\mathbf{u}'(\mathbf{x}, t) \times \mathbf{u}'(\mathbf{x} + \boldsymbol{\xi}, t + \tau)) \cdot \hat{\mathbf{g}}}. \qquad (2.40)$$

Due to assumptions introduced here, (2.39) provides for a value of α_1 independent of $\hat{\mathbf{g}}$ and $\hat{\boldsymbol{\omega}}$. In contrast to the coefficient α given by (2.17), the coefficient α_1 depends not only on correlations described by the function h or, what is the same, by the function h^*, but is a more complicated quantity. The corresponding expressions for α_2, α_3, and α_4, which are rather complicated too, will not be given here.

Departing from various conceptions on the structure of the fluctuating motions several investigations have been carried out which provide for special results on $\alpha_1, \alpha_2, \ldots$ or related quantities. The motions in the solar convection zone are influenced in a high degree by the stratification of the medium. Investigations which include the effect of stratification were performed by Steenbeck et al. (1966) and by Krause (1967). Furthermore, with other assumptions on the motions, special results were presented by Rädler (1969a), Moffatt (1970a), Krause and Rädler (1971), and Roberts and Soward (1975).

The electromotive forces connected with derivatives of $\bar{\mathbf{B}}$ given in the last three lines of (2.37) have not been investigated in detail up to now.

2.4.5. So far we have taken into account the electromotive force $\boldsymbol{\mathscr{E}}^B$ only, but have not considered $\boldsymbol{\mathscr{E}}^0$ which can be unequal zero under conditions which allow $\mathbf{B}'$ to be unequal zero even though $\bar{\mathbf{B}}$ vanishes. Since in the solar convection zone the magnetic Reynolds number responsible for fluctuations, R_m is much larger than unity we cannot exclude without further investigation that some parts of the fluctuating motion give rise to 'local dynamos', i.e. dynamos restricted to the scales of fluctuations, which contribute to $\mathbf{B'}$ also in absence of $\bar{\mathbf{B}}$. This is related to ideas by Batchelor (1950) and Kasantsev (1967). According to the fact that the fluctuating motions vary rapidly, each such dynamo will have a short lifetime only, and the corresponding part of $\mathbf{B'}$ excited by some kind of seed will reach only small values and then decay again. Of course, $\mathscr{E}^0$ depends on assumptions about the seeds. For simplicity we suppose the seeds to be isotropically distributed. From the assumptions used above for the fluctuating motions including weak influence of Coriolis forces we can now conclude

$$\mathscr{E}^0 = \kappa \hat{\mathbf{g}} + \lambda \hat{\mathbf{g}} \times \hat{\boldsymbol{\omega}} \tag{2.41}$$

with some coefficients κ and λ. Taking into account arbitrary strength of Coriolis forces, a term proportional to $\hat{\boldsymbol{\omega}}$ has to be added. Unfortunately there does not exist any investigation which provides for more detailed information on $\mathscr{E}^0$ or coefficients κ and λ.

For the solar cycle we conjecture that the electromotive force $\mathscr{E}^B$, the magnitude of which depends on that of $\bar{\mathbf{B}}$, is always large compared with $\mathscr{E}^0$ so that the latter can be neglected. This seems to follow already from the argument given above in favour of the small efficiency of the local dynamos producing $\mathbf{B'}$.

We are faced with a completely different situation when $\bar{\mathbf{B}}$ is zero at a given time. Then $\mathscr{E}^B$ is zero too, and if the local dynamos lead to a non-vanishing $\mathscr{E}^0$, we have to expect that $\bar{\mathbf{B}}$ grows. Especially the part of $\mathscr{E}^0$ proportional to $\hat{\mathbf{g}} \times \hat{\boldsymbol{\omega}}$ gives rise to a $\bar{\mathbf{B}}$ field of dipole type. Thus, as a consequence of small scale magnetic fields, a large-scale field can occur.

3. Kinematic Dynamo Theory of the Solar Cycle

3.1. BASIC FEATURES

In order to discuss the general principle of the dynamo theory of the solar cycle we consider the Sun as a sphere of electrically conducting matter with a mean motion like differential rotation and with turbulent motions in the outer layers, and we suppose the mean motion as well as the distribution of the turbulence to be symmetric with respect to both the rotation axis and the equatorial plane, and to be stationary. Furthermore, we understand the solar cycle, which should be comprehended theoretically here, substantially as an interplay between a poloidal and a toroidal mean magnetic field, both supposed to be axisymmetric too but antisymmetric with respect to the equatorial plane. In a rough picture, the poloidal field is of dipole type, and the toroidal field consists of two oppositely oriented belts, one in each hemisphere. Both fields generate and attenuate each other so that an oscillation with a period of 22 years results. Compared with the poloidal field the toroidal one reaches much higher amplitudes and is responsible, e.g., for the appearance of sunspots.

We have first to discuss how a poloidal field can generate a toroidal one and vice versa. Provided a poloidal field exists, a toroidal one must already occur due to differential rotation. Accepting that for the generation of the toroidal field the differential rotation is more effective than the turbulence, we may neglect all effects of turbulence except that represented by the turbulent conductivity. Departing from a toroidal field, however, a poloidal one can be generated neither by differential rotation nor by any other axisymmetric motion but only by turbulence. Parker (1955) was the first to recognize that the poloidal field can occur as a consequence of cyclonic convection. Within the frame of mean-field electrodynamics the effect of cyclonic convection he pointed out is reflected by the α-effect.

Let us consider in more detail this crucial point of the dynamo theory of the solar cycle, the generation of a poloidal field from a toroidal one. Having in mind that a poloidal field is connected with toroidal currents we have to ask for electromotive forces caused by turbulence which induce toroidal currents from toroidal fields. Only three of the electromotive forces occurring in (2.37) show this property. With respect to (2.32) and (2.38) we note them in the form

$$\mu\beta_1\hat{\boldsymbol{\omega}}\times\bar{\mathbf{j}}, \qquad -\alpha_1(\hat{\mathbf{g}}\cdot\hat{\boldsymbol{\omega}})\bar{\mathbf{B}}, \qquad \mu\alpha_9(\hat{\mathbf{g}}\cdot\hat{\boldsymbol{\omega}})\hat{\mathbf{g}}\times\bar{\mathbf{j}}. \tag{3.1}$$

Even if we relinquish linearity in $\hat{\boldsymbol{\omega}}$ no other terms occur here.

In addition to differential rotation at least one of the electromotive forces given in (3.1) is required to have indeed an interaction between poloidal and toroidal magnetic fields. It remains to be scrutinized whether this interaction actually allows a dynamo, i.e., prevents that the magnetic fields vanish in course of time, and whether and under which conditions alternating magnetic fields occur. Unfortunately, the questions arising here can be answered only on the basis of either analytical calculations for extremely simplified models or lengthy numerical calculations for more realistic models.

As the first of the electromotive forces listed in (3.1) we consider that given by $-\alpha_1(\hat{\mathbf{g}}\cdot\hat{\boldsymbol{\omega}})\bar{\mathbf{B}}$, i.e. the idealized α-effect. In conjunction with differential rotation it actually allows dynamos, called $\alpha\omega$-type dynamos, for both stationary and alternating magnetic fields. The dynamo mechanism suggested by Parker (1955) already before the elaboration of mean-field electrodynamics can be understood as being of $\alpha\omega$-type, generating alternating fields; however, he did not investigate spherical dynamo models. Departing from the conception of mean-field electrodynamics Steenbeck and Krause (1969) elaborated spherical dynamo models of the $\alpha\omega$-type for alternating fields. They presented numerical results on excitation conditions, space and time structure of the fields etc. and compared them with observational material as the butterfly diagram. Meanwhile numerous investigations of such models with alternating fields have been carried out; see, e.g., Deinzer and Stix (1971), Roberts and Stix (1972), Roberts (1972), Ruzmaikin and Ivanova (1975), Jepps (1975). Also non-axisymmetric fields were considered; see Stix (1971), Krause (1971), Roberts and Stix (1972). It has been proved by Levy (1972) that dynamos of the $\alpha\omega$-type are also able to generate stationary fields; see also Stix (1973) and Deinzer et al. (1973, 1974).

Next we consider the electromotive force described by $\mu\beta_1\hat{\boldsymbol{\omega}}\times\bar{\mathbf{j}}$, i.e. the $\boldsymbol{\omega}\times\mathbf{j}$-effect. In combination with differential rotation this effect too allows dynamos. In other words, differential rotation and a special type of anisotropic electric conductivity can lead to dynamos. In the first investigations of models of that kind carried out by Rädler (1969c, 1970) and by Roberts (1972) only stationary fields were found. Dealing with the possibility of alternating fields Roberts (1972) got numerical results which suggest that such fields exist too, but because of some convergency difficulties he regarded these results as not quite convincing. Recent results by Rädler (1975) confirm that alternating fields can occur. Some details will be given in the appendix.

As the last of the electromotive forces listed in (3.1) we consider that described by $\mu \alpha_9(\hat{\mathbf{g}} \cdot \hat{\boldsymbol{\omega}})\hat{\mathbf{g}} \times \bar{\mathbf{j}}$. Since it is very similar to $\mu \beta_1 \hat{\boldsymbol{\omega}} \times \bar{\mathbf{j}}$ we have good reason to assume that it together with differential rotation can give rise to dynamos too.

3.2. Remarks concerning the further elaboration of models of the solar cycle

Many efforts have been devoted to the elaboration of dynamo models for the solar cycle which reflect as many features of observations as possible. Suppose the basis is correct, one may, when fitting the models to observations, draw some conclusions on parameters inaccessible for direct observations, e.g. on the radial dependence of angular velocity or on parameters characterizing the fluctuating motions.

In this context we must have in mind that each of the electromotive forces occurring in Ohm's law (2.37) can influence the excitation conditions and the space and time behaviour of the magnetic fields. We have to scrutinize carefully which of these electromotive forces should be taken into account and which of them could be neglected. In most of the hitherto discussed dynamo models for the solar cycle only the idealized α-effect and the turbulent conductivity were involved, i.e. only effects which are already known from the case of homogeneous isotropic turbulence, and all other effects of turbulence were cancelled. The possible influence of these effects should be discussed in more detail, otherwise the conclusions of the kind mentioned above are only of restricted value.

Considering again the generation of the poloidal from the toroidal field we have to clarify which of the electromotive forces listed in (3.1) should be taken into account. It seems that the α-effect plays the most important part indeed. As for orders of magnitude, the ratio of the idealized α-effect to the $\boldsymbol{\omega} \times \mathbf{j}$-effect is given by $\alpha_1 \lambda_B / \beta_1$, where λ_B is a characteristic length for the variation of the magnetic field. Provided (2.35a) and (2.39) determine the orders of magnetic of α_1 and β_1 even when only h is involved and l_1 and m_1 are cancelled, this ratio can be replaced by λ_B / λ_K with λ_K as given in (2.22). It is to be expected that $\lambda_B / \lambda_K \gg 1$; for according to the above-mentioned estimates λ_K is hardly bigger than 10^4 m. In this way the $\boldsymbol{\omega} \times \mathbf{j}$-effect turns out to be of minor importance, and the same can be concluded for the related effect proportional to $(\hat{\mathbf{g}} \cdot \hat{\boldsymbol{\omega}})\hat{\mathbf{g}} \times \bar{\mathbf{j}}$. These arguments, however, should undergo a more detailed examination. Some investigations of dynamo models in which beside the differential rotation both the α-effect and the $\boldsymbol{\omega} \times \mathbf{j}$-effect are included have been carried out by Rädler (1975).

The behaviour of the toroidal field can be influenced by some of the electromotive forces occurring in (2.37) even without the assistance of the poloidal field. Using

(2.25) we write these electromotive forces in the form

$$-\gamma\hat{\mathbf{g}}\times\bar{\mathbf{B}}, \qquad -\mu(\gamma_1-\gamma_2)\hat{\mathbf{g}}(\hat{\mathbf{g}}\cdot\bar{\mathbf{j}}), \qquad -\mu\gamma_2\bar{\mathbf{j}}. \tag{3.2}$$

The first of them, which can act in the same way as a diamagnetism of the matter, can strongly influence the dynamo mechanism. This becomes clear from the dynamo models investigated by Ruzmaikin and Ivanova (1975) in which diamagnetism is included, even though in a manner which does not correspond exactly to the conception of mean-field electrodynamics outlined here. The last two electromotive forces in (3.2), which can be interpreted in terms of an anisotropy of the turbulent conductivity, have not been investigated in this respect up to now.

Finally we have to consider the generation of the toroidal from the poloidal field and that influence on the poloidal field which occurs without the assistance of the toroidal one. All the electromotive forces due to turbulence which have to be discussed here, are induced by the poloidal field. As long as the poloidal is small compared with the toroidal field, they should be small compared with those given in (3.1) and (3.2). Therefore, they are presumably of minor importance for dynamo models as considered here.

As indicated above, our considerations are restricted to axisymmetric magnetic fields. If non-axisymmetric ones should be included some points have to be rediscussed.

4. The Influence of the Mean Magnetic Field on the Motions and its Consequences for the Dynamo Mechanism

4.1 BASIC IDEAS

Up to now we have dealt with the question of how mean electromagnetic fields in electrically conducting matter are influenced by the motions of the matter but we did not pay attention to the fact that the motions can be influenced by the electromagnetic fields. Within the frame of such considerations, dynamo mechanisms turn out to be possible which allow the mean magnetic fields to grow incessantly. In accordance with that, the kinematic dynamo theory of the solar cycle allows arbitrary mean-field amplitudes. However, if a magnetic field grows we have to expect that also its back reaction to the motions due to Lorentz forces becomes stronger and, finally, provides for a limitation of the field strength. In this way also the mean-field amplitude in the solar cycle is controlled.

Already this point gives rise to extend the foregoing considerations so that this influence of the electromagnetic fields on the motions is included. Doing so we overstep the bounds of mean-field electrodynamics and are confronted with much more complicated aspects of mean-field magnetohydrodynamics. Only some of the problems arising here have been tackled up to now.

To begin with, we have to complete the basic electrodynamic Equations (2.1) by hydrodynamic equations. We want to demonstrate only some basic features. For this purpose we restrict ourselves to incompressible media. Then only the Navier–Stokes equation and the continuity equation have to be added which read

$$\rho\left(\frac{\partial \mathbf{u}}{\partial t} + (\mathbf{u} \cdot \mathrm{grad})\mathbf{u}\right) = -\mathrm{grad}\, p + \rho\mathbf{f} + \mathbf{j} \times \mathbf{B} \tag{4.1a}$$

$$\mathrm{div}\, \mathbf{u} = 0. \tag{4.1b}$$

Here ρ is the mass density, which is to be regarded as a constant, p is the hydrodynamic pressure, and $\mathbf{f}$ denotes viscous or other body forces. Finally, $\mathbf{j} \times \mathbf{B}$ represents the Lorentz force which is responsible for the influence of the electromagnetic fields on the motion. In the same way as with the Equations (2.1) we take the average of the Equations (4.1) thus obtaining

$$\rho\left(\frac{\partial \bar{\mathbf{u}}}{\partial t} + (\bar{\mathbf{u}} \cdot \mathrm{grad})\bar{\mathbf{u}}\right) = -\mathrm{grad}\, \bar{p} + \rho\bar{\mathbf{f}} + \bar{\mathbf{j}} \times \bar{\mathbf{B}} - \rho\overline{(\mathbf{u}' \cdot \mathrm{grad})\mathbf{u}'} + \overline{\mathbf{j}' \times \mathbf{B}'} \tag{4.2a}$$

$$\mathrm{div}\, \bar{\mathbf{u}} = 0. \tag{4.2b}$$

Just as observed with Ohm's law, the formal correspondence between the Navier–Stokes equation for the original and the mean fields is disturbed by additional forces, $-\rho\overline{(\mathbf{u}' \cdot \mathrm{grad})\mathbf{u}'}$ and $\overline{\mathbf{j}' \times \mathbf{B}'}$, appearing with the mean fields. The first of them is already known from the theory of hydrodynamic turbulence; it can be interpreted in terms of Reynolds stresses.

This consideration shows that within the mean-field magnetohydrodynamics of incompressible media in addition to the electromotive force, $\overline{\mathbf{u}' \times \mathbf{B}'}$, the two forces, $-\rho\overline{(\mathbf{u}' \cdot \mathrm{grad})\mathbf{u}'}$ and $\overline{\mathbf{j}' \times \mathbf{B}'}$, have to be investigated. As far as $\overline{\mathbf{u}' \times \mathbf{B}'}$ is concerned all general relations discussed above hold also when $\bar{\mathbf{u}}$ and $\mathbf{u}'$ and, consequently, K_{ij} are specified to depend on $\bar{\mathbf{B}}$. Of course, then $\overline{\mathbf{u}' \times \mathbf{B}'}$ is no longer linear in $\bar{\mathbf{B}}$. A part of the methods by which $\overline{\mathbf{u}' \times \mathbf{B}'}$ was determined can be used also to determine $-\rho\overline{(\mathbf{u}' \cdot \mathrm{grad})\mathbf{u}'}$ and $\overline{\mathbf{j}' \times \mathbf{B}'}$.

If we remove the restriction to incompressible media, some details become more complicated; e.g. correlations between ρ' and $\mathbf{u}'$ have to be considered.

4.2 SOME SPECIAL RESULTS

The dynamo models of the solar cycle are determined by assumptions on both the mean motion and the electromotive force caused by the fluctuating motions. Since there is no observational evidence for a substantial variation of the mean motion during the cycle it is reasonable to suppose $\bar{\mathbf{u}}$ to be independent of $\bar{\mathbf{B}}$. There are, however, several reasons to assume that the fluctuating motions are considerably influenced by the magnetic field, so that for the electromotive force, $\overline{\mathbf{u}' \times \mathbf{B}'}$, a non-linear dependence on $\bar{\mathbf{B}}$ has to be expected.

There are several investigations of $\overline{\mathbf{u}' \times \mathbf{B}'}$ in which this non-linear dependence on $\bar{\mathbf{B}}$ is considered. Unfortunately, in most cases suppositions were used which scarcely apply to the solar convection zone; nevertheless, the results may be suggestive here too.

Especially the α-effect was considered which occurs with an originally homogeneous isotropic non-mirror-symmetric turbulence under the influence of a homogeneous mean magnetic field. As long as this influence is weak one may easily conclude that α, now understood as a function of $\bar{B}$, has the form

$$\alpha = \alpha_0(1 - a\bar{B}^2) \tag{4.3}$$

with some coefficients α_0 and a; see Krause and Rädler (1971). Using second order correlation approximation and supposing $R_m \ll 1$ as well as incompressibility of the medium, Rüdiger (1974) computed the value of a. Within the same frame he discussed α for large $\bar{B}$ too and found that it vanishes as $\bar{B}^{-3}$ if $\bar{B} \rightarrow \infty$. Interesting relations for α have been derived by Vainshtein and Zeldovich (1972).

Furthermore, $\overline{\mathbf{u}' \times \mathbf{B}'}$ has been studied in detail, especially with respect to its dependence on $\bar{\mathbf{B}}$, for a random superposition of inertial waves in a rotating fluid by Moffatt (1970b, 1972) and Soward (1975). Comprehensive calculations for an inhomogeneous turbulence in a rotating fluid have been carried out by Roberts and Soward (1975).

Finally, first investigations on the consequences of an α-effect which is non-linear in $\bar{\mathbf{B}}$ for dynamo mechanisms of the $\alpha\omega$-type have been presented by Stix (1973) and Jepps (1975).

Appendix to Section 3.1

We give here an example of a dynamo model which works on the basis of differential rotation and the $\omega \times \mathbf{j}$-effect and allows alternating magnetic fields. The model consists of a sphere of conducting moving matter. The mean motion is supposed to be a differential rotation with an angular velocity, ω, given by

$$\omega = \begin{cases} \omega_0 & \text{for } 0 \leqslant x \leqslant 0.3 \\ \omega_0 + \dfrac{\Delta\omega}{4}(2 + 3v - v^3) & v = 5(x - 0.5) \quad \text{for } 0.3 \leqslant x \leqslant 0.7 \\ \omega_0 + \Delta\omega & \text{for } 0.7 \leqslant x \leqslant 1. \end{cases} \tag{A.1}$$

As for the electromotive forces induced by turbulence, only the $\omega \times \mathbf{j}$-effect is involved with a coefficient, β_1, specified by

$$\beta_1 = \begin{cases} 0 & \text{for } 0 \leqslant x \leqslant 0.6 \\ \dfrac{\beta_{1\max}}{16}(8 + 15v - 10v^3 + 3v^5) & v = 5(x - 0.8) \quad \text{for } 0.6 \leqslant x \leqslant 1 \\ \beta_{1\max} & \text{for } x = 1. \end{cases} \tag{A.2}$$

Here ω_0, $\Delta\omega$ and $\beta_{1\max}$ are constants, and x is the normalized radius of the sphere with $x = 1$ at its surface. Figure 2 shows the profiles of ω/ω_0 and $\beta_1/\beta_{1\max}$. The surroundings of the sphere are supposed to be a vacuum. Again, only axisymmetric magnetic fields are considered. For the generation of the toroidal field from the

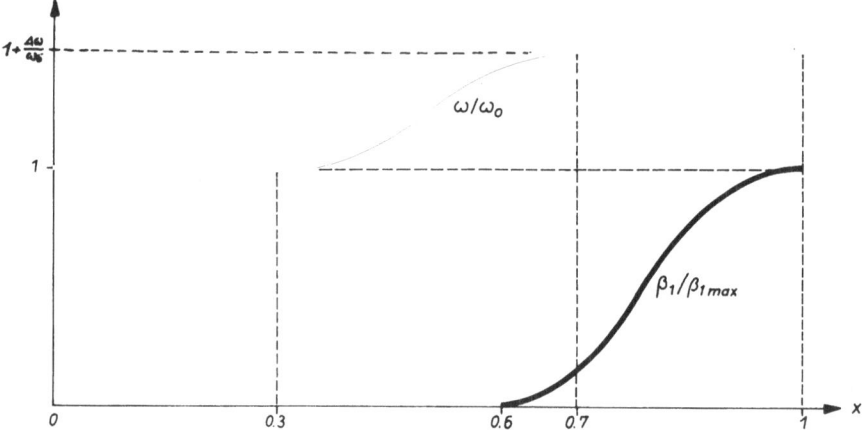

Fig. 2.   Profiles of the normalized angular velocity, ω/ω_0, and the normalized $\boldsymbol{\omega}\times\mathbf{j}$-coefficient, $\beta_1/\beta_{1\text{max}}$.

poloidal one only the differential rotation is taken into account, and the $\boldsymbol{\omega}\times\mathbf{j}$-effect is neglected. The generation of the poloidal field from the toroidal one is, however, due to the $\boldsymbol{\omega}\times\mathbf{j}$-effect. As one can easily see from the equations governing this model, the condition of excitation of magnetic fields can be expressed by a dynamo number, P, given by

$$P = \mu^2\sigma^2\Delta\omega\beta_{1\text{max}}R^2 \tag{A.3}$$

where R is the radius of the sphere. Since $\beta_{1\text{max}}$ may be supposed to be positive, $P>0$ corresponds to $\Delta\omega > 0$, i.e. an inward decreasing angular velocity, and $P<0$ corresponds to $\Delta\omega < 0$, i.e. an inward increasing angular velocity. Furthermore, a dimensionless frequency, Ω, of the mean fields is used for which the time is measured in units of $\mu\sigma R^2$.

By means of a representation of the magnetic field by spherical harmonics the partial differential equations governing this model were reduced to an infinite set of ordinary differential equations. But only the first harmonics up to a certain order, say N, were taken into account, and the solutions of the corresponding truncated sets of equations were computed numerically for different N. In the case of convergency of these solutions for growing N, the solutions for sufficiently high N were regarded as such of the original equations.

TABLE I

Marginal dynamo numbers, P, dimensionless frequencies, Ω, and the types of the respective solutions

P	Ω	Type of solution
$-2\,347$	16.6	dipole
$-2\,216$	6.5	quadrupole
$+4\,914$	35.5	quadrupole
$+19\,268$	73.4	dipole

For both $P > 0$ and $P < 0$ solutions were found which correspond to magnetic fields of dipole as well as of quadrupole type. In Table I the marginal values of P which were found within the region $-2500 \ldots +20\,000$ and the respective values of Ω were listed. Table II shows one example illustrating the convergency of the solutions of the truncated equations.

TABLE II

Marginal dynamo numbers, P, and dimensionless frequencies, Ω, for related solutions of the truncated equations in dependence on the truncation parameter, N

N	P	Ω
4	5 014.7	35.48
6	4 912.5	35.47
8	4 913.5	35.47
10	4 913.5	35.47

References

Batchelor, G. K.: 1950, *Proc. Roy. Soc.* **A201**, 406.
Braginski, S. I.: 1964, *Zh. Eksper. Theoret. Fiz.* **47**, 1084.
Csada, I. K.: 1951, *Acta Phys. Hung.* **1**, 235.
Deinzer, W. and Stix, M.: 1971, *Astron. Astrophys.* **12**, 111.
Deinzer, W., v. Kusserow, H.-U., and Stix, M.: 1973, *Mitteilungen Astron. Ges.* **34**, 155.
Deinzer, W., v. Kusserow, H.-U., and Stix, M.: 1974, *Astron. Astrophys.* **36**, 69.
Drobyshevski, E. M. and Yuferev, V. S.: 1974, *J. Fluid Mech.* **65**, 33.
Jepps, S. A.: 1975, *J. Fluid. Mech.* **67**, 625.
Kasantsev, A. P.: 1967, *Zh. Eksper. Theoret. Fiz.* **53**, 1806.
Krause, F.: 1967, Habilitationsschrift Univ. Jena.*
Krause, F.: 1971, *Astron. Nachr.* **293**, 187.
Krause, F. and Rädler, K.-H.: 1971, in *Handbuch der Plasmaphysik und Gaselektronik 2* (ed. by R. Rompe and M. Steenbeck), Akademie-Verlag, Berlin.
Krause, F. and Roberts, P. H.: 1973a, *Astrophys. J.* **181**, 977.
Krause, F. and Roberts, P. H.: 1973b, *Mathematika* **20**, 24.
Krause, F.: 1975, this volume, p. 305.
Krause, F. and Rüdiger, G.: 1975, *Solar Phys.* **42**, 107.
Levy, E. H.: 1972, *Astrophys. J.* **171**, 621.
Moffatt, H. K.: 1970a, *J. Fluid Mech.* **41**, 435.
Moffatt, H. K.: 1970b, *J. Fluid Mech.* **44**, 705.
Moffatt, H. K.: 1972, *J. Fluid. Mech.* **53**, 385.
Parker, E. N.: 1955, *Astrophys. J.* **122**, 293.
Rädler, K.-H.: 1966, Dissertation Univ. Jena.
Rädler, K.-H.: 1968a, *Z. Naturforsch.* **23a**, 1841.*
Rädler, K.-H.: 1968b, *Z. Naturforsch.* **23a**, 1851.*
Rädler, K.-H.: 1969a, *Geodät. Geophys. Veröffentl. Potsdam*, Reihe II, Heft 13, 131.
Rädler, K.-H.: 1969b, *Monatsber. Dtsch. Akad. Wissensch. Berlin* **11**, 195.*
Rädler, K.-H.: 1969c, *Monatsber. Dtsch. Akad. Wissensch. Berlin* **11**, 272.*
Rädler, K.-H.: 1970, *Monatsber. Dtsch. Akad. Wissensch. Berlin* **12**, 468.
Rädler, K.-H.: 1974, *Astron. Nachr.* **295**, 85.
Rädler, K.-H.: 1975, *Astron. Nachr.* (in preparation).
Roberts, P. H.: 1971, in W. H. Reid (ed.), *Mathematical Problems in the Geophysical Sciences*, American Mathematical Society, Providence, R. I., p. 129.

Roberts, P. H.: 1972, *Phil. Trans. Roy. Soc. London* **A272**, 663.

Roberts, P. H. and Stix, M.: 1971, 'The Turbulent Dynamo', NCAR Technical Note TN/IA-60.

Roberts, P. H. and Stix, M.: 1972, *Astron. Astrophys.* **18**, 453.

Roberts, P. H. and Soward, A. M.: 1975, *Astron. Nachr.* **296**, 49.

Rüdiger, G.: 1974, *Astron. Nachr.* **295**, 275.

Ruzmaikin, A. and Ivanova, B.: 1975, *Astron. Zh.* (in print).

Soward, A. M.: 1975, *J. Fluid Mech.* **69**, 145.

Steenbeck, M. and Krause, F.: 1969, *Astron. Nachr.* **291**, 49.

Steenbeck, M., Krause, F. and Rädler, K.-H.: 1963, *Sitzungsber. Dtsch. Akad. Wiss. Berlin, Klasse Math.-Phys.-Tech.*, Heft 1.

Steenbeck, M., Krause, F., and Rädler, K.-H.: 1966, *Z. Naturforsch.* **21a**, 369.*

Stix, M.: 1971, *Astron. Astrophys.* **13**, 203.

Stix, M.: 1973, *Astron. Astrophys.* **24**, 275.

Stix, M.: 1974, *Astron. Astrophys.* **37**, 121.

Stix, M.: 1975, this volume, p. 367.

Sweet, P. A.: 1950, Monthly Notices Roy. Astron. Soc. **110**, 69.

Vainshtein, S. I. and Zeldovich, Ja. B.: 1972, *Usp. Fiz. Nauk* **106**, 431 (*Soviet Phys.-Usp.* **15**, 159 (1972)).

Articles marked with an asterisk have been translated into English by Roberts and Stix (1971).

DISCUSSION

Vainshtein: There is α-effect *without* rotation caused only by magnetic forces. It is a nonlinear effect. I think that you must take it into account. In this case $\alpha \sim \mathbf{B} \cdot \mathrm{curl}\ \mathbf{B}$.

Krause: In a paper by Roberts and Soward (Roberts, P. H. and Soward, A. M.: 1975, *Astron. Nachr.* **296**) it was shown that in the first approximation no α-effect is caused by the helicity of a mean magnetic field. More detailed, writing down $\alpha = a \cdot (\bar{\mathbf{B}} \cdot \mathrm{curl}\ \bar{\mathbf{B}})$, the factor a will prove to be zero in the first approximation in $\bar{\mathbf{B}}$. I checked this result to be correct. My question is, in what approximation you found the constant a to be unequal zero.

Vainshtein: The term $(\mathbf{B} \cdot \mathrm{curl}\ \mathbf{B})$ appears in the first nonlinear approximation.

Stix: If the coefficients of the various electromotive forces are difficult to obtain, can you at least obtain information about the *sign* of the coefficients? In particular what is the sign of β_1, the coefficient of the $\boldsymbol{\omega} \times \mathbf{j}$ term?

Rädler: In all calculations I know the coefficient β_1 turned out to be positive. For a certain range of suppositions, departing from Bochner's theorem, it can be shown that β_1 must necessarily be positive.

Notes added in proof concerning chapter 3

Strictly speaking, the factor α_1 in (3.1) has to be replaced by $\alpha_1 - (\alpha_9 + \alpha_{10})/r$, and γ in (3.2) by $\gamma - (\gamma_2 + \gamma_3)/r$, where r is the radial coordinate. With respect to the foregoing discussions, however, this is of minor importance.

The above-mentioned estimation of the ratio $\alpha_1 \lambda_B/\beta_1$ implies a special assumption on h. Obviously, α_1 depends only on the symmetric and β_1 only on the antisymmetric part of h where symmetry with respect to $(\hat{\boldsymbol{\omega}} \cdot \boldsymbol{\xi})$ is considered. Only if both parts are of the same order of magnitude $\alpha_1 \lambda_B/\beta_1$ may be replaced by λ_B/λ_K. However, this assumption and, consequently, the conclusion on the minor importance of the $\boldsymbol{\omega} \times \mathbf{j}$-effect are questionable.

DYNAMO IN THE PRESENCE OF
DIFFERENTIAL ROTATION

S. I. VAINSHTEIN

Sibizmiran, U.S.S.R.

It is well know that the 'dynamo' theory has a number of vetoes; e.g. axisymmetric, two-dimensional, central-symmetric, etc. dynamo are impossible. In principle, the problem is essentially three-dimensional in any coordinate system. This is the main difficulty of both the theory itself and its possible applications. In fact, one prefers to believe that, for example, a non-rigid body-rotating star or convection in the Earth's nucleus possesses axis symmetry. However, due to the above vetoes one has to add finer effects (Coriolis strength, density, inhomogeneity) to create asymmetrical convection. On the other hand, the authors try to find the most simple movements with minimum deviations from axial symmetry. Thus, the Herzenberg's dynamo (Herzenberg, 1958) is realized by two rotating cylinders, axes of which are parallel to each other (see also Galaitis, 1973; Galaitis and Freinberg, 1974), the Lortz's dynamo-spiral movement (Lortz, 1968; Ponomarenko, 1973). Nevertheless, the mentioned vetoes possess a common feature, the assumption regarding the symmetry extends both to the movement and to the field. Hence, it makes sense to raise a question whether symmetric movements are able to generate an asymmetric field. A positive answer to this question, in particular, is given by Tverskoy's model (Tverskoy, 1966) – the toroidal vortex. The latter possesses axial symmetry. Nevertheless, the toroidal vortex is a complex motion; we will proceed along the path of a minimum simplification.

Is dynamo possible in differential rotation? This motion is the simplest and likely the most wide-spread in nature. Besides, is dynamo possible in two-dimensional motion of the differential rotation type, that is, in the cylindrical system

$$v_r = v_z = 0, \qquad v_\varphi \neq 0, \qquad v_\varphi = v_\varphi(r)?$$

On the face of it, the answer must be negative as Zeldovich's system (Zeldovich, 1956) exists which rules out a two-dimensional dynamo in the arbitrary (that is, not necessary two-dimensional) magnetic field. In this problem, however, the velocity is even one-dimensional as **v** depends only on r. However, the mentioned theory is proved for the unlimited conductive medium. If the conductive medium is situated at $r < R$, and at $r > R$ the medium will be unconductive (in particular, vacuum), then the veto can be removed and dynamo is possible. We will notice that no unlimited bodies exist in nature.

Let us explain why the veto is taken away from the two-dimensional dynamo in the presence of vacuum. From induction equation

$$\frac{\partial \mathbf{H}}{\partial t} = \mathrm{rot}\,[\mathbf{vH}] + \nu_m \, \Delta \mathbf{H} \tag{1}$$

Bumba and Kleczek (eds.), Basic Mechanisms of Solar Activity, 345–351. *All Rights Reserved.*
Copyright © 1976 by the IAU.

equation for H_z results in given geometry

$$\frac{\partial H_z}{\partial t} + (\mathbf{v}\nabla)H_z = \nu_m \Delta H_z .$$ (2)

In view of the fact that the behaviour H_z is described by the heat conductivity equation and $H_z \to 0$ at $r \to \infty$ in the unlimited fluid, $H_z \to 0$ at $t \to \infty$, that is, H_z is damping.

Then, assuming that $H_z = 0$, it is easy to prove that the equation for the vector-potential of removed field components has also the form of conductivity equation. Hence, both H_x and H_y are suppressed.

Now, let the vacuum be at $r > R$. Then, assuming that the permanence for media $\mu = I$, we will have the boundary conditions: field continuity, while the field in vacuum is potential, as well as

$$\frac{1}{r} \cdot \frac{\partial H_z}{\partial \varphi} - \frac{\partial H_\varphi}{\partial z} = 0$$ (3)

on the boundary. Condition (3) corresponds to zero reduction of the current component, normal toward the boundary. Now, conductivity Equation (2) with boundary conditions (3) must yield no field H_z damping! In general, all theorems of dynamo impossibility are proved when one of the field components separates from the other, that is, it behaves independently of them. At the same time, H_z is related to H_φ through boundary condition (3). It will be shown below that just this very fact takes away the veto from the one-dimensional dynamo.

It seems that one might object that the vacuum boundary conditions on the sky body boundary are not very topical. The Sun, for instance, is surrounded by a highly conductive corona which directly passes into the solar wind. Are vacuum boundary conditions topical here? However, H_z field component does not separate from the other already in the presence of electric conductivity dependence upon r (thus ν_m is axisymmetric too). In fact, the equation for H_z has in this case the form:

$$\frac{\partial H_z}{\partial t} + (\mathbf{v}\nabla)H_z = \nu_m\left(\frac{1}{r} \cdot \frac{\partial}{\partial r} r \frac{\partial H_z}{\partial r} + \frac{1}{r^2} \cdot \frac{\partial^2 H_z}{\partial \varphi^2}\right) - \frac{\partial \nu_m}{\partial r} \cdot \frac{\partial H_r}{\partial z} .$$ (4)

It can be seen from (4) that to conclude on damping is impossible again. Field generation exists in case (4) too. The assumption of inhomogeneous $\nu_m(r)$ is naturally associated with the sky body boundary itself; in the particular case of vacuum ν_m it changes sharply, in vacuum $\nu_m = \infty$.

1. One-Dimensional Problem Solution in Vacuum

Let us express the field $v_\varphi(r)$ in this way (Vainshtein, 1975):

region I: $v_\varphi = \omega_0 r$ at $r < r_0$;

region II: $v_\varphi = 0$ at $r > r_0, r < R$.   (5)

The rigid body rotation at $r < r_0$ simplifies the calculations, although it is not important, in principle, whether there is a jump in angular velocity $\omega(r)$ or if the transition is smooth. Then, the equations for H_r, H_φ, H_z have the form:

$$\frac{\partial H_z}{\partial t} = -\omega_0 \frac{\partial H_z}{\partial \varphi} + \nu_m \Delta H_z ;$$

$$\frac{\partial H_r}{\partial t} = -\omega_0 \frac{\partial H_r}{\partial \varphi} + \nu_m \left(\Delta H_r - \frac{1}{r^2} H_r - 2\frac{1}{r^2} \frac{\partial H_\varphi}{\partial \varphi} \right) ;$$

$$\frac{\partial H_\varphi}{\partial t} = -\omega_0 \frac{\partial H_\varphi}{\partial \varphi} + r H_r \cdot \frac{\partial \omega_0}{\partial r} + \nu_m \left(\Delta H_\varphi - \frac{1}{r^2} H_\varphi + 2\frac{1}{r^2} \cdot \frac{\partial H_r}{\partial \varphi} \right)$$

(6)

The solution of the system (6) is to be naturally sought in the form:

$$H_{r,\varphi,z} = f_{r,\varphi,z}(r) \exp\left[Et + i(m\varphi + kz) \right] .$$

(7)

Similarly to (5), introduce the function $f_\pm = f_r \pm i f_\varphi$ and we obtain Bessel equations for regions I, II and vacuum (region III):

$$\frac{\partial^2 f_z}{\partial p^2} + \frac{1}{p} \cdot \frac{\partial f_z}{\partial p} - \left(1 + \frac{m^2}{p^2} \right) f_z = 0 ;$$

$$\frac{\partial^2 f_\pm}{\partial p^2} + \frac{1}{p} \cdot \frac{\partial f_\pm}{\partial p} - \left(1 + \frac{(m \pm 1)^2}{p^2} \right) f_\pm = 0 ;$$

(8)

where in region I

$$p = \beta r , \qquad \beta = \sqrt{(E + \nu_m k^2 + i m \omega_0)/\nu_m} ;$$

in region II

$$p = \varkappa r , \qquad \varkappa = \sqrt{(E + \nu_m k^2)/\nu_m} ;$$

in region III

$$p = kr .$$

We are looking for the solution of set (8) in the form:

region I:

$$f_z = A I_m(p), \qquad f_\pm = B_\pm I_\pm(p) \text{ (limited in zero)} ;$$

region II:

$$f_z = C I_m(p) + D K_m(p), \qquad f_\pm = L_\pm I_\pm(p) + M_\pm K_\pm(p) ;$$

region III:

$$f_z = F K_m(p), \qquad f_\pm = i F K_\pm(p) \text{ (disappearing in infinity)} ,$$

where

$$I_\pm = I_{m\pm1}, \qquad K_\pm = K_{m\pm1} .$$

In region III we have made use of condition rot $\mathbf{H} = 0$.

The unification of solutions in the III regions yields a set of algebraic equations for coefficients.

The coupling condition is as follows: the continuity of all solutions and

$$\frac{\partial f_z}{\partial r}\bigg|_I = \frac{\partial f_z}{\partial r}\bigg|_{II} ;$$

$$\left(\frac{\partial f_\pm}{\partial r} \pm \frac{i r_0 \omega_0}{2 \nu_m}(f_+ + f_-)\right)\bigg|_I = \frac{\partial f_\pm}{\partial r}\bigg|_{II} .$$

(9)

Nine equations and 10 coefficients one can obtain. The Equation (10) is obtained by using the condition div $\mathbf{H} = 0$ (condition (3) is derived from the written ones). In order to derive this equation, take the divergence from (6), and we obtain:

$$E\gamma - i m \omega_0 \gamma = \nu_m \Delta\gamma ; \qquad \gamma = \operatorname{div} \mathbf{H} .$$

(10)

Further, in order $\gamma \equiv 0$, it is sufficient γ to be equal to zero on the boundary with vacuum (this results from uniqueness of solution (10) in given boundary conditions). Writing div $\mathbf{H} = 0$ on the boundary of the regions II–III and taking into account the fact that the field is continuously crossing the boundary, it is easy to understand that $\partial H_\varphi/\partial\varphi$, $\partial H_z/\partial z$ are also continuous on the boundary; from this

$$\frac{\partial H_r}{\partial r}\bigg|_{II} = \frac{\partial H_r}{\partial r}\bigg|_{III} .$$

(11)

Condition (II) is the last equation to be found. The determinant of the tenth order is presented in the form of a two (co-)factor product, one of which does not yield dynamo.

Let us assume that βr_0, $\varkappa r_0$, $kR \gg 1$.

Below we will see that this situation corresponds to the great Reynolds number $R_m = \omega_0 r_0^2/\nu_m \gg 1$. Now $p \gg 1$ on the boundaries, and one can use asymptotes

$$I_m(p) = (1/2\pi p)^{1/2}[\exp(p) + \exp(-p - (m + 1/2)\pi i)] ,$$

$$K_m(p) = (\pi/2p)^{1/2} \cdot \exp(-p) .$$

(12)

The second factor of the determinant is simplified, considering that th $\varphi \approx \pm 1$, $\varphi = \beta r_0 + i(\pi/2)(m + \frac{1}{2})$, which is fulfilled with exponential accuracy. In this case, the given factor is divided, in its turn, into two factors, one of which yields the equation

$$\operatorname{th} \varkappa \Delta r = \mp \frac{\varkappa}{\beta} ; \qquad \Delta r = R - r_0 ,$$

(13)

which does not yield dynamo. To the second factor corresponds the equation:

$$\operatorname{th} \varkappa \Delta r = \varkappa \frac{k \mp \beta}{\varkappa^2 \mp \beta k}$$

(14)

(The upper sign corresponds to th $\varphi = +I$) or in the non-dimensional form ($q = \varkappa \Delta r$)

$$\operatorname{th} q = \frac{D - \sqrt{q^2 + iC}}{q^2 - D\sqrt{q^2 + iC}} .$$

(15)

If in (15) we assume $D = k\,\Delta r$, $C = m\omega_0(\Delta r)^2/v_m$, $Re(q^2 + iC)^{1/2} > 0$. $\omega_0 = 0$, that is, $C = 0$ (absence of rotation, the trivial case), then (15) has no roots and one must use other factors of the determinant giving field damping. If $m = 0$ (the purely axisymmetric case, that is both the velocity and $\bar{H}$ are axisymmetric), then $C = 0$, so that it is tantamount to $\omega_0 = 0$ and the dynamo is impossible (theorem by S. I. Braginsky, 1964).

If $k = 0$, that is $D = 0$ (the purely two-dimensional case), then the equation is analogous to (13), only on the right side instead of $\varkappa/\beta$ there is $\beta/\varkappa$. This equation also has no growing solutions (Ya. B. Zeldovich theorem (Zeldovich, 1956)).

At $\Delta r \to 0$ we obtain the rotating cylinder as a rigid-body in vacuum. It is natural that the dynamo is impossible ($D = C = 0$). It can be seen therefore that the field will be essentially three-dimensional and will possess all the three components.

It can be easily seen that for the dynamo-solution, the root must lie in the region $Re(q^2 - k^2) > 0$. Solution (15) should be looked for in C-D-plane by giving q. Assuming $q = 1.00 - i\,0.50$, we obtain graphically $C = 0.11$, $D = -0.55$.

It is not difficult to verify that the given root corresponds to accepted assumptions and yields dynamo. In fact $\varkappa\,\Delta r \approx \beta\,\Delta r \approx k\,\Delta r \approx 1$ is not in contradiction with the use of asymptotes (12), if $r_0 \gg \Delta r$; the latter condition corresponds to $Rm = Cr_0^2/m(\Delta r)^2 \gg 1$, i.e. to the great Reynolds number. Further

$$E = (\varkappa^2 - k^2)v_m = \frac{v_m}{(\Delta r)^2}(0.47 - i\,1.00) \tag{16}$$

By using the condition $C \approx I$, we obtain $Re\,E \sim m\omega_0$, which is quite natural.

2. Discussion

The considered example supports the following arguments: dynamo occurs in all the cases when it is impossible to prove the contrary, using some standard rules. It seems that this affirmation is unlikely to be proved, nevertheless, it is practically always helpful.

In order to be sure of the existence of the dynamo in a determined situation, the following procedure should be adopted:

(1) Write the induction equation in the natural curvilinear coordinate system.

(2) Verify whether one of the field components is not prevalent; in the affirmative, this would lead to damping and the dynamo is impossible, as the remaining components are sure to be damped (which can be proved by transition to the equation for vector-potential).

Theorems on the impossibility of the dynamo were proved in just this way.

The astonishing simplicity and symmetry of the model under consideration permits one to hope that, practically, the veto for the dynamo will be taken away for all the symmetric models. But then, more complex asymmetric motions will be the more so dynamo-instable. Thus we might suggest a thesis-assumption: all motions in nature are instable with respect to magnetic fluctuations. The dynamo problem can be, therefore, turned upside down by looking for such motions in real limited bodies without generation.

Since the dynamo-solution has a character of instability in relation to the magnetic field fluctuations, the latter will be excited at least by thermodynamical fluctuations no matter whether it corresponds to observational data or not. Therefore one may ask why the solar cycle period ≈ 22 years, when $\omega_0 \sim 10^{-6}\,\mathrm{s}^{-1}$ – i.e., the field growth time caused by differential rotation is less than one month – and why all the celestial bodies do not have a magnetic field, etc. On the other hand the model considered is similar to the model of a rotating galaxy; the increment $\sim r\,\partial\omega/\partial r \sim \omega$, therefore the field is growing in the time period of a galaxy rotation, i.e. rather rapidly. And finally, both the occurrence of the field and its strengthening on the solar surface (for instance sunspots) as well as its disappearance can be easily explained by differential rotation only, taking into account either the simplest dynamo, or 'antidynamo' (Vainshtein, 1973). The analysis of the most simple shift motions the differential rotation and the shift (Vainshtein, 1973), showed that even such simple motions could cause a reduction of the field's scale and later on its strengthening or effective destruction. That is why the idea of the 'frozen-in' is a very simplified picture. Taking into account the increasing contribution of electromagnetic forces with decreasing field scale one can come to the following dilemma:

(1) If the electroconductivity is not very high (e.g. turbulent) and if in spite of the scale reduction, the electromagnetic forces remain insignificant, then the idea of the 'frozen-in' field is very often inadequate.

(2) If the electromagnetic forces contribution is essential, then the idea of the 'frozen-in' field is valid, however, the field of velocities itself acquires a small scale structure and becomes complicated.

References

Braginskij, S. I.: 1964, *Zh. Exper. Teoret. Fiz.* **47**, 1084.
Galaitis, A.: 1973, *Magnitnaya Gidrodinamika* **4**, 12.
Galaitis, A. and Freinberg, Ya.: 1974, *Magnitnaya Gidrodinamika*, No. 1, 37.
Herzenberg, A.: 1958, *Phil. Trans. Roy. Soc. A.* **250**, 543.
Lortz, D.: 1968, *Plasma Phys.* **10**,, No. 11, 967.
Ponomarenko, Yu. B.: 1973, *Prikladnaya Mehanika i Tehnicheckaya Fizika*, No. 6, 47.
Tverskoy, B. A.: 1966, *Geomagnetizm i Aeronomija* **6**, 11.
Vainshtein, S. I.: 1973, *Zh. Exper. Teoret. Fiz.* **65**, 550.
Vainshtein, S. I.: 1975, *Zh. Exper. Teoret. Fiz.* **62**, 997.
Zel'dovich, Ya. B.: 1956, *Zh. Exper. Teoret. Fiz.* **31**, 154.

DISCUSSION

Gilman: Your magnetic field produced by the dynamo must be non-axisymmetric. What form does it take? In particular are there wave numbers in the directions perpendicular to the shear of rotation for which dynamo occurs at the lowest magnetic Reynolds number?

Vainshtein: There is dependence on r, z and φ. The dependence on z is proportional to I^{ikz} where k is the wave number in the direction z the dependence on φ is proportional to $I^{im\varphi}$, $m = 1, 2, \ldots$; Reynolds number must be great.

Deinzer: Are these magnetic fields at all if there is an axisymmetric flow? I think they would be excluded by Cowling's theorem.

Vainshtein: Cowling's theorem is proved for the case when both magnetic field and velocity field are axisymmetric. In my case the magnetic field is not axisymmetric.

Stix: I agree with Dr Weiss that the model presented here is not in conflict with Cowling's theorem. But there is another theorem that says that no toroidal motion can give rise to dynamo action. Rotation, no matter how differential, is such a toroidal motion. The theorem has recently been in an even more restrictive form by Busse who found that the magnetic Reynolds number based on the radial velocity component must exceed a critical value. So my question is whether the model contradicts this theorem. Another question is: What are your boundary conditions at infinity?

Vainshtein: One cannot prove that the dynamo mechanism cannot act if the differential rotation only is present and conductivity is inhomogeneous.

Stix: I was referring to the magnetic Reynolds number based on the *radial* velocity component, not to the one based on the differential rotation.

Vainshtein: At infinity the magnetic field tends to zero. The generation of magnetic field can be in the presence of differential rotation without radial motion.

ON FAST MAGNETIC FIELD RECONNECTION

E. R. PRIEST

Dept. of Applied Mathematics, The University, St. Andrews, Scotland

and

A. M. SOWARD

School of Mathematics. The University, Newcastle-upon-Tyne, England

Abstract. The first model for 'fast' magnetic field reconnection at speeds comparable with the Alfvén speed was put forward by Petschek (1964). It involves one shock wave in each quadrant radiating from a central diffusion region and leads to a maximum reconnection rate dependent on the electrical conductivity but typically of order 10^{-1} or 10^{-2} of the Alfvén speed. Sonnerup (1970) and Yeh and Axford (1970) then looked for similarity solutions of the magnetohydrodynamic equations, valid at large distances from the diffusion region; by contrast with Petschek's analysis, their models have two waves in each quadrant and produce no sub-Alfvénic limit on the reconnection rate.

Our approach has been, like Yeh and Axford, to look for solutions valid far from the diffusion region, but we allow only one wave in each quadrant, since the second is externally generated and so unphysical for astrophysical applications. The result is a model which qualitatively supports Petschek's picture; in fact it can be regarded as putting Petschek's model on a firm mathematical basis. The differences are that the shock waves are curved rather than straight and the maximum reconnection rate is typically a half of what Petschek gave. The paper is a summary of a much larger one (Soward and Priest, 1976).

1. Introduction

The basic principles for two-dimensional magnetic field reconnection in a highly conducting plasma were established many years ago by Dungey (1953), Sweet (1958) and Parker (1963). They consider oppositely-directed magnetic field lines which are carried towards one another by a converging plasma flow, as shown in Figure 1. The magnetic field is assumed frozen to the plasma, save in a 'diffusion' region of dimensions l and L (and L out of the plane of the figure). There the electric current density is so large that the field lines can slip through the plasma. They enter the diffusion region at a speed v_i, are reconnected at the neutral point N, where the magnetic field strength vanishes, and are carried out of the region with speed v_0. The outflow magnetic field strength B_0 is less than the inflow strength B_i, so that some of the magnetic flux is annihilated in the diffusion region. The corresponding fall in magnetic energy appears partly as heat, through ohmic dissipation, and partly as an increase in the kinetic energy of the plasma.

For steady, incompressible flow Sweet and Parker derived the following order of magnitude relationships between the input and output parameters:

$$v_0 = v_A \equiv B_i/(\mu\rho)^{1/2}, \tag{i}$$

$$v_i = \eta/l, \tag{ii}$$

$$v_i L = v_0 l, \tag{iii}$$

where $\eta = (\mu_0\sigma)^{-1}$ is the magnetic diffusivity and σ the electrical conductivity. The first arises because the plasma is accelerated away from N by the magnetic tension

Bumba and Kleczek (eds.), Basic Mechanisms of Solar Activity, 353–366. *All Rights Reserved.*

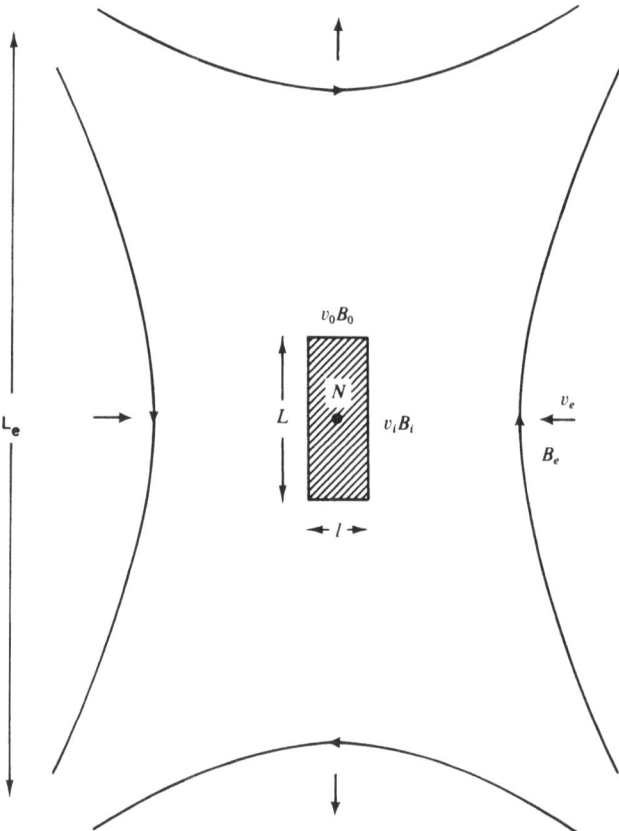

Fig. 1.   The overall magnetic configuration for the reconnection process. Oppositely directed magnetic fields of strength B_e are carried towards a magnetic neutral point with speed v_e from an 'external region' whose scale length is L_e. Surrounding N is a 'diffusion region' with dimensions l and L where the electric current density is strong. Subscripts i and o refer to inflow and outflow values at the edge of the diffusion region.

B_i^2/μ and an excess gas pressure of the same order. The second expresses the fact that flux is being carried in at the speed with which it diffuses through the plasma. The final equation results from the principle of mass conservation.

For given input values v_i, B_i, (i)–(iii) determine v_0 and the dimensions of the diffusion region. In particular one finds

$$L = v_A \eta / v_i^2 \,,$$

which must be compared with a typical overall dimension L_e for variations in the magnetic field. If

$$v_i \gg (v_A \eta / L_e)^{1/2} \,,$$

then L is much less than L_e in value and the diffusion region occupies only a small part of the flow. Recent attention has been concentrated on such a situation, which we refer to as 'fast' reconnection and which is relevant for most astrophysical and geophysical applications. In particular, the object has been to study the 'external' region which surrounds the diffusion region and to determine the maximum allowable value of the speed v_e with which magnetic flux (of strength B_e at a distance L_e) can be carried towards the neutral point and reconnected. It is necessary to distinguish between the flow speeds v_e and v_i, since, as the plasma approaches the diffusion region it may well be greatly accelerated.

In the next section we summarise critically the present fast reconnection models of the external region; the plasma is assumed perfectly conducting so the diffusion region is regarded as a point at the origin or a line through it. In Section 3 a brief account is given of an asymptotic similarity treatment which we have recently completed (Soward and Priest, 1976).

2. Previous Models of Fast Reconnection

2.1. PETSCHEK

The first to set up a qualitative mechanism for fast reconnection was Petschek (1964). He noted that, as plasma crosses a slow magnetohydrodynamic shock wave (a finite Alfvén wave in the incompressible limit), the magnetic field direction rotates towards the normal. This fact enabled him to construct a model of the external region with one shock in each quadrant of the x, y plane. The first quadrant is shown in Figure 2a, and the configuration in the rest of the x, y plane can be constructed from it by assuming symmetry about both the x- and y-axes. The wave remains stationary at the position OA while plasma and magnetic flux passes through it. In the process, magnetic energy is converted into kinetic energy and heat. Both the wave and the field lines which have passed through it are assumed to be straight.

Petschek estimates the magnetic field strength to the right of the diffusion region (considered as a straight line rather than a point) by assuming OA to coincide with the vertical axis. He then uses Equations (i)–(iii) to link the external region to the diffusion region and gives a qualitative argument for expecting the inflow speed v_e to possess a maximum value. The maximum depends weakly on the magnetic Reynolds number $R_{me} = v_{Ae}L_e/\eta$ and is typically 0.1 or 0.01 times the Alfvén speed $v_{Ae} = B_e/(\mu\rho)^{1/2}$. In view of the qualitative nature of Petschek's estimate, Roberts and Priest (1975) recently attempted to estimate the maximum inflow speed or, loosely speaking, 'reconnection rate' more quantitatively. They solve for the magnetic field in the region to the right of OA, assuming OA to be inclined at an angle α to the vertical. After matching to the diffusion region, they plot the reconnection rate v_e as a function of α and find that v_e does indeed possess a maximum value. Typical ones are 0.1 v_{Ae} for $R_{me} = 10^2$ and 0.02 v_{Ae} for $R_{me} = 10^6$.

But Petschek's mechanism was not generally accepted. For instance, Sweet and Green (1966) and Priest (1972) showed that, if the shock OA is straight, the field lines to the left of OA are bowed away from O, rather than being straight as indicated in Figure 2a. (However, Vasyliunas (1975) has recently claimed that the resulting transition from convex to concave field lines as the plasma leaves the diffusion region presents no problem.) Even more doubt on the validity of Petschek's analysis was cast by the appearance of the alternative solutions due to Sonnerup (1970) and Yeh and Axford (1970).

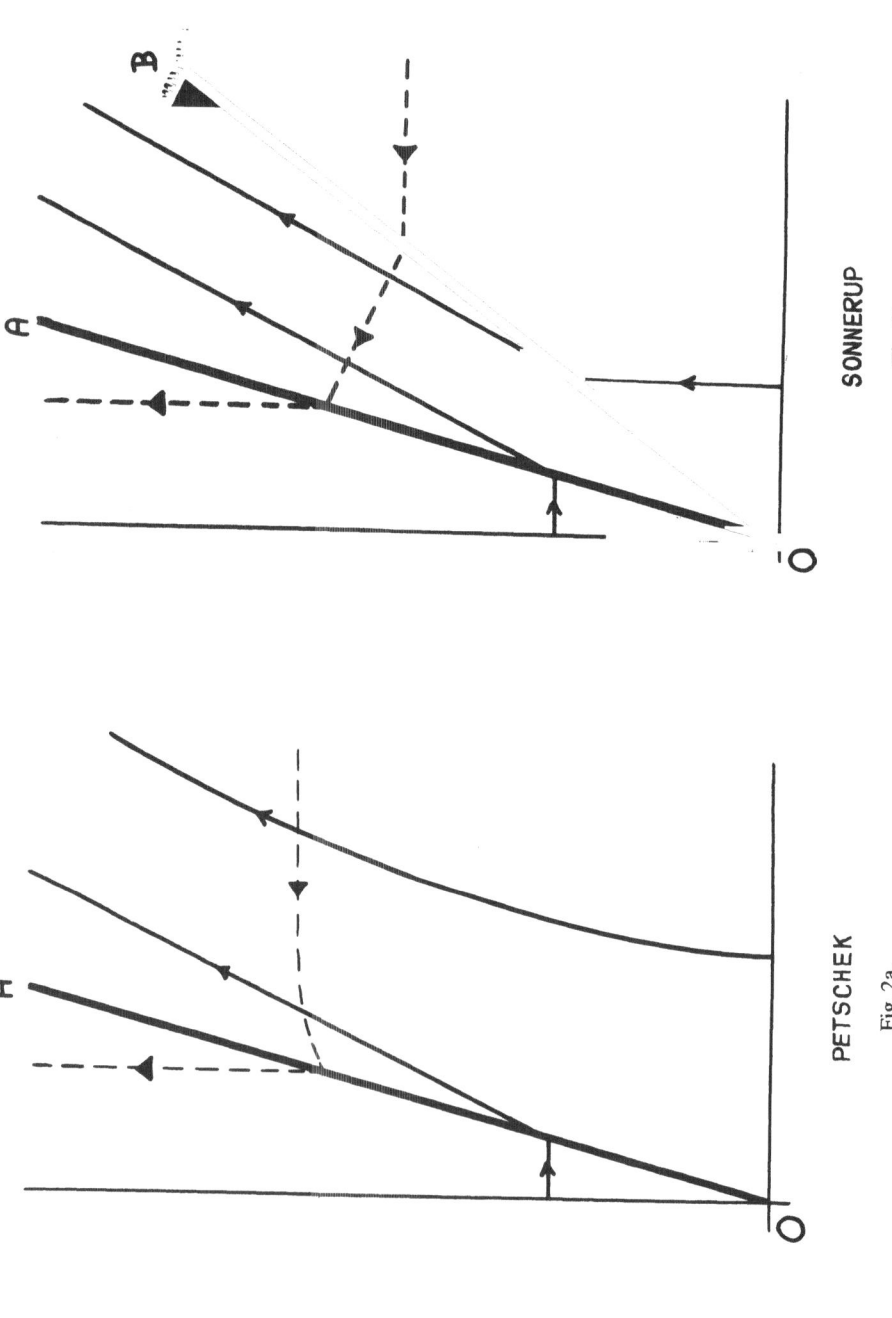

PETSCHEK

Fig. 2a.

SONNERUP

Fig. 2b.

Figs. 2a–d.   Typical magnetic field lines (———) and streamlines (– – – – –) in the first quadrant of the external region for various fast reconnection models. (a) Petschek's mechanism has one discontinuity OA. (b) Sonnerup's model contains a second discontinuity OB. (c) Yeh and Axford's model. (d) The present model. The flow passes smoothly through the Alfvén line OB, shown dotted.

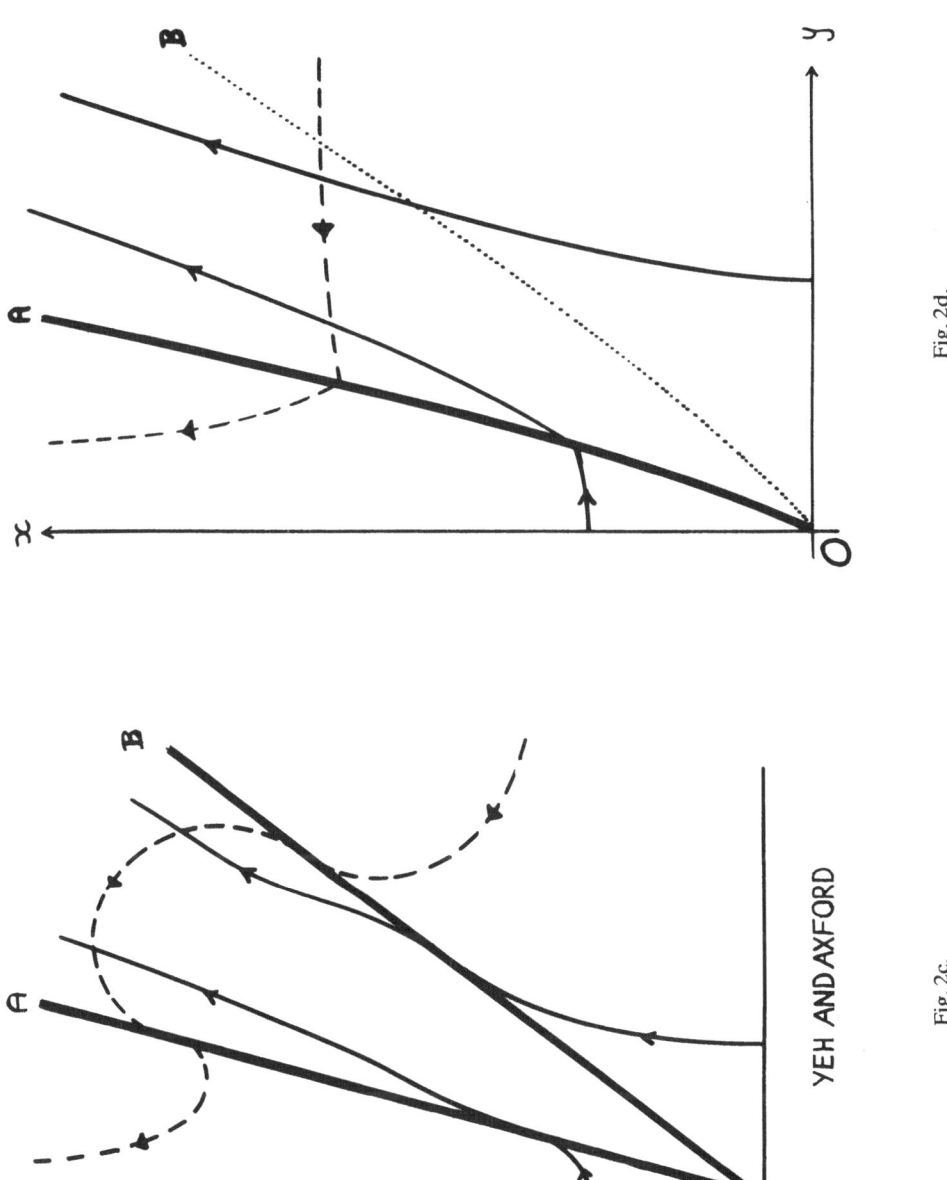

Fig. 2d.

Fig. 2c.

YEH AND AXFORD

2.2. SONNERUP

The configuration proposed by Sonnerup (1970) is shown in Figure 2b. There are now two discontinuities in each quadrant. They are, however, different in nature. In the compressible case, *OA* becomes a slow magnetohydrodynamic shock-wave, but *OB* becomes an expansion fan. Furthermore, the wave *OA* is generated at *O*, whereas *OB* is generated at the corner *B*, a fact which makes the model inapplicable in detail to astrophysical problems. In spite of this fault, the model proves to be most useful because of its simplicity, which arises from the fact that the plasma velocity and magnetic field are both uniform in each of the three regions bounded by the *x*- and *y*-axes, *OA* and *OB*. The conservation relations across *OA* and *OB* thus enable one to relate the values in the three regions analytically.

The elegance of the model has made it easy for Cowley (1974a, b) to generalize it in two ways. Firstly, he includes magnetic field and plasma velocity components normal to the plane of Figure 2b and finds that it is possible to choose the inflow and outflow values of the normal components arbitrarily and independently. Secondly, he is able to construct solutions in which the two inflowing magnetic fields differ in magnitude; for a given set of inflow parameters the solution is unique (Priest and Cowley, 1975). The generalization to include compressibility has not, however, been performed in a self consistent way; the most that has been done so far is a partially consistent treatment due to Yeh and Dryer (1973).

It may be argued that the wave *OB* represents a lumping together for mathematical convenience of the magnetohydrodynamic interaction which exists to the right of *OA* in Petschek's mechanism. But, in view of the fact that Sonnerup's model appears to yield no upper limit on the reconnection rate, by contrast with Petscheks model, such a representation is unlikely to be exact. It is therefore of importance to analyse Petschek's mechanism in more detail.

2.3. YEH AND AXFORD

Yeh and Axford (1970) adopted a different philosophy to that of Petschek for seeking the external region flow. They argue that, on some scale length intermediate between the size of the diffusion region and the distance between the magnetic field sources, there is no natural scale length and the magnetohydrodynamic variables are of self-similar form. They express the plasma velocity **v** and magnetic induction **B** in terms of a stream function ψ and vector potential A defined by

$$(v_r, v_\theta) = \left(\frac{1}{r}\frac{\partial\psi}{\partial\theta}, -\frac{\partial\psi}{\partial r}\right),$$

$$(B_r, B_\theta) = \left(\frac{1}{r}\frac{\partial A}{\partial\theta}, -\frac{\partial A}{\partial r}\right).$$

The equations to be solved for steady, two-dimensional, incompressible flow are

$$\rho(\mathbf{v}\cdot\nabla)\mathbf{v} = -\nabla(p + B^2/2\mu) + (\mathbf{B}\cdot\nabla)\mathbf{B}/\mu, \tag{1a}$$

$$\mathbf{E} + \mathbf{v}\wedge\mathbf{B} = 0, \tag{1b}$$

$$\nabla\cdot\mathbf{B} = \nabla\cdot\mathbf{v} = 0, \tag{1c}$$

where ρ and E are constants. The form for the solution which Yeh and Axford assume is

$$\psi = rg(\theta), \qquad A = rf(\theta); \tag{2}$$

in other words the components of **v** and **B** depend on θ alone. The resulting ordinary differential equations for $g(\theta)$ and $f(\theta)$ are solved subject to the boundary conditions

$$\psi = \frac{\partial A}{\partial \theta} = 0 \quad \text{on} \quad \theta = 0, \pi/2, \tag{3}$$

where, as indicated in Figures 2a and 3, we take for convenience the x-axis as vertical and measure θ from it. The conditions (3) arise from the symmetry assumptions that a streamline is traced by the x- and y-axes while the magnetic field lines intersect it normally.

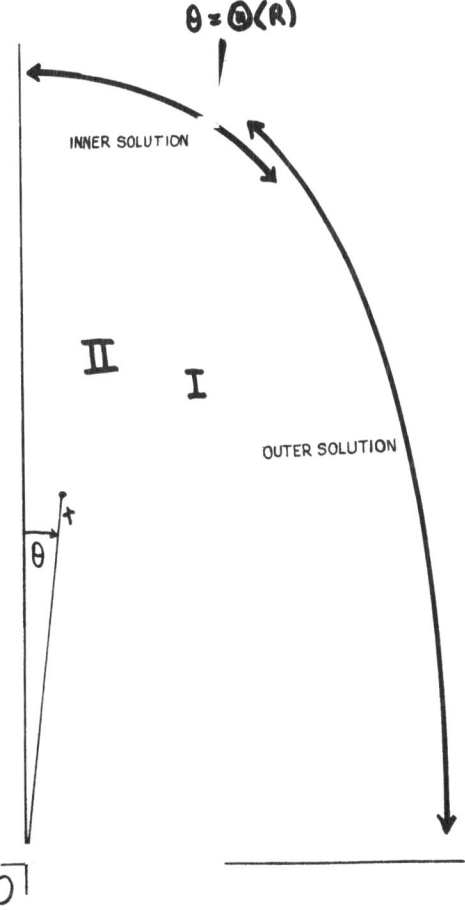

Fig. 3.   Different domains for the solution. The discontinuity OA (finite Alfvén wave) is situated at $\theta = \Theta(R)$, where $R = \log_e (r/l) + \pi/(8\,M_i)$; it separates the first quadrant into regions I and II. The outer solution is valid from $\theta = \pi/2$ down to small values of θ, whereas the inner solution applies for θ/Θ between zero and some large value.

The resulting form for the streamlines and field lines is shown in Figure 2c. They are curved, except in the special case of a uniform total pressure (gas plus magnetic), when the model reduces to that of Sonnerup. As in Sonnerup's model, there are two straight discontinuities and again the wave OB is subject to the criticism of being externally generated. More unluckily, however, the model does not work at all, because a detailed investigation of the nature of OB (Vasyliunas, 1975) shows it to be unphysical; it is not possible to construct a solution which joins thé flow on both sides except in the special case of Sonnerup's solution. The concept of looking for similarity solutions such as (2) is, however, important.

3. An Asymptotic Similarity Solution for Fast Reconnection

For reasons given in the previous section, Petschek's mechanism is the only fast reconnection model directly relevant to an astrophysical application. But a lingering doubt about it remains, because of both the ad-hoc way in which it was constructed and the semi-quantitative nature of Petschek's analysis. For these reasons, Soward and Priest (1976) have looked for two-dimensional solutions of the steady, incompressible magnetohydrodynamic equations (1) which are asymptotic in form and therefore valid at large distances from the diffusion region in the same spirit as Yeh and Axford's analysis. (The solutions, it transpires, have the added bonus of remaining valid right up to the diffusion region provided that the inflow Alfvén Mach number $M_i \equiv v_i(\mu\rho)^{1/2}/B_i$ is much less than unity.) We have asked the question "Just what is the distant external region flow for fast reconnection in practice with no externally generated discontinuities OB present?" Is it qualitatively like Petschek's picture or is there, as suggested by Coppi and Friedland (1971), no need for any discontinuities at all?

There are two curves OA, OB in each quadrant passing through the origin O (Figure 2d) along which the normal components of plasma and Alfvén velocity are the same in magnitude. It does indeed prove possible to construct solutions which are continuous across OB, but across OA the tangential components are forced to suffer discontinuities.

We first of all look for similarity solutions more general than (2), namely of the form

$$\psi = r^{1-n}g(\theta), \qquad A = r^{1+n}f(\theta),$$

where $n > 0$, so that the inertia forces are negligible as $r \to \infty$. But due to the singular behaviour of ψ and A at OB we find similar difficulties at OB to those unearthed by Vasyliunas in Yeh and Axford's case. The difficulties, however, become less severe in the limit as n approaches zero. This leads us to search instead for solutions close to (2) in form but containing a weak (logarithmic) dependence on r. Specifically, we put

$$\psi = rg(R, \theta), \qquad A = rf(R, \theta), \tag{4}$$

where

$$R = \log_e(r/L) + R_0 \tag{5}$$

and R_0 is a constant, chosen for convenience (to make $B = B_i$ when $r = l$) as $\pi/(8\, M_i) - \log_e M_i$.

Series solutions for f and g of the following form are then assumed:

$$f = R^{1/2}f_0(\theta) + R^{-1/2}(f_{11}(\theta) \log_e R + f_1(\theta)) + \dots,$$
$$g = R^{-1/2}g_0(\theta) + R^{-3/2}(g_{11}(\theta) \log_e R + g_1(\theta)) + \dots. \tag{6}$$

The largest powers of R are determined to be $+\frac{1}{2}$ and $-\frac{1}{2}$ in order to make both the product fg of order unity (see 1b) and the Lorentz force to dominate the inertial term in (1a). Having substituted into (1) we derive ordinary differential equations for the unknown functions of $\theta, f_0, f_{11}, f_1, g_0, g_{11}, g_1, \dots,$ by equating the coefficients of powers of R in (1) to zero. The equations are then solved for $0 < \theta \le \pi/2$ subject to the conditions

$$g = \partial f/\partial \theta = 0 \quad \text{at} \quad \theta = \pi/2 \tag{7}$$

which result from (3). But the resulting solutions are found not to obey the remaining boundary conditions, namely

$$g = \partial f/\partial \theta = 0 \quad \text{at} \quad \theta = 0.$$

The form (4) and (5) is therefore not valid right down to $\theta = 0$; it represents only an 'outer' solution, as indicated diagrammatically in Figure 3.

The 'inner' solution, which holds for small values of θ, is constructed by changing the variable from θ to

$$\xi = \theta/\Theta(R),$$

where

$$\Theta(R) = \Theta_o R^{-1} + (\Theta_{11} \log_e R + \Theta_1)R^{-2} + \dots$$

is, at this stage, just a typical small value of θ. We write as the inner solution

$$\psi = rG(R, \xi), \qquad A = rF(R, \xi),$$

where

$$F = R^{-1/2}F_0(\xi) + R^{-3/2}(F_{11}(\xi) \log_e R + F_1(\xi)) + \dots$$

and

$$G = R^{-1/2}G_0(\xi) + R^{-3/2}(G_{11}(\xi) \log_e R + G_1(\xi)) + \dots. \tag{8}$$

Again, having substituted for ψ and A in (1) and equated the coefficients of powers of R, we find a series of ordinary differential equations in ξ for the unknown functions $F_o, F_{11}, F_1, G_o, G_{11}, G_1 \dots$. They are solved for $0 \le \xi < \infty$, subject to the boundary conditions

$$G = \partial F/\partial \xi = 0 \quad \text{at} \quad \xi = 0. \tag{9}$$

Finally, certain free parameters in the inner and outer asymptotic expansions (6) and (8) are determined after 'matching' the expansions by means of the conditions

$$\lim_{\theta \to o} g = \lim_{\xi \to \infty} G$$

and   (10)

$$\lim_{\theta \to 0} f = \lim_{\xi \to \infty} f$$

to all orders.

The details of the above procedure for determining the inner and outer expansions are in practice far from simple and are given by Soward and Priest (1976). To lowest order we find Equation (1) for the outer region becomes

$$\frac{d^2 f_o}{d\theta^2} + f_o = 0 ,$$

$$f_o \frac{dg_o}{d\theta} - g_o \frac{df_o}{d\theta} = 1 ,$$

with solution, subject to the boundary conditions (7),

$$f_o = -a \sin \theta , \qquad g_o = a^{-1} \cos \theta ,$$ (11)

where a is a constant.

In the inner region, on the other hand, (1) reduces to

$$(G_0^2 - F_0^2) \frac{d^2 G_o}{d\xi^2} = 0 ,$$ (12)

$$F_o \frac{dG_o}{d\xi} - G_o \frac{dF_o}{d\xi} = \Theta_o ,$$ (13)

which are to be solved for $\xi \geq 0$ subject to the boundary conditions

$$G_o = dF_o/d\xi = 0 \quad \text{at} \quad \xi = 0 ,$$

resulting from (9). Near $\xi = 0$, the solution of (12) is

$$G_o = \text{const} \times \xi$$

but it cannot hold for all values of ξ since it is not possible to match it to the solution (11) in the sense of (10). It can be seen that the most general continuous solution of (12) is a linear function of ξ with changes in gradient at the places where $G_o = F_o$ and $G_o = -F_o$, which correspond to the Alfvén curves OA and OB, respectively. Without loss of generality, we assume the curve $\xi = 1$ to be situated where $G_o = F_o$, so that, as indicated in Figure 3, OA is given by $\theta = \Theta(R)$. Furthermore, we suppose that F_o and G_o do suffer changes in gradient at $\xi = 1$ but not at the curve where $G_o = -F_o$, since a discontinuity at OA is generated from O and so acceptable but one at OB is generated from some external point B and so is not acceptable on physical grounds. The resulting lowest order inner region solution of (12) and (13), which is capable of being matched to (11), is then

$$F_o = \sqrt{\Theta_o}, \qquad G_o = \sqrt{\Theta_o}\,\xi \quad \text{in} \quad \text{II} \ (0 \le \xi < 1),$$
$$F_o = \sqrt{\Theta_o}(2 - \xi), \quad G_o = \sqrt{\Theta_o} \quad \text{in} \quad \text{I} \ (\xi > 1).$$

Finally, the matching condition (10) implies that

$$a = \Theta_o^{-1/2}.$$

The next order terms in the asymptotic expansions (6) and (8) are found in a straightforward manner and the matching condition leads to a value for Θ_o of $\pi/8$, so that to lowest order the discontinuity OA is located at

$$\theta = \Theta \equiv \pi/(8\ R).$$

At this order, however, a weak singularity in the behaviour of $F_1(\xi)$ and $G_1(\xi)$ near $\xi = 1$ appears. In order to find a continuous transition through OA, it proves necessary to determine the inner and outer expansions to the next highest order (to terms of order $R^{-5/2}$, for instance, in F and G) and to take account of finite conductivity effects near OA. The reader is referred to Soward and Priest (1976) for the details. It suffices here to state that having written $B' = BB_i^{-1}M_i^{-1/2}$, $v' = vv_i^{-1}M_i^{1/2}$ we find that the lowest order contributions to each of the components of plasma velocity and magnetic induction are

$$\left. \begin{array}{ll} B'_x = -(8\ R/\pi)^{1/2}, & B'_y = (\pi R/2)^{-1/2}(\pi/2 - \theta), \\[2mm] v'_x = -(\pi/32)^{1/2}R^{-3/2}(\pi/2 - \theta), & v'_y = -(8\ R/\pi)^{-1/2}, \end{array} \right\} \text{in I} \qquad (14)$$

and

$$\left. \begin{array}{l} B'_x = -(\pi R/2)^{-1/2}\log_e\left[(1+\xi)/(1-\xi)\right], \ B'_y = -(8\ R/\pi)^{-1/2}, \\[2mm] v'_x = (8\ R/\pi)^{1/2}, \qquad v'_y = -(\pi/32)^{1/2} \\[2mm] \qquad \times R^{-3/2}\{\log_e\left[(1+\xi)/(1-\xi)\right] - \xi\}. \end{array} \right\} \text{in II}$$

The dominant components are thus B_x, v_y in I and v_x in II. The form for the resulting field lines and streamlines is shown in Figure 2d, where the curve OB, through which there is a continuous transition, is shown dotted. The position of OB has been exaggerated; both OB and OA are in fact situated close to Ox in the inner region. Also it should be noted that the curvature of the magnetic field lines in II allays the fears of Green and Sweet (1966) and Priest (1972) that Petschek's mechanism would not work.

Now, from (14), the inflow speed on the y-axis a distance L_e from the origin is

$$v_e = v_i(8\ R_e M_i/\pi)^{-1/2} \qquad (15)$$

where, according to (5), the value of R at $r = L_e$ is

$$R_e = \pi/(8\ M_i) + \log_e\left[L_e/(LM_i)\right]. \qquad (16)$$

Further, the inflow and outflow speeds v_i and v_o, indicated on Figure 1, are related to v_e by

$$v_i B_i = v_o B_o = v_e B_e \qquad (17)$$

due to the constancy of the electric field. Equations (i), (ii), (iii), (15), (16), (17) may therefore be combined to yield the following relation between the Alfvén Mach numbers $M_i \equiv v_i(\mu\rho)^{1/2}/B_i$ and $M_e \equiv v_e(\mu\rho)^{1/2}/B_e$ on the y-axes at distances l and L_e, respectively, from the neutral point:

$$\frac{\pi}{8\,M_e} = \frac{\pi}{8\,M_i} + \tfrac{1}{2}\log_e (R_{me}^2 M_i M_e),\tag{18}$$

where $R_{me} = L_e v_{Ae}/\eta$ is the magnetic Reynolds number based on the external conditions and $v_{Ae} \equiv B_e(\mu\rho)^{-1/2}$ is the Alfvén speed. Equation (18) is strictly valid only where $M_i \ll 1$, since, otherwise, the external region solution does not hold right down to the diffusion region. Nevertheless, we have sketched the solution in Figure 4

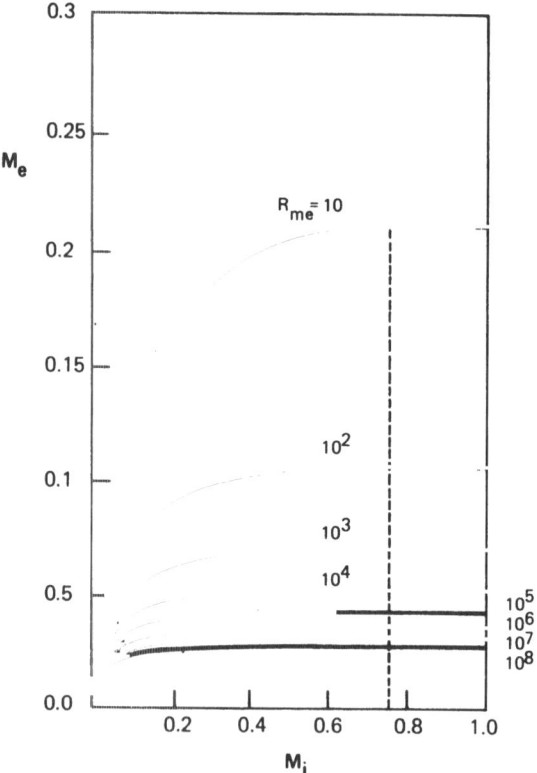

Fig. 4. The external Alfvén Mach number M_e is plotted against the inflow Alfvén Mach number M_i for various values of the magnetic Reynolds number $R_{me} = v_{Ae}L_e/\eta$. The maximum values of M_e, namely $M_{e\,\mathrm{max}}$, are located at the intersection of the curves with the broken line.

for values of M_i right up to unity. M_e, plotted as a function of M_i alone, possesses a maximum value $M_{e\,\mathrm{max}}$, which occurs at $M_i = \pi/4$, and which is sketched as a function of the magnetic Reynolds number in Figure 5. $M_{e\,\mathrm{max}}$ decreases from 0.2, when $R_{me} = 10$ to 0.02 when $R_{me} = 10^6$ and, as can be seen from the figure, is somewhat less than the value derived from Petschek's original analysis.

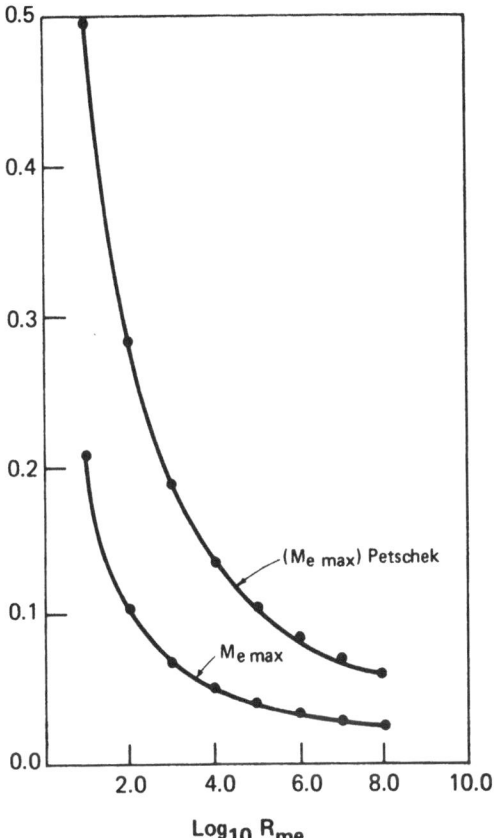

Fig. 5. $M_{e\,max}$ from our analysis as a function of R_{me}. For comparison the corresponding value from Petschek's simpler analysis is also shown.

In conclusion, Petschek's mechanism has been shown to be the only workable model for fast reconnection in practice. It now has a sound mathematical basis. Reconnection can occur at any rate up to typically a tenth or a hundredth of the Alfvén speed, depending on the magnetic Reynolds number. By contrast with Petschek's original analysis, the shocks, magnetic field lines and streamlines are all in general curved in the manner indicated by Figure 2d.

References

Coppi, B. and Friedland, A. B.: 1971, *Astrophys. J.* **169**, 379.
Cowley, S. W. H.: 1974a, *J. Plasma Phys.* **12**, 319.
Cowley, S. W. H.: 1974b, *J. Plasma Phys.* **12**, 341.
Dungey, J. W.: 1953, *Phil. Mag.* **44**, 725.
Green, R. M. and Sweet, P. A.: 1966, *Astrophys. J.* **147**, 1153.
Parker, E. N.: 1963, *Astrophys. J. Suppl. Ser.* **77**, 8, 177.
Petschek, H. E.: 1964, *AAS-NASA Symposium on the Physics of Solar Flares* (ed. W. N. Hess), NASA SP-50, p. 425.

Priest, E. R.: 1972, *Monthly Notices Roy. Astron. Soc.* **159**, 389.
Priest, E. R. and Cowley, S. W. H.: 1975, *J. Plasma Phys.* **14**, 271.
Roberts, B. and Priest, E. R.: 1975, *J. Plasma Phys.* **14**, 417.
Sonnerup, B. U. O.: 1970, *J. Plasma Phys.* **4**, 161.
Soward, A. M. and Priest, E. R.: 1976, submitted for publication.
Sweet, P. A.: 1958, *Nuovo Cimento Suppl.* **8**, Ser. X, 188.
Vasyliunas, V. M.: 1975, *Rev. Geophys. Space Phys.* **13**, 303.
Yeh, T. and Axford, W. I.: 1970, *J. Plasma Phys.* **4**, 207.
Yeh, T. and Dryer, M.: 1973, *Astrophys. J.* **182**, 301.

DISCUSSION

Gilman: How is your magnetic Reynolds number defined?

Priest: $R_{me} = v_{Ae}L_e\mu_0\sigma$ (see text), but one could just as easily express the results in terms of a magnetic Reynolds number based on v_e rather than v_{Ae}.

Smith: It seems to me that your results will be applicable to crude models of a solar flare only if the conductivity σ is taken as a turbulent conductivity so that R_m is reduced from 10^6, a number typical of the nonturbulent corona.

Priest: The reconnection model works whether you take a Coulomb or a turbulent value for σ. For a flare, there is probably an initial stage with the Coulomb value, which is then triggered to a main phase with the turbulent value. In each case (and for any astrophysical application), one needs to determine whether some additional consideration gives rise to an upper limit on M_e which is less than $M_{e\,max}$. For instance, one may require that the width l of the diffusion region be greater than a collision mean-free path or an ion gyroradius. Whether this is a stringent limitation or not in practice depends crucially on how high in the solar atmosphere the reconnection is taking place.

DYNAMO THEORY AND THE SOLAR CYCLE

M. STIX*

High Altitude Observatory, National Center for Atmospheric Research,† Boulder, Colo. 80303, U.S.A.

Abstract. In this paper solutions of the mean field induction equation in a spherical geometry are discussed. In particular, the 22-year solar magnetic cycle is considered to be governed by an axisymmetric, periodic solution which is antisymmetric with respect to the equatorial plane. This solution essentially describes flux tubes travelling as waves from mid-latitudes towards the equator. In a layer of infinite extent the period of such dynamo waves solely depends on the strength of the two induction effects, differential rotation and α-effect (cyclonic turbulence). In a spherical shell, however, mean flux must be destroyed by turbulent diffusion, so the latter process might actually control the time scale of the solar cycle.

A special discussion is devoted to the question of whether the angular velocity *increases* with increasing depth, as the dynamo waves seem to require, or whether it *decreases*, as many theoretical models concerned with the Sun's differential rotation predict. Finally, theories for the sector structure of the large scale photospheric field are reviewed. These describe magnetic sectors as a consequence of the sectoral pattern in the underlying large scale convection, as non-axisymmetric solutions of the mean field induction equation, or as hydromagnetic waves, modified by rotational effects.

1. Introduction

The subject of 'mean field electrodynamics', which was discussed during this morning's session, has led us to a mean-field induction equation. In the present lecture I intend to demonstrate and discuss solutions of this mean-field induction equation. In a spherical geometry, in particular, there is an axisymmetric solution, which is periodic in time and antisymmetric with respect to the equatorial plane; the cyclic behaviour of the large scale magnetic field of the Sun usually is ascribed to this solution. I shall discuss mean fields of this type in Section 2. Before doing so, however, let us recall some of the conditions which should be satisfied for the mean field induction equation to be a good approximation.

One of these conditions has to do with the neglect of all terms in the induction equation for the *fluctuating* part of the magnetic field which contain products of fluctuations. As compared to the linear terms, these second order terms are small if either the magnetic Reynolds number is small or if the fluctuating velocity is 'slow' in the sense that

$$v\tau \ll l, \tag{1}$$

where v, τ and l are rms velocity, correlation time and correlation length of the turbulent convection. Both these conditions are *not* satisfied in the solar convection zone. It is usually assumed that $v\tau \approx l$ (and observations of granules and supergranules confirm this assumption); and the magnetic Reynolds number is very large indeed. As shown in Figure 1, it increases from $\approx 10^3$ in the photosphere to $\approx 10^{10}$ at a depth of 10^5 km. Now the magnetic Reynolds number is the ratio of the free decay time, $\mu\sigma l^2$, of a magnetic field fluctuation to the lifetime of the convective eddy which

---

* On leave from Universitäts-Sternwarte, Göttingen, Germany.
† The National Center for Atmospheric Research is sponsored by the National Science Foundation.

Bumba and Kleczek (eds.), Basic Mechanisms of Solar Activity, 367–388. All Rights Reserved.
Copyright © 1976 by the IAU.

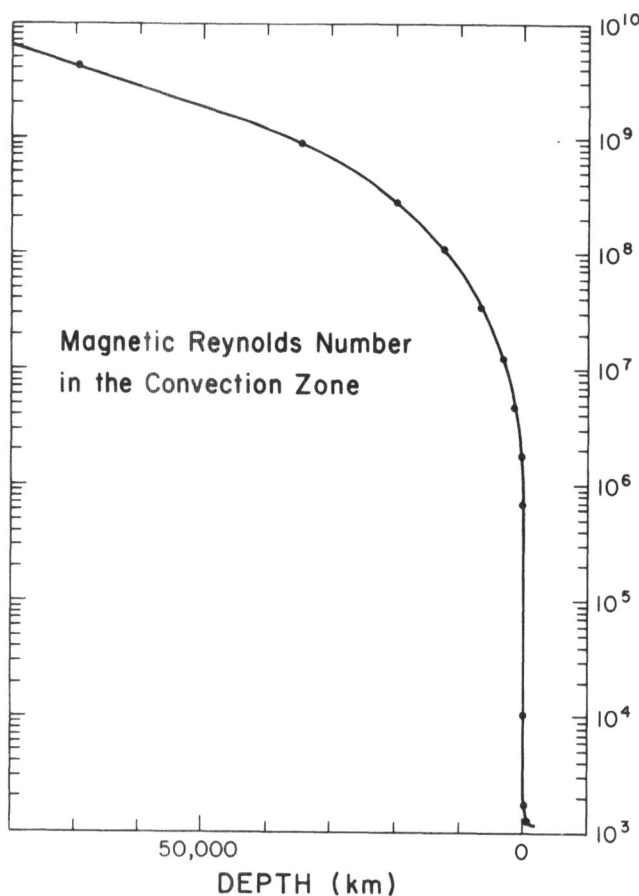

Fig. 1.   The magnetic Reynolds number, $\mu\sigma v l$, as a function of depth, z, in the solar convection zone. The conductivity, σ, is taken from Kopecký and Soytürk (1971) for the four values closest to the surface, and computed from $\sigma = 0.003 \times T^{3/2}$ mho m^{-1} (Spitzer, 1962; Piddington, 1975), where the temperature, T, is taken from the table of Baker and Temesváry (1966). The convection velocity, v, is taken from the same table, and $l = z/2$ is assumed. For the three photospheric values I used $v = 1$ km s^{-1} and $l = 100$ km.

created that fluctuation. Since this ratio is large, more magnetic field fluctuations will be created before the old ones can decay, so that, in an equilibrium state, the rms magnetic field will be large as compared to the mean field. It therefore appears that in the solar case the above-mentioned second-order terms should *not* be neglected. I think that this is essentially the point where the criticism of Piddington (1971, 1972, 1973, 1975) of the solar dynamo theory becomes relevant.

How is the 'mean field' defined? In the mean field electrodynamics, as it is developed by Steenbeck *et al.* (1966), Krause (1968), Rädler (1968a, b) and Krause and Rädler (1971), mean values originally are understood as ensemble averages. As the authors point out, they may be replaced by averages over space or time. Such averages are of course required for comparison of observations with predictions of the theory. But what is an appropriate volume or an appropriate time span of integration? On the Sun, there is no length scale which is large compared to *all* convection cells but still small compared to the solar radius. This problem appears to be particularly intriguing since it is the *largest* convection cells which contribute most to the α-effect which is so essential for the regeneration of the mean field (Stix, 1974).

After these somewhat pessimistic introductory remarks concerning but two of many problems related to the solar dynamo (see e.g. Stix, 1974) I shall nevertheless proceed to illustrate the properties of the mean-field induction equation. The capability of this equation to explain the solar cycle will be demonstrated in the following section, and some problems related to special forms of differential rotation will be discussed in Section 3. Section 4 will contain some remarks concerning the sector structure of the large scale solar field. Whenever it is convenient, I shall comment on the above-mentioned problems and, at least in some cases, speculate how they might eventually be solved.

2. The Axisymmetric Periodic Mean Field

The induction equation for the mean magnetic field, **B** (with the overbar omitted since only mean fields will be discussed), was first, in a simplified form, given by Parker (1955), and subsequently derived under more general circumstances by Steenbeck *et al.* (1966), Krause (1968), Rädler (1968a, b; 1969), Iroshnikov (1970a), Parker (1970), Moffat (1970a, b), Yoshimura (1972), Gubbins (1974a), and Deinzer and Stix (1975). In its most often used form it is

$$\frac{\partial \mathbf{B}}{\partial t} = \text{curl } (\mathbf{v} \times \mathbf{B} + \alpha \mathbf{B}) - \eta \text{ curl curl } \mathbf{B} , \tag{2}$$

where **v** is the mean velocity field, $\alpha\mathbf{B}$ the additional electric field caused by the helicity of the fluctuating velocity field, and η the turbulent electro-magnetic diffusivity. The 'semi-empirical' equation derived by Leighton (1969) can be written essentially in the same form (Yoshimura, 1972; Stix, 1974). The large magnetic Reynolds number dynamos of Braginskiĭ (1964; see also Soward, 1971, 1972; Gubbins, 1973a) also lead to equations of the form (2).

Let us suppose that the mean velocity is a pure rotation, and is *symmetric* with respect to the equatorial plane, i.e., in spherical polar co-ordinates (r, θ, ϕ),

$$\mathbf{v} = (0, 0, r\omega(r, \theta) \sin \theta) , \tag{3}$$

where $\omega(r, \pi - \theta) = \omega(r, \theta)$. Let us further suppose that $\alpha(r, \theta)$ is *antisymmetric* with respect to the equatorial plane, i.e. $\alpha(r, \pi - \theta) = -\alpha(r, \theta)$. This is plausible since the helicity of the convection, which causes the α-effect and is itself caused by the Coriolis force, is also antisymmetric; and all explicit expressions for α indeed have this antisymmetry. Whenever the induction effects have these symmetries the solutions of Equation (2) can be divided into two uncoupled sets of solutions, namely the 'antisymmetric' (dipole-type, odd) and the 'symmetric' (quadrupole-type, even) sets. In general, both sets contain steady and time-dependent modes, which can be axially symmetric or ϕ-dependent. Only the axisymmetric modes will be considered in this section.

In order to proceed further I introduce the two magnetic Reynolds numbers

$$R_\alpha = \frac{\alpha_0 r_\odot}{\eta}, \qquad R_\omega = \frac{\Delta\omega r_\odot^2}{\eta}; \tag{4}$$

α_0 is a typical value of α in the northern hemisphere, $r_\odot$ is the radius of the Sun and $\Delta\omega$ is a characteristic value of differences in the angular velocity. The condition

$$|R_\alpha| \ll |R_\omega| \tag{5}$$

then defines the so-called $\alpha\omega$-dynamo, where, in the ϕ-component of Equation (2), the α-effect can be neglected in comparison to the effect of differential rotation. The ratio of the poloidal and toroidal field components is then $\sim |R_\alpha/R_\omega|^{1/2}$ (Steenbeck and Krause, 1969). In the solar case we have $B_{pol}/B_{tor} \approx 0.1$ (Yoshimura, 1975c)*, and may therefore use the limit (5). In this limit, the frequencies and growth rates of the magnetic field solutions depend on the product, P, of the two magnetic Reynolds numbers defined by Equations (4); P is called the *dynamo number*.

In order to find solutions of Equation (2) in a spherical geometry a number of numerical models have been described in recent years. In these, Equation (2) is either reduced to an eigenvalue problem (Steenbeck and Krause, 1969; Deinzer and Stix, 1971; Roberts and Stix, 1972; Levy, 1972; Köhler, 1973; Stix, 1973; Deinzer *et al.*, 1974) or it is solved as an initial value problem (Leighton, 1969; Jepps, 1975; Yoshimura, 1975a). Although different functions $\omega(r, \theta)$ and $\alpha(r, \theta)$ have been employed in these models, some general properties are common to them. These are: *If $\alpha \cdot \partial\omega/\partial r$ is negative in the northern and positive in the southern hemisphere (i.e. $P < 0$) the most preferred mode has dipole-type symmetry and is oscillatory.* With most preferred I mean that the growth rate of the mode becomes positive at the smallest $|P|$. Moreover, *for the same sign of $\alpha \cdot \partial\omega/\partial r$ the oscillatory modes of both symmetries, dipolar and quadrupolar, travel from higher latitudes toward the equatorial plane. For the opposite sign of $\alpha \cdot \partial\omega/\partial r$ the travel direction is reversed and the quadrupolar oscillatory mode is preferred.* Exceptions to the first of these rules occur sometimes, in particular if the spatial distribution of the induction effects is such that high order harmonics play only a minor role in the magnetic field. The preferred modes would then be steady and quadrupolar for $P < 0$ and steady and dipolar for $P > 0$ (Stix, 1973; Deinzer *et al.*, 1974). However, these cases have a very large spatial separation of the inducing effects and are thus very probably irrelevant to the Sun where the dynamo is believed to be confined to the convection zone (see below, in part. Figure 5). Another possibility to make steady modes preferred is by means of a sufficiently strong large scale meridional circulation (Roberts, 1972), but this is not observed on the Sun.

The migration of the oscillatory magnetic fields can be understood in terms of a propagating wave ('dynamo wave', 'Parker wave'). This becomes particularly clear if local cartesian co-ordinates are introduced. In this way the latitude migration of the mean solar field was first explained by Parker (1955). More recently, Yoshimura (1975b) pointed out that *the dynamo waves generally propagate along the surfaces of*

---

* This value is an upper limit; higher spatial resolution of the magnetograph would probably lead to a smaller ratio.

isorotation. The direction of propagation is given by the vector

$$\alpha \cdot \nabla\omega \times \mathbf{e}_\phi, \tag{6}$$

where $\mathbf{e}_\phi$ is the unit vector in the azimuthal direction. Thus, if $\alpha > 0$ in the northern hemisphere and the angular velocity increases with increasing depth, the migration is equatorwards, in agreement with the numerical results mentioned before.

A schematic illustration of an antisymmetric oscillatory $\alpha\omega$-dynamo is given in Figure 2. Consider an antisymmetric toroidal mean field as shown in Figure 2a. The α-effect, with $\alpha_{north} > 0$ and $\alpha_{south} < 0$, will cause a symmetric toroidal mean current, $\sigma\alpha\mathbf{B}$, and an associated poloidal field as indicated in Figure 2b. This poloidal field is subject to the differential rotation. For $\partial\omega/\partial r < 0$ the toroidal field pattern shown in Figure 2c will result. Add this to the original field of Figure 2a; the result is an enhancement of the original toroidal flux tubes on their equatorial side, and a destruction on their polar side, with the net effect that the tubes are moved closer to the equatorial plane (Figure 2d). At the same time weak flux tubes of opposite polarity are formed at higher latitudes, indicating the advent of the next cycle. The argument is now repeated: symmetric currents are again caused by the α-effect and again they generate a poloidal field (Figure 2e). From this, differential rotation winds up the toroidal field shown in Figure 2f, which resembles the original one, but is reversed in sign.

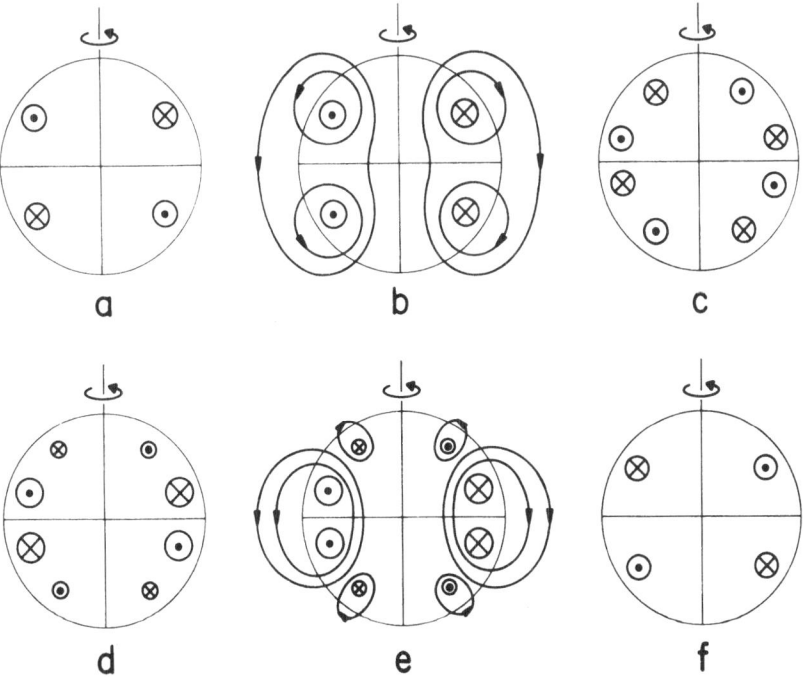

Fig. 2. Oscillatory dynamo action, with $\alpha_{north} > 0$, $\alpha_{south} < 0$, and $\partial\omega/\partial r < 0$. The toroidal field (a), together with the α-effect, sets up toroidal currents and so generates a poloidal field (b). This is wound up by differential rotation (c), with the result that the original field is moved equatorwards (d). Again, toroidal currents and associated poloidal fields are caused by the α-effect (e), so that differential rotation finally can reverse the field (f). An encircled dot (cross) indicates a vector pointing out of (into) the plane of the figure.

The scheme outlined in Figure 2 differs from the schemes described by Steenbeck and Krause (1969, Figure 2) and Krause and Rädler (1971a, Figure 4). There the field reversal depended on a phase lag caused by diffusion and a spatial separation of the two induction effects, α-effect and differential rotation. The numerical models show however that such a spatial separation is not necessary in order to obtain oscillatory modes (Roberts, 1972). On the contrary, as already mentioned, it seems to favour the steady modes (Deinzer *et al.*, 1974).

It is illustrative to show oscillatory magnetic fields in form of a movie, and 12 frames of such a movie are shown in Figure 3. The field is a numerical solution of the

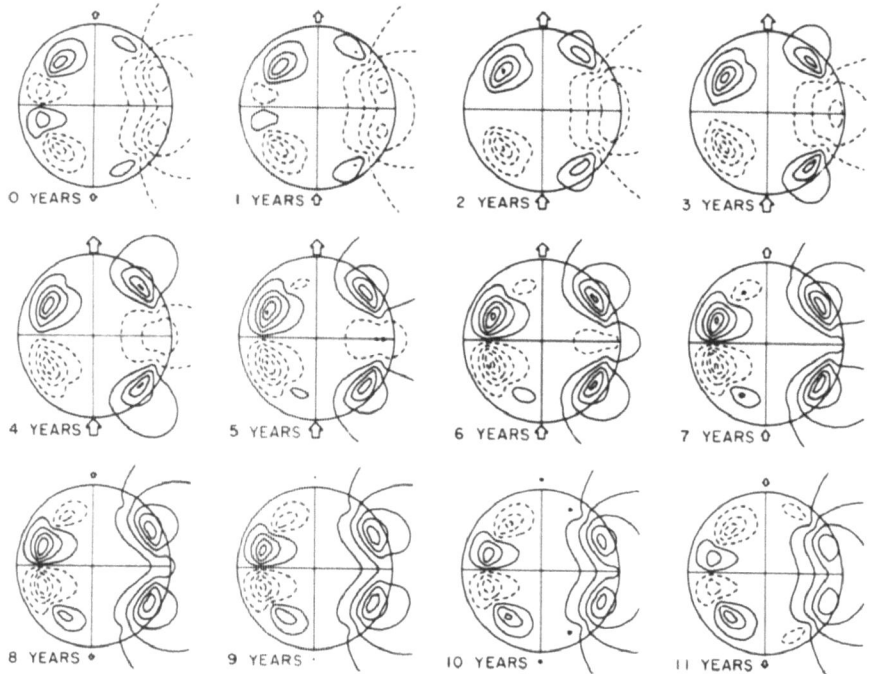

Fig. 3. A numerical solution of an antisymmetric, oscillatory $\alpha\omega$-dynamo. The shear is radial, with $\partial\omega/\partial r < 0$, and $\alpha \sim \cos\theta$. Each frame is a meridional cross-section through the solar model. The contours of constant toroidal field strength are on the left and the poloidal lines of force are on the right. The levels of the curves are ± 0.1, ± 0.3, ... ± 0.9 times the maximum values (over the whole cycle) of the toroidal field and the poloidal flux function respectively. Positive values (solid curves) indicate toroidal fields pointing out of the figure, and clockwise poloidal field lines; negative values (dahsed curves) indicate the opposite. The direction and magnitude of the field at the poles is indicated by vertical arrows. The time scale is adjusted so that 11 years cover one half-cycle.

model described by Deinzer and Stix (1971). I would like to emphasize two features of this solution. First, the toroidal mean field, which is shown on the lefthand side of each frame in the form of isogauss contours, migrates towards the equatorial plane, and two flux tubes are visible during most of the time interval shown. This, of course, means that two consecutive cycles overlap. How much of this effect would be visible on the solar surface in form of overlapping butterfly wings cannot however be predicted by this model; the process of field eruption depends on the magnitude of the field which cannot be obtained from a linear theory. The second feature to which I would like to draw your attention is the harmonic structure of the field, in particular of the poloidal field which is shown on the righthand side of each frame in form of its lines of force. The octupole harmonic is clearly discernible during the entire cycle so that the mean poloidal field never looks like a pure dipole. Moreover, the poloidal field, at least at low latitude, participates in the *equatorward* migration of the toroidal field, as it should do according to the scheme of Figure 2. The harmonic structure of the field seems to be essential for the oscillatory (or 'migratory') nature of the $\alpha\omega$-dynamo. This is confirmed by a model of Steenbeck and Krause (1966) which had steady solutions when the field expansion was truncated after the first poloidal and toroidal harmonics, but yielded oscillatory fields without such a truncation (Krause and Rädler, 1971b; Roberts, 1972). For the field shown in Figure 3 6 poloidal and as many toroidal harmonics were retained, and no changes occurred at higher truncation levels.

There is observational evidence that the solar mean poloidal field does migrate towards the equator. Figure 4a shows contours of the radial field component in a

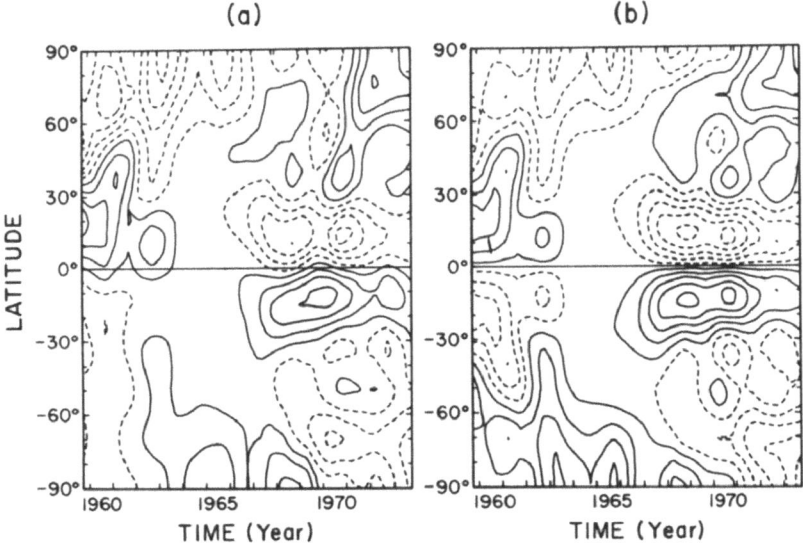

Fig. 4. Contours of the radial magnetic field component, 1959–1973. The total field (a) and the antisymmetric part (b) are shown. The contour levels are ± 0.1, ± 0.3, ... ± 0.9 times the maximum values, which are 1.65 (a) and 1.19 (b) G. Solid curves indicate positive levels, dashed curves indicate negative levels. The smoothing constant in time is 1 yr, and the smoothing in latitude is determined by the truncation of the expansion (7) after the 9th harmonic.

time-latitude diagram, computed according to

$$B_r(t, \theta) = \sum_{n=1}^{9} (n+1) g_n^0(t) P_n(\theta),$$ (7)

where $P_n(\theta)$ are zonal surface harmonics and the g_n^0 are zonal expansion coefficients.

The latter were obtained by Altschuler *et al.* (1974) from the photospheric magnetic field data, measured by the Mt. Wilson magnetograph. Such a 'butterfly diagram' of the poloidal field was also computed by Stenflo (1972), but his data extended only through 1969. For this reason and because of too little smoothing in time his diagram did not show the features relevant to the solar cycle as clearly as Figure 4a. These features are even more pronounced in Figure 4b where only the odd harmonics of the sum (7) were retained; this figure is antisymmetric and therefore the exact observational counterpart of the antisymmetric theoretical poloidal field, as shown on the right of the frames of Figure 3. The butterfly wings in the latitude range $\pm 35°$ during 1965–1973 almost exactly coincide with the wings of the classical sunspot butterfly diagram of cycle 20 (Yoshimura, 1975c), which represents the subsurface toroidal field. As shown in Figure 3, this coincidence is also present in the theoretical model.

I would like to add here a remark concerning the averaging problem mentioned in the introduction: At least as long as we are not interested in ϕ-dependent mean fields we may simply average over all longitudes and over a sufficiently long period of time; the time smoothing constant in Figure 4 is 1 yr.

Several features of the solar cycle are not well resembled by the field shown in Figure 3. For example, the field penetrates too far into the interior of the Sun. This is, of course, a consequence of the constant magnetic diffusivity used in the model. Since this diffusivity is the *turbulent* diffusivity, it should be replaced by a much smaller value in the radiative core. The skin effect would then prevent the oscillatory field from entering the core. A model of this type was computed by Roberts and Stix (1972), and Figure 5 shows four phases of the resulting field. The field expulsion from the solar core can alternatively be simulated by an inner boundary condition, as in the models of Köhler (1973) and Yoshimura (1975a).

Another difference of the field shown in Figure 3 and the real solar cycle is that the field is, on the average, located at too high latitudes. Also, the poloidal field at high latitudes virtually does not migrate towards the poles, as it should do in order to resemble the observed poloidal field (Figure 4a) and to reproduce the poleward rush of the high latitude prominence zones. Both these defficiencies do not however occur in more sophisticated models. For example, a concentration of the α-effect and shear regions to lower lattitudes brings the butterfly wings closer to the equator and at the same time allows the polar field to diffuse freely towards the poles (Stix, 1974; Yoshimura 1975a). Figure 6 shows how the neutral line of the radial field takes longer to drift towards the poles as the shear is more concentrated to low latitudes. This result is obtained from the model of Köhler (1973; see also Stix, 1974); the radial shear is proportional to $\sin^n\theta$. A very similar result was found earlier by Leighton (1969, Figure 4).

Let us finally discuss the most important question concerning the oscillatory $\alpha\omega$-dynamo, namely the question how the period is defined. According to the scheme outlined in Figure 2 non-uniform rotation and α-effect are responsible for the destruction and amplification of mean toroidal flux tubes. The strength of these two induction effects should therefore decide how long it would take these flux tubes to travel from midlatitudes toward the equator. And, indeed, the dynamo waves discussed in rectangular co-ordinates by Parker (1955), Iroshnikov (1970b), and Yoshimura (1975b) have periods which depend only on this strength (and the wave

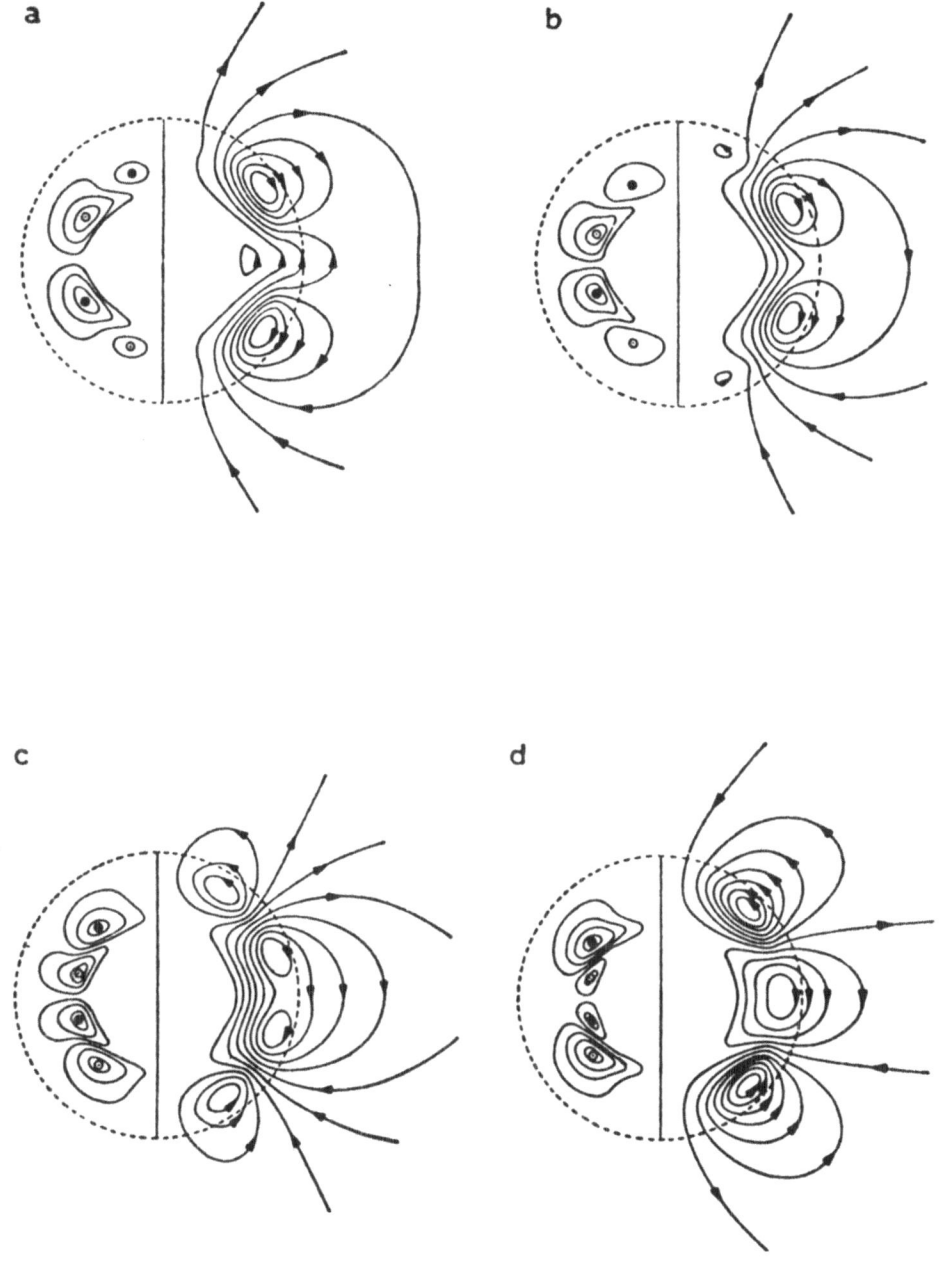

Fig. 5.   Field expulsion from the radiative solar interior. The diffusivity decreases from its turbulent value in the convection zone to the very small molecular diffusivity in the core, where the field cannot penetrate due to the skin effect (after Roberts and Stix, 1972).

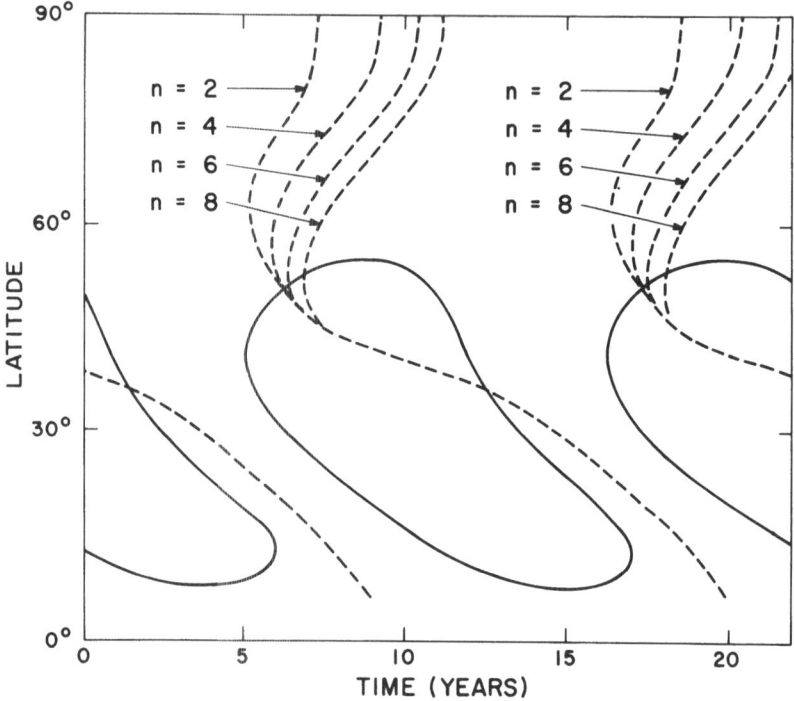

Fig. 6. The poleward drift of the neutral line of the radial field component (dashed curves) and the 50% level of the toroidal fields (solid curve). The latter is drawn only for the case $n = 6$ in order to avoid confusion. The radial shear is $\sim \sin^n\theta$.

number, k). The frequency is

$$\Omega = \left(\frac{|kH\Gamma|}{2}\right)^{1/2}, \tag{8}$$

(Parker, 1955), where H is the shear and $\Gamma \ (= \alpha)$ measures the strength of the cyclones causing the α-effect. Using $k = 2/r_\odot$ (i.e. one flux tube in each hemisphere) and $H \approx r_\odot \, \partial\omega/\partial r$ we see that (8) is equivalent to

$$\Omega = \left|\alpha\frac{\partial\omega}{\partial r}\right|^{1/2} \tag{9}$$

in a spherical geometry with radial shear. Unfortunately, it is difficult to estimate the values of α and $\partial\omega/\partial r$ in the solar convection zone. If we assume that the total variation of the angular velocity in depth is about the same as its total variation in latitude, and if we assume 2×10^8 m as the depth of the convection zone we obtain $|\partial\omega/\partial r| \approx 5 \times 10^{-15} \ \mathrm{m}^{-1} \, \mathrm{s}^{-1}$. For the α-effect Steenbeck and Krause (1969) made an estimate of $0.26 \ \mathrm{m \, s}^{-1}$; the values of Yoshimura (1972) and Leighton (1969) are 0.5 and $0.7 \ \mathrm{m \, s}^{-1}$ respectively (see Yoshimura, 1972). With $\alpha = 0.5 \ \mathrm{m \, s}^{-1}$ we obtain a frequency of $5 \times 10^{-8} \ \mathrm{s}^{-1}$ which is somewhat larger than the frequency of the real solar cycle $(9 \times 10^{-9} \ \mathrm{s}^{-1})$. The frequency becomes still larger if we directly apply Krause's (1968) formula for α and data obtained from the mixing-length theory

(Köhler, 1973), even if we correct for the effect of rotation angles larger than $\pi/2$ which occur in convection cells in the case of a small Rossby number (Stix, 1974). We obtain $\alpha \approx 50$ m s^{-1} (see Figure 7) and, accordingly, a frequency of 5×10^{-7} s^{-1}. On the other hand we may use an *a-posteriori* argument, namely that the ratio of the poloidal and toroidal field strengths should match the observed ratio of, say, 0.01 and so obtain an α between 0.01 and 0.05 m s^{-1} (Köhler, 1973). The frequency (9) would then agree with the observed one.

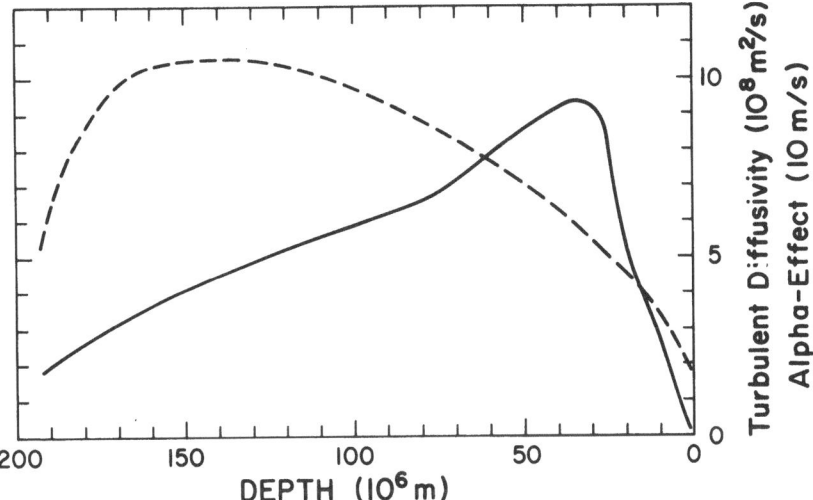

Fig. 7. Alpha-effect (α, solid) and turbulent diffusivity (η_t, dashed) in the solar convection zone. α is computed according to an expression derived by Krause (1968; see Steenbeck and Krause, 1969, Equation (23)), but is multiplied by the Rossby number, Ro, in the case $Ro < 1$. For the diffusivity I used $\eta_t = \frac{1}{3}vl$, where v and l are velocity and size of convection cells. Data of Baker and Temesváry (1966) were used to compute α and η_t.

Turbulent diffusion plays no role in the foregoing interpretation of the solar cycle period. Only the growth rate depends on the diffusivity. This interpretation, which has been strongly emphasized by Yoshimura (1975a, b), contrasts with earlier work where the period of the cycle had been identified essentially with the time the mean field needs to diffuse over a characteristic distance, d, in latitude (Leighton, 1969) or depth (Steenbeck and Krause, 1969). According to this picture, the frequency is

$$\Omega = 2\pi \left/ \frac{2d^2}{\eta_t} \right. \tag{10}$$

where η_t is the turbulent magnetic diffusivity and the factor 2 appears in the diffusion time since fields of both polarities must diffuse in order to complete a full cycle.

Using $\eta_t = 10^9$ m^2 s^{-1} (Figure 7) and $d = 2 \times 10^8$ m we find $\Omega = 8 \times 10^{-8}$ s^{-1}, which again is somewhat larger than the frequency of the real solar cycle. The value of η_t used here might however be too large by an order of magnitude (Stix, 1974).

Is the period of the solar cycle a wave period or a diffusion time? We have already seen (Figure 2) that it is the *wave* character of the $\alpha\omega$-dynamo which causes the latitude migration of the mean toroidal fields. And we have also seen that this is confirmed by the predicted (Figure 3) and observed (Figure 4) participation of the poloidal field in this latitude migration. But on the other hand, there is also no doubt that turbulent diffusion plays an important role in the numerical models mentioned earlier in this section: without diffusivity, such models fail altogether to produce reasonable results (e.g. Yoshimura, 1975a, b). At least partially, this might be a consequence of the geometry of the models. In a slab of infinite extent the dynamo waves may easily propagate, but in a spherical geometry tubes of mean flux must dissipate as they reach the boundaries or the equatorial plane. This process of turbulent dissipation might well be so slow that it dictates the time behaviour of the entire dynamo.

I shoud like to emphasize once more that the diffusion discussed here is *turbulent* diffusion; it is relevant to the *mean* field, or the small wave number end of the magnetic spectrum (Krause, 1968; Rädler, 1968b; Parker, 1971). Thus, since I consider the solar cycle to be governed by a periodic mean field, I disagree with Dr Piddington who argued that turbulent diffusion is irrelevant to the solar cycle (Piddington, 1971, 1972, 1973, 1975). The difficult question, then, is of course what happens at small scales, i.e. at the large wave number end of the spectrum. Does the drift across the spectrum proceed to sufficiently large wave numbers so that ohmic dissipation can destroy the magnetic fluctuations? The large magnetic Reynolds number mentioned in the introduction seems to indicate that this is not the case. Possible solutions of this problem are that convective motions of smaller, so far unresolved, scale exist on the Sun, or (and) that *dynamic* mechanisms accelerate the dissipation of magnetic flux (e.g. Sweet, 1958; Petschek, 1964; Parker, 1972), or that the small scale magnetic flux is simply lost by eruption through the solar surface (e.g. Parker, 1973, 1975; Stix, 1974). The 'ephemeral active regions' recently described by Harvey *et al.* (1975) make the latter idea particularly attractive.

3. The Role of the Angular Velocity

One of the main facts discussed in the preceding section was that the dynamo waves propagate along surfaces of isorotation. Now some of the theoretical work on solar differential rotation – for a review see Gilman (1974) – suggests that in the convection zone these surfaces are cylinders, and that the angular velocity increases with increasing distance from the axis of rotation. With $\alpha_{\text{north}} > 0$ and $\alpha_{\text{south}} < 0$ the direction of wave propagation (expression (6)) is then such as depicted in Figure 8a, away from the equatorial plane. Of course, mean toroidal flux propagating in this way can never produce the observed butterfly diagram as it errupts to the solar surface.

We may consider three possible answers to this puzzle. Firstly, the Sun might not be an $\alpha\omega$-dynamo, so that the toroidal flux would not necessarily propagate in the direction given by the vector product (6). Secondly, we may accept the concept of a solar $\alpha\omega$-dynamo but question the use of a positive α in the northern and negative α in the southern hemisphere. The propagation of the dynamo waves would then be opposite to the arrows of Figure 8a, and butterfly wings like the observed ones would

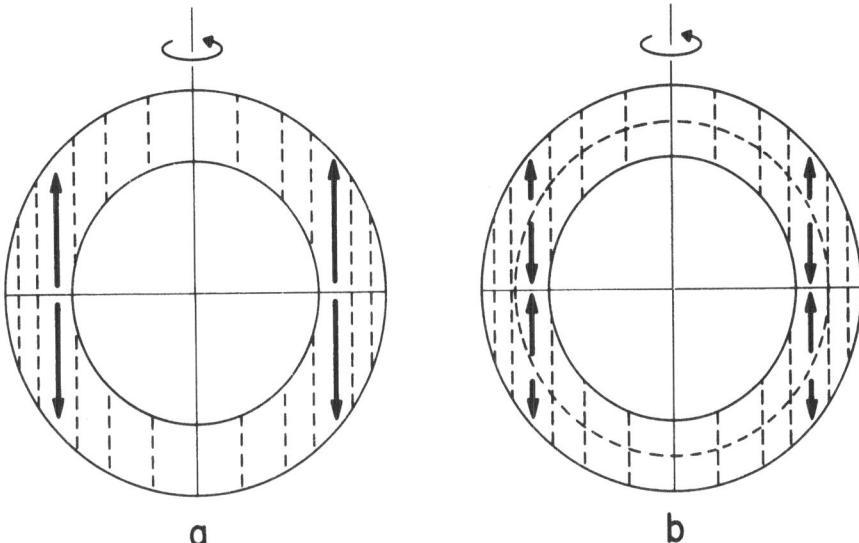

Fig. 8. Propagation of dynamo waves (heavy arrows) in the solar convection zone. The cylindrical surfaces of isorotation are indicated by the vertical dashed lines. $\alpha_{\text{north}} > 0$ and $\alpha_{\text{south}} < 0$ in (a) and outside the dashed circle of (b). Inside the dashed circle of (b) the sign of α is opposite.

result. Thirdly, the theory of the Sun's differential rotation might be insufficient. The puzzle would then simply not exist.

Let me first discuss the second possibility. The only case where a negative α in the northern hemisphere (and a positive α in the southern) has been reported is the work of Yoshimura (1972). Using the Boussinesq approximation he solved the linearized equations of motion and induction explicitly and found that α should be positive in the outer part of the convection zone in the northern hemisphere, and negative in the inner part, and should have the opposite signs in the southern hemisphere. Along cylindrical isorotation surfaces the dynamo waves would then travel in the directions indicated in Figure 8b. This possibility will also be discussed in a forthcoming paper by Durney (1976). The flux producing the butterfly diagram should, according to this model, erupt from the deeper part of the convection zone. This idea is consistent with recent estimates of Parker (1975) of the time of rise of the magnetic flux tubes. Only in the deeper part of the convection zone would these tubes stay long enough so that the dynamo could operate.

There are two arguments against this type of solar dynamo. Firstly, the α-effect used here depends on the special type of averaging. If Yoshimura (1972) had computed his mean quantities not only as longitudinal averages, but also as depth averages, the result would have been $\alpha \equiv 0$. The contributions from the upper layer would have cancelled those from the lower layer. The only way to obtain a non-zero α-effect when averaging over a volume is to make use of anisotropies in addition to the one introduced by rotation. Steenbeck *et al.* (1966) and Krause (1968) essentially used the density gradient and thus obtained their α which is positive in the entire northern and negative in the entire southern hemisphere. Secondly, there is a definite phase relation between the poloidal and toroidal field components. Only the case

$\partial\omega/\partial r<0$ *and* $\alpha_{\text{north}}>0$, $\alpha_{\text{south}}<0$ seems to be consistent with a negative (positive) radial field in the region of the northern (southern) butterfly wing of cycle 20 (see Figure 4 and Stix, 1976).

Since the α-effect is caused by the helicity, $\overline{\mathbf{v}\cdot\text{curl}\,\mathbf{v}}$, of the convective flow, $\mathbf{v}$, we may wonder whether it is possible to obtain the helicity from numerical models of the convection zone and then compute α according to

$$\alpha = -\frac{1}{3}\overline{\mathbf{v}\cdot\text{curl}\,\mathbf{v}}\,\tau, \tag{11}$$

where τ is the correlation time of the velocity field (Steenbeck and Krause, 1969). For example, Gilman (1972) has simulated the convection zone in a rotating annulus, and Figure 9 shows the helicity which he obtained for Taylor numbers between 10^2 and 10^6. We see that, for all Taylor numbers, left-handed helicity dominates in the northern hemisphere, i.e. $\alpha_{\text{north}}>0$, according to Equation (11). If at all a change in sign occurs within a hemisphere, the helicity tends to be right-handed in the upper and left-handed in the lower part of the northern hemisphere, so that the structure of α would be opposite to that obtained by Yoshimura (1972). This difference must be caused by the different approximations used by the two authors (although both used the Boussinesq approximation and longitudinal averaging, and considered linear modes characterized by a single azimuthal wave number): e.g., Yoshimura used a thin shell with hydrostatic equilibrium in the vertical direction, which Gilman did not, and his model was spherical whereas Gilman's was an annulus. Before we can use Equation (11) to compute α we probably have to wait for more numerical models.* These models most desirably should include such features as the nonlinear superposition of different modes and a variation of the boundary conditions. And, after all, they should be non-Boussinesq since, as we know, the density gradient plays such an important role in the determination of the α-effect.

Let me now discuss the first of the possibilities mentioned above. Mean field dynamo action without helicity, i.e. without α-effect, was first proposed by Rädler (1969a). He found that an additional mean electric field of the form

$$\beta\boldsymbol{\omega}\times\mathbf{j} \tag{12}$$

should exist in turbulent flows under the influence of rotation, where $\boldsymbol{\omega}$ is the rotation vector and $\mathbf{j}$ is the mean current density. However, numerical results of Rädler (1969b, 1970) and Roberts (1972) and an analytical treatment in cartesian coordinates by Gubbins (1974) indicate that only *steady* dynamo action can be obtained from the use of expression (12). For oscillatory modes the convergence was only 'suggestive, but not convincing' (Roberts, 1972). I have extended Roberts' search for oscillatory modes to higher truncation levels in the spherical harmonic expansion, but was still unable to find satisfactory convergence† (Stix, 1976). This was particularly so when cylindrical surfaces of isorotation were considered.

---

* P. A. Gilman has recently informed me that his new spherical shell model essentially leads to the same helicity as the annulus model.
† As reported by Dr Rädler at this Symposium, oscillatory $\boldsymbol{\omega}\times\mathbf{j}$-dynamos can be obtained from a slightly different model.

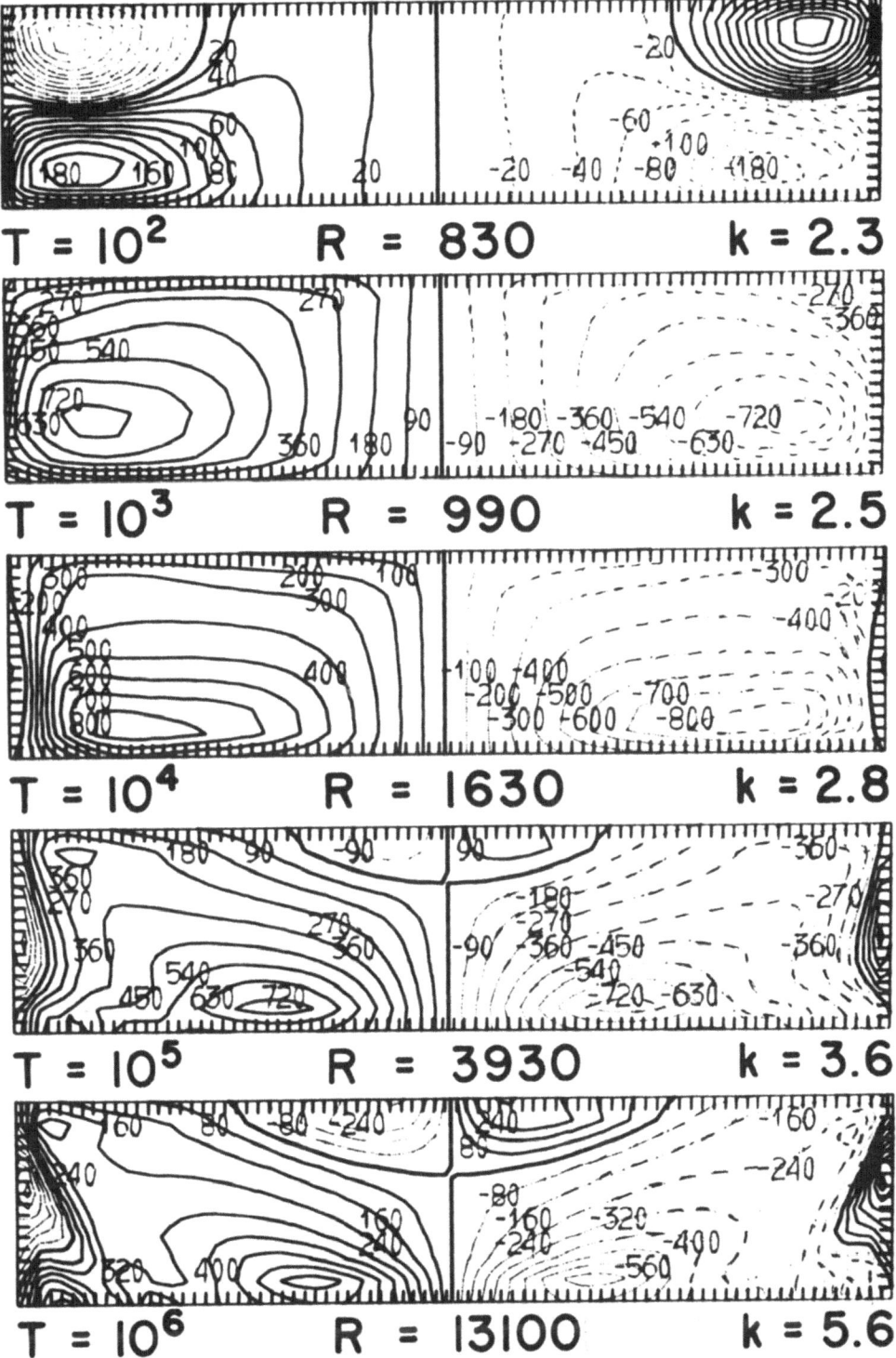

$T = 10^2$ $R = 830$ $k = 2.3$

$T = 10^3$ $R = 990$ $k = 2.5$

$T = 10^4$ $R = 1630$ $k = 2.8$

$T = 10^5$ $R = 3930$ $k = 3.6$

$T = 10^6$ $R = 13100$ $k = 5.6$

Fig. 9.   Normalized helicity of the most unstable linear convection modes, according to the rotating annulus model of Gilman (1972). T, R and k denote the Taylor number, the Rayleigh number, and the azimuthal wave number. Solid contours indicate positive (right-handed) helicity; dotted contours indicate negative (left-handed) helicity. The upper (lower) edge of each box is the top (bottom) of the convection zone, and north is to the right, south to the left (courtesy P. A. Gilman).

Thus, the prospects of a solar $\omega \times \mathbf{j}$-dynamo are not very promising, and we must conclude that either a completely different (i.e. not mean-field-electrodynamics type) theory of the solar cycle has to be developed or we have to revert to the $\alpha\omega$-dynamo. I prefer the latter; after all, the Sun *does* rotate and there *is* therefore helicity in the solar convection zone, giving rise to an α-effect.

The unsuccessful attempts to solve our puzzle by means of the first two possibilities leave us with the third. Are the surfaces of isorotation really cylinders? I think we should very carefully reconsider all assumptions and models which lead to such a law of rotation. I will however not do so in this lecture which is on solar dynamo theory and not on solar differential rotation.

4. Magnetic Sectors

In a recent paper (Stix, 1974) I have discussed possible theories of the sector structure of the large scale photospheric magnetic field. I will, therefore, spend only a small portion of this lecture on solar sectors. In particular, I shall not review all the observational evidence. I shall, however, discuss one observational fact in some detail, namely the fact that large-scale features on the Sun apparently do not, or at least not completely, participate in the differential rotation of the solar plasma. This is true for the photospheric magnetic field itself (Wilcox and Howard, 1970) but also for related features such as the electron corona (Hansen *et al.*, 1969), the coronal green line emission (Antonucci and Svalgaard, 1974) and coronal holes (Wagner, 1975; Timothy *et al.*, 1975).

A behaviour like this can probably most easily be explained as a *wave* phenomenon, and the three theoretical approaches which I will consider in this section have this interpretation in common.

There are first the models which ascribe the magnetic sectors to a corresponding pattern in the large scale velocity field. For example, models of the solar convection zone by Busse (1970, 1973), Durney (1970, 1971), Yoshimura and Kato (1971), Gilman (1972, 1975) and Yoshimura (1974) predict large scale motions in the form of rolls aligned parallel to the axis of rotation. These rolls are not destroyed by the mean differential rotation. In fact, they *produce* the differential rotation as they transport angular momentum towards the equatorial plane. At the same time, these rolls are damped on one of their sides (in longitude) and built up on the other, which is typical for a wave phenomenon. How such a velocity field can cause a similar, i.e. sectorial, structure in the magnetic field has been discussed e.g., by Yoshimura (1971, 1972). The longitudinal wave number, m, obtained from this theory is comparatively large: For a convection zone thickness of $0.2 \, r_\odot$, Durney (1970) found that $m = 10$ marked the most unstable mode, and Gilman (1975) obtained $9 \le m \le 21$ for Taylor numbers between 0 and 10^6. The resulting magnetic features should, therefore, be narrow and elongated. Perhaps the elongated coronal hole reported by Timothy *et al.* (1975) is a prominent example for such a feature (Figure 10).[*]

---

[*] Magnetic neutral lines derived from $H\alpha$ observations indicate however that the magnetic sector underlying this coronal hole covered a longitude interval of approx 45°, at least twice as wide as the hole itself (McIntosh *et al.*, 1975). The longitudinal wave number, m, was therefore only 4.

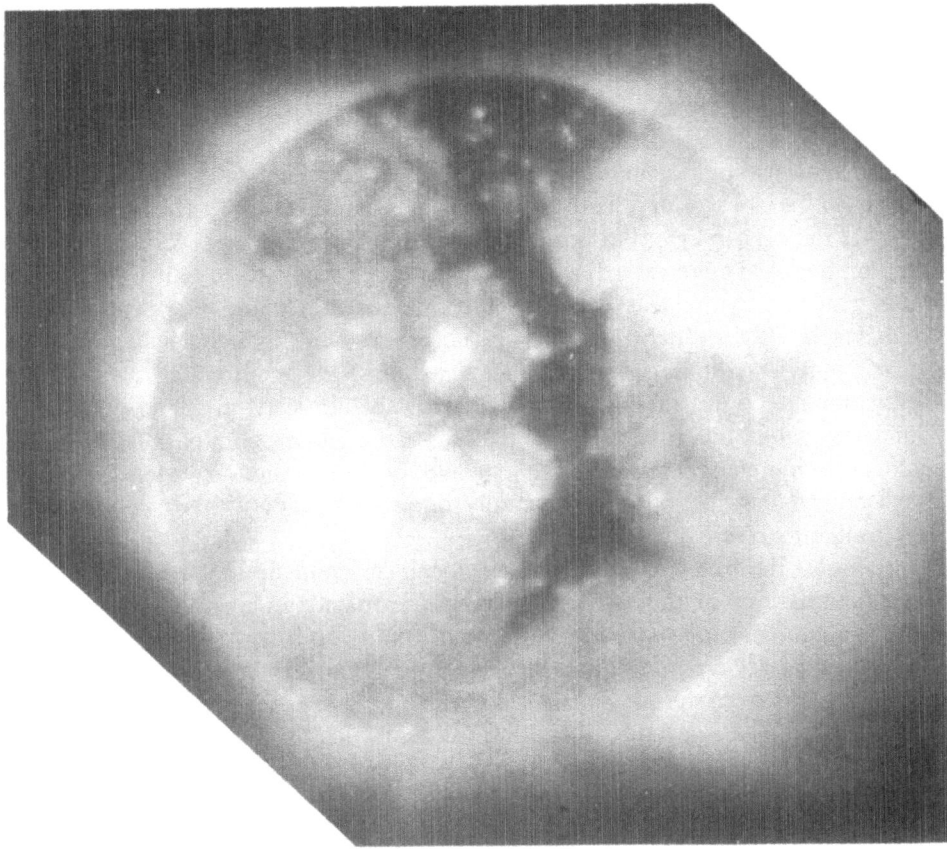

Fig. 10. An elongated coronal hole. This coronal image covers the wave-length intervals 3–32 Å and 44–54 Å; it was obtained on 1 June 1973 by the AS&E X-ray telescope on the Apollo Telescope Mount (courtesy A. S. Krieger, American Science and Engineering, Inc.).

In the observed photospheric magnetic field the dominant longitudinal wave number is usually much smaller than 10. This is particularly conspicuous in the spherical harmonic analysis of the Mt. Wilson data of Altschuler *et al.* (1974), who found that harmonics with $m = 1$ and $m = 2$ were dominant during most of the period from 1959 to 1972. The theoretical approach leading to such small longitudinal wave numbers is to consider the magnetic sectors as ϕ-dependent solutions of the mean field induction equation. These solutions are proportional to

$$\exp\left(i(\Omega t + m\phi)\right),\qquad(13)$$

and have been described by Stix (1971, 1974), Krause (1971) and Roberts and Stix (1972). Again, they are essentially waves, and their longitudinal propagation velocity, as computed by Stix (1974), is of the order 10 deg of longitude per year. This is about as fast as the sector structure inferred by Svalgaard and Wilcox (1975) propagates. According to these authors, a four-sector pattern persisted during the past five sunspot cycles, drifting slowly westwards during the first half and eastwards during the second half of each cycle. Since only *linear* modes have been considered so far in theory, such a dependence of the drift on the phase of the cycle has not yet been predicted. It could only be obtained from a non-linear coupling of the sectoral modes (13) with the axisymmetric periodic mean field described in Section 2. Little is however known about the nature of this non-linear coupling. At this time I can only say that both eastwards and westwards propagating linear modes exist, i.e. the real part of Ω in (13) can be either positive or negative (Stix, 1974).

As for the axisymmetric mean fields, the mean fields with $m > 0$ can be either symmetric or antisymmetric with respect to the equatorial plane. It appears that the *symmetric* fields are slightly preferred, i.e. excited at slightly smaller absolute dynamo numbers (Stix, 1971, 1974; Krause, 1971), a result which seems to be observationally confirmed by Wilcox and Howard (1968) who found that sector boundaries generally cross the solar equator rather than being parallel to it. An analogous symmetry rule is, incidentally, valid for the convection pattern described above: antisymmetric convection generally requires a larger critical Rayleigh number than symmetric convection (Gilman, 1975).

As long as the non-linear coupling between different modes is negligible, the sector boundaries maintain their shape as they propagate in longitude (Stix, 1974), which is a quite natural behaviour for a wave.

Finally, I would like to mention that the linear mean field modes described here may drift relatively to the mean plasma flow not only because they are waves, but also because they are *mean* fields. As such, they are subject to the *turbulent* magnetic diffusivity, i.e. they are *not* frozen-in fields.

Of course, linear modes with $m \neq 0$ could also be obtained from the exact induction equation, i.e. without the α-effect. Contrary to the axisymmetric modes they are not in conflict with Cowling's theorem (Cowling, 1934). Examples of such modes have been presented by G. O. Roberts (see P. H. Roberts, 1971) and Gubbins (1972, 1973b). However, the point here is that we need the mean field approach in order to explain the mean *axisymmetric* field, so I think it is only natural to use the same equation, namely the mean field induction equation, for *all* modes, $m = 0$ *and* $m > 0$. In order to compute the $m > 0$ modes described above I have indeed used the *same* forms of differential rotation and α-effect which led to the axisymmetric solution of Section 2 (Figure 3).

The third theoretical explanation that has been proposed for the origin of solar magnetic sectors is that they are a manifestation of hydromagnetic waves travelling in the azimuthal direction along a subsurface toroidal field (Kato and Nakagawa, 1970; Suess, 1971, 1975; Roberts and Stix, 1972). The angular velocity of these waves is of the order $Ro_m v/r_\odot$, where v is the Alfvén velocity corresponding to the mean toroidal field and Ro_m is the magneto-hydrodynamic Rossby number, $Ro_m \sim v/r_\odot \omega$ (Roberts and Stix, 1972; Suess, 1975). Reasonable propagation velocities (~ 1 deg of longitude per year) can be obtained in this way, but the estimate is very uncertain since v is uncertain; we do not know exactly the magnitude of the mean toroidal field, nor do we know the depth (i.e. density) we should use in order to compute v. Also, important effects, such as shear, stratification and the geometry of the spherical shell, have not been taken into account. We may therefore consider the hydromagnetic wave theory as the most speculative of the three theories offered in this section.

Acknowledgements

This lecture was prepared when I was a visitor at the High Altitude Observatory of the National Center for Atmospheric Research in Boulder, Colorado. M. D. Altschuler, B. R. Durney, P. A. Gilman, P. S. McIntosh, D. E. Trotter and H. Yoshimura all contributed helpful comments and made some of their unpublished material available to me.

References

Altschuler, M. D., Trotter, D. E., Newkirk, G., and Howard, R.: 1974, *Solar Phys.* **39**, 3.
Antonucci, E. and Svalgaard, L.: 1974, *Solar Phys.* **34**, 3.
Baker, N. and Temesváry, S.: 1966, *Tables of Convective Stellar Envelope Models*, 2nd ed., NASA, New York.
Braginskij, S. I.: 1964, *Zh. Eksper. Teoret. Fiz.* **47**, 1084 (= *Soviet Phys. JETP* **20**, 726).
Busse, F. H.: 1970, *Astrophys. J.* **159**, 629.
Busse, F. H.: 1973, *Astron. Astrophys.* **28**, 27.
Cowling, T. G.: 1934, *Monthly Notices Roy. Astron. Soc.* **94**, 39.
Deinzer, W. and Stix, M.: 1971, *Astron. Astrophys.* **12**, 111.
Deinzer, W. and Stix, M.: 1975, preprint.
Deinzer, W., v. Kusserow, H.-U., and Stix, M.: 1974, *Astron. Astrophys.* **36**, 69.
Durney, B. R.: 1970, *Astrophys. J.* **161**, 1115.
Durney, B. R.: 1971, *Astrophys. J.* **163**, 353.
Durney, B. R.: 1976, *Astrophys. J.* **204**, 589.
Gilman, P. A.: 1972, *Solar Phys.* **27**, 3.
Gilman, P. A.: 1974, *Ann. Rev. Astron. Astrophys.* **12**, 47.
Gilman, P. A.: 1975, *J. Atmospheric. Sci.* **32**, 1331.
Gubbins, D.: 1972, *Nature Phys. Sci.* **238**, 119.
Gubbins, D.: 1973a, *Geophys. J. Roy. Astron. Soc.* **33**, 57.
Gubbins, D.: 1973b, *Phil. Trans. Roy. Soc. London* **A274**, 493.
Gubbins, D.: 1974a, *Studies Appl. Math.* **L111**, 157.
Gubbins, D.: 1974b, *Rev. Geophys. Space Phys.* **12**, 137.
Hansen, R. T., Hansen, S. F., and Loomis, H. G.: 1969, *Solar Phys.* **10**, 135.
Harvey, K. L., Harvey, J. W., and Martin, S. F.: 1975, *Solar Phys.* **40**, 87.
Iroshnikov, R. S.: 1970a, *Astron. Žh.* **47**, 726 (= *Soviet Astron.* **14**, 582).
Iroshnikov, R. S.: 1970b, *Astron. Žh.* **47**, 1253 (= *Soviet Astron.* **14**, 1001).
Jepps, S. A.: 1975, *J. Fluid Mech.* **67**, 625.
Kato, S. and Nakagawa, Y.: 1970, *Solar Phys.* **14**, 138.
Köhler, H.: 1973, *Astron. Astrophys.* **25**, 467.
Kopecký, M. and Soytürk, E.: 1971, *Bull. Astron. Inst. Czech.* **22**, 154.
Krause, F.: 1968, Habilitationsschrift, Univ. Jena.*
Krause, F.: 1971, *Astron. Nachr.* **293**, 187.
Krause, F. and Rädler, K.-H.: 1971a, in R. Howard (ed.), 'Solar Magnetic Fields', *IAU Symp.* **43**, 770.
Krause, F. and Rädler, K.-H.: 1971b, in R. Rompe und M. Steenbeck (eds.), *Ergebnisse der Plasmaphysik und der Gaselektronik*, Akad. Verlag, Berlin, Vol. 2, p. 1.
Leighton, R. B.: 1969, *Astrophys. J.* **156**, 1.
Levy, E. H.: 1972, *Astrophys. J.* **171**, 621.
McIntosh, P. S., Krieger, A. S., Nolte, J. T., and Vaiana, G. S.: 1975, AAS-SPD Meeting, San Diego.
Moffat, H. K.: 1970a, *J. Fluid Mech.* **41**, 435.
Moffat, H. K.: 1970b, *J. Fluid Mech.* **44**, 705.
Parker, E. N.: 1955, *Astrophys. J.* **122**, 293.
Parker, E. N.: 1970, *Astrophys. J.* **162**, 655.
Parker, E. N.: 1971, *Astrophys. J.* **163**, 279.
Parker, E. N.: 1972, *Astrophys. J.* **174**, 499.
Parker, E. N.: 1973, *Astrophys. Space Sci.* **22**, 279.
Parker, E. N.: 1975, *Astrophys. J.* **198**, 205.
Petschek, H. E.: 1964, in W. N. Hess (ed.), *AAS-NASA Symposium on the Physics of Solar Flares*, NASA Sp-50, p. 409.

Piddington, J. H.: 1971, *Proc. Astron. Soc. Australia* **2**, 7.
Piddington, J. H.: 1972, *Solar Phys.* **22**, 3.
Piddington, J. H.: 1973, *Astrophys. Space Sci.* **24**, 259.
Piddington, J. H.: 1975, *Astrophys. Space Sci.* **35**, 269.
Rädler, K.-H.: 1968a, *Z. Naturforsch.* **23a**, 1841.*
Rädler, K.-H.: 1968b, *Z. Naturforsch.* **23a**, 1851.*
Rädler, K.-H.: 1969a, *Monatsber. Dt. Akad. Wiss. Berlin* **11**, 194.*
Rädler, K.-H.: 1969b, *Monatsber. Dt. Akad. Wiss. Berlin* **11**, 272.*
Rädler, K.-H.: 1970, *Monatsber. Dt. Akad. Wiss. Berlin* **12**, 468.*
Roberts, P. H.: 1971, in W. H. Reid (ed.), *Mathematical Problems in the Geophysical Sciences*, American
    Mathematical Society, Providence, R.I., p. 129.
Roberts, P. H.: 1972, *Phil. Trans. Roy. Soc. London* **A272**, 663.
Roberts, P. H. and Stix, M.: 1971, *The Turbulent Dynamo*, NCAR Technical Note TN/IA-60.
Roberts, P. H. and Stix, M.: 1972, *Astron. Astrophys.* **18**, 453.
Soward, A. M.: 1971, *J. Math. Phys. N.Y.* **12**, 1900.
Soward, A. M.: 1972, *Phil. Trans. Roy. Soc. London* **A272**, 431.
Spitzer, L.: 1962, *Physics of Fully Ionized Gases*, 2nd ed., New York, Interscience.
Steenbeck, M. and Krause, F.: 1966, *Z. Naturforsch.* **21a**, 1285.*
Steenbeck, M. and Krause, F.: 1969, *Astron. Nachr.* **291**, 49.*
Steenbeck, M., Krause, F., and Rädler, K.-H.: 1966, *Z. Naturforsch.* **21a**, 369.*
Stenflo, J. O.: 1972, *Solar Phys.* **23**, 307.
Stix, M.: 1971, *Astron. Astrophys.* **13**, 203.
Stix, M.: 1973, *Astron. Astrophys.* **24**, 275.
Stix, M.: 1974, *Astron. Astrophys.* **37**, 121.
Stix, M.: 1976, *Astron. Astrophys.* **47**, 243.
Suess, S. T.: 1971, *Solar Phys.* **18**, 172.
Suess, S. T.: 1975, *Am. Inst. Aeron. Astron. J.* **13**, 443.
Svalgaard, L. and Wilcox, J. M.: 1975, *Solar Phys.* **41**, 461.
Sweet, P. A.: 1958, in B. Lehnert (ed.), 'Electromagnetic Phenomena in Cosmical Physics', *IAU Symp.* **6**,
    123.
Timothy, A. F., Krieger, A. S., and Vaiana, G. S.: 1975, *Solar Phys.* **42**, 135.
Wagner, W. J.: 1975, *Astrophys. J.* **198**, L141.
Wilcox, J. M. and Howard, R.: 1968, *Solar Phys.* **5**, 564.
Wilcox, J. M. and Howard, R.: 1970, *Solar Phys.* **13**, 251.
Yoshimura, H.: 1971, *Solar Phys.* **18**, 417.
Yoshimura, H.: 1972, *Astrophys. J.* **178**, 863.
Yoshimura, H.: 1974, *Publ. Astron. Soc. Japan* **26**, 9.
Yoshimura, H.: 1975a, *Astrophys. J. Suppl. Ser.* **29**, No. 294.
Yoshimura, H.: 1975b, *Astrophys. J.* **201**, 740.
Yoshimura, H.: 1975c, this volume, p.137.
Yoshimura, H. and Kato, S.: 1971, *Publ. Astron. Soc. Japan* **23**, 57.

* Articles marked with an asterisk have been translated into English (Roberts and Stix, 1971).

DISCUSSION

Roxburgh: Our picture of the formation of the Sun suggests that at an early stage it was completely convective and therefore probably a dynamo operated throughout the whole Sun producing a magnetic field in the central regions. What do you think, is the connection between the present central field and the present dynamo field produced in the solar convective zone?

Stix: I think if the central field exists it is not connected to the dynamo field because the time scale of the central field is much longer.

Gilman: Suppose that the bottom of the convection zone is rotating solidly at nearly the equatorial rate, as one interpretation of sector and coronal hole rotation indicates. Then the radial shear is much stronger near the pole than the equator. What sort of α-effect dynamo would then occur, and would it be in agreement with the important observations?

Stix: I would expect toroidal fields generated at too high latitudes.

Yoshimura: I would like to comment about two questions. One is about the determination of the sign of α and gradient of ω. You said that it is possible to determine them by determining the phase shift between the poloidal and toroidal field observed at the surface. However, this is possible only if the radial gradient of ω is dominant. According to the diagrams of the fields of myself, there is virtually no phase shift observed. So that it is rather questionable that the determination is possible. Moreover, if the radial gradient of ω is dominant, there is no branching of the two wings in each hemisphere in the diagram of the poloidal field which has been observed and presented in this symposium. The other question is about the interpretation of the sector structure of the magnetic field. You said the interpretation by the global-scale nonaxisymmetric velocity fields has some drawbacks, i.e. the theories predict wave numbers of 10 to 20. However, the theories sofar studied have used many approximations. So, it is safe to say that the correct determination of a wave number has not yet been done. Moreover, I would like to make one more comment about Dr Busse's fluid mechanical convection experiment. He obtained some interesting results even in the experiment of plane parallel case. That is, besides the ordinary Bénard cell convection, he observed some coherent laminar flow cells whose scale is much larger than the Bénard cells and also larger than the depth of the system. In the case of the Sun also, it is possible that similar kind of larger scale coherent convective flow exists whose wave number may be as well as 2 or 3, or 5 or 6 besides the supergranular or granular convection. So we should be very careful about objecting to the existence of the global convection. If it exists, it surely can explain the sector structure of the magnetic field.

Weiss: It is difficult to predict the horizontal scale of plane form of convection. The azimuthal wavenumber of 10 or 11 was originally put forward on the assumption that the horizontal width would be comparable with the depth of the convecting layer in a Boussinesq fluid.

Gilman: Dr Yoshimura comments that current spherical shell convection models are not good enough to give reliable prediction of the longitudinal wave numbers. I am inclined to agree, and point out that strong enough shear in the differential rotation could significantly reduce the wave number predicted by current models, which are Boussinesq and based on initial states of solid rotation. Induced $\mathbf{j} \times \mathbf{B}$ forces will also tend to reduce the longitudinal wave number. Compressibility should also contribute.

Stix: I am looking forward to seeing convection models with smaller longitudinal wave number.

Stenflo: I would like to comment on the sector structure of solar magnetic fields and the modes in longitude. Power spectra of the observed variations of magnetic fields with longitude show that the power is quite high out to wave numbers of 20 or even more.

Wilcox: The term 'magnetic sector' was first used in the description of a large-scale structure observed in the interplanetary magnetic field. These sectors were then found in the photospheric field observed with Babcock magnetographs, and are seen with particular clarity in observations of the mean solar magnetic field (the Sun seen as a star). If defined in this way there are nearly always either four or two sectors. It is these sectors that have been shown to have the rigid rotation properties discussed by Stix and other theorists.

Deinzer: It still seems that an angular velocity increasing inwards is required to produce the desired sort of butterfly diagrams. Imagine you drive the solar dynamo by differential rotation increasing inwards and being constant on spheres.

Is it entirely inconceivable that the electromagnetic (Lorentz) forces could drive the observed equatorial acceleration? Maybe this is more a question to Dr Gilman.

Gilman: Dr Deinzer asks whether the presence of $\mathbf{j} \times \mathbf{B}$ forces could produce an equatorial acceleration when they are induced by a dynamo with rotation increasing with depth. This is conceivable, but I would suspect it would produce a fairly large dependence of the differential rotation on the magnetic cycle. Observations indicate this occurs weakly, if at all.

Roxburgh: If the central regions of the Sun are magnetically isolated from the surface convective zone then just beneath the convective zone there will be a region where the angular velocity increases inwards. As the surface regions are slowed down by the angular momentum loss in the solar wind there will be an angular velocity gradient between the more rapidly spining core and the envelope. This region is probably weakly turbulent due to Goldreich-Schubert-Fricke instabilities and could be the site of the solar dynamo.

Stix: The dynamo could operate there only if this weak turbulence still provides sufficient turbulent diffusivity so that the mean field can diffuse down into the shear region. Perhaps convective overshooting would also help to do this.

Mestel: I would like to take up Dr Deinzer's query. If a dynamo does operate, then presumably the field amplifies until magnetic forces are strong enough to react back in at least one essential part of the motion. This morning we heard some ideas and preliminary results on magnetic back-reaction on the α-term; maybe there is also an effect on the non-uniform rotation which drives the whole process. There is in fact one simple model – in which angular momentum is transported by magnetic stresses and large-scale meridian circulation only, but not by turbulence – which can give both equatorial acceleration and Ω increasing inwards. This result is no more than a hint: one needs to include turbulence, and the circulation field should emerge from the theory, instead of being postulated. But perhaps the possibility should be borne in mind, especially if the purely hydrodynamic models of differential rotation continue to give trouble. Dr Gilman's point about a non-observed cycle-dependent Ω is of course very important: to be acceptable the theory would need to show that the time of redistribution of angular momentum when the field changes is long compared with the solar cycle.

Durney: I think that we all agree that the mixing length theory is logically highly unsatisfactory. However, it could well give results that are not too bad. It gives good results for the Sun because the parameters are adjusted to do so. Nevertheless it also predicts accurately the spectral type separating stars with fast and slow rotations, that is it predicts accurately the spectral type of stars having appreciable surface convection zones.

SOLAR MAGNETIC FIELDS AND CONVECTION

VII. *A Review of the Primordial Field Theory*

J. H. PIDDINGTON

National Measurement Laboratory, CSIRO, Sydney, Australia 2008

Abstract. We review the primordial field theory of solar magnetic fields (Papers I–VI) whose three main features are, first, a permanent dipole-like magnetic field, second a mainly toroidal field formed by shearing and rolling into individual, helically twisted ropes as suggested by Babcock, and third a mechanism for reversing the toroidal field. The theory explains numerous observational effects where the dynamo theory fails.

(i) An active region forms when a rope section emerges and expands layer by layer to form a rotating arch filament system and then spots. Only a rope model explains the radial inflow of magnetic elements to build up a spot, as well as the spiral structure and other features pointed out by Babcock. (ii) The model explains umbral and penumbral structures, the Wilson depression, Evershed flow, the sunspot energy deficit and the very slow loss of flux fragments by some sunspots. (iii) The model is then extended to background magnetic fields to show that surface magnetic fields are like the upppermost branches of a magnetic 'tree' whose trunk is a flux rope. This explains unipolar magnetic regions, 'pepper and salt', and ephemeral active regions. Tension in the submerged flux ropes accounts for the observed migrations of magnetic regions, active longitudes and magnetic longitudes. (iv) On a smaller scale spicules, mottles and other network elements are explained in terms of the tree structure. (v) The mechanism of reversal of the toroidal field system is explained and the manner of disposal of old toroidal fields. (vi) The basic error in the dynamo theory is discussed briefly. We point out that radical changes in dynamo theory have been suggested by Parker, but appear to have been ignored by others.

1. Introduction

In a series of papers (Piddington, 1971a, b, 1974, 1975a, d, e) I have developed a theory of solar magnetic fields which might be termed the *primordial field theory*. One of its important features is that submerged magnetic fields are in the form of helically twisted flux ropes as shown in Figure 1a, an idea proposed originally by Babcock (1961). A puzzling feature of Babcock's paper is that it is often quoted as though it also provided observational support for the *dynamo theory* (see, for example, Parker, 1970a; Stix, 1974) which is the very antithesis of the primordial field theory. Hence, before reviewing the latter it is desirable to explain this misunderstanding.

The dynamo theory describes two separate and distinct phenomena as follows.

(i) As pointed out some decades ago (see Elasser, 1956; Cowling, 1953), non-uniform rotation $\omega(\theta)$ of a conducting sphere will draw out a poloidal field to form a stronger, mainly toroidal field. This ω-effect operates equally well with or without the added feature of twisted flux ropes and in Figure 1b the result is illustrated *with* twists. There is no argument about this ω-effect, which is common to both theories and is readily tailored to explain the butterfly diagram and the polarity of sunspot magnetic fields for an 11-yr cycle.

(ii) The second dynamo phenomenon is 'cyclonic turbulence' (Parker, 1955, 1970b) illustrated in Figure 1c. A rising convective cell carries some flux upwards and Coriolis forces rotate the flux loop to provide one of many elements of a reversed poloidal field. These elements are merged by turbulence to provide the α-effect, whose very essence is dominance of the convective motions which create new

Bumba and Kleczek (eds.), Basic Mechanisms of Solar Activity, 389–407. All Rights Reserved.
Copyright © 1976 by the IAU.

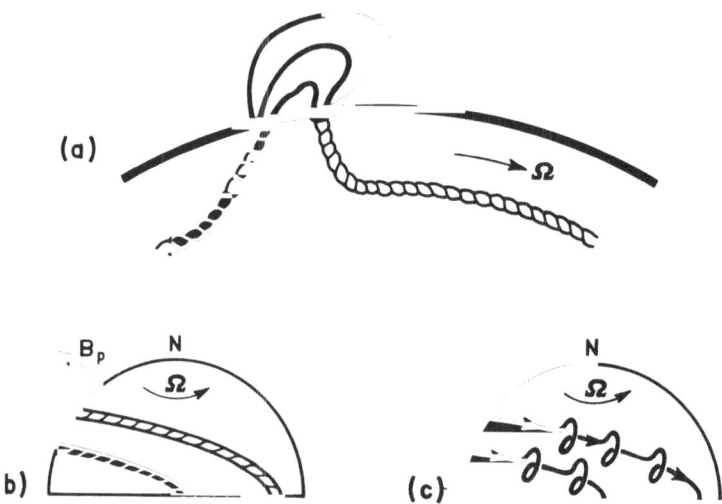

Fig. 1.   Solar magnetic fields according to different theories. (a) A section of one of Babcock's flux ropes which has erupted to provide a sunspot pair. (b) Submerged flux ropes wound into a mainly toroidal field. (c) A more-or-less uniform toroidal field system erupts in numerous cyclonic loops to provide the α-effect of the $\omega\alpha$ solar dynamo.

small-scale magnetic fields and then tangle and merge them to form new large-scale fields. The α-effect and the flux-rope model are thus completely incompatible because the submerged flux ropes are powerful and cohesive and unaffected by convective motions.

The explanation of the misunderstanding is now clear. Babcock's flux ropes represent a form of the ω-effect and he attempted to use them to provide an α-effect as well. While 99% of the ropes left the Sun, the remaining 1% were retained to merge and form a new poloidal field. This part of his model is not acceptable (Piddington, 1972a), and if we allow 100% of the ropes to leave the Sun then the model is divorced from the dynamo theory and no longer supports that theory by its agreement with observations.

The success of Babcock's model of the ω-effect in explaining several features of spot groups was widely acclaimed, and in Sections 2 and 3 we review the extension of this comparison with more recently discovered observational features. In Section 4 we then show how the flux rope model may be made part of a 22-yr cycle by replacing the α-effect by a primordial field together with a meridional oscillation which causes a small change in the $\omega(r, \theta)$ pattern. Finally, for completion, we comment in Section 5 on the solar dynamo theory in its more recent and more divided versions.

2. Sunspots

Figure 1a shows how Babcock (1961) accounted for a pair of sunspots with powerful, twisted magnetic fields. This model was widely acclaimed, yet most recent reviews (see, for example, Meyer *et al.*, 1974) state that "after the flux emerges through the photosphere it is concentrated by supergranule convection to form a sunspot". We show below that this view is incorrect; meanwhile we explain how the flux of a powerful, twisted rope is *temporarily* scattered and so *appears* to be controlled by gas motions.

In Figure 2a an inverted U section of a rope is seen approaching the surface, the magnetic field being contained by a higher external gas pressure p_e given by

$$p_e = p_i + B^2/8\pi, \tag{1}$$

where p_i is the internal gas pressure and B the field strength. At a critical level (depth $\gtrsim 1000$ km) p_e is too small to satisfy this equation and activity starts. The diameter of a substantial flux rope is $D \approx 10^4$ km while the scale height is only ≈ 300 km. Thus, when a layer of the rope of thickness only $D/30$ projects through the critical level it expands upwards while the remainder of the rope remains intact. This is the first arch filament, and as the rope continues to rise layer after layer erupt in their turns to provide an Arch Filament System (AFS) discussed further below.

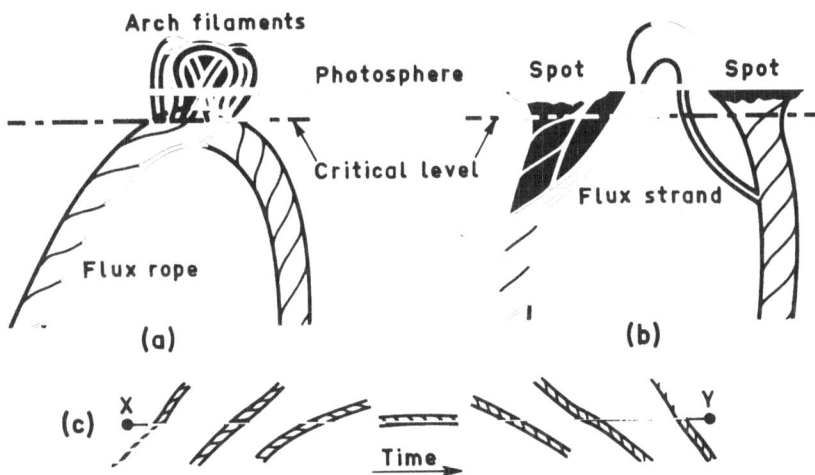

Fig. 2.   Some details of the formation of an active region and spot pair. (a) A flux rope rises through the critical level and erupts layer by layer to provide a rotating Arch Filament System. (b) The two parts of the rope have separated and become nearly vertical. Some flux strands have unwound and separated from the rope. (c) A plan view of a steadily rotating Arch Filament System; rotation is caused by the twist in the rope.

Figure 2b shows the two rope sections after the horizontal section has emerged, a process which may take a couple of days corresponding to a velocity of $\approx$ 0.06 km s^{-1}. A submerged flux rope is strongly cohesive and virtually indestructable, and we now see why scattered flux moves radially inwards towards a spot (Vrabec, 1974). It is because most flux remains connected to the submerged rope sections and is gathered together as those sections approach the surface. However, at this stage the ropes will start to unwind by the upward propagation of Alfvén waves; they will then 'fray' and lose some strands as shown in Figure 2b.

Returning to the AFS, we note that the model predicts a puzzling sequence illustrated in Figure 2c. The outer layer of a twisted rope will be strongly tilted from the axis of the rope whose projection on the surface is shown by the line XY. Thus the first arch filament is tilted as shown by the left-hand filament. The tilt decreases to zero as the axis of the rope emerges and then increases in the opposite direction. Just such a remarkable sequence has been observed by Frazier (1972), to provide further confirmation that sunspot fields are indeed helically twisted. It will be seen that the effect may now be used to estimate the pitch of the helically twisted field.

In Paper I of this series we described several other features of a spot group which seem explicable only in terms of submerged helically twisted flux ropes. These include the marked asymmetry of spot groups, the ordered, large-scale proper motions of sunspots and the very long lives of some preceding (but not following) sunspots. For brevity, further discussion of these effects is omitted.

2.1. THE MAGNETIC STRUCTURE OF A SUNSPOT

Returning to the problem of the mode of creation of sunspots, we show in Figure 3 the left-hand section of an emerging flux rope. The horizontal part of the rope has been torn apart by internal pressure and the rope has partially untwisted and frayed into 'strands' which, as shown in the following section, are also helically twisted. Some of these cut the surface and so account for pores and smaller magnetic elements, while others marked S are still submerged. The rope below the section XX is intact and continues to move upwards as the rope section swings into a vertical position. It is easily seen that this upward movement must cause the pores and other surface magnetic elements to *move radially inwards* towards the small spot of central pore shown in Figure 3.

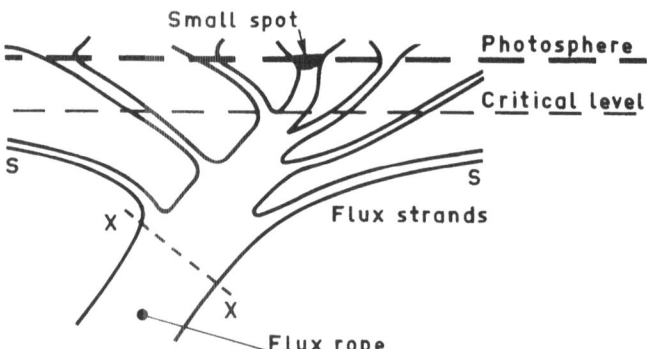

Fig. 3.   A flux rope rising and gathering its frayed strands to provide a growing sunspot.

This process has been observed by Vrabec (1974) who also attributed it to magnetic control as in the model shown. The evidence for magnetic control (as opposed to the traditional idea of supergranule control) is as follows.

(i) Magnetic stresses make a rope strongly cohesive so that its submerged flux is not dispersed.

(ii) Over an area of more than 30 supergranule cells Vrabec (his Figure 2) observed ordered, convergent flow of magnetic features. Polarities of both signs are rather mixed yet only features of like polarity move towards the spot; adjacent features of opposite polarity do not move.

(iii) The paths of the moving magnetic features sometimes show marked curvature corresponding to the spiral structure of the spot penumbra. This effect is expected from the fact that the flux strands have *partially* unwound from the rope and so are curved.

(iv) The speeds of the magnetic features range up to 1 km s^{-1} which is >2 times that of supergranules. it is easily accounted for if some of the strands are nearly horizontal.

We have shown how flux accumulates above a rope end, and now consider why this causes pores and spots to be dark. It has been shown (Stenflo, 1973; Harvey and Hall, 1974) that even in quiet regions and even for magnetic elements of size $\gtrsim 200$ km the field strength is always ≈ 2000 G. On the other hand pores are known to form when the average field strength is only ≈ 1400 G. Accordingly (Paper IV) we suggest that pores are loosely twisted bundles of flux fibres whose total photospheric area is about 2/3 that of the pore, the remaining 1/3 being non-magnetic. The pores are dark because there is no convection within the individual strands, and also because convection between the strands is partly or wholly suppressed. One might ask why are the individual flux strands not observed as tiny pores, and the answer seems to be that they *are*. There are few really good white-light photographs, but Danielson's (1961, Figure 8 and others) Stratoscope I photos show pores which are so tiny that their lower limit cannot be distinguished from the dark intergranule regions.

According to this model a sunspot magnetic field has the form shown in Figure 4. It is made up of hundreds or thousands of flux strands whose flux may be typically $\approx 4 \times 10^{18}$ Mx and field strength ≈ 4000 G falling to 2000 G at the photosphere

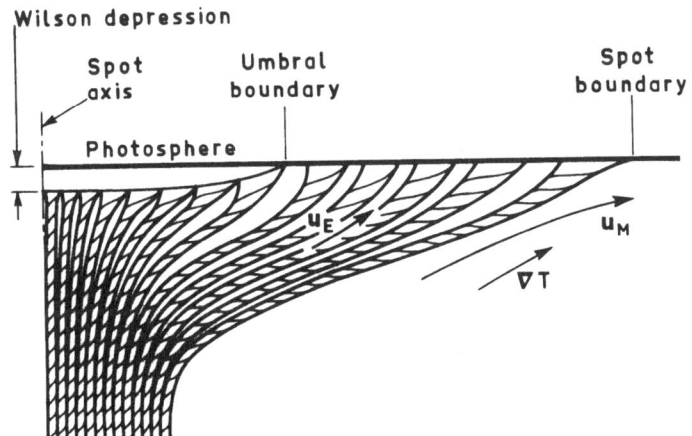

Fig. 4.   A section of a sunspot magnetic field showing the individual twisted strands but, for simplicity, not showing the overall rope twist. The field in the subsurface strands is everywhere ≈ 4000 G. At the surface it is generally pressure limited to ≈ 2000 G or less when the strand is tilted as in the penumbra. However, near the centre of a large spot the subsurface flux strands cannot expand and the full 4000 G is seen at the surface. The outward Evershed flow $\mathbf{u}_E$ is a combination of flows of dark plasma within the flux strands and bright plasma between. The enhanced temperature gradient ∇T drives a new convective motion, the 'moat' convection $\mathbf{u}_M$, which accounts for the sunspot energy deficit.

where it is pressure limited, except at the centres of large spots where its full value is observed. The twisted strands are twisted together (not shown) to form a rope with varying amounts of non-magnetic plasma between the fibres. Other features of the model are as follows.

(i) In the umbra the 2000–4000 G strands completely fill the photosphere and we see only the colder magnetic plasma, thus accounting for the umbral darkness.

(ii) In the penumbra the dark magnetic plasma forms lanes between which the hot non-magnetic plasma convects. This is a modified granule convection in the form of long striae with characteristic period a few times the granule period. A similar convective pattern has been observed by Danielson and by Bray and Loughhead outside sunspots in regions between pores (see Paper IV).

(iii) The Wilson depression is explained in terms of completely isolated magnetic atmosphere described in Section 4 below. It is cooler than the main non-magnetic atmosphere, and at every level it adjusts its pressure according to Equation (1).

(iv) The rope model explains the origin of huge spots, some with diameters $\gtrsim 10^5$ km (Newton, 1955) and some with central fields of $\gtrsim 4000$ G. It does not seem that these features could possibly be accounted for in terms of the compression of weak fields by the supergranule convective motions which are both too weak and of too small a scale.

2.2. MOTIONS IN AND AROUND SUNSPOTS

The fraying flux rope model of sunspot magnetic fields accounts for a number of plasma motions in and around spots. It explains, in the first place, why some spots vanish within a few days often accompanied by fragmentation while others last for weeks or months (Bumba, 1963). Spot fields without twists must disintegrate immediately by the flute instability, a fate which would overtake any spot formed in the traditional manner by supergranule motions. Twisted fields last as long as the twists and then decay as explained below. Other motions are as follows.

(i) In Paper I we discussed the proper motions of sunspots which are sometimes highly ordered over large distances and quite different for preceding and following spots. The motions are clearly the result of magnetic forces, and are not consistent with the traditional idea of supergranule control. They relate closely to the observed asymmetry of spot groups explained by Babcock (1961) in terms of flux ropes.

(ii) The model opens interesting possibilities for the explanation of light bridges, umbral granulation and umbral flashes. The submerged flux strands are separated by non-magnetic plasma which is *not* thermally isolated from the plasma outside the flux rope, and so is hotter than that in the strands. Upwellings of this plasma between the strands explain these curious effects.

(iii) In Paper IV we show how the Evershed outflow is related to the 'moat convection' and how each is explained. Partial suppression of convection beneath the 'umbrella' (Figure 4) will lead to an enhanced upward temperature gradient in the non-magnetic plasma and hence to a slow flow of hot gas $\mathbf{u}_E$ as shown. The gas within the flux tubes is cooler, thus accounting for the bright and dark curved striae of the penumbra; this gas provides an Evershed flow of cool gas within the individual, twisted flux strands. This flow is caused by the general upward movement of the whole rope system discussed above and in Section 4.

(iv) An extension of these ideas leads to simple explanations of the 'moat convection' and as yet unexplained sunspot energy deficit. Sheeley (1972) observed a new type of motion radially away from sunspots with velocities up to 1 km s^{-1} to distances of $\approx 2 \times 10^4$ km, and Vrabec (1974) pointed out that this differs from supergranule motions in four major respects. We consider that this is a new non-random convective motion $\mathbf{u}_M$ (Figure 4) forced by the enhanced temperature gradient ∇T which develops beneath the penumbra.

In Paper IV we have only shown that this convective motion accounts readily for the sunspot energy deficit. A temperature difference (across the moat) of only $\gtrsim 2\%$ at a depth of ≈ 3000 km and a flow at speed 0.75 km s^{-1} carries away energy at a rate of 2×10^{29} erg s^{-1} from a spot of radius 2×10^4 km. This equals the energy deficit for such a spot, and the depth is appropriate because the thermal energy above that level is enough to provide the photospheric emission for ≈ 2 days which seems a reasonable time scale.

(v) Finally we consider the motions which represent the slow decay of a sunspot (Harvey and Harvey, 1973; Vrabec, 1974). These moving magnetic features are isolated elements, which suggests that their submerged sections are twisted flux strands as depicted in Figure 5. They move independently, some overtaking others

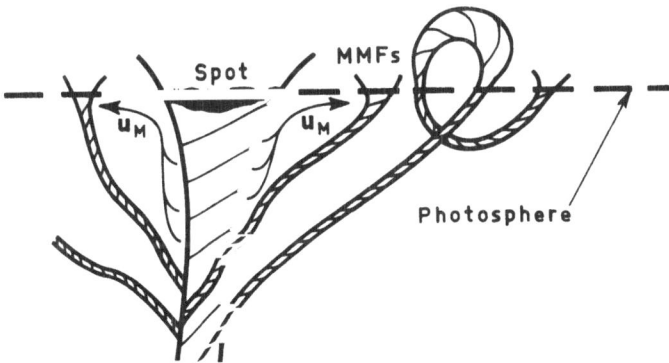

Fig. 5.   A section of a flux rope below a decaying spot. Flux strands unwind to ever-increasing depths and are then carried across the moat as moving magnetic features (MMFs) by the moat convection $\mathbf{u}_M$. Strands are twisted and so the kink instability creates loops as shown.

and some moving along curved paths which match the penumbral filaments, which is explained, like the penumbral filaments, by flux strands which are unwinding from the flux rope. The total flux (the sum of the fluxes of both signs) transported is an order of magnitude more than the spot flux, but the net flux is just that of the spot. This effect is readily explained by the kink instability which will cause a succession of kinks to propagate up a flux strand as shown. The same effect accounts for the sudden appearance of features in the middle of the moat. Finally, we recall that long-lived spots decay by losing flux at a uniform rate, and explain the effect of the unwinding at a uniform rate of the outermost layer of flux strands.

We suggest that these various phenomena in and around sunspots can only be explained in terms of the rope-strand model and confirms the features of the model inferred from earlier observations.

3. Background Magnetic Fields

After the spots have vanished, active regions expand and decay. The magnetic elements are clearly affected by the supergranule motions, but it is easily seen that this control is not complete as in the dynamo and other diffuse field theories. There are many reasons for this conclusion, one being the observations (Stenflo, 1973; Harvey and Hall, 1974) of magnetic elements of field strength ≈ 2000 G and diameters only a few hundred kilometers. As shown in Paper IV these are only just explicable in terms of pressure limitation (Equation (1) above) in the low photosphere, and their preservation requires that the flux bundle concerned is twisted and so cohesive. Other evidence for a highly ordered subsurface magnetic structure is seen in the further development of the flux-rope-strand model as follows.

(i) A notable feature of the background magnetic fields is the unipolar magnetic region, which is a huge ($\gtrsim 10^5$ km) area with photospheric magnetic field of one strongly predominant polarity (see, for example, Livingston and Harvey, 1971, Figure 4). If solar magnetic fields could be manipulated by the convective eddies a very different picture would be seen, with the churned up fields having completely mixed polarities. The flux-rope model explains the result in terms of the structure shown in Figure 6, where the rope has untwisted down to the level Z_0. Above that level the rope has broken into smaller ropes and frayed into strands with flux typically $\approx 4 \times 10^{18}$ Mx, so that a rope of average flux 10^{21} Mx has about 200 strands and large ropes up to $\gtrsim 10^4$ strands. From considerations of energy equipartition we have

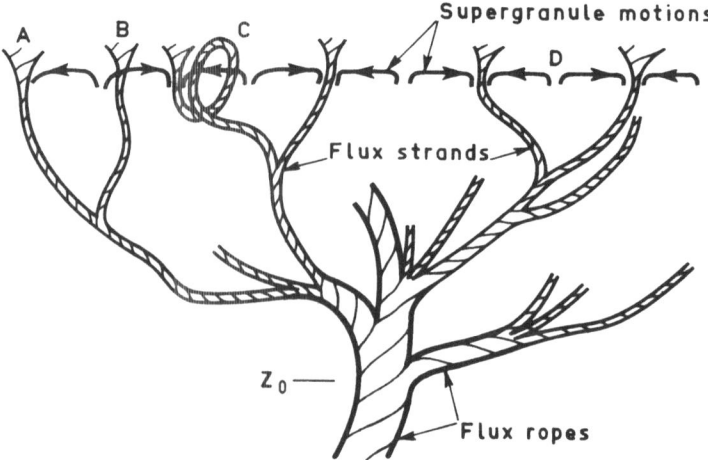

Fig. 6.   A magnetic flux tree which is the structure beneath a unipolar magnetic region. The original rope (below the level Z_0) has split into smaller ropes and frayed into hundreds or even thousands of flux strands. Most of these are carried to the supergranule boundaries giving the illusion of control. Some are in the cell interiors, and some are looped.

calculated that the strands may be bent to a radius roughly equal to that of a supergranule cell as shown. Only with this model does it seem possible to explain how the fields remain essentially vertical, yet are shuffled about by the supergranule motions and concentrated in the cell boundary regions (strand A, and others).

(ii) The model explains why some field concentrations are observed (Livingston and Harvey, 1975) in the interiors of cells as shown by strand B. This, and the fact that no weak fields are observed, shows that the flux cannot be concentrated by the supergranule motions as in the dynamo and similar diffuse-field theories.

(iii) Being helically twisted some strands will develop a loop (strand C) as a result of the kink instability, thus accounting for a small proportion of flux of opposite polarity in otherwise unipolar regions ('pepper and salt') and for tiny bipolar magnetic regions which appear and vanish (Frazier, 1970; Harvey *et al.*, 1975). The loops also provide a likely explanation of the X-ray bright points.

(iv) On occasions a supergranule cell may become surrounded by flux strands (cell D) whose total rigidity is then sufficient to control the convection. In this way we account for the observations (Livingston and Orrall, 1974) of cells with lifetimes up to 7 days.

3.1. MIGRATIONS OF MAGNETIC REGIONS

Unipolar magnetic regions migrate over large distances and while proper motions are difficult to measure, Bumba and Howard (1965) have observed motions of about 0.1 km s^{-1}. This figure is in reasonable agreement with that of Section 2 above of 0.06 km s^{-1} for the rate of upward motion of the ropes which must result in the 'peeling' of the ropes out of the Sun and their dissipation in space. In Paper V we have made a rough estimate of the rate of peeling for a rope of field strength 4000 G lying at a depth of 10^5 km; the result is 0.13 km s^{-1}. Thus observational and theoretical estimates lead to a figure of about 0.1 km s^{-1} or about 0.7 solar circumference per year.

This result explains a number of puzzling phenomena which do not agree with Leighton's (1964) model of dispersion of magnetic fields by a combination of differential rotation and random walk by supergranule motions. These phenomena are now explained in terms of the peeling of flux ropes having the configuration shown in Figure 7a.

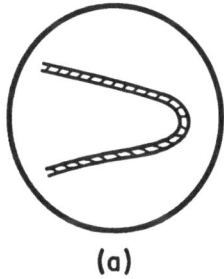
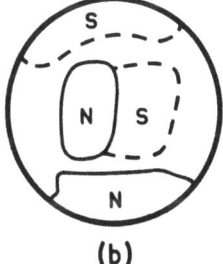

(a)   (b)

Fig. 7. Large-scale solar magnetic fields. (a) A submerged flux rope. (b) Surface background magnetic fields, illustrating the accumulation of f-region flux near the poles, and 'magnetic latitudes'.

(i) Following (f) magnetic regions migrate eastwards and polewards to cause flux accumulations at high latitudes (Figure 7b) and apparent reversal of a general field which, as seen in Section 4, is probably only remnants of f-region flux. Babcock (1961) estimated that this flux would provide polar fields 100 times stronger than observed, and so there must be a dissipative mechanism. This was confirmed by Bumba and Howard (1965) who observed that most of migrating fields simply disappear by a process of expansion and weakening.

This effect is not explicable by the dynamo mechanism because huge areas of a *single* polarity simply fade away. We suggest that it is due to the peeling of a flux rope out of the Sun and its entire loss, as described in Section 4.

(ii) The synodic period of 27 days for recurrent geomagnetic storms and lower latitude solar background fields (see Newkirk, 1971) seems to indicate that the Sun has a rigidly rotating core which somehow controls surface features. However, the flux-rope model provides a more plausible interpretation of such *uniform rotation*.

The preceding (p) magnetic regions migrate westwards and equatorwards with speeds of $\approx 0.1 \text{ km s}^{-1}$ and along paths as delineated by the flux rope of Figure 7a. The p regions which originate at latitudes near $\pm 40°$ move mainly westward, and by adding their proper motion to the rotational surface speed of $\approx 1.3 \text{ km s}^{-1}$ the synodic period is reduced from 29 days to ≈ 27 days. Near the equator the proper motion is equatorwards and so does not change the synodic period which is 27 days, so that there is a strong tendency for a general rotation at this period.

(iii) The remarkable feature of active longitudes, or the appearance of new active regions within or close to old p regions, is explained by the flux-rope model. As a p rope section is peeled from the Sun it disturbs an adjacent rope and causes it to erupt; the ropes may have some magnetic connection, or the effect may be due to turbulence. Since f regions migrate polewards they traverse latitudes which were active earlier in the 11-yr cycle and so have lost their ropes; thus active longitudes tend to involve p rather than f regions.

(iv) Finally, there is the puzzling phenomenon of *magnetic longitudes* which stretch across the equator as shown in Figure 7b (see Stenflo, 1972; Svalgaard *et al.*, 1974). This is simply an extension of active longitudes, for the case where the flux rope is peeled out across the equator so that the p magnetic regions invade the opposite hemisphere. It is easily seen that if two active longitudes develop on opposite sides (front and back) of the Sun and in opposite hemispheres they will turn the Sun into a dipole with its axis in the equatorial plane as sometimes observed (Wilcox, 1971).

It may now be seen that the combination of polar fields and magnetic longitudes shown in Figure 7b is just that invoked by Svalgaard *et al.* (1974) to explain large-scale coronal and interplanetary magnetic fields.

3.2. THE PHOTOSPHERIC AND CHROMOSPHERIC NETWORKS

Let us turn from the large-scale magnetic features to those which are known to be responsible for the facular network. The structural features of the network have been reviewed recently by Bray and Loughhead (1974) and Michard (1974), and the main ones are shown in Figure 8a which follows earlier sketches of Beckers and of

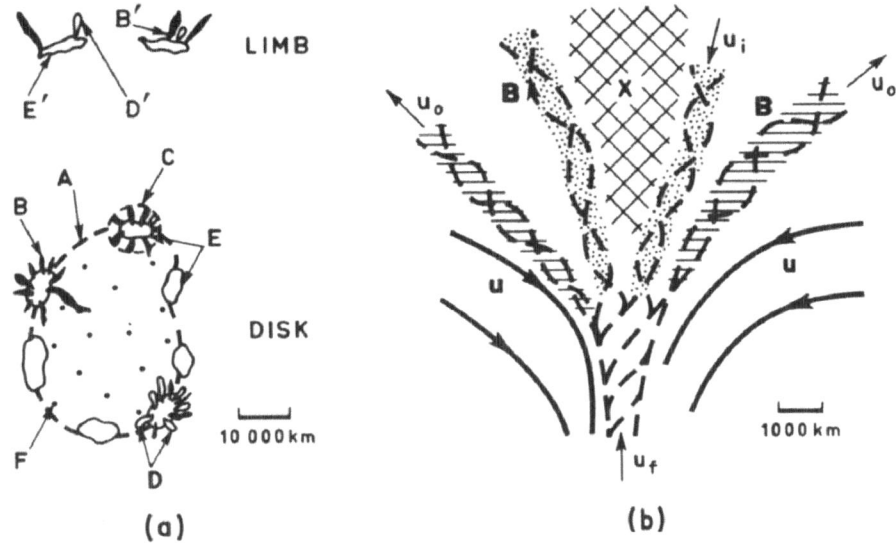

Fig. 8. The chromospheric network. (a) The various network elements comprising dark and bright mottles, rosettes and bushes, spicules and grains (after Beckers). (b) Their structure is explained in terms of a flux strand which frays in the chromosphere into ≈ 10 flux fibres.

Michard. We explain these various elements in terms of the individual flux strands shown in our earlier Figure 6, and now shown in further upward projection in Figure 8b. The base of this strand is in or below the photosphere, and the strand has untwisted and frayed into a dozen or so flux 'fibres'. Since a typical strand has a flux of $\approx 4 \times 10^{18}$ Mx, a typical fibre has a flux of $\approx 4 \times 10^{17}$ Mx. The models of the various elements are as follows.

(i) Around the boundary of a supergranule cell A are distributed about ten network elements each attributed to a single flux strand. The various fibres of each strand delineate the *dark mottles B*, a group of which may form a *rosette C*. Another common network element is the *chain* (or double chain), which is not shown but is explained as a particularly loose bundle of fibres which has been compressed and flattened by supergranule motions.

(ii) On the limb rosettes become *bushes* and it seems likely that dark mottles appear as *spicules B'* which are intermittent emissions of plasma (u_0) caused by the buffeting of a flux fibre by the granule motion u (Piddington, 1972b). The twisted fibres account for the observed rotation of spicules and for the fact that they are long and slender. Some ejected plasma will fall back and some may traverse the corona and fall into a distant flux strand (u_i).

(iii) As well as dark mottles there are *bright mottles D, D'* which form a structure E, E' within and below the rosette or bush. These elements might be attributed to hot non-magnetic gas (X in Figure 8b), heated by transverse oscillations of the flux fibres and perhaps identical with Giovanelli's (1974, Figure 5) 'diffuse' component. The vibrations of the flux fibres (Alfvén waves) are, in turn, attributed to granule motions at lower levels.

This rather tentative magnetic model of the photospheric and chromospheric network gives some idea of the great difficulties which will be met in attempting a

detailed plasma model. One-or two-component models seem unlikely to satisfactor-
ily explain the main features which include hot and cold non-magnetic plasma and
inward and outward moving magnetic plasma.

(iv) As seen in Figure 6, every flux strand is connected to its flux rope to form a
tree-like structure, with the outer strands all tending to be tilted away from the axis of
the tree trunk. This might explain the ordered 'porcupine' structure of some fields of
spicules (Bray and Loughhead, 1974, p. 37). Again, as the whole magnetic structure
or 'tree' is migrating, one might expect to see the upper branches of the tree all tilted
away from the direction of migration. In this way we explain the puzzling 'wheat field'
pattern of spicules.

(v) In the interior of the cells we see the *fibrils*, some of which may be unusually
long dark mottles. Some, however, which occur in active regions are a different
phenomenon, caused by a flux fibre which loops across one or two supergranule cells
rather than rising into the corona as do the great majority of flux fibres.

(vi) Also in the cell interior are the dark and bright *grains* (F in Figure 8b) which
are attributed to the stray magnetic elements known to occur within the cells
(Livingston and Harvey, 1975). The grains are then smaller examples of the network
elements C and E.

4. The Poloidal and Toroidal Field Systems

The primordial field, or flux-rope, model of solar magnetic fields is based firmly on
the observations of surface fields described by Babcock (1961) and greatly extended
in later papers summarized above. All of these effects are caused by toroidal fields,
and so we require an explanation of the origin of these toroidal fields and the reason
why they reverse every 11 years. We have suggested earlier (Piddington, 1971a) that
they are wound from a non-reversing and presumably *primordial poloidal field*. The
observational evidence for the existence of such a field is discussed in Paper IX (in
preparation) of this series and is summarized briefly here.

(i) The observed reversals of the polar fields during 1957–58 (Babcock, 1959)
were taken as proof of the reversal of a 'general' poloidal field, an effect which only
seems explicable in terms of a dynamo field. Today the situation has changed
drastically, and it is known that the highly erratic behaviour of the polar fields is
caused primarily by the migration to those regions of a large amount of flux of the **f**
magnetic regions. There is no evidence that these fields have merged with α-effect
(dynamo) fields at lower latitudes to provide a 'general' poloidal field. On the other
hand it is quite possible that the **f**-region fields are so strong that they usually hide a
weaker primordial field. A choice between these possibilities must be made on the
basis of other evidence.

(ii) Bumba *et al.* (1968) report the observation of magnetic fields of a new cycle
forming in one broad longitude zone, in an area where weak fields from the old cycle
were still visible. This result is explicable only in terms of new toroidal fields which
are wound below the old, presumably from a permanent poloidal field.

(iii) It is reasonably well established that solar magnetic fields are often delineated
by plasma structures, and during the eclipse of 1954, June 30 the coronal structure

revealed an almost flawless dipole field. Waldmeier (1955) termed this a *new, undisturbed form* as distinct from the so-called minimum form which always reveals relics of active region fields. We suggest that on this unique occasion we were allowed a glimpse of a true, deep-seated dipole-type field, the polarity being positive or outwards in the northern hemisphere.

The alternative explanation, that the field is a dynamo field, is most unlikely because such a field is patched together from cyclonic loops which, observationally, are identified with the fields of active regions. These and their remnants are always highly irregular in structure, and a highly ordered and symmetric field seems possible only when the background fields have left the Sun entirely as required in the primordial field model.

(iv) It has often been suggested that a primordial field would have been lost during the Hayashi phase of turbulence. This argument is based on the concept of turbulent or eddy diffusion which has been shown in Papers II and III to be false.

4.1. THE TOROIDAL FIELD SYSTEM

All important theories of solar magnetic fields agree on one process: that a toroidal field arises from the progressive winding up of a poloidal field by the latitude variation $\partial\omega/\partial\theta$ of the non-uniform rotation. They also agree that after attaining a certain stage of development the submerged fields will erupt to provide active regions and spot groups. By adjusting the various parameters it is then a simple matter to account for the observed butterfly diagram, so that this feature is common to all theories and provides no particular support for the dynamo theory.

A major addition to these ideas was Babcock's (1961) concept of rolling the toroidal flux into flux ropes whose cohesive forces made them immune to convective motions. Such twists account for so many observational features that they must be an accepted feature of solar fields. However, twisted flux ropes cannot provide the dynamo α-effect and so must be divorced from dynamo theory. The subsurface field system inferred from observations is now totally different from that of the dynamo fields, which comprise a generally diffuse and tangled field system which is dominated by the convective motions to provide the α-effect.

Submerged flux ropes have field strength of ≈ 4000 G, and if the surface fields have a strength averaged over the whole sum of, say, 20 G then the vertical ropes occupy only $\approx \frac{1}{2}\%$ of the convection zone. The remaining $99\frac{1}{2}\%$ is non-magnetic and the two plasmas provide two entirely separate and mainly isolated atmospheres. The isolation is imposed, of course, by the very large electrical conductivity in the convection zone (Paper II) and the consequent very slow rate of diffusion of plasma across field lines. The isolation allows the plasma pressure inside the flux-rope system to adjust itself to satisfy Equation (1) at all levels. This is not possible if strong fields are formed in the traditional manner by the compression of weak fields.

The next problem is that of the elimination of the flux ropes of one 11-yr cycle before, or during, their replacement by those of the following cycle. We suggest that all toroidal fields must leave the Sun and float away, and most may be lost simply by the peeling process discussed in Section 3. If this process continues until the ropes are peeled back to their junctions with the poloidal field, then most of the plasma within the flux ropes flows back into the shearing regions where it existed before the ropes were created.

It is possible, however, that some ropes are broken in two or more places and that we must account for the disposal of the large amount of plasma within these rope sections. Such a section will become shorter and shorter as it is peeled from both ends, but its plasma cannot escape. However, the section must eventually lose all of its twists and fragment into fibres of diameter <100 km. In a thin layer within the photosphere the electrical conductivity falls to a sufficiently low value for these fibres to be separated from their plasma within a few days and so they are also lost from the Sun.

4.2. REVERSAL OF THE TOROIDAL FIELDS

If the Sun does have a significant primordial field then, somehow or other, the field and the differentially rotating plasma must have interacted and evolved so that, averaged over long periods, the magnetic field lines lie on surfaces of isorotation. This does not mean that at all times they lie on those surfaces, otherwise there would be no ω-effect and no toroidal fields at all. For 11 years the field lines of a poloidal field must tilt one way relative to the $\omega = \text{const.}$ surfaces, and for the next 11 years they must tilt in the opposite way. A simple form of such a model is shown in Figure 9, which was inferred from observations of meridional oscillations by Richardson and Schwarzschild (1953), and also deduced as an essential alternative to the untenable α-effect (Piddington, 1971a).

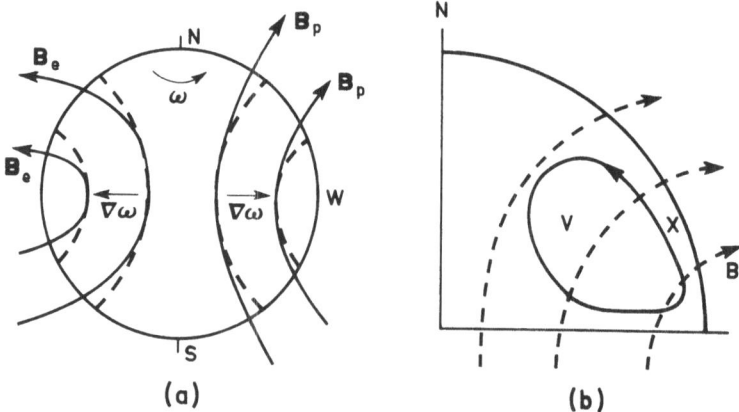

Fig. 9.   Illustrating the mechanism of reversal of the solar toroidal magnetic fields. (a) A dipole-like magnetic field has its lines of force lying, on an average, on the surfaces of isorotation; for 11 years they are tilted equatorwards of these surfaces (B_e) and for 11 years polewards (B_p). Consequently two opposed sets of toroidal fields are created by the shears $\nabla\omega$. (b) A meridional oscillation which could change the field B from the equatorial form B_e to the polar form B_p.

Figure 9a shows a solar polar section with hypothetical surfaces of rotation as dashed lines. It will be seen that $\nabla \cdot \omega$ is outwards which is a requirement needed to account for the twists observed in sunspot magnetic fields (Babcock, 1961), and is, incidentally, the opposite to that required by dynamo theory (Stix, 1974). The figure also shows two halves of a permanent poloidal field as they appear during the two halves of a 22-yr cycle. On the left is the 'equatorial' form B_e, with the emerging field lines tilted equatorwards of the isorotation surfaces. On the right is the 'polar' form B_p with the opposite direction of tilt. In Figure 9b is shown a single quadrant of a meridional oscillation V which, with speeds of only ≈ 100 cm s^{-1}, will transform the B_e field to the B_p form. The toroidal fields wound by the ω-effect from B_p is opposite that wound from B_e; each successive set of toroidal ropes is wound below its predecessor and helps to push the latter out of the Sun. Thus, at least kinematically we have explained the solar 22-yr cycle in terms of a primordial field and a meridional oscillation.

While the dynamo theory is limited to kinematics, we have been able to provide a simple dynamical picture of the above hydromagnetic oscillator (Piddington, 1971b). In the region X of Figure 9b the magnetic field is sheared and so resists and eventually reverses the motion to provide a simple harmonic oscillator. Reasonable values of the various parameters lead to a period of 18 yr in good agreement with that observed. There is even the possibility of explaining the excitation of the meridional oscillation in terms of meridional forces exerted by the toroidal fields.

There is an interesting alternative to the simple model of Figure 9, which requires no change in the poloidal field. Instead of the above V oscillation we invoke an oscillation which involves an ordered transfer of angular momentum $\omega(r, \theta)$. In this way the surfaces of isorotation oscillate back and forth across the stationary field lines. The study of these different hydromagnetic oscillators should provide interesting problems and perhaps a true picture of solar magnetic fields.

5. Discussion of the dynamo Theory

It has often been stated that the "main features of the dynamo that produces the solar cycle are now generally accepted" (Weiss, 1971). Yet as shown in Papers I, IV and V and in the above review most observational evidence is opposed to a dynamo theory and theoretical arguments given in Papers II and III appear to raise fatal difficulties. We believe that the sequence of events leading to this disagreement may be summarized as follows.

(i) Cosmic dynamo theory opens fascinating possibilities, and the winding of a toroidal field (ω-effect) provided a major piece of the mechanism which Parker (1955) completed with his cyclonic turbulence or α-effect. However, apart from the solar differential rotation, the model lacked observational support until Babcock's (1961) model appeared to provide such support. Overlooked was the fact that Babcock's model is concerned mainly with the ω-effect, which is not in dispute, and his attempt to reverse the poloidal field was neither necessary nor satisfactory (Section 4 above). The flux-rope concept may be adopted without difficulty by the primordial field model, thereby removing the only major flaw in Babcock's model.

(ii) The next major event in this sequence was the discovery of the supergranulation and of its *apparent* control of the magnetic fields. At this stage the concepts of cyclonic turbulence, flux ropes and supergranule control were taken as complementary and as providing a sound observational basis for dynamo theory. In fact, the reverse is true: as seen in Sections 2 and 3 flux ropes evolve almost entirely under the influence of magnetic stresses and are quite incompatible with the α-effect.

(iii) In turning to the theoretical side, we must again go back two decades to Elsasser's (1956) assumption that just as turbulence disperses smoke, heat, or any other passive property of a fluid, so it will *eliminate* a passive magnetic field. This assumption has gained general recognition and it is difficult to overestimate its pervasive influence in hydromagnetic turbulence theory.

The relevant equation for the magnetic field **B** is

$$\frac{\partial \mathbf{B}}{\partial t} = \nabla(\mathbf{V} \times \mathbf{B}) + \eta \nabla^2 \mathbf{B}, \tag{2}$$

where **V** is the plasma velocity and η the diffusivity. The point is that turbulence V_t is part of the V term and represents transport of the field lines *with* the plasma, while only the η term represents transport *through* the plasma. There are two similar terms in the case of a scalar quantity (Paper III) but here the physical results of both are much the same, leading to Elsasser's error. Using V_t as a diffusion effect leads to a new diffusivity (Leighton, 1964; Rädler, 1968; Parker, 1970a, 1971; Vainshtein and Zel'dovich, 1972; Nakagawa and Priest, 1973; Stix, 1974; Meyer *et al.*, 1974)

$$\eta_t \approx 0.2 \, V_t L, \tag{3}$$

where L is the characteristic eddy dimension. If we now insert η_t in place of η in Equation (2) we increase the diffusion rate in the convection zone by a factor of $\gtrsim 10^7$. The electrical resistivity is supposed to be similarly increased.

The above procedure is an essential part of modern dynamo theory, yet it is *quite unjustified* (Piddington, 1972a, 1973, Paper III; Parker, 1973a) and invalidates solar dynamo theory.

(iv) The above procedure not only grossly overestimates the diffusivity, but hides an even more serious error in dynamo theory. By taking V_t away from the **V** term of Equation (2) we lose sight of the fact that these motions cause rapid amplification of the magnetic fields by shear, so that instead of eliminating unwanted fields they build them up. Parker (1973a, b) has attempted to meet this difficulty by two *drastic revisions* of the dynamo theory. First, instead of the magnetic loops formed by cyclonic turbulence merging rapidly, the fields are drawn out into long, thin filaments. When a filament is thin enough it is able to merge with an oppositely directed neighbour and vanish. Second, in order to accelerate merging over the rate allowed by molecular diffusion, the Petschek mechanism is invoked.

This new dynamo shows even less agreement with observations and, in our opinion, less plausibility than the old. Additionally, it has been analysed (Paper III, Section 4) and appears to fail on theoretical grounds; fields either halt convection or become chaotic.

(v) We have seen above the diffusivity used in the dynamo is far too large, that turbulence creates new fields faster than it destroys them, and that Parker has, accordingly, proposed a radically new 'thin filament' dynamo. Yet in a recent review, Stix (1974) has virtually ignored these developments. Without giving any physical arguments he repeats the claim that transport of magnetic field *with* the gas is equivalent to diffusion *through* the gas.

References

Babcock, H. D.: 1959, *Astrophys. J.* **130**, 364.
Babcock, H. D.: 1961, *Astrophys. J.* **133**, 572.
Bray, R. J. and Loughhead, R. E.: 1974, *The Solar Chromosphere*, Chapman and Hall, London.
Bumba, V.: 1963, *Bull. Astron. Inst. Czech.* **14**, 91.
Bumba, V. and Howard, R.: 1965, *Astrophys. J.* **141**, 1502.
Bumba, V., Howard, R., Martres, M. J., and Soru-Iscovici, I.: 1968, in R. Howard (ed.), 'Structure and Development of Solar Active Regions', *IAU Symp.* **35**, 13.
Cowling, T. G.: 1853, in G. P. Kuiper (ed.), *The Sun*, University Chicago Press, p. 532.
Danielson, R. E.: 1961, *Astrophys. J.* **134**, 312.
Elsasser, W. M.: 1956, *Rev. Mod. Phys.* **28**, 135.
Frazier, E. N.: 1970, *Solar Phys.* **14**, 89.
Frazier, E. N.: 1972, *Solar Phys.* **26**, 130.
Giovanelli, R. G.: 1974, *Solar Phys.* **38**, 117.
Harvey, J. and Hall, D.: 1974, *Bull. American Astron. Soc.* **6**, 81.
Harvey, J. and Harvey, K.: 1973, *Solar Phys.* **23**, 61.
Harvey, K. L., Harvey, J. W., and Martin, S. F.: 1975, *Solar Phys.* **40**, 87.
Krause, F. and Rädler, K.-H.: 1971, in R. Howard (ed.), 'Solar Magnetic Fields', *IAU Symp.* **43**, 770.
Leighton, R. B.: 1964, *Astrophys. J.* **140**, 1547.
Leighton, R. B.: 1969, *Astrophys. J.* **156**, 1.
Livingston, W. and Harvey, J.: 1971, in R. Howard (ed.), 'Solar Magnetic Fields', *IAU Symp.* **43**, 51.
Livingston, W. and Harvey, J.: 1975, American Astron. Soc., Solar Phys. Div., Meeting Boulder, Colorado.
Livingston, W. C. and Orrall, F. Q.: 1974, *Solar Phys.* **39**, 301.
Meyer, F., Schmidt, H. U., Weiss, N. O., and Wilson, P. R.: 1974, *Monthly Notices Roy. Astron. Soc.* **169**, 35.
Michard, R.: 1974, in R. G. Athay (ed.), 'Chromospheric Fine Structure', *IAU Symp.* **56**, 3.
Nakagawa, Y. and Priest, E. R.: 1973, *Astrophys. J.* **179**, 949.
Newkirk, G.: in R. Howard (ed.), 'Solar Magnetic Fields', *IAU Symp.* **43**, 547.
Newton, H. W.: 1955, *Vistas Astron.* **1**, 666.
Parker, E. N.: 1955, *Astrophys. J.* **122**, 293.
Parker, E. N.: 1970a, *Ann. Rev. Astron. Astrophys.* **8**, 1.
Parker, E. N.: 1970b, *Astrophys. J.* **162**, 665.
Parker, E. N.: 1971, *Astrophys. J.* **163**, 279 and **164**, 491.
Parker, E. N.: 1973a, *Astrophys. Space Sci.* **22**, 279.
Parker, E. N.: 1973b, *Astrophys. J.* **180**, 247.
Piddington, J. H.: 1971a, *Proc. Astron. Soc. Australia* **2**, 7.
Piddington, J. H.: 1971b, *Solar Phys.* **21**, 4.
Piddington, J. H.: 1972a, *Solar Phys.* **22**, 3.
Piddington, J. H.: 1972b, *Solar Phys.* **27**, 402.
Piddington, J. H.: 1973, *Astrophys. Space Sci.* **24**, 259.
Piddington, J. H.: 1974a, in R. G. Athay (ed.), 'Chromospheric Fine Structure', *IAU Symp.* **56**, 269.
Piddington, J. H.: 1975a, *Astrophys. Space Sci.* **34**, 347, Paper I.
Piddington, J. H.: 1975b, *Astrophys. Space Sci.* Paper II **35**, 269.
Piddington, J. H.: 1975c, *Astrophys. Space Sci.* Paper III **38**, 157.
Piddington, J. H.: 1975d, *Astrophys. Space Sci.* Paper IV.
Piddington, J. H.: 1975e, *Astrophys. Space Sci.* Paper V.
Rädler, K. H.: 1968, *Z. Naturforsch* **23a**, 1851.
Richardson, R. S. and Schwarzschild, M.: 1953, Academia Lincei, Convegno 11, Rome 1952, p. 228.
Sheeley, N. R.: 1972, *Solar Phys.* **25**, 98.
Simon, G. W. and Leighton, R. B.: 1964, *Astrophys. J.* **140**, 1120.
Stenflo, J. O.: 1972, *Solar Phys.* **23**, 307.
Stenflo, J. O.: 1973, *Solar Phys.* **32**, 41.
Stix, M.: 1974, *Astron. Astrophys.* **37**, 121.
Svalgaard, L., Wilcox, J. M., and Duvall, T. L.: 1974, *Solar Phys.* **37**, 157.
Vainshtein, S. I. and Zel'dovich, Ya. B.: 1972, *Usp. Fiz. Nauk.* **106**, 431; *Soviet Phys. Usp.* **15**, 159.
Vrabec, D.: 1974, in R. G. Athay (ed.), 'Chromospheric Fine Structure', *IAU Symp.* **56**, 201.
Waldmeier, M.: 1955, *Z. Astrophys.* **36**, 275.
Weiss, N. O.: 1971, in R. Howard (ed.), 'Solar Magnetic Fields', *IAU Symp.* **43**, 757.
Wilcox, J. M.: 1971, *Publ. Astron. Soc. Pacific* **83**, 516.

DISCUSSION

Stenflo: Can you explain how the twisting of the toroidal magnetic flux is produced in the first place?

Newkirk: What supplies the energy to twist the ropes in Piddington's theory?

Giovanelli: I think that it is differential rotation, which Piddington does not attempt to explain but takes as an observed fact. You will recall that Babcock, whom Piddington follows, also invoked twisting due to buffeting or shear by convective eddies.

Parker: I would like to point out the origin of Piddington's divergent view, the 'Piddington heresy', in contrast to the orthodox dogma that the solar magnetic fields are the result of a hydromagnetic dynamo. The basis is that Piddington does not believe in turbulent diffusion. Without turbulent diffusion the magnetic lines of force are locked into the solar gas because of the high electrical conductivity. The resistive decay times are 10^5 yr or more, instead of 10 yr. Magnetic lines of force do not reconnect, and the hydromagnetic dynamo cannot operate. Piddington's disbelief in turbulent diffusion is based on the fact that the turbulent eddies wrap and stretch the magnetic field into filaments, with the field strength in each filament growing exponentially in time. Thus the mean square field quickly reaches the equipartition value and halts the turbulence. Piddington argues that turbulent diffusion of the mean fields does not proceed beyond this point.

Piddington has put his finger on a weak spot in turbulence theory. There is no formal deductive answer to his objection to turbulent diffusion. All 'theories' of turbulent diffusion of the mean field merely take it for granted that there is no problem with the growth of the *mean square* field. Yet the elementary calculation shows the exponential growth.

The answer to Piddington's objection to turbulent diffusion is, I think, the rapid reconnection of the lines of force of neighboring strands of field. This reconnection takes place at a fraction of the Alfvén speed (as pointed out by Priest this morning). The strands are continuously cut to pieces and the exponential growth does not occur. Thus I believe that the turbulent decay of a weak mean field goes on relatively unhindered by the tensions in the individual small-scale strands. Indeed, on the basis of a simple statistical hypothesis, it is possible to show that the mean magnetic field diffuses in the same way as a scalar field (such as smoke) carried in the same fluid.

Piddington has developed his ideas of the behaviour of magnetic flux in the Sun in response to his view that it is permanently frozen into the fluid.

If I may comment now on one aspect of Piddington's picture of the magnetic tree whose emergence carries the active region to the surface, it is to note that an essential part of his idea is the unwrapping of the twisted flux rope at its apex where it emerges through the surface. It is a theoretical fact that magnetic buoyancy pushes the twisted coils upward along this tube, concentrating them in the apex. Thus, the apex is the place of *maximum* twisting, rather than the place of least twisting suggested by Piddington. I do not understand how the twisted rope is to fray, coming at its apex to form the many separate features of the active region.

Giovanelli: The tube expands greatly when it reaches the outer solar layers where the external gas pressure is too low to confine the magnetic field. The twisting is unable to prevent this expansion. There would then be a discontinuity at the surface, which is eliminated by the upward propagation of the twists from just under the surface, the process which Piddington associates with 'fraying' in the case of a rope consisting of many strands.

A participant: I should expect that the ropes would simply untwist completely with the Alfvén velocity as soon as they break through the photosphere and that the flux concentration dilemma still remains.

Priest: The sunspot field which you assume is, in my view, unlikely. If one forms a flux tube by bringing together many twisted fibres, rapid reconnections will occur, leaving a simply twisted field. It is easy to reconnect field lines which are aligned arbitrarily and not just at 180°.

Further the amount of twist in your diagrams is greatly exaggerated. Raadu showed that, as soon as field lines are twisted more than once around the axis of a finite tube, the field becomes unstable to the kink instability. Can the observers tell us how common are spiral penumbral structures?

Giovanelli: In white light the penumbral fibrils are fairly straight and radial, but at times fibrils seem to cross at angles, contrary to what Dr Priest has suggested (but there may be a vertical separation between them). The superpenumbral fibrils seen in Hα are much longer, and many show curved spiral structure.

Stenflo: Hale noticed the Hα vortex pattern around sunspots already in 1904. Later statistical analysis by Richardson showed that a large fraction of all sunspots have a clear twist as seen in Hα. Measurements of the transverse magnetic fields made at Crimea show that the magnetic field is generally twisted down in the photosphere as well.

Bumba: There is evidence about the spiral structures around the sunspots in the chromosphere as has been pointed out by Dr Stenflo. But you may observe such structures around sunspots very rarely in the photosphere. We have succeeded in obtaining a series of good photographs of three regular sunspots demonstrating the development of such spiral organization of penumbral fibrils. Studying these series of photographs we come to the conclusion that the observed spiral structures are possibly not connected with the kinematics of the discrete sunspots. On the other hand Knoška's investigation of sunspot rotation shows clearly that many sunspots rotate around their vertical axis, although the greater part of them rotate in the same direction in both solar hemispheres.

Stix: There are indirect indications of polar field reversals prior to the one of 1957/58, e.g. poleward migration of filament zones, or numbers of polar formulae. But even if we forget these, there was still another observed polar field reversal occurring in 1971/72, and I wonder whether you or maybe Dr Howard could comment on the reality of this last reversal.

Howard: There is no doubt that the polar field of the Sun reversed around the time of the last solar maximum – in 1969 and 1972. My paper in *Solar Physics* gives more details of this.

Giovanelli: But your own results (*Solar Phys.* **25**, 5 (1972); **38**, 59, 283 (1974)) show that the polar fields vary irregularly, with little correspondence between north and south. Under these circumstances it is rather difficult to speak confidently about reversals of polar fields being associated with solar maxima; to me, it seems so far that they have been quite irregular. Observations will be needed over a much longer period to establish the pattern.

Vandakurov: I want to note that the problem of turbulent diffusion in the presence of a magnetic field may also be considered from the point of view of stability theory. The estimate turbulent transport characteristics, we use the hypothesis of Kadomtsev and Pogutse which was cited in my paper, published recently in *Solar Physics*. This seems to favour usually quoted values of the turbulent diffusion. It should be emphasized however that this approach is valid only if the magnetic configuration is stationary. But as far as the convection zone with large convective motion is concerned, the fulfilment of the stationary condition is rather a difficult problem.

Chvojková: Wouldn't it be possible to join these two theories? Dr Stix's report, for example, has finished with the evolution of a poloidal field, while Babcock had begun by a poloidal field which became amplified by winding around the Sun due to differential rotation. Babcock's mechanism could, perhaps, additionally amplify the field given by Stix. It seems that in most phenomena the joint model would be very similar to Piddington's.

A MODEL OF THE SOLAR CYCLE DRIVEN BY THE DYNAMO ACTION OF THE GLOBAL CONVECTION IN THE SOLAR CONVECTION ZONE

HIROKAZU YOSHIMURA

Hale Observatories, Carnegie Institute of Washington,
California Institute of Technology, 813 Santa Barbara Street,
Pasadena, Calif. 91109, U.S.A.

Extensive numerical studies of the dynamo equations due to the global convection are presented to simulate the solar cycle and to open the way to study general stellar magnetic cycles. The dynamo equations which represent the longitudinally-averaged magnetohydrodynamical action (mean magnetohydrodynamics) of the global convection under the influence of the rotation in the solar convection zone are considered here as an initial boundary-value problem. The latitudinal and radial structure of the dynamo action consisting of a generation action due to the differential rotation and a regeneration action due to the global convection is parameterized in accordance with the structure of the rotation and of the global convection. This is done especially in such a way as to represent the presence of the two cells of the regneration action in the radial direction in which the action has opposite signs, which is typical of the regeneration action of the global convection. The effects of the dynamics of the global convection (e.g., the effects of the stratification of the physical conditions in the solar convection zone) are presumed to be all included in those parameters used in the model and they are presumed not to alter the results drastically since these effects are only to change the structure of the regeneration action topologically. However, since the structure of the differential rotation is not known precisely, several typical cases of the differential rotation are examined. A nonlinear process is included by assuming that part of the magnetic field energy is dissipated away when magnetic-field strength exceeds some critical value, simulating the formation of active regions and subsequent dissipation. By adjusting the parameters within a reasonable range, oscillatory solutions (the dynamo waves) are obtained to simulate the solar cycle with periods of the right order of magnitude and with patterns of evolution of the latitudinal distribution of the toroidal component of the magnetic field near the surface similar to the observed Butterfly Diagram of sunspots. In those cases, which simulate the solar cycle well, a slight radial gradient of differential rotation increasing inward plays a role, but the latitudinal equatorial acceleration dominates the differential rotation at least near the surface where sunspots are assumed to be formed. The evolution of the latitudinal distribution of the radial (poloidal) component of the magnetic field shows patterns similar to the Butterfly Diagram but having two branches of different polarity in each hemisphere, predicting that the solar general magnetic field has quadropole-like structure. The development of the radial structure of the magnetic field associated with the solar cycle is presented to explain the simultaneous presence of active regions with the polarity of the new cycle in a region of magnetic field with the polarity of the old

Bumba and Kleczek (eds.), Basic Mechanisms of Solar Activity, 409–413. All Rights Reserved.
Copyright © 1976 by the IAU.

Fig. 1.   The evolution of the latitudinal distribution of the *toroidal* general magnetic field near the surface for the standard case of this study in which the latitudinal gradient of the differential rotation (equatorial acceleration) plays the dominant role with the presence of slight radial gradient (rotation increasing inwards). Abscissa is sin (latitude) from south to north pole and the ordinate is time evolving from upward to downward. Note that this diagram well represents the observed Butterfly Diagram of sunspots and that there are slight poleward propagating branches.

Fig. 2. The same as Figure 1, but of the *poloidal* (*radial*) general magnetic field. Note that there are two branches (wings), in each hemisphere. This predicts that the Sun behaves quadrupole-like not dipole-like. This prediction has been confirmed by Yoshimura (1975c, 1976a, b). This branching is due to the fact that the two (northern and southern) hemispheres behave rather independently. The branching is sensitive to the rotational law of the convection (dynamo) zone of the Sun.

cycle. The importance of the poleward migrating branch of the Butterfly Diagrams of the toroidal and poloidal fields, barely seen for sunspots but clearly seen for polar prominences and for the observed general magnetic field, is emphasized in relation with the relative importance of the role of the latitudinal and radial shears of the differential rotation. Furthermore, by studying general cases of the differential rotational law, the evolution of the magnetic field inside the Sun is found to show that the dynamo waves of the magnetic field propagate along isorotation surfaces (this was done by making a movie showing the evolution of the field); e.g., in the case of purely latitudinal differential rotation, the waves propagate radially, while in the case of purely radial differential rotation, they propagate latitudinally. The direction of the propagation and the phase shift between the evolutions of the poloidal and toroidal fields depend upon the sign of the regeneration action as well as on the rotational law. The importance of the explicit recognition of these results, which hold for any law of the differential rotation, is stressed with regard to the problem of inferring the actual rotational law inside the Sun and stars in general. Further, the general dependence of the characteristics of the dynamo waves responsible for the solar cycle on those parameters adopted, i.e., on the structure of the generation and regeneration actions, is discussed in order to understand the nature of the solar cycle and to parameterize the basic characteristics of the solar cycle. The present formulation of the dynamo equations can be applied to further studies of the solar and stellar cycles.

References

Yoshimura, H.: 1975a, *Astrophys. J. Suppl. Series* **29**, 467.
Yoshimura, H.: 1975b, *Astrophys. J.* **201**, 740.
Yoshimura, H.: 1975c, this volume, p. 137.
Yoshimura, H.: 1976a, *Solar Phys.* (in press).
Yoshimura, H.: 1976b, *Solar Phys.* (in press).

DISCUSSION

Gilman: If I understand correctly, you included one particular differential rotation profile in your calculation of regeneration action produced by convection, but then you used this regeneration action form for dynamo calculations with other differential rotation profiles. This seems inconsistent to me, if the regeneration action is sensitive to the presence of differential rotation.

Yoshimura: I think you misunderstand. I did not assume one particular form of the differential rotation in my calculation of the regeneration action of the dynamo processes. In my model of the solar cycle, I adopted and tested various forms of the regeneration action with general properties of the action which are not influenced much by the form of the differential rotation as far as it is equatorial acceleration. I parameterized the general properties so that the regeneration action changes its form only topologically and searched for the correct forms of the regeneration action and of the differential rotation which well explain the observed characteristics of the solar cycle. The exact consistency of the two forms should be examined by studying the dynamics of the convection zone. However, this is not the present approach. My approach is opposite to this; i.e., the purpose is to determine the forms of the regeneration action and of the differential rotation, which are useful when we study the dynamics of the convection zone, by comparing the solutions of the model with the observed characteristics of the solar cycle. The flexibility of the formulation of the present model, in my opinion, can well serve for this purpose.

Gilman: Please summarize the condition for the last case in the movie, in which propagation of the dynamo wave is toward the equator.

Yoshimura: In the last case, the radial gradient of the differential rotation predominates the process. And the rotation increases inwards. The structure of the regeneration action is such that the upper part of the action predominates the process, which means the action with positive (negative) sign in the northern (southern) hemisphere predominates the dynamo process. Then the dynamo waves, the amplitude of which is considerable only in the upper part, propagates along isorotation surfaces which are parallel to the surface toward the equator.

Durney: Does your model rule out an angular velocity decreasing with depth in the lower part of the convection zone?

Yoshimura: No, not necessarily, if the radial gradient decreasing inwards does not predominate the process. In order to answer this question correctly, there are many problems to be solved. If the active regions originate in the upper part of the convection zone, the Butterfly Diagram of sunspots should not reflect the distribution of the toroidal field of the lower part. In this case, the form of the differential rotation in the lower part could not be determined by observing the surface phenomena as far as the differential rotation in the lower part does not affect the field in the upper part so much. If the active regions originate in the lower part, the phenomena observed at the surface would greatly be affected by the form of the differential rotation in the lower part which directly affects the field in the lower part. In this case, the angular velocity in the lower part may be decreasing inwards since this brings about the equatorial propagation of the dynamo waves in the lower part. However, to make a comment on Dr Durney's question, I think he has in mind that the iso-rotation surfaces of the differential rotation may be cylindrical, which means that the rotation rate is decreasing inwards if we fit it to the equatorial acceleration observed at the surface. I tested this case using my model of the solar cycle and found that it is difficult to explain the various observed characteristics of the solar cycle by this case of cylindrical form of the differential rotation which has been predicted by some investigators of the differential rotation.

Roxburgh: What would happen if you had many cells in the radial direction as I would expect in a compressible medium. Have you done or is it possible to do calculation for a multi-cell model?

Yoshimura: Do you mean a multi-cell structure in the radial direction with global horizontal scale? If so, I think the model is quite unlikely because we have already difficulties to explain the elongated form of the global convection of one cell. Moreover, various observed surface phenomena of magnetic field and activity can well be explained by one-cell model of the global convection. To answer this question correctly, we should study the dynamics of the convection zone. At present, we have no correct answer. Even for the case that the global convection has a multi-cell structure, it is possible to solve the dynamo equations of my model of the solar cycle if the number of cell is not so large. However, I have not done it yet.

Stix: If there are a number of large scale circulation cells above each other in the convection zone, it appears to me that a meaningful regeneration term can only be obtained by averaging vertically over a number of them, in addition to the longitudinal averaging employed by Dr Yoshimura. But then the Boussinesq approximation would yield no regeneration. We would need the density stratification, as has been emphasized by Dr Krause and Dr Rädler in their papers.

(Note added in proof by H. Yoshimura): If we average the regeneration term vertically, we should also average the original MHD equations vertically. Then the resulting equations have only one space coordinate, i.e., the latitude, as in the case of Leighton's model of the solar cycle. In this case, we have some difficulties if the vertical space to be averaged is comparable to the depth of the convection zone. For, in this case, the differential rotation would not depend on the vertical coordinate. Moreover, the latitudinal gradient of the rotation rate would bring about the vertical propagation of the dynamo waves, the phenomenon which cannot be described by one space coordinate model. Only in the case that the vertical space to be averaged is much smaller than the depth of the system, various physical quantities (as well as the dynamo equations) can have vertical structure which is essential to describe the behavior of the dynamo waves. But, in order to assume this, the fluid motions which are responsible for the dynamo processes, should have much smaller scale than that of the system. In the case of the global convection, this seems to be very unlikely even though the global convection has a multi-cell structure in the radial direction.

ON THE DISTRIBUTION OF
ANGULAR VELOCITY IN THE SUN

JAAKKO TUOMINEN

University of Helsinki, Finland

Abstract. The angular velocity distribution as a function of heliographic latitude is directly observed at the solar surface. Also the dependence of the angular velocity on the distance from the solar centre can be studied observationally for the outermost layers of the Sun. But within the Sun the study of angular velocity depends very much on assumptions. One possibility to study the problem is given by the fact that in a bipolar magnetic group the leader is compact while the follower is dispersed over a large area. In Leighton's theory, following Babcock, this phenomenon is explained, in principle, in the following way: Slide 1. When a magnetic rope, as a result of magnetic buoyancy, rises to the surface, it is twisted by differential rotation. Dashed lines in the figure represent *isotachial* surfaces. When the rope has risen to the surface, the two ends are twisted in opposite directions. The surface differential rotation continues to twist one of the now free ends of the rope, while it untwists the other free end. Both twisting and untwisting are, in fact, very slow. They are fastest at the latitude of 35°, where the magnetic rope is rotated once in 26 days. We understand that the distribution of angular velocity within the Sun is involved in this picture. The twisting in the interior can be written equal to the untwisting of the follower spot produced by the surface differential rotation. Of course there are many solutions. Slide 2 gives two of them. The lines represent calculated *isotachial* surfaces and the numbers give the observed angular velocities in radians per day. In the left-hand side distribution the velocity decreases inwards, while in the right-hand side distribution it increases inwards. The right-hand side distribution is also consistent with the conclusion derived from observations that the tilt of sunspot axes from the vertical is relatively small, and with the observation that at the equatorial plane, near the surface, the rotational velocity does not vary with depth.

Reference

Tuominen, I. V. and Tuominen, J.: 1968, *Astrophys. Letters* **1**, 95.

DISCUSSION

Kasinskij: I would like to comment on Dr Tuominen's report by presenting some evidence about angular velocity changes in the chromosphere. The considerations are based on the vector-shifting butterfly diagrams which I have constructed for the mean flare positions relative to sunspots. This vector-diagram disclosed that at the beginning of a cycle there are strong 'east' and at the end of the cycle visible 'west' displacements. Thus, by interpreting this phenomenon in terms of angular velocity changes one can say that the southern hemisphere rotates more slowly at the beginning and faster at the end of a cycle than the northern hemisphere.

Some synchronized form of rotation takes place near the maximum. Therefore one can conclude that in the chromosphere there exists some kind of torsion oscillation which may be predicted by some theories.

Bumba and Kleczek (eds.), Basic Mechanisms of Solar Activity, 415. *All Rights Reserved.*
Copyright © 1976 by the IAU.

ON THE DIPOLE-LIKE PROGRESSIVE WAVE
IN THE PHOTOSPHERE

I. K. CSADA

Konkoly Observatory, Budapest, Hungary

Abstract. Statistical analysis of the photospheric magnetic field averaged over 27 day time intervals shows periodical variations. The results of such an analysis are usually considered to be a pure mathematical representation of the observation material and additional analysis is necessary to infer the physical meaning of the results. Some ideas for this analysis are given by the mathematical method of the 'non-axisymmetric dynamo' theories which suggest the representation of the field by the series

$$H_x = \sum_n \sum_{m=1}^n \mathscr{F}_{nm}(r, v)\, e^{i(\Omega + m\varphi)}$$

i.e. the field variation is equivalent to the interference of progressive wave moving parallel to the equator with velocity

$$|v| = \Omega/m .$$

If $n = 1$ the variation is equivalent to a progressive dipole field circulating over the photosphere by a time $2\pi/\Omega$ and this period can be derived from the 27-day averages. The superimposition of the circulation to the rotation will result in the apparent 'field rotation' being observable as recurrence time of the largest magnetic areas. Observational evidences are found for a progressive dipole circulating in a great circle with a period of $4\tfrac{1}{6}$ yr.

The higher multipole field components (if $n \geq 2$) have no characteristic variations but their mean effect results in statistical spread in the 'multipole field rotation'.

Bumba and Kleczek (eds.), Basic Mechanisms of Solar Activity, 417. *All Rights Reserved.*

PART 4

STELLAR ACTIVITY OF THE SOLAR TYPE

STELLAR ACTIVITY OF THE SOLAR TYPE
OBSERVATIONAL ASPECTS

G. GODOLI

Catania Astrophysical Observatory and Astronomical Institute, University of Catania, Italy

1. Introduction

By the expression 'stellar activity of the solar type' we mean the manifestations of stellar activity determined by a mechanism of the same kind as that producing the solar activity.

As we already know, this mechanism should consist in the interplay of solar poloidal magnetic fields, differential rotation and convection. Gravitational actions by planets and interactions with diffuse matter could also play as triggering agents. It is evident that if this mechanism acts on different stars in different physical conditions it could give rise to quantitatively and qualitatively different phenomena. Vice versa, from the analysis and interpretation of these phenomena, knowing the physical conditions in which they take place, we can hope to deduce new information on the mechanism involved.

Gershberg (1975) notes that since we have no complete physical theory either for the totality of events that take place in a solar active region, or for the mechanism involved in solar activity as a whole, somebody could argue that analogies between solar and stellar events do not contain new information. Wittily Gershberg points out that also Newton found only (!) some analogies between the falling of an apple and the motion of the Moon!

The problem of stellar activity of the solar type was thoroughly investigated by Unsöld in the second edition of his *Physik der Sternatmosphären* (1955) and was successively taken up again by Smith and Smith in their book on solar flares (1963) and by Bray and Loughhead in their book on sunspots (1964). In 1967 a wide research programme on stellar activity of the solar type was undertaken at the Catania Astrophysical Observatory (Godoli, 1967). In 1968 a review of the problem and preliminary results collected at Catania were presented by the writer at the Nobel Symposium. Problems connected with stellar activity of the solar type were reviewed by Oster in the volume recently dedicated to Albrecht Unsöld (1975). Finally, a thorough analysis on the analogies between stellar and solar flares and between stellar and solar flare activities was presented by Gershberg (1975) at the IAU Symposium No. 67 on 'Variables in Relation to the Evolution of Stars and Stellar Systems'.

We can assert that in the last few years the problem of stellar activity of the solar type has been an aspect of the increasing convergence in the fields of stellar and solar physics.

In this review we shall summarize the latest results in this field.

For the sake of completeness we shall also briefly recall information reported in the quoted literature.

Bumba and Kleczek (eds.), Basic Mechanisms of Solar Activity, 421–446. All Rights Reserved.
Copyright © 1976 by the IAU.

2. The Sun as a Variable Star

2.1. INTRODUCTION

Before trying to describe stellar observational facts that could be determined by a mechanism of the same kind as that producing the solar activity, let us briefly consider how the manifestation of solar activity would appear at stellar distances.

2.2. SOLAR ROTATION AND MAGNETIC FIELD

As Howard (1972) pointed out, if the Sun were observed as a distant star its rotation rate could be detected only at the very highest stellar spectroscopic dispersion and the 5% variation observed in the Sun rotation rate would pass entirely undetected.

In 1967 Bumba *et al.*, using, instead of the solar image, a parallel beam from the coelostat mounting falling on the magnetograph slit, showed that the magnetic field of the Sun observed as a star is about two orders of magnitude weaker than the smallest fields that could be detected in stellar observations at that time.

In 1967 Severny started more or less systematic observations of the magnetic field of the Sun observed as a star. The results of these observations and their relationships to the sector structure of the interplanetary magnetic field are well known to all solar physicists. We only recall the explanation given by Tuominen (1970) of the fact that the magnetic flux from sunspots is in antiphase with the fluctuations of the measured total field (Severny, 1969). According to Tuominen, since the following part of a bipolar magnetic region disintegrates much faster than the leading one, the following part has a greater influence on the observed large scale fields. On the contrary, when determining the magnetic flux from sunspots, the leader spots are more effective. Tuominen also suggests that the long lifetime of the large scale magnetic region can be explained on the basis of the occurrence of the activity complexes (Bumba and Howard, 1965).

2.3. SOLAR LUMINOSITY

The problem of the variations of the so-called solar constant is very old and intricate. We recall here only the attempts made for collecting information on the variation of the solar constant from the amount of light reflected by the planets Uranus and Neptune (Johnson and Iriarte, 1959; Serkowski, 1961; Jerzykiewicz and Serkowski, 1966; Albrecht *et al.*, 1969).

2.4. SUNSPOTS

The largest sunspot group ever recorded, observed in 1947, had a maximum area of a little more than a hundredth of the solar disk. If we consider this group completely dark, it would give rise, owing to the solar rotation, to a light variation of about one hundredth of magnitude. This value falls within the possibility of photoelectric photometry.

2.5. PHOTOSPHERIC AND CHROMOSPHERIC FACULAE

A variation of the same order as above would be given by photospheric faculae.

Since the presence of magnetic fields in the solar atmosphere is related to the H_2 and K_2 emission component of the H and K lines of Ca II it is obvious to expect variation of these emissions if we observe the Sun as a star.

Actually Sheeley (1967) and Bumba and Růžičková-Topolová (1967) found, independently, with different methods, variations of the intensity of K_2 emission averaged on the disc of as much as 15% during a solar rotation and of 40% during a solar cycle, whereas Bappu (1967) found these values too optimistic.

2.6. SOLAR FLARES

Optical white light flares would give rise to variations at the limit of the possibilities of photoelectric photometry. However, as early as 1950, Greenstein pointed out that bright solar flares could double the brightness of an M dwarf: moreover a flare on such a star would be more prominent in the violet because of the steep drop in the stellar continuum.

Some years later, in 1959, Schatzman showed that the flux of solar type II bursts could be detected with the most sensitive radiotelescope of that time even at stellar distances.

3. Stellar Flares

3.1. INTRODUCTION

The task of picking out, from the multiplicity of observational stellar facts, those that could be determined by a mechanism of the same kind as that producing the solar activity is extremely intricate.

We shall first analyse stellar flares since these phenomena have many common features with solar flares. Speaking of stellar flares we will be in the wake of the problems concerning photospheric, chromospheric and coronal stellar activity.

Many reviews are available on the problem of stellar flares. We recall, among the most recent ones those of Gershberg (1971, 1975), Lovell (1971), Ambartsumian and Mirzoyan (1971), Gershberg and Pikelner (1972), Kunkel (1975), Gershberg and Lund (1975). Therefore we shall only summarize the most relevant information from our point of view (for more details and references we refer to the quoted reviews) and present the latest results. We shall confine ourselves to only UV Ceti-type stars although we know that flare activity can be present in stars of other types too.

UV Ceti-type stars are red dwarfs with spectra showing emission lines (dKe and dMe). Surface temperatures are less than 3500 K; luminosities are of the order of 10^{-5} solar luminosity; masses are of the order of 10^{-1} solar masses.

These stars undergo the most rapid variations of all eruptive variables. In a time interval ranging from a fraction of a second to several minutes, their integrated brightness sometimes increases by more than five magnitudes. The decay time is of the order of 1–60 min. Very intense and complex events have been observed to last even a few hours. Optical observations from the U- to the V-band show that flare amplitude decreases in going from shorter to longer wavelengths.

Typical frequency is 10^{-1} events per hour.

Traditional methods of photographic and photoelectric spectroscopy cannot be used to get information on spectra of stellar flares since exposure time must be longer than the duration of flares. Image tubes must be used as light detectors.

During a flare, spectra undergo radical changes. The overall characteristics of flare spectra almost completely coincide with the characteristics of the spectra of T Tauri type stars. Bright intense lines and strong continuous emission are present especially in the ultraviolet. At the beginning of the flare, hydrogen and neutral helium lines increase strongly. Simultaneously strong emission in the Balmer continuum appears. In some strong flares also the 4686 Å line of the ionized helium and the 5184 Å line of neutral magnesium appear. Ca II emissions increase somewhat later and to a lesser extent than the emission of the Balmer lines. After flare maximum the decay of the continuum emission is much more rapid than the decay of the line emission and the helium lines fade quicker than the hydrogen lines. Intensified calcium emission has been observed even several hours after strong flares.

We notice that spectral features, as TiO bands and the Ca I 4227 Å line, characteristic of the cold stellar atmosphere, remain during flares.

We have already pointed out that as early as 1959 Schatzman showed the possibility of observing radio bursts from flare stars.

This possibility directed the attention of Lovell to these objects. Investigation began at the Mark 1250-ft radiotelescope at Jodrell Bank in 1958, September 28 and was continued at intervals until 1960, April 14. 474 h of observations were made in the meter waveband at 240 MHz with a technique suitable for rejecting spurious bursts of noise arising from terrestrial sources. When all doubtful cases had been excluded, 13 events with durations of several minutes were noticed. For UV Ceti, rates of occurrence of radio and optical events were closed. Notwithstanding these preliminary results simultaneous optical and radio observations were needed in order to ascertain the nature of the recorded radio events.

Many national and international campaigns have been organized since then also under the auspices of Commission 27 of the International Astronomical Union.

Up-to-date radio emission during stellar flares has been recorded in the range 20 cm–15 m.

With a radio apparatus sensitivity of 0.3 f.u. radio flare occurrence is close to optical flare occurrence.

Observable X-ray emission during stellar flares was predicted by Gurzadian (1965, 1966b), Grindlay (1970), Gurzadian (1971) and Crammel et $al.$ (1974).

An estimation of the upper limit of the X-ray flux from flares of UV Ceti type stars was carried out by Gershberg et $al.$ (1969) studying the correlation between stellar flares and sudden cosmic noise absorption. No correlation was found. Supposing that the effect is smaller than the fluctuations of cosmic noise an upper limit of 10^{-4} erg cm^{-2} s^{-1} was determined.

Arrangements had been made to observe YZ CMi in January and February 1971 simultaneously in the optical, radio and X-range with the satellite SAS-A launched in 1970, December 12. Unfortunately flare star observation from SAS-A had to be cancelled due to some instrumental problems.

At the 14th International Cosmic Ray Conference, Grindlay and Heise (1975) presented a paper describing observations that allowed the detection of X-ray emission from flare stars for the first time (Heise et $al.$, 1975).

Observations were made by soft and hard X-ray detectors of the Space Research Laboratory at Utrecht and the Center for Astrophysics at Cambridge Mass. which were on board the Astronomical Netherlands Satellite (ANS). Utrecht detectors are sensitive in the ranges 0.2–0.28 keV and 1–7 keV; Cambridge detectors in the range 1–30 keV.

The stars YZ CMi and UV Cet were observed for the periods 1974 October 19–22 and 1975 January 3–9 respectively.

A total of 300 min of good quality data with low background was obtained on YZ CMi with the satellite continuously pointed to the source for typically 5–12 min per orbit. Optical and radio monitoring was conducted for only about a third of the period of X-observation. No optical or radio flares were detected.

On October 19 at 20 05 UT an X-ray flare was observed with the soft and medium energy detectors. No increase in count rate was observed in the Cambridge detector, probably for pointing reasons that cannot be verified. Unfortunately no optical or radio coverage was available.

The total duration of the flare was 6 min at 0.2–0.28 keV and 1.5 min at 1–7 keV. The total energy released in the range 0.2–0.28 keV was $4.2 \pm 0.3 \times 10^{31}$ erg with a peak luminosity of $2.5 \pm 0.4 \times 10^{29}$ erg s^{-1} and in the range 1–7 keV was $1.9 \pm 0.4 \times 10^{32}$ with a peak luminosity of $3.6 \pm 0.7 \times 10^{30}$ erg s^{-1}.

A total of 316 min of good quality data with low background was obtained on UV Cet. During this time four events were observed in the optical range.

On January 8, at 01 17 12 UT a large optical flare was observed by Kunkel and Zarate (1975): its peak luminosity was $>1.5 \times 10^{30}$ erg s^{-1}. X-ray measurements started 28^s later at 01 17 40 with a strong enhancement in the 0.2–0.28 keV. X-ray flares lasted 48 s. The minimal total energy released in this range was $2.9 \pm 0.6 \times 10^{30}$ erg with an average luminosity of at least $6.1 \pm 1.3 \times 10^{28}$ erg s^{-1}. During this time the upper limit in the Utrecht medium energy detector and in the Cambridge detector was respectively 1.8×10^{30} and 1.3×10^{30} erg s^{-1}.

During the other three optical events X monitoring was conducted with negative results. Only the upper limit of the 4σ level for the X-ray flux was given.

New observations of UV Ceti had been planned for the period 1975, July 4–9 (Grindlay, 1975). Unfortunately this period is not suitable for ground-based observations.

Ground-based optical and radio observations were organized by Moffett (1975a, 1975b) for autumn 1975–early winter 1976 in coordination with X-ray observations from MIT/SAS-C satellite. Observations of the flare stars UV Cet, YZ CMi, AD Leo were planned.

Nowadays we know of hundreds of UV Ceti-type stars (including BY Dra and UVn stars) of which 39 (single stars or multiple systems) are within a distance of 25 pc from the Sun (Rodonò 1975a). As far as I know, the latest discovery of a flare star was made at Catania (Cristaldi and Rodonò, 1976) last July in the large proper motion pair G. 208-44/45 (Giglas et al., 1967) recently recognised by Harrington et al. (1974) as a nearby binary system located at 4.7 pc.

Photoelectric patrol for more than 10 h has been carried out, generally in the B band, for 34 stars (Rodonò, 1975a) mainly at the observatories of Armagh, Boyden, Catania, Cerro Tololo, Crimea, McDonald, Tokyo. Systematic sequential and simultaneous *UBV*, radio, and polarimetric observations have also been executed.

Among the 39 objects within a distance of 25 pc from the Sun, 60% are binaries or multiple systems (Rodonò, 1975b). Although Rodonò (1975b) found no evidence of mutual flare triggering between components of binary systems, it is stimulating to note that a percentage of 60% could be higher than the average one (Batten, 1973) and that many of these systems have very eccentric orbits. Not so long ago there had been a belief that only the fainter components of these systems were flare stars. Nowadays flare activity on the more luminous component of visual binaries and on both of the components has been discovered in several instances.

Since dK and dM stars are intrinsically faint, UV Ceti type stars can be detected only in the neighbourhood of the Sun. Because dM stars constitute 80% of the galactic stellar population and dMe stars 5% of the dM type stars, we can argue, following Gershberg, that the flares of UV Ceti-type stars, are the most widespread kind of stellar variability.

Flare stars have not been discovered only in the neighbourhood of the Sun but also in hundreds in stellar aggregates (associations and clusters). Photographic patrol vigorously pursued particularly by Ambartsumian, Haro, Rosino and collaborators is generally used for discovery and study of flares in aggregates.

The multiple exposure technique has been employed with a time resolution of 5–10 min, inadequate to detect faint short-lived events. Recently also photoelectric methods were successfully employed by Rodonò (1974).

According to Ambartsumian and Mirzoyan (1971) T Tauri-type stars, flare stars in stellar aggregates and UV Ceti-type stars form a wide class of comparatively young, related, non-stable objects.

According to Gershberg (1975) in the light of up-to-date observations we would seemingly reject the hypothesis that UV Ceti-type stars are the later stage of T Tauri-type stars. It would be better to think that the variables of both types have originated in the same volume but have different behaviours due to mass differences. For flare activation, the existence of a deep convection is important, not the age itself.

3.2. Properties of stellar flares

3.2.1. *Extension*

We have various proofs that stellar flares are events localized in small regions of the stellar atmospheres.

(1) As we shall see, spectroscopic and photometric methods show that the temperature of the perturbed region must be of the order of several 10^4 K. If this temperature increase involves the whole stellar atmosphere, we should expect an increase in the visual surface brightness greater by one–two orders of magnitude than that observed.

(2) We have already noticed that some spectral features, which are characteristic of the cold stellar atmospheres, remain unchanged during flares. We must deduce that either the flares are localized in a small part of the stellar surface or they develop in high atmospheric layers without influencing the photosphere.

(3) Flare stars have sizes of 1–2 light seconds, whereas several flare structures have been observed to approach maxima in less than a second.

3.2.2. *Physical Properties*

The rates of energy production during a great stellar flare are of the order of approximately 10^{25} erg s^{-1}, over a bandwidth of 400 MHz, in the radio spectrum and of the order of 10^{30} erg s^{-1}, over a bandwidth of 6.7×10^8 MHz in the optical spectrum.

Spectroscopic and photometric methods show that the temperature of the perturbed region must be of the order of several 10^4 K and the electron density of the order of 10^{12}–10^{14} cm^{-3}.

Assuming localized emission, brightness temperature in radio flares at metric wavelength is of the order of 10^{16}–10^{18} K.

Spectrographic evidences of mass motions during a flare exist in literature although they are very scanty. The most convincing observation is that of Wolf 359 (the fourth nearest star or the third nearest system) by Greenstein and Arp (1969). In an image tube spectrogram, taken in an unknown phase of a flare, the K line was shifted by -59 km s^{-1} and the mean of about 9 hydrogen lines was changed by -23 km s^{-1}. Arp and Greenstein ascribed the negative shift to a hot cloud, ejected towards the observer, at velocity like that in solar spicules and less than that in eruptive prominences.

3.2.3. *Statistical Properties*

Generally the duration of a flare increases with the increase of the amplitude.

In many radio optical correlations there has been evidence that the maximum phase of the radio flare follows that of the optical flare by several minutes.

In 1972, October 11 simultaneous radio and optical recordings of a large flare on UV Ceti were obtained by Lovell *et al.* (1974). In this event we have a delay of about 500 s between the onset of the flare at optical and radio frequencies and moreover the maximum phase of the radio flare follows that of the optical flare by 11 ± 1 min.

Statistical investigations show that, although flares of UV Ceti-type stars occur in time, generally following the Poisson law, the number of the occurrences of very close flares is higher than expected according to the Poisson law. Also for stellar flares we could therefore have sympathetic phenomena. New information by Rodonò (1976) on this matter will be reported in this section. Also homologous phenomena could be present.

In a given star the flare frequency increases as the amplitude decreases.

The existence of a periodical or quasi-periodical time variation in the frequency of flares in a given star is much discussed.

3.2.4. *Relationship with the Parent Star*

There appears to be in each star an upper limit to the energy radiated by a single flare and to the time averaged energy radiated by all the flares. According to Cristaldi and Rodonò (1975) the first limit is, in the B band, 60 times the energy emitted by the quiet star in one minute; the second one is, also according to Kunkel (1970, 1973), one percent of the energy radiated by the quiescent atmosphere.

Flare frequency decreases as the stellar luminosity increases but more powerful flares are observed in more luminous stars, although there are indications that flare luminosity increases slower than star luminosity (Cristaldi and Rodonò, 1975). Due to these indications, it seems difficult to deduce information on the relationship between flare activity and luminosity of the parent star from the relationship between flare frequency and stellar luminosity.

3.3. PHYSICAL NATURE OF STELLAR FLARES

A complete model of stellar flares should describe:
(1) the mechanism of storage and dissipation of the flare energy;
(2) typical behaviour of the flare light curves, colours and spectra; and
(3) regularities observed by flare statistics.

As far as storage and dissipation of the flare energy is concerned, we have three main models: the nebular model, the fast electrons model, and the shock wave model.

The nebular model elaborated by Gershberg (1967) and also by Kunkel (1967) assumes that an optical flare is due to the radiation of a hot ionized and rapidly emitting gas mass ejected by the star.

Ambartsumian showed as early as 1954 that the continuum emission, usually present in spectra of T Tauri-type stars and appearing in the spectra of flare stars during the flares, should not be of thermal nature. Afterwards further evidence was obtained in favour of the non-thermal and non-synchrotron nature of the flare continuous emission.

As early as 1965 Gurzadyan assumed that 'fast electrons' i.e. non thermal, but not extremely relativistic electrons, whose energy is about 1–2 MeV, can be formed in the stars' atmospheres directly above the photosphere. The scattering of photons at such electrons is accompanied by an increase in the frequency of the photon scattered after collision (inverse Compton effect). The flare itself should be (Gurzadian, 1966a) a sudden appearance of fast electrons over the photosphere of the star. These electrons should cause a transition of the infrared quanta of the photosphere into visual and ultraviolet quanta. Naturally, assuming identical energy parameters for the electrons, the amplitude of the flare increases, in agreement with the observations, when the stellar temperature decreases. In this model polarization of the flare light is to be expected and radio emission during the flare is possible. Finally, Gurzadyan (1972) has recently pointed out that non-thermal bremsstrahlung produced by the fast electrons is practically of no significance, except for the very brightest flares observed.

According to the shock wave model of Klimishin (1970), Korovyakovskaya (1972) the flare could be due to a de-excitation of ionized hydrogen gas behind a shock front propagating in the stellar atmosphere. The Greenstein and Arp observation already discussed would suggest that a shock wave is responsible for the heating of the matter that emits radiation during the flare. Klimishin, assuming that the matter density is constant in the star atmosphere, obtained flare rise and decay times that are not in agreement with the observations. On the other hand, Korovyakovskaya was able to explain many photometric and spectroscopic characteristics of the flares assuming a stellar atmosphere with a density gradient and obtaining differences in the relaxation times and acceleration of the shock front when it moves into a region of lower density.

Unfortunately, till now, the available observations (mainly flare colours) do not allow a selection of a model (Cristaldi and Rodonò, 1975).

3.4. COMPARISON OF UV CETI-FLARES WITH SOLAR FLARES

3.4.1. *The Atmosphere of the Sun and of the UV Ceti-Stars*

Although we shall discuss afterwards the problems of the activity of stellar photospheres, chromospheres and coronae we must now point out some facts.

(1) In flare stars, as in the Sun, photospheric dark short-lived spots exist.

(2) In flare stars, as in the Sun, spot appearances show periodicities of several years.

(3) During stellar flares, as during solar flares, mass motion can develop.

(4) The flare star spectra at the minimum, i.e. outside the flares, differ from the spectra of normal dwarf stars by the presence of emission lines: as for the Sun, active stellar chromospheres are correlated with the capacity of producing flares.

(5) According to Kahn (1974) the delay between the onset of the flare at optical and radio frequencies shows that the disturbance coming from the optical flare must have spread to a corona of the star. The energy required to heat the corona would come from the mechanical energy of the convective motions in and near the photospheric layers.

3.4.2. *Analogies between Flares of UV Ceti-Type Stars and Solar Flares*

Flares of UV Ceti-type stars and solar flares have many important common features.

(1) Both phenomena are transient events.

(2) Both phenomena refer only to a small region of the stellar surface.

(3) For stellar and solar flares the ratios between rise and decay times are of the same order.

(4) Variations of spectra during stellar and solar flares are analogous.

(5) Optical thickness in $H\alpha$-lines for stellar and solar flares are similar.

(6) Electron density of stellar flares is of the same order as electron density of solar flares spreading to some higher densities.

(7) Spectral behaviours of radio stellar flares and temporal relationships between optical and radio stellar flares are similar to the ones of the solar radio noise storm and sometimes of the solar type II burst. Typical velocities are $1000 \, \mathrm{km \, s^{-1}}$.

(8) The ratio between the rates of energy production in the optical spectrum and in the radio spectrum is 10^5 for stellar and solar flares.

(9) Homologous and sympathetic stellar flares can exist; we must deduce that also in flare stars persistent morphological features of the active region (most probably its magnetic field) control the flare process and that triggering agents can radiate from a flare.

10. Periodical or quasi-periodical time variations in the frequency of stellar flares could be present as in the frequency of solar flares.

(11) According to Cristaldi and Rodonò (1975) the Sun might be regarded as a low-activity flare star.

3.4.3. *Differences between Flares of UV Ceti-Type Stars and Solar Flares*

We also have differences between flares of UV Ceti-type stars and solar flares.

(1) Although stellar flares have been observed with rise time greater than 10 min and with total duration greater than 1 h, generally the flares of UV Ceti-type stars have a lifetime of one order of magnitude smaller than solar flares.

(2) While the total energy output in a stellar flare is of the same order of the undisturbed stellar continuum, the total energy output in a solar flare is about a millionth of the normal quiescent emission.

(3) Temperatures and probably densities of stellar flares are greater than temperatures of solar flares.

3.4.4. *The Problem of the Continuous Emission*

As is well known optical continuum emission from solar flares has been observed only in a few cases, mainly for very strong events known as white light flares. On the contrary, optical continuum emission during the flash phase is typical for all stellar flares. Certainly this difference is in part due to different detection conditions in the Sun and in the flare stars. On the other hand, as also pointed out by Haupt and Schlösser (1974), the ratio between the energy radiated in the optical continuum and total optical energy is much smaller for solar flares (0.01) than for stellar flares (0.5).

Two main mechanisms have been proposed for interpreting the optical continuum emission of solar flares.

According to Švestka (1970) in a limited volume of the flare region a strong impulsive process occurs shortly after the flare onset, accelerating protons and electrons up to the relativistic range for electrons and, in very strong events, up to the relativistic range of protons. Protons accelerated to energies above 20 MeV penetrate down to the lowest chromospheric and upper photospheric layers and produce a heating of the atmosphere in the limited bombarded region which can be observed as a shortlived increase of the continuum in the optical spectral region.

According to Hudson (1971) decelerate electrons, diffusing out of the region where they produce the hard X-rays of the flash phase, could produce the optical continuum emission.

If we assume that one of these two mechanisms is responsible for the optical continuum emission of flare stars, we deduce that in the atmospheres of flare stars a larger number of particles must be accelerated.

Since during proton solar flares intensive proton and heavy element emission into interplanetary space is observed it is sound to explore if flare stars could be the source of the soft cosmic ray component (Lortet-Zuckermann, 1965).

Actually considering the rate of energy released during a stellar flare and adopting the factor, which has been determined empirically in the case of the Sun, for the conversion of flare energy to cosmic ray energy, Lovel (1974) showed that M type flare stars may be the major source of the galactic cosmic rays for energies from 10^6 to

3×10^8 eV and that K type stars may contribute to one fifth of the total cosmic-ray energy up to 10^9 eV.

4. Photospheric Stellar Activity

4.1. INTRODUCTION

The hypothesis that non-uniform distribution of regions of constant brightness, coupled with rotation, could be responsible for stellar photometric variations, was proposed as early as the seventeenth century (see Ledoux and Walraven, 1958).

The hypothesis of variable superficial regions was introduced in the nineteenth century. Analogies with sunspots were called upon by Wolff and Secchi (l.c.).

According to Vardya (1970), the first plausible proof of the existence of stellar spots was found by Jaschek and Malaroda (1970) in the A2p star 73 Dra (HD 196502). These authors have found in 73 Dra, a star of effective temperature ~9000 K, lines of CN and CH, corresponding to an effective temperature of ~6600 K. These molecular features, that seem variable, would indicate the presence of a cold spot.

As also pointed out by Piotrowsky et al. (1974) we could actually interpret any light curve assuming the existence of convenient spots on the photosphere of a rotating star. Therefore we must accept this assumption only if it either describes other known phenomena too or if photometric variations can not be implied in the frame of the known phenomenology.

Quite recently Friedemann and Gürtler (1975) calculated the photometric variations determined by the presence of one circular spot on the photosphere of a rotating star. Naturally a light curve depends (1) on the size of the spot, (2) on the temperature difference between the spot and the photosphere, (3) on the position of the spot on the stellar surface, (4) on the orientation of the rotational axis in relation to the observer. Colour indices exhibit rather small changes ($\leq 0\overset{m}{.}05$) especially for large temperature differences between the photosphere and the spot. Spots also lead to a shift of the star in the colour magnitude diagram.

Main classes of objects showing photospheric phenomena that could be and have indeed been interpreted in terms of photospheric activity are binary, magnetic and flare stars.

4.2. PHOTOSPHERIC ACTIVITY ON BINARY STARS

In 1930 Sitterly assumed that some distortion in the light curve of the binary system RS Canum Venaticorum could be due to photospheric spots. The same assumption was made by Kron (1947) and Struve (1952) for AR Lacertae and by Binnendijk (1970) for certain W Ursae Majoris systems. Quite recently it was taken up again and re-elaborated by Walter (1971) for SW Cygni and by Hall (1972) for RS Canum Venaticorum.

Due to the methodological reasons discussed, it might be more advisable to try to describe these distortions by taking into account gas-streams between components (see e.g. Catalano and Rodonò, 1974; Piotrowsky et al., 1974) required to explain other properties of these systems too.

Notwithstanding this some authors still prefer a more elaborated modified spot assumption (Arnold and Hall, 1973; Hall, 1975; Mullan, 1975).

We notice that quite recently Weiler (1975a, 1975b) found for RS Canum Venaticorum and for other five binaries of the same type significant variations in the emission line intensities of the H and K line of Ca II and Hα. At least for two systems maximum emission correlated with assumed spot activity was found.

4.3. PHOTOSPHERIC ACTIVITY ON MAGNETIC STARS

Very comprehensive and complete reviews are available on magnetic and related stars (e.g. Cameron, 1967; Pikelner and Khokhlova, 1971, 1972; Preston, 1971).

We recall here that all the observed magnetic fields, ranging from the limit of detectability of hundreds of gauss to 30 kG, seem to be variable with periods generally ranging from about 0.5 to 200 days.

Generally, stars with observable magnetic fields have peculiar spectra. Vice versa variable magnetic fields have been observed in all Ap stars having lines narrow enough for measurement of Zeeman splitting.

Also the other properties of these stars (brightness, colour, intensity of the spectral lines) generally show correlated periodic variations. Irregular variations of a short period are observed too.

Among the proposed models two are relevant for our discussion: the solar-cycle model and the oblique-rotator model.

The solar cycle model is based on the assumption that phenomena of the same type as those associated with the solar cycle, enhanced by orders of magnitude, could produce the observed variations. Although the variations observed in some stars could be actually interpreted with such a model (Blanco *et al.*, 1972) some theoretical difficulty arises mainly connected with the high velocity required by the process of dissipation and reversal of magnetic fields naturally assumed to be superficial.

The oblique rotator model, having the axis of a static magnetic field inclined to the axis of rotation of the star, can describe fairly well the period vs line-width relation deduced by Deutsch for spectrum variables and the observed regular variations including the photometric one (Kodaira, 1973; Catalano, 1975). Notwithstanding this some difficulty arises for long period variables (Preston and Wolff, 1970; Catalano and Strazzulla, 1975). Moreover, active photospheric regions are required in this model too, in order to describe both short period irregular variations and the crossover effect.

Severny (1970), Borra and Landstreet (1973a, 1973b) and Boesgard (1974) opened the very promising possibility of observing weak stellar magnetic fields down to 15–20 G. Unfortunately results are still very scanty and no M dwarfs have been observed. It is interesting to note that some of the stars with weak magnetic fields have Ca II emission components.

4.4. PHOTOSPHERIC ACTIVITY ON FLARE STARS

Existence of slow, quasi periodic photometric variations of small amplitude (several hundredths of magnitude) in flare stars outside flare activity has been reported for many years.

More recently Chugainov (1966) observed brightness variations in BY Dra (K6 V) with a period of 3.8 days and proposed to interpret these variations with the existence of a spot on the surface of a rotating star.

Following Chugainov's discovery, Krzeminsky and Kraft (1967) and Krzeminsky (1969) searched for similar brightness variations among six emission line objects (including BY Dra) and three non-emission line objects in the range of spectral types dK 7 to dM 3.5. Four of the emission line objects showed variability in the range of 0.06–0.1 mag. Variability and period of BY Dra were confirmed; a period of 2.2 days was found for another of these four variable objects (FF And). The remaining emission line and the non-emission line stars showed no variability in excess of 0.02–0.04 mag.

Independently of the Krzeminsky and Kraft programme, the Catania group undertook in 1967, in the frame of its research project on stellar activity of the solar type, the study of luminosity variations between flares with the design of interpreting them assuming the presence of active regions (Godoli, 1967; Cristaldi et al., 1969; Godoli, 1968).

The observation of photometric variations of flare stars outside flare activity has been recently pursued by several authors (Andrews, 1968; Ferraz Mello and Torres, 1971; Torres et al., 1972; Chugainov, 1973, 1974; Vogt, 1973, 1975; Robinson and Kraft, 1974; Martins, 1975).

We notice that Robinson and Kraft (1974) observed Pleiades' and Hyades' dwarfs in the spectral-type range K3 v–M0 v finding photometric variations outside flare activity in the Pleiades but not in the older Hyades. Actually Kraft and Greenstein (1969) found that the Pleiades contain more dMe stars than the Hyades.

Inconstancy of the amplitude of the photometric variations and of the mean brightness and colour was observed (Chugainov, 1973, 1974; Vogt, 1973). In BY Dra a steady increase in the mean light level is accompanied by a steady decrease in the amplitude of quiescent variations (Chugainov, 1973, 1974; Vogt, 1973).

Changes of period were observed by Chugainov (1973, 1974) although Vogt (1973, 1975) threw some doubt on this matter.

Evidence that flare activity may be almost the same independently of the periodical brightness variations was finally obtained by Chugainov (1974).

Evans (1971) attempted to interpret photometric variations of binary flare stars outside stellar activity, taking into account obscuring material temporarily located near Lagrangian points of the binary systems. But, stability considerations showed that the possible lifetime of circumstellar material near these points would be too short (Bopp and Evans, 1973).

Let us interpret photometric variations of flare stars outside flare activity as modulation by rotation of a spotted star (Kron, 1950a, 1950b, 1950c; Chugainov, 1966, 1973, 1974; Krzeminsky and Kraft, 1967; Krzeminsky, 1969; Torres et al., 1972; Bopp and Evans, 1973; Evans, 1973; Torres and Ferraz Mello, 1973; Vogt, 1973, 1975; Mullan, 1974).

Spots several hundredths of the stellar hemisphere large, extended to high latitude (up to 60°) with low effective temperature are needed. Estimations of the size of the spots can be made when the star belongs to a binary system.

Inconstancy of the amplitude of the photometric variations and of the mean brightness could be due to variations of photospheric activity. For BY Dra Chugainov (1973) suggests an activity cycle of 8–9 yr.

Period changes could be due either, as on the Sun (Albrecht *et al.*, 1969), to a migration in latitude of the spots in regime of differential rotation or to the presence of not constant, preferential longitudes.

The independence of flare activity from the periodical brightness variations, if confirmed, could point out that flares are not associated with photospheric activity. It would also rule out the possibility that brightness variations could be determined by bright active regions consisting of clumps of microflares.

Finally we note that equatorial rotational velocities of the order of 10–15 km s^{-1} would turn up. Such velocities, greater than expected for the lower end of the main sequence, may point, together with the presence of emission lines that we shall discuss further on, to the youth of the flare stars, although, as we have seen, some doubt exists on this topic.

If active regions really exist on flare stars strong magnetic fields associated with starspots ought to be present. The flare mechanism could consist in the release of energy by magnetic field collapse (Evans, 1973; Worden, 1974).

4.5. CONCLUDING REMARKS ON PHOTOSPHERIC STELLAR ACTIVITY

Some of the assumptions on the existence of star spots, that we have described, are extremely naïve.

We should always take into account not only spots but also other forms of activity that, on grounds of the solar analogy, ought to be present in stellar active regions. But for such an extension we need more observational details although some considerations have recently been made in this direction by Arnold and Hall (1973) for RS Canum Venaticorum-type binaries, by Mullan (1975) for W Ursae Majoris-type binaries and by Martins (1975) for flare stars.

5. Chromospheric Stellar Activity

5.1. INTRODUCTION

A section of the colloquium held in München in 1969 on the spectrum formation in stars with steady state extended atmospheres was dedicated to the chromospheres and coronae of stars (Groth and Wellman, 1970). Moreover an entire colloquium held at Goddard Space Flight Center in 1972 was dedicated to the problem of stellar chromospheres (Jordan and Avrett, 1973). For this reason we shall very briefly summarize here only the most important observational results and present the latest ones.

Arguments for the existence of stellar chromospheres and coronae are by now very convincing.

In the solar type situation we have a strong convection, kept by subphotospheric hydrogen ionization, which causes a host of waves. Among them the acoustic modes still appear to be the most important ones. As these waves propagate upward they steepen into shocks which dissipate into chromospheres and coronae.

As an example of convective models we recall those by de Loore (1970) although it was pointed out at the Goddard Colloquium that the de Loore's models exaggerate the mechanical flux when convective zones are thinner than the mixing lengths. De Loore calculated photospheric models for 90 stars with effective temperatures ranging from 2500 K to 41 600 K and acceleration of gravity ranging from 10 to 10^5 cm s^{-2}. He deduced that all the investigated stars contain unstable layers, including the hottest. Nevertheless, according to de Loore, only stars with the effective temperature of 8300 K or less contain layers where the convective energy transport is important. For main sequence stars, the largest fluxes would be generated in F2 III stars.

But, as pointed out by Thomas at the Münich Colloquium, we could have situations, other than the solar one, where chromospheres and coronae do exist as a consequence of mechanical instability different than acoustic waves. So we must not expect chromospheres and coronae only in stars of spectral class F0 and later ones (Lamers and de Loore, 1974).

Moreover the Catania group (Blanco et al., 1974) showed that for stars with $T_{eff} < 5000$ K the theoretical acoustic energy flux is inadequate, by two orders of magnitude for the coolest stars, to account for the observed chromospheric emission fluxes. New information by Blanco et al. (1975) on this matter will be reported in this section.

5.2. STELLAR CHROMOSPHERES

Ca II resonance doublet emission (Ca II H and K at 3933.7 and 3968.5 Å, $4s\,^2S - 4p\,^2P°$) which is one of the best known indicators of stellar chromospheres was first reported by Eberhard and Ludendorff in Arcturus (α Boo, K1 III) (Eberhard and Schwarzschild, 1913). In their famous 1949 list Joy and Wilson reported 445 stars known to have H and K emissions (novae were excluded). The list indicated that, with few exceptions, calcium emission occurs only in stars of G, K and M types. Although supergiants, giants, subgiants are well represented in the Joy and Wilson list, the best represented are the dwarfs. Other stars with H and K emission components were found afterwards (Greenstein, 1952). A longer list was published by Bidelman (1954). In 1969 Warner (1969) added a list of 200 southern G, K and M stars with emission components. In 1970 Wilson and Woolley presented a list of calcium emission intensities of 325 main sequence late type stars.

The ultraviolet magnesium counterpart of the Ca II resonance doublet emission (Mg II resonance doublet at 2795.5 and 2802.7 Å, $3s\,^2S - 3p\,^2P°$), was first reported in the solar spectrum by Durant et al. in 1949. In late type stars it was reported by Doherty (1971, 1972), Gurzadian (1972), Kondo (1972), Kondo et al. (1972), Linsky and Basri (1974), Moos et al. (1974), Kondo et al. (1975); in intermediate type stars by Kondo (1972), Dupree (1974); in a wide range of spectral types by Oganesyan (1974).

Chromospheric Ly-α emission was observed from Arcturus by Rottman et al. (1971), Moss and Rottman (1972), Moss et al. (1974) and from an intermediate type star by Dupree (1974). O I λ 1304 was observed from Arcturus by Moss and Rottman (1972). O II λ 1334.5 was observed in an intermediate type star by Dupree

(1974). Other ultraviolet lines were observed in the spectrum of Procyon (α CMi F5 IV) (Evans *et al.*, 1975).

Rosendhal (1973) found changes in the strength or structure of the Hα as well as in the strength of other prominent lines in the red region of the spectrum of 13 early type supergiants among the 20 examined. This behaviour could be due to possible coupling between mass loss and turbulence.

Linsky (1973) reported that in the Sun the profile of the Ca II triplet line at λ 8498 Å shows a definite double reversal in the weak and slightly stronger plages. Therefore Linsky suggested that the 8498 Å line could be a very sensitive indicator of the activity of stellar chromospheres. Anderson (1973) obtained photoelectric profiles of this line for 30 late-type stars but at the resolution and precision of his study none of the stars observed show any evidence of emission although several have very strong K line emission.

Vaughan and Zirin (1968) pointed out that the λ 10 830 Å line of neutral helium is a better indicator for the existence of stellar chromospheres than Ca II lines: on the Fraunhofer solar spectrum, indeed, the λ 10 830 Å line is the only one in the range from 3000 Å to 11 000 Å that originates solely in the chromosphere; moreover H and K lines of Ca II and Hα line are excited in a wide range of temperature and may show either enhanced emission or absorption depending upon physical conditions. These authors have found in a wide research on 86 stars the λ 10 830 Å line of neutral helium in absorption in a substantial number of G and early K stars and in emission in five stars. According to Vaughan and Zirin the He line would probably not be present, at least with appreciable strength in either F or M stars. However we notice that, as far as M stars are concerned, only supergiants, bright giants and giants have been observed.

As early as 50 years ago, Stratton realized that Ca II resonance doublet emissions were correlated with the luminosity of the late type stars. After many years Wilson (1954, 1959, 1967, 1970) and Wilson-Bappu (1957) gave the relationship between the Ca II emission line width and the absolute visual magnitude. Since this relationship is valid for over a range of more than 15 mag and it does apply to the Sun too, we can expect that the mechanism which produces the line widths is the same for all the stars including the Sun.

Further analysis of this effect has been carried out by Reimers (1973).

A relationship between H and K absorption lines and luminosities was also reported in the range of spectral types from about F5 to K2 (Lutz *et al.*, 1973).

Mg II resonance doublet emissions also follow the Wilson Bappu relationship (Kondo, 1972; Kondo *et al.*, 1975). According to Kondo *et al.* (1972) the widths of the Mg II emission lines are between two and three times greater than the widths of Ca II emission lines for a given star.

A relationship between Ca II and Mg II resonance doublet absorption and luminosity was recently investigated by Ayres *et al.* (1975).

Considering that, on the Sun, the absorption structures at the edges of the core of Hα line correspond to the emission structures of K_2 (McMath *et al.*, 1956), Kraft *et al.* (1964) searched for and found in the spectra of late type stars an increase in the width of the Hα core in absorption with increasing luminosities. Further measurements were made by LoPresto (1971) (McMath solar telescope) and by Caplan (1973).

Bonsack and Culver (1966) found such a behaviour also for weak absorption lines formed in the deeper atmosphere of K type stars. According to these authors, the profiles of the weak lines in K stars are dominated by the motions of large elements of gas. Therefore the velocities of these elements must be correlated with stellar luminosity in the same way as the phenomena at higher levels which determine the widths of the K line emission and $H\alpha$ absorption.

According to Anderson (1973) the central intensity as well as the equivalent width of the Ca II triplet line at λ 8498 Å show no evidence of the Wilson Bappu effect in thirty late-type stars observed.

The existence of chromospheres of the UV Ceti-type stars has been suspected since the first spectra of these stars were obtained outside flare activity.

Discussing their list, Joy and Wilson pointed out that the strongest K emissions appear in T Tauri stars and in the dMe dwarfs of extremely low luminosity.

In 1960 Wilson (1961) obtained at Palomar a sprectrogram of EV Lac covering the photographic region with a dispersion of 9 Å mm^{-1}. He observed emission lines of H, He I, Si I, Ca I, Fe I, Ca II. Since the excitation requirements for these lines are different by one order of magnitude, Wilson suggested a stratification of chromospheric layers. From the sharpness of the metallic lines and the lack of differential velocities he assumed that thermal motion is probably the principal source of broadening for the hydrogen lines. With this assumption he deduced a temperature of about 14 000 K for the chromospheric zone giving rise to the hydrogen emission.

On the other hand, Gershberg (1970) found that the number of hydrogen atoms per unit area of the chromosphere in UV Ceti-type stars is greater by a factor of 3–50 than the corresponding number for solar chromosphere.

Jennings and Dyck (1972) and Jennings (1973) showed that H and K reversals and other heavy-element emission lines, especially those of ionized species, tend to vanish in stars which exhibit intrinsic polarization and infrared excess. The infrared excess, measured by the 0–L colour (the magnitude at 11 μ minus the magnitude at 3.5 μ) has been interpreted by thermal reradiation from circumstellar mineral grains. Since the infrared excesss is strongly correlated with polarization, scattering by particle must be responsible for observed intrinsic polarization. According to Jennings and Dyck the disappearance of emission lines implies that grains are responsible for the weakening of the chromospheres through a mechanism involving thermodynamic or indirect dynamic effects or a combination of the two.

5.3. CHROMOSPHERIC STELLAR ACTIVITY

According to some authors (e.g. Unsöld, 1964) we could consider intensities of H and K emission components not only as chromospheric indicator S but also as chromospheric activity indicators.

We must point out that solar H and K emission components, in integrated sunlight, are below the limit of detectability at those dispersions generally used in the observations of stellar H and K emissions. We can therefore deduce that solar type stars with observed H and K emissions must have more active chromospheres than the Sun.

Ca II H and K emission component intensities (and hence the general degree of the chromospheric activity) and the angular velocity of a main sequence star, decay as the inverse square root of the age (Wilson, 1963; Wilson and Skumanich, 1964; Skumanich, 1972; Boesgaard and Hagen, 1974). Assuming that the factor which determines the intensity of H and K emissions in the chromosphere of a star is simply the average magnetic field strength over its surface, one can predict that the magnetic field strength also gradually decays as the star ages and moreover that angular velocity must be greater when the magnetic field is stronger.

Since late type giants possess a more intense chromosphere than their counterparts on the main sequence it would be necessary to explain how the magnetic field could be reactivated when the star begins to evolve upwards to the giant branch in the H-R diagram (Matsushima, 1974). According to Matsushima (1974) a decrease of metal content could produce an increase of the surface convective flux.

According to the Catania group (Blanco *et al.*, 1974) the Sun as a whole does not fit in with the relations between H and K emission component intensities and age unless only active regions are considered. Therefore we must consider stellar H and K emission components as indicators of chromospheric stellar activity rather than, simply, as indicators of stellar chromospheres.

Apart from active regions, we know nowadays that the brightness of knots of the solar chromospheric network at the centre of the Sun and at the northern and southern poles changes with the phase of the solar cycle (Tsap and Labe, 1973).

Therefore, solar analogies would suggest that it is better to search for chromospheric stellar activity by analysing stars with variable emission components.

Many stars, in which the H and K emission components are probably variable, are already indicated in the Joy and Wilson list.

For a period of about 5 years (from 1950 to 1954) Popper (1956) found no definite intensity change in H and K emissions of five main sequence K and M stars.

As far as I know, definite variation in the intensity of H and K emissions was first observed in 1962 by Kandel in HD 119850 (spectral type dM 2.5) (Kandel, 1962, 1966). In 1963 Griffin discovered variations in Arcturus (K1 III). Deutsch (1967) found large changes in the profiles of Ca II H and K in two giants (α Tauri, K5 III and γ Aquilae K3 II) and smaller changes in most other K giants examined. Large profile changes have been found in α Tauri only after an interval of 100 days or more. The changes are not regular, but they tend to reoccur after about 1000 days, probably the rotational period of the star.

Nine stars in the spectral range G2-M0, observed by Liller (1968) in the period September 1964 – August 1965, showed, almost continuously, small random variations of the K line emission component in both intensity and shape. The stars α Tauri and, possibly, λ Andromedae exhibit greater fluctuations sometimes of flare-like nature with a recovery time of the order of half an hour. No evidence of cyclic variations was detected.

Boesgaard (1969) found fairly strong H and K emission components in a M3 S star in which no emission had been observed for many years.

Vaughan and Zirin (1968) observed temporal changes of the neutral helium λ 10 830 emission line in two of the five stars showing this line in emission.

Since 1965 photoelectric flux measurements at the centre of stellar H and K lines have been systematically performed at Mount Wilson with the main goal of collecting data which may ultimately lead to the detection and study of stellar analogues of the solar cycle (Wilson, 1968). According to Wilson (1968) no undoubted variations were found during the first year observations.

Soon afterwards, Wilson (1969) announced that flux variations in both components of 61 Cygni had been found during the period June–December 1967.

Nowadays we know that for several of the stars observed by Wilson it is likely that a cycle has nearly been completed. Up to now none of these stars have been reported to begin a new cycle (Hale Observatories, Annual Report 1973–74). New information by Wilson (1976) will be reported in this section.

Active chromospheres in eclipsing binary stars were observed for a long time (Wright, 1973).

Chromospheric activity on flare stars was observed by Popper in 1953 on BY Dra (K6 v) when this star was not yet known to be a flare star. On a spectrogram obtained on 1953 July 21, 37 UT he observed hydrogen emission lines many times stronger than on the other plates. H and K lines were probably also strengthened.

Recently Bopp (1974a, b) discovered variations of up to a factor of three in the equivalent widths of the emission lines of seven flare stars (among nine observed) at a time when no detectable flare was photometrically in progress. The time scale of this variation is of the order of one day or less. As Bopp (1974a) also points out, such a time scale shows that quiescent chromospheric emission must be localized: a variable emission produced by a uniform chromosphere would in fact imply a corresponding variation of the convective transport over the entire stellar surface. According to Bopp the observations are best explained if one assumes that active regions are present on a rotating star with the rotation axis inclined at a considerable angle to the plane of the sky in such a way that the active regions, although partially visible in all their lifetime are observed with varying degrees of foreshortening and limb darkening. Rotational velocities from five to twenty times the solar rotational velocity are found, as obtained from photospheric activity features.

6. Coronal Stellar Activity

6.1. INTRODUCTION

Stellar coronal parameters have been computed by many authors under the main assumption that the mechanical energy flux from convective layers is responsible for the heating of the corona (de Jager and Neven, 1961; Kuperus, 1965; Bierman, 1969; Nariai, 1969; de Loore, 1970; de Jager and de Loore, 1971). According to coronal models constructed by de Loore (1970) for the Sun and for stars with effective temperatures between 5350 K and 8320 K, the stars with $T_{eff} = 7130$ K and $\log g = 4$ possess the hottest and densest coronae with a computed temperature of 3.7×10^6 K and $\log N_e = 10.4$.

Expected stellar radio fluxes (Weimann and Chapman, 1965; Oster, 1971) and X fluxes (Bierman, 1969; de Jager, 1971; de Loore and de Jager, 1970; Landini and Monsignori Fossi, 1973) have been computed.

A radio candidate was

> Betelgeuse (α Ori, Me I);

X candidates were

> Arcturus (α Boo, K1 IIIp)
> Procyon (α CMi, F5 IV-V)
> β Cas (F2 IV)
> α Cen (G2 V–K0 V)

Since the gravitational potential of giants and particularly of supergiants is smaller than that of main sequence stars, one could expect no coronas of solar type in these stars (Bierman and Lust, 1960) or perhaps, better, since a large supply of acoustic energy is available, a stellar wind much more powerful than the solar wind. Actually, attempts were made in interpreting mass loss in red giants by a stellar wind mechanism (Weymann and Chapman, 1965; Fusi Pecci and Renzini, 1975).

6.2. STELLAR CORONAE

While observing chromospheric Ly-α emission from β Gem (K0 III) with the 'Copernicus' Satellite, McClintock et al. (1974) detected an emission line at 1218.4 ± 0.2 Å identified with the $2s^2 \ {}^1S_0 - 2p \ {}^3P_1$ intercombination line of O V. This line was first detected and identified in the solar limb spectra obtained by Burton et al. in 1965 (Burton et al., 1967). This identification was subsequently confirmed by laboratory measurements by Edlén et al. (1969) and by new observations by Burton and Ridgeley (1970). In the solar spectrum this line has a strength smaller than 0.03 that of Ly-α. In β Gem it has a strength ≈ 0.3 that of Ly-α. According to McClintock et al. (1974), calculations indicate that in β Gem the line is formed in a corona at temperatures near 260 000 K rather than in a transition region as in the solar case. Other high-excitation lines are being searched for in β Gem. The O V line is not observed in Aldebaran (α Tau, K5 III), Arcturus (α Boo, K1 III), or ε Eri (K2 V) which have been similarly studied by the Copernicus satellite.

If stellar coronae do exist, it is to be expected that the scattering of light by free electrons would cause the apparent angular distribution of light across the star to depend upon the plane of polarization. Hanbury Brown et al. (1974) made observations of the hot supergiant β Ori (B8 Ia) with the stellar intensity interferometer at Narrabri using linearly polarized light. The angular diameter was measured at two base lines with light polarized parallel and perpendicular to the base line and the results were used to derive two independent measures of the angular size of the star. No significant change of angular diameter with polarization was observed. From computation of the effects of the expected polarization the authors concluded that to obtain positive results it would be necessary to improve the sensitivity of the measurements by a factor in a range of 10 to 100.

6.3. CORONAL STELLAR ACTIVITY

Significant flux density of 0.11 ± 0.03 flux units from Betelgeuse was detected at a wavelength of 1.9 cm by Kellermann and Pauliny-Toth in 1966, February 21 at the

National Radio Astronomical Observatory. No significant flux was found in the next or in ten other nights in February and March. Low (1965) had found strong infrared radiation from this star at 10 μ, variable in intensity, but unfortunately no infrared measurements are available during the period of radio measurements.

Probable significant fluxes from Betelgeuse (M2 ib) and π Aurigae (M3 ii) were detected at a λ of 2.85 cm by Seaquist (1967) at the Algonquin Radio Observatory on 1966 October 30 and August 13 respectively. No significant flux was found from δ Orionis (O 9.5 ii) and ψ Aurigae (M0 iab) on 1960 October 30.

Significant flux density of 0.005 ± 0.001 flux units from Antares (α Scorpii) was detected at a wavelength of 11.1 cm by Wade and Hjellming (1971) at the NRAO on 1970, June 4, and November 12. No significant flux was detected at 3.7 cm. According to Oster (1971) what Wade and Hjellming observed may have been a stellar type iv burst or a series of them.

Further observations (Hjellming and Wade 1971a, b) showed that Antares' radio source is variable, detectable at both 3.7 and 11.1 cm and associated with the B3 v companion to the red supergiant M1 ib. According to Hjellming and Wade this radio emission could be related either to the infall of matter from the red supergiant or to particles' acceleration and ejection from the surface of the B star itself. This second possibility is the interesting one in our framework.

Many other objects of this kind have been discovered in the last three years.

Radioemission was found from a Bep star by Braes et al (1972). Also this star has an infrared excess, probably due to ejected circumstellar material (Geisel, 1970). Other objects of this kind have been discovered.

Catura et al. (1974, 1975) recently detected X-ray emission in the range from 0.2 to 1.6 keV from an area of the sky which contains the binary star system Capella (α Aurigae). Capella is a spectroscopic binary at 15 pc consisting of an F8-G0 giant of 2.9 M and a G5 giant of 3.0 M. The system rotates in a nearly circular orbit with a period of 104.023 days. The separation between the stars is of about 1 AU. No radio emission from the Capella system has been reported.

The detector was pointed at Capella in 1974, April 5 for a $1\overset{s}{.}2$ period. The number of counts obtained when Capella was within the field of view was about 10 times the average counting rate. The authors are of the opinion that it is very improbable that this signal is a random fluctuation in the background counting rate. The fact that the spectrum shows no indication of a turnover at 0.25 keV from X-ray absorption by interstellar matter, is consistent with a nearby source. If this identification is correct, an X-luminosity of Capella during the observation comes out ranging from 10^{31} to 10^{34} erg s^{-1}.

Fitting the data with a function describing thermal bremsstrahlung the authors found temperatures ranging from 5 to 15×10^6 K.

This source was not observed during a previous survey of the same region in 1972. The authors recall other soft X-ray sources, identified with nearby stellar objects, which other observations of comparable or better sensitivity have failed to detect. According to the authors one must conclude that either many spurious observations have been reported, or these sources of soft X-ray emission are strongly variable in time.

Comparison of these events with those observed on YZ CMi and UV Ceti is stimulating.

7. Concluding Remarks

I should like to conclude this review on the observational aspects of stellar activity of the solar type with some remarks.

Although, as we have seen, information on Ap stars can be extremely stimulating in our framework, it is advisable to look for objects with much weaker magnetic fields. The recently opened possibility of observing stellar weak magnetic fields must therefore be pursued.

As we have seen, there is a lot of information that could be collocated in our framework. Notwithstanding this, we have not yet ascertained the existence of stellar activity cycles and we are just beginning to understand some associations among different known manifestations of stellar activity.

What we absolutely need to substantially improve the situation is an international cooperative effort that must exploit all the different available techniques and must be concentrated for a long time on a few objects of different types. We cannot hope the stars will help us in understanding the Sun unless we observe them as we observe the Sun.

Acknowledgements

It is a pleasure to thank F. Catalano, S. Catalano and M. Rodonò for helpful discussions and S. Motta and M. Rodonò for reading the manuscript.

References

Albrecht, R., Maitzen, H. M., and Rakos, K. D.: 1969, *Astron. Astrophys.* **3**, 236.
Ambartsumian, V. A. and Mirzoyan, L. V.: 1971, *IAU Colloq.* **15**, 98.
Anderson, C. M.: 1973, *Bull. Am. Astron. Soc.* **5**, 413.
Andrews, A. D.: 1968, IAU Comm. 27, *Inf. Bull. Var. Stars* **273**.
Arnold, C. N. and Hall, D. S.: 1973, IAU Comm. 27, *Inf. Bull. Var. Stars* **843**.
Ayres, T. R., Linsky, J. L., and Shine, R. A.: 1975, *Astrophys. J.* **195**, L121.
Babcock, H. W.: 1968, *Ann. Rep. Director Mt. Wilson Palomar Obs. 1966–67*, p. 262.
Babcock, H. W.: 1974, *Ann. Rep. Director Hale Obs. 1973–74*, p. 136.
Bappu, M. K. V.: 1967, *Kodaikanal Obs. Repr. No. 37*.
Bateson, F. M.: 1971, *Southern Stars* **24**, 23.
Batten, A. H.: 1973, *Binary and Multiple Systems of Stars*, Pergamon Press, Oxford.
Bidelman, P.: 1954, *Astrophys. J. Suppl.* **1**, 175.
Biermann, L. and Lüst, R.: 1960, *Stars Stellar Systems* **6**, 260.
Biermann, L.: 1969, *Proc. Roy. Soc. London* **A313**, 357.
Binnendijk, L.: 1969, *Astron. J.* **74**, 1031.
Binnendijk, L.: 1970, *Vistas Astron.* **12**, 217.
Blanco, C., Catalano, F. A., Godoli, G., and Vaccari, S.: 1972, *Mem. Soc. Astron. Ital.* **43**, 655.
Blanco, C., Catalano, S., Marilli, E., and Rodonò, M.: 1974, *Astron. Astrophys.* **33**, 257.
Blanco, C., Catalano, S., and Marilli, E.: 1976, this volume, p. 473.
Boesgaard, A. M.: 1969, *Publ. Astron. Soc. Pacific* **81**, 283.
Boesgaard, A. M.: 1974, *Astrophys. J.* **188**, 567.
Boesgaard, A. M. and Hagen, W.: 1974, *Astrophys. J.* **189**, 85.
Bonsack, W. K. and Culver, R. B.: 1966, *Astrophys. J.* **145**, 767.
Bopp, B. W. and Evans, D. S.: 1973, *Monthly Notices Roy. Astron. Soc.* **164**, 343.
Bopp, B. W. and Moffett, T. J.: 1973, *Astrophys. J.* **185**, 239.

Bopp, B. W.: 1974a, *Monthly Notices Roy. Astron. Soc.* **166**, 79.
Bopp, B. W.: 1974b, *Monthly Notices Roy. Astron. Soc.* **168**, 255.
Bopp, R. W. and Feckel, F. Jr.: 1974, *Publ. Astron. Soc. Pacific* **86**, 978.
Borra, E. F. and Landstreet, J. D.: 1973a, *Astrophys. J.* **185**, L139.
Borra, E. F. and Landstreet, J. D.: 1973b, *Astrophys. J.* **185**, L145.
Braes, L. L. E., Habing, H. J., and Schoenmaker, A. A.: 1972, *Nature* **240**, 230.
Bray, R. J. and Loughhead, R. E.: 1964, *Sunspots*, Chapman and Hall Ltd., London, p. 282.
Bumba, V. and Howard, R.: 1965, *Astrophys. J.* **141**, 1502.
Bumba, V., Howard, R., and Smith, S. F.: 1967, in R. C. Cameron (ed.) *Magnetic and Related Stars, Mono Book Corporation, Baltimore*, p. 131.
Bumba, V. and Růžičková-Topolová, B.: 1967, *Solar Phys.* **1**, 216.
Burton, W. M., Ridgeley, A., and Wilson, R.: 1967, *Monthly Notices Roy. Astron. Soc.* **135**, 207.
Burton, W. M. and Ridgeley, A.: 1970, *Solar Phys.* **14**, 3.
Cameron, R. C.: 1967, *The Magnetic and Related Stars*, Mono Book Corporation, Baltimore.
Caplan, J. G.: 1973, *Astron. Astrophys.* **28**, 213.
Catalano, F. A.: 1975, *Mem. Soc. Roy. Sci. Liège VI Series* **7**, 117.
Catalano, F. A. and Strazzulla, G.: 1975, *IAU Colloq.* **32**, (to be published), preprint.
Catalano, S. and Rodonò, M.: 1974, *Publ. Astron. Soc. Pacific* **86**, 390.
Catura, R. C., Acton, L. W., and Johnson, H. M.: 1974, *Bull. Am. Astron. Soc.* **6**, 445.
Catura, R. C., Acton, L. W., and Johnson, H. M.: 1975, *Astrophys. J.* **196**, L47.
Chugainov, P. F.: 1966, Comm. 27 IAU, *Inf. Bull. Var. Stars* **122**.
Chugainov, P. F.: 1973, *Izv. Krymsk. Astrofiz. Obs.* **48**, 3.
Chugainov, P. F.: 1974, *Izv. Krymsk. Astrofiz. Obs.* **52**, 3.
Crannell, C. J., McClintock, J. E., and Moffett, T. J.: 1974, *Nature* **252**, 659.
Cristaldi, S., Godoli, G., Narbone, M., and Rodonò, M.: 1969, in '*Non-Periodic Phenomena in Variable Stars*', IAU Colloq., Academic Press, Budapest, p. 149.
Cristaldi, S. and Rodonò, M.: 1975, *IAU Symp.* **67**, 75.
Cristaldi, S. and Rodonò, M.: 1976, *Astron. Astrophys.* **48**, 165.
Deutsch, A. J.: 1967, *Publ. Astron. Soc. Pacific* **79**, 431.
Doherty, L. R.: 1971, *Phil. Trans. Roy. Soc. London* **A270**, 189.
Doherty, L. R.: 1972, *Astrophys. J.* **178**, 495.
Dupree, A. K.: 1974, *Bull. Am. Astron. Soc.* **6**, 446.
Durand, E., Oberly, J. J., and Tousey, R.: 1949, *Astrophys. J.* **109**, 1.
Durney, B. R. and Stenflo, J. O.: 1972, *Astrophys. Space Sci.* **15**, 307.
Eberhard, G. and Schwarzschild, K.: 1913, *Astrophys. J.* **38**, 292.
Edlén, B., Palenius, H. P., Bockasten, K., Hallin, R., and Bromander, J.: 1969, *Solar Phys.* **9**, 432.
Evans, D. S.: 1971, *Monthly Notices Roy. Astron. Soc.* **154**, 329.
Evans, D. S.: 1973, *Bull. Am. Astron. Soc.* **5**, 400.
Evans, R. G., Jordan, C., and Wilson, R.: 1975, *Nature* **253**, 612.
Ferraz Mello, S. and Torres, C. A. O.: 1971, IAU Comm. 27, *Inf. Bull. Var. Stars* **577**.
Friedemann, C. and Gürtler, J.: 1975, *Astron. Nachr.* **296**, 125.
Fusi Pecci, A. and Renzini, A.: 1975, *Astron. Astrophys.* **39**, 413.
Geisel, S. L.: 1970, *Astrophys. J.* **161**, L105.
Gerola, H., Linsky, J. L., Shine, R., McClintock, W., Henry, R. C., and Moos, H. W.: 1974, *Astrophys. J.* **193**, L107.
Gershberg, R. E.: 1967, *Astrophysics* **3**, 64.
Gershberg, R. E., Neshpor, Y. I., and Chugainov, P. F.: 1969, *Izv. Krymsk. Astrofiz. Obs.* **39**, 140.
Gershberg, R. E.: 1970, *Astrophysics* **6**, 92.
Gershberg, R. E.: 1971, *Flares of Red Dwarf Stars* (transl. from the Russian by D. J. Mullan), Armagh Obs., N. Ireland.
Gershberg, R. E. and Pikelner, S. B.: 1972, *Comments Astrophys. Space Sci.* **4**, 113.
Gershberg, R. E.: 1975, *IAU Symp.* **67**, 47.
Gershberg, R. E. and Lund, L.: 1975, Proc. III European Astron. Meeting.
Giglas, H. L., Burnham, R. Jr., and Thomas, N. G.: 1967, *Lowell Obs. Bull. No. 138*.
Gliese, W.: 1969, *Veroeffentl. Astron. Rechen. Inst. Heidelberg No. 22*.
Godoli, G.: 1967, *Oss. astrofis. Catania Pubbl. No. 115*.
Godoli, G.: 1968, in Y. Ohman (ed.) *Mass Motions in Solar Flares and Related Phenomena*, Wiley Interscience Division, New York, p. 211.
Greenstein, J. L.: 1952, *Publ. Astron. Soc. Pacific* **64**, 71.
Greenstein, J. L. and Arp, H.: 1969, *Astrophys. Letters* **3**, 149.
Griffin, R. F.: 1963, *Observatory* **83**, 255.

Grindlay, J. E.: 1970, *Astrophys. J.* **162**, 187.

Grindlay, J.: 1975, Private communication.

Grindlay, J. and Heise, J.: 1975, 'Observation of X Ray from Flare Stars with ANS' (preprint).

Groth, H. G. and Wellmann, P.: 1970, *Spectrum Formation in Stars with Steady State Extended Atmospheres*, Nat. Bureau Standard, Washington, Special Publ. 332.

Gurzadian, G. A.: 1965, *Astrophysics* **1**, 170.

Gurzadian, G. A.: 1966a, *Astrophysics* **2**, 109.

Gurzadian, G. A.: 1966b, *Dokl. Akad. Nauk SSSR* **166**, 53.

Gurzadian, G. A.: 1971, *Astron. Astrophys.* **13**, 348.

Gurzadian, G. A.: 1972, *Astron. Astrophys.* **20**, 145.

Gurzadian, G. A.: 1972, *Sky Telesc.* **43**, 350.

Hall, D. S.: 1972, *Publ. Astron. Soc. Pacific* **84**, 323.

Hall, D. S.: 1975, Preprint.

Hanbury Brown, R., Davis, J., and Allen, L. R.: 1974, *Monthly Notices Roy. Astron. Soc.* **168**, 93.

Harrington, R. S., Dahn, C. C., and Guetter, H. H.: 1974, *Astrophys. J.* **194**, L87.

Haupt, W. and Schlosser, W.: 1974, *Astron. Astrophys.* **37**, 219.

Heise, J., Brinkman, A. C., Schrijver, J., Mewe, R., Gronenschild, E., Den Boggende, A., and Grindley, J.: 1975, *Astrophys. J.* **202**, 73.

Hjellming, R. M. and Wade, C. M.: 1971a, *Astrophys. J.* **168**, L115.

Hjellming, R. M. and Wade, C. M.: 1971b, *Science* **173**, 1087.

Howard, R.: 1972, *Science* **177**, 1157.

Hudson, H. S.: 1972, *Solar Phys.* **24**, 414.

Jager, C. de and Neven, L.: 1961, *Mem. Soc. Roy. Sci. Liège*, V. Serie, **4**, 552.

Jager, C. de: 1971, *Phil. Trans. Roy. Soc. London* **A270**, 175.

Jager, C. de and Loore, C. de.: 1971, *Astrophys. Space Sci.* **11**, 284.

Jaschek, M. and Malaroda, S.: 1970, *Nature* **225**, 246.

Jennings, M. C. and Dyck, H. M.: 1972, *Astrophys. J.* **177**, 427.

Jennings, M. C.: 1973, *Astrophys. J.* **185**, 197.

Jerzykiewicz, M. and Serkowski, K.: 1966, *Lowell Obs. Bull.* **6**, 295.

Johnson, H. L. and Iriarte, B.: 1959, *Lowell Obs. Bull.* **4**, 99.

Jordan, S. D. and Avrett, E. H.: 1973, *Stellar Chromospheres*, NASA, Washington, SP-317.

Joy, A. H. and Wilson, R. E.: 1949, *Astrophys. J.* **109**, 231.

Kahn, F. D.: 1974, *Nature*, **250**, 125.

Kandel, R.: 1962. *Compt. Rend. Acad. Sci. Paris* **255**, 1575.

Kandel, R.: 1966, in M. Hack (ed.), *Colloquium on Late Type Stars*, Oss. Astron., Trieste, p. 146.

Kellermann, K. I. and Pauliny Toth, I. I. K.: 1966, *Astrophys. J.* **145**, 953.

Klimishin, I. A.: 1970, *Tsirk. ShA. O.* **6**, 13.

Kodaira, K.: 1973, *Astron. Astrophys.* **26**, 385.

Kondo, Y.: 1972, *Astrophys. J.* **171**, 605.

Kondo, Y., Giuli, R. T., Modisette, J. L. and Rydgren, A. E.: 1972, *Astrophys. J.* **176**, 153.

Kondo, Y., Morgan, T. H., and Modisette, J. L.: 1975, *Astrophys. J.* **196**, L125.

Korovyakovskaya, A. A.: 1972, *Astrophysics*, **8**, 148.

Kraft, R. P., Preston, G. W., and Wolff, S. C.: 1964, *Astrophys. J.* **140**, 235.

Kraft, R. P. and Greenstein, J. L.: 1969, in S. Kumar (ed.), *Low Luminosity Stars*, Gordon and Breach, London, p. 65.

Kron, G. E.: 1947, *Publ. Astron. Soc. Pacific* **59**, 261.

Kron, G. E.: 1950a, *Publ. Astron. Soc. Pacific* **62**, 141.

Kron, G. E.: 1950b, *Astron. J.*, **55**, 69.

Kron, G. E.: 1950c, *Astron. Soc. Pacific Leaflet* **6**, No. 257.

Krzeminsky, W. and Kraft, R. P.: 1967, *Astron. J.* **72**, 307.

Krzeminsky, W.: 1969, in S. Kumar (ed.), *Low Luminosity Stars*, Gordon and Breach, London, p. 57.

Kunkel, W. E.: 1967, *Astron. J.* **72**, 810.

Kunkel, W. E.: 1970, *Publ. Astron. Soc. Pacific* **82**, 1341.

Kunkel, W. E.: 1973, *Astrophys. J. Suppl.* **25**, 1.

Kunkel, W. E.: 1975, *IAU Symp.* **67**, 15.

Kunkel, W. E. and Zarate, N.: 1975, Preprint.

Kuperus, M.: 1965, *Rech. Astron. Obs. Utrecht* **17**, No. 1.

Lamers, H. J. G. L. M. and Loore, C. de: 1974, Preprint.

Landini, M. and Monsignori Fossi, B. C.: 1973, *Astron. Astrophys.* **25**, 9.

Ledoux, P. and Walraven, Th.: 1958, *Handbuch der Physik* **51**, 353.

Liller, W.: 1968, *Astrophys. J.* **151**, 589.

Linsky, J. L.: 1973, in S. D. Jordan and E. H. Avrett (eds.), *Stellar Chromospheres*, NASA, Washington, SP-317, p. 48.
Linsky, J. L. and Basri, G.: 1974, *Bull. Am. Astron. Soc.* **6**, 458.
Lo Presto, J. C.: 1971, *Publ. Astron. Soc. Pacific* **83**, 674.
Loore, C. de: 1970, *Astrophys. Space Sci.* **6**, 60.
Loore, C. de and Jager, C. de: 1970, *IAU Symp.* **37**, 238.
Lortet-Zuckermann, M. C.: 1965, *Kleine Veroeffentl. Remeis Sternw. Bamberg* **4**, 30.
Lovell, B.: 1971, *Quart. J. Roy. Astron. Soc.* **12**, 98.
Lovell, B.: 1974, *Phil. Trans. Roy. Soc. London* **A277**, 489.
Lovell, B., Maridis, L. N., and Contadakis, M. E.: 1974, *Nature* **250**, 124.
Low, F.: 1965, *IAU Circ.* **1884-5**.
Lutz, T. E., Furenlid, I., and Lutz, J. H.: 1973, *Astrophys. J.* **184**, 787.
McClintock, W., Linsky, J., Gerola, H., Shine, R., Henry, R. C., and Moos, H. W.: 1974, *Bull. Am. Astron. Soc.* **6**, 315.
McMath, R. R., Mohler, O. C., Pierce, A. K., and Goldberg, L.: 1956, *Astrophys. J.* **124**, 1.
Martins, D. H.: 1975, *Publ. Astron. Soc. Pacific* **87**, 163.
Matsushima, S.: 1974, *Sci. Rep. Tôhoku Univ. First Ser.*, **57**, 23.
Moffett, T. J.: 1974, *Sky Telesc.* **48**, 94.
Moffett, T. J.: 1975a, Comm. 27 IAU, *Inf. Bull. Var. Stars* **995**
Moffett, T. J.: 1975b, Private communication.
Moos, H. W. and Rottman, G. J.: 1972, *Astrophys. J.* **174**, L73.
Moos, H. W., Linsky, J. L., Henry, R. C., and McClintock, W.: 1974, *Astrophys. J.* **188**, L93.
Morton, D. C.: 1967, *Astrophys. J.* **150**, 535.
Mullan, D. J.: 1974, *Astrophys. J.* **192**, 149.
Mullen, D. I.: 1975, *Astrophys. J.* **198**, 563.
Nariai, K.: 1969, *Astrophys. Space Sci.* **3**, 150.
Oganesyan, D. B.: 1974, *Soviet Astron.* **17**, 617.
Oster, L.: 1971, *Astrophys. J.* **169**, 57.
Oster, L.: 1975, in B. Baschek, H. W. Kegel, and G. Traving (eds.), *Problems in Stellar Atmospheres and Envelopes*, Springer-Verlag, Berlin, p. 301.
Pikelner, S. B. and Khokhlova, V. L.: 1971, *Comments Astrophys. Space Phys.* **3**, 190.
Pikeluer, S. B. and Khokhlova, V. L.: 1972, *Soviet Phys. Uspekhi* **15**, 395.
Piotrowski, S. L., Rucinski, S. M., and Semeniuk, I.: 1974, *Acta Astron.* **24**, 389.
Popper, D. M.: 1953, *Publ. Astron. Soc. Pacific* **65**, 278.
Preston, G. W. and Wolff, S. C.: 1970, *Astrophys. J.* **160**, 1071.
Preston, G. W.: 1971, *Publ. Astron. Soc. Pacific* **83**, 571.
Reimers, D.: 1973, *Astron. Astrophys.* **24**, 79.
Robinson, E. L. and Kraft, R. P.: 1974, *Astron. J.* **79**, 698.
Rodonò, M.: 1974, *Astron. Astrophys.* **32**, 337.
Rodonò, M.: 1975a, Private communication.
Rodonò, M.: 1975b, Preprint.
Rodonò, M.: 1976, This volume, p. 475.
Rosendhaì, J. D.: 1973, *Astrophys. J.* **182**, 523.
Rottman, G. J., Moos, H. W., Barry, J. R., and Henry, R. C.: 1971, *Astrophys. J.* **165**, 661.
Schatzman, E.: 1959, *IAU Symp.* **9**, 552.
Seaquist, E. R.: 1967, *Astrophys. J.* **148**, L23.
Serkowski, K.: 1961, *Lowell Obs. Bull.* **5**, 157.
Severny, A.: 1969, *Nature* **224**, 53.
Severny, A.: 1970, *Astrophys. J.* **159**, L73.
Sheeley, N. R. Jr.: 1967, *Astrophys. J.* **147**, 1106.
Sitterly, B. W.: 1930, *Princeton Univ. Obs. Contr.* **III**, No. 11, p. 21.
Skumanich, A.: 1972, *Astrophys. J.* **171**, 565.
Smith, H. J. and Smith, E. V. P.: 1963, in *Solar Flares*, The McMillan Company, New York, p. 270.
Struve, O.: 1952, *Publ. Astron. Soc. Pacific* **64**, 20.
Švestka, Z.: 1970, *Solar Phys.* **13**, 471.
Torres, C. A. O., Ferraz Mello, S., and Quast, G. R.: 1972, *Astrophys. Letters* **11**, 13.
Torres, C. A. O. and Ferraz Mello, S.: 1973, *Astron. Astrophys.* **27**, 231.
Tsap, T. T. and Laba, I. S.: 1973, *Izv. Krynsk. Astrofiz. Obs.* **48**, 73.
Tuominen, J.: 1970, *IAU Symp.* **43**, 754.
Unsöld, A.: 1955, in *Physik der Sternatmosphären*, 2nd edition, Springer-Verlag, Berlin, p. 606.
Unsöld, A.: 1964, *Observatory* **84**, 152.

Vardya, M. S.: 1970, *Observatory* **90**, 155.
Vaughan, A. H. and Zirin, H.: 1968, *Astrophys. J.* **152**, 123.
Vogt, S. S.: 1973, *Bull. Am. Astron. Soc.* **5**, 399.
Vogt, S. S.: 1975, *Astrophys. J.* **199**, 418.
Wade, C. M. and Hjellming, R. M.: 1971, *Astrophys. J.* **163**, L105.
Walter, K.: 1971, *IAU Colloq.* **15**, 497.
Warner, B.: 1969, *Monthly Notices Roy. Astron. Soc.* **144**, 333.
Weiler, E. J.: 1975a, IAU Comm. 27, *Inf. Bull. Var. Stars* **1014**.
Weiler, E. J.: 1975b, *Bull. Am. Astron. Soc.* **7**, 267.
Weymann, R. and Chapman, G.: 1965, *Astrophys. J.* **142**, 1268.
Wilson, O. C.: 1954, *Proc. N.S.F. Conference on Stellar Atmospheres*, Indiana Univ. Press, Bloomington.
Wilson, O. C. and Bappu, M. K. V.: 1957, *Astrophys. J.* **125**, 661.
Wilson, O. C.: 1959, *Astrophys. J.* **130**, 499.
Wilson, O. C.: 1961, *Publ. Astron. Soc. Pacific* **73**, 15.
Wilson, O. C.: 1963, *Astrophys. J.* **138**, 832.
Wilson, O. C. and Skumanich, A.: 1964, *Astrophys. J.* **140**, 1401.
Wilson, O. C.: 1967, *Publ. Astron. Soc. Pacific* **79**, 46.
Wilson, O. C.: 1968, *Astrophys. J.* **153**, 221.
Wilson, O. C.: 1969, in S. Kumar (ed.), *Low Luminosity Stars*, Gordon and Breach, London, p. 103.
Wilson, O. C.: 1970, *Publ. Astron. Soc. Pacific* **82**, 865.
Wilson, O. C. and Woolley, R.: 1970, *Monthly Notices Roy. Astron. Soc.* **148**, 463.
Wilson, O. C.: 1976, this volume, p. 447.
Wordern, S. P.: 1974, *Publ. Astron. Soc. Pacific* **86**, 595.
Wright, K. O.: 1973, *IAU Symp.* **51**, 117.

DISCUSSION

Mestel: Did I understand you to say that one needs photospheric activity in an Ap star to understand the cross-over effect? Isn't this effect simply explained in the oblique rotator model as a geometrical consequence of the rotation of contiguous regions of opposite polarity?

Godoli: Yes, in order to describe the cross-over effect, we require magnetic regions of opposite polarity and the magnetic activity is the main manifestation of active regions.

Mestel: The difficulty seems to be largely semantic: you mean by photospheric activity in an Ap star just the existence of observable emerging field-lines.

Ruždjak: Which is the highest Balmer line seen in emission in flare stars? Has the Balmer recombination continuum been observed.?

Godoli: Balmer series has been observed to H_{10}–H_{14}. As I have already said, strong emission in the Balmer continuum appears at the beginning of flares.

Schröter: I would like to add to your list of observable effects of stellar activity and its implication in interpretation one other aspect which concerns the problem of element abundances. The interaction of solar magnetic fields with photospheric and chromospheric plasma occurs apparently in two different modes; In one mode magnetic fields suppress radiation flux (sunspots, pores); in the other mode they lead to enhanced radiation (plages, network, faculae), etc. Let us apply this (so far not well understood) fact to stars with magnetic fields.

Imagine that in a stellar atmosphere many spots are present. Lines of elements with low ionization potential will occur much strengthened in these spots. Determining the abundance of elements like Li and ignoring the presence of such activity regions may well lead to an apparent overabundance of these elements. Let us now assume the surface of the star is covered considerably by features like network of faculae in which the magnetic fields cause locally enhanced radiation. Consider once again the problem of determination of the abundance of Li (or for example, C). Ignoring the presence of these active regions shall this time lead to an apparent 'under abundance' of Li (or overabundance of C). To my knowledge the effect of the presence of such structures has so far not been taken in account in determining element abundances (or in interpreting deviations observed in magnetic active stars).

Zwaan: Have attempts been made to find cyclic variations in the rate of flare occurrence in flare stars?

Godoli: Yes; variations have been found but not yet periodic variations.

Roxburgh: Could Dr Schröter quantify his argument on abundance determinations? If we looked at the Sun as a star what would be the error in the determination of the Lithium abundance?

Schröter: To answer your question, let us look at the behaviour of the resonance line of Li on the Sun. This line is almost absent in the photosphere ($W_\lambda \leq 1$ Å) and has an equivalent-width of > 200 mÅ in sunspots. How large the error in abundance may be when ignoring the contribution of spot-like areas to the observed spectrum depends strongly on the procedure of spectrum analysis. I can imagine that errors up to factor 10 may well arise.

CHROMOSPHERIC VARIATIONS
IN MAIN SEQUENCE STARS

O. C. WILSON

*Hale Observatories, Carnegie Institution of Washington,
California Institute of Technology*

The work described here is still in progress; hence this paper will be a brief sampling of the results to date.

There are several observable indices of the more or less periodic variations in solar activity known generally as the solar cycle. It has long been known that one of these indices is the total flux in the chromospheric emission components of the H and K lines of Ca II. In searching for analogous variations in stars, this same quantity appears to be the most easily observable one, provided suitable instrumentation is available.

About ten years ago a coude scanner was completed for the 100-in. telescope on Mt. Wilson. This is a two channel machine, with two photomultipliers and two sets of pulse counting electronics. It is arranged so that one channel receives the signal from a slit whose width can be chosen, and across which the spectrum can be moved in the direction of dispersion. The other channel receives the combined signals from two 25 Å windows on either side of the region through which the spectrum can be scanned. In the present work the scan mode was not used. Taking into account the total radial velocity of a star at a given time, the second slit can be set at the centers of the H and K lines with a precision of two or three hundredths of an ångström. Both the entrance and exit slits were set for one Ångström width at the focal plane.

An observation consists of four integrations, two in K and two in H, with the counts in both channels started by pushing a button and terminated simultaneously by the machine at a pre-set value of the count in the one Ångström channel. For most stars under normal conditions the latter is chosen to be 10^4. The counts at the line centers are added and divided by the corresponding counts in the monitor channels, and these numbers are then divided by the same ratios obtained from a standard lamp observed each day. In this way the effect of drifts and changes in the electronics are eliminated. Standard stars are also observed each night. These are stars with minimal H-K fluxes which probably have little or no chromospheric emission and their constancy bears out this assumption.

All the stars in the program are main sequence objects. They extend from above the Sun, at spectral types about F5, down to M0. Each star is observed on every night when it is available, and every effort is made to obtain a dozen or more observations per season. This is essential since many of the stars appear to have fluctuations on a variety of time scales and as many observations as possible are necessary to give a good representation of the longer term variations.

Figure 1 shows the results on one of the standard stars and the scatter here, of the order of ±2%, is probably due almost entirely to the unavoidable errors of observation. This star is typical of the eighteen standards around the sky, all of which have fluxes in the range 1300 to 1500 instrumental units.

Bumba and Kleczek (eds.), Basic Mechanisms of Solar Activity, 447–452. All Rights Reserved.
Copyright © 1976 by the IAU.

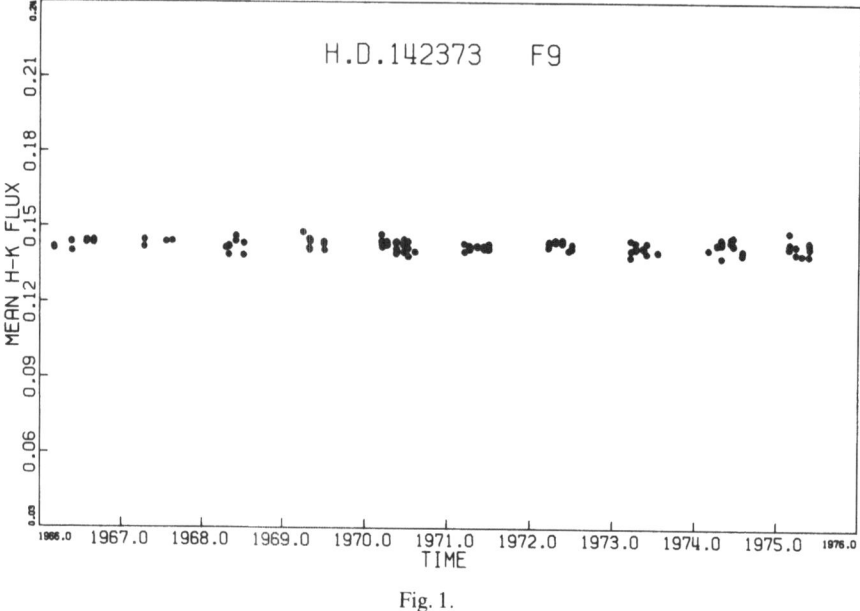

Fig. 1.

The second and third slides show the data on two stars near the Sun on the main sequence. These are typical of such objects and also of F-type stars; there are obvious variations in their fluxes, but there is no evidence for periodicity. Note that these two stars have rather strong chromospheric emissions and are therefore young. Similar stars with fluxes intermediate between the preceding ones and the standard stars, show behavior similar to that of Figures 2 and 3, but on a reduced scale.

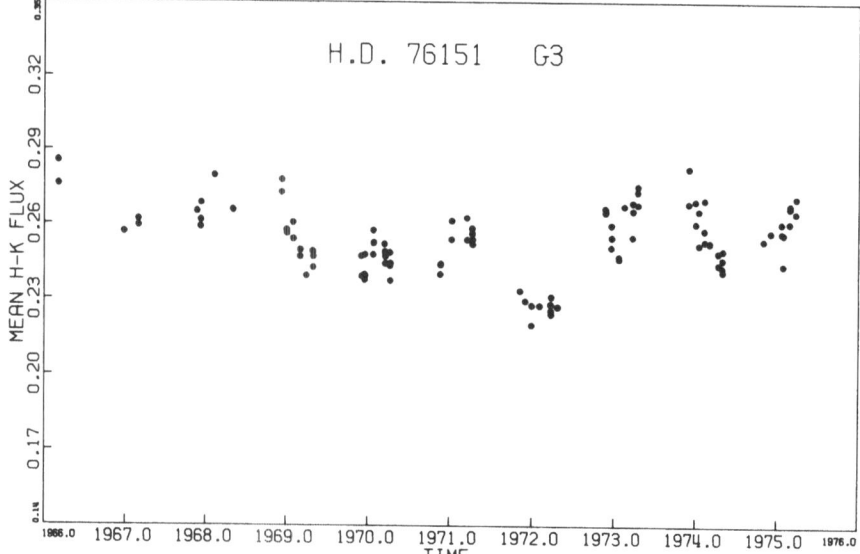

Fig. 2.

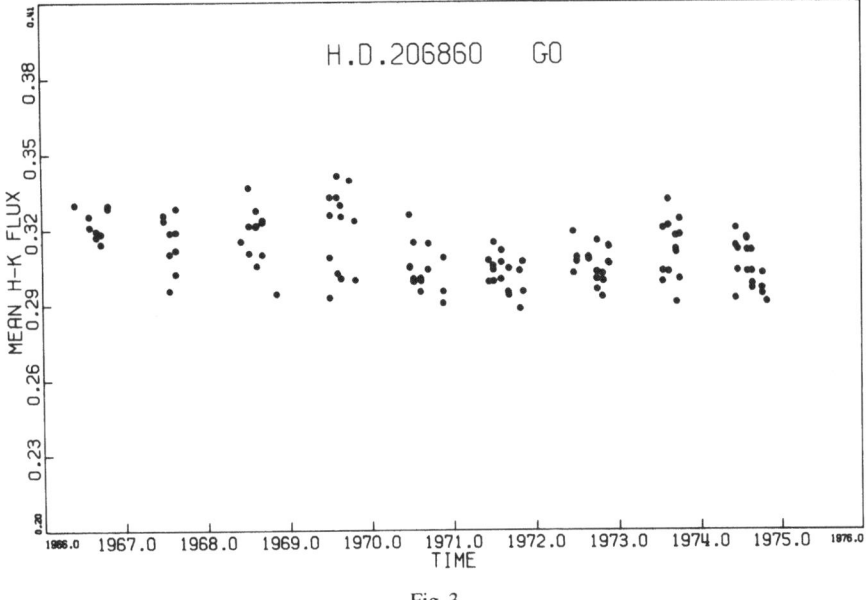

Fig. 3.

In Figures 4, 5, and 6 are shown results for three K-type stars with moderately strong emission. All three are probably undergoing cyclical variations which may well be analogous to the solar cycle. Note that the rise time seems to be shorter than the decay time in all three cases as is true for the Sun. This difference appears in exaggerated form for HD 32147 where the next rise had not yet begun even at the end of the 1974–75 observing season. The scatter in the seasonal point groups is

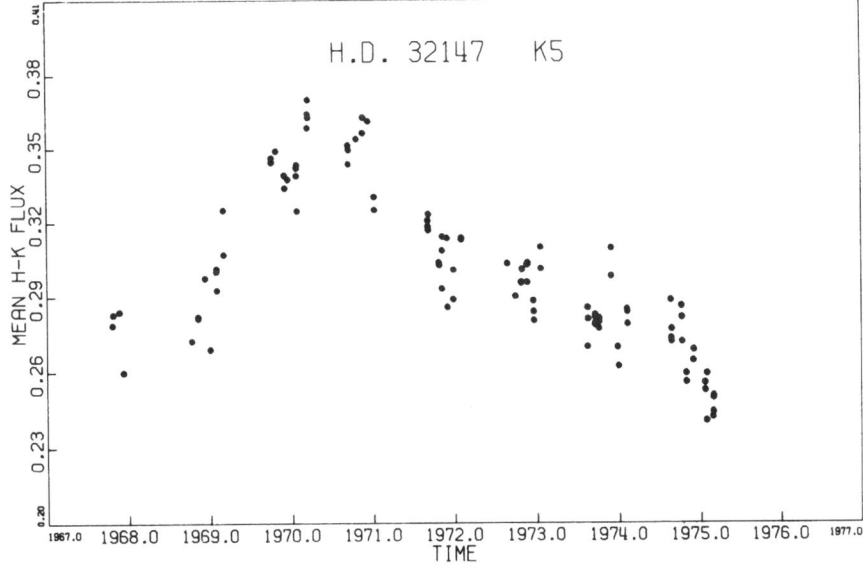

Fig. 4.

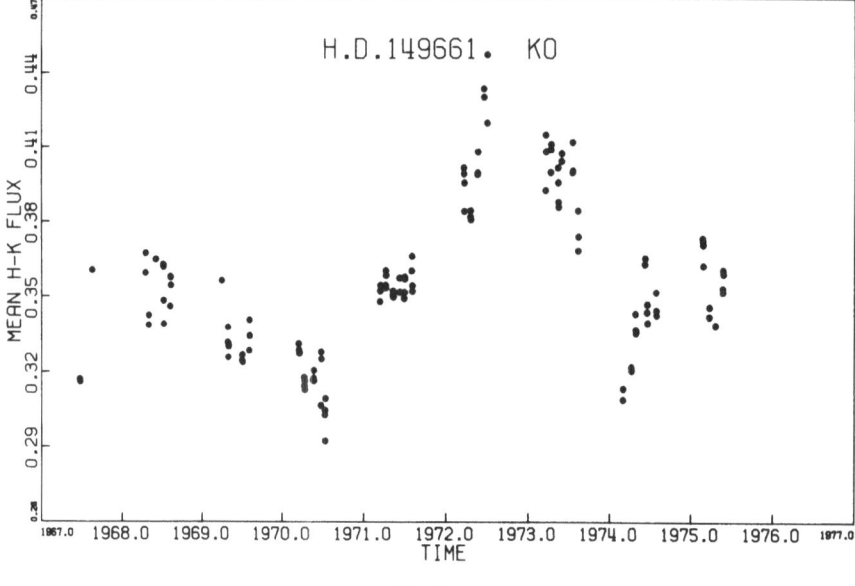

Fig. 5.

partly due to the overall change during an observing season, but a major part of it is caused by intrinsic variations of unknown source and time scale. Rotational modulation is a highly probable source for some of it, but more continuous observations would be needed to disentangle this effect from other possibly sporadic ones.

Figure 7 gives the results for another K-type star, HD 22049, which has significantly more chromospheric flux than the three preceding ones. Here there is no

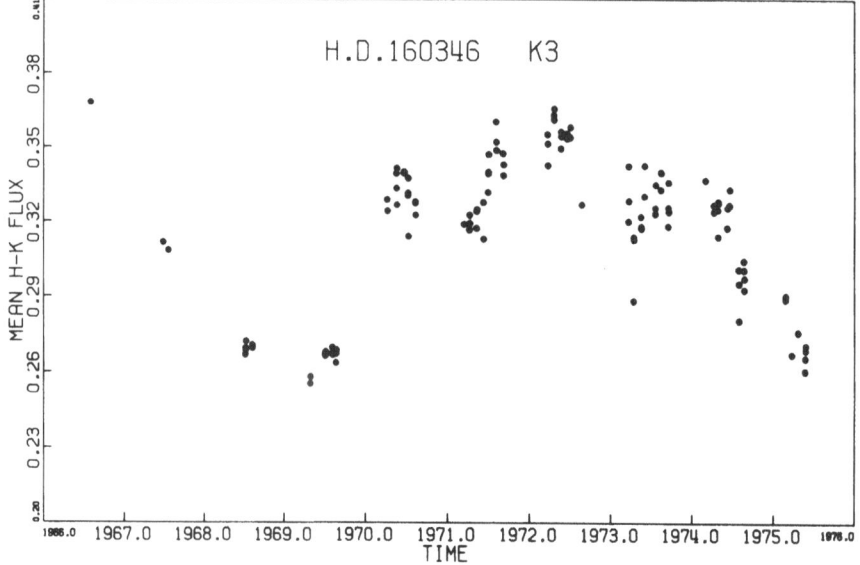

Fig. 6.

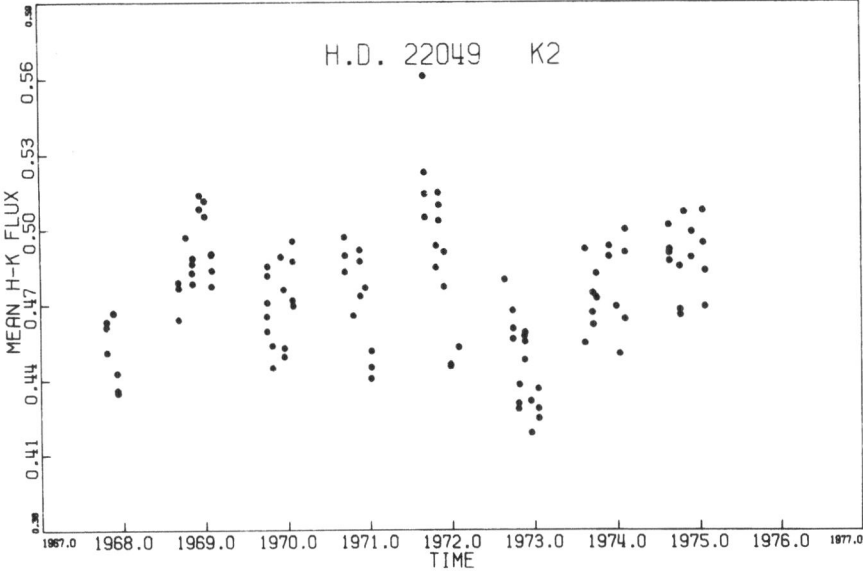

Fig. 7.

obvious cyclical variation, but there is a great deal of 'noise' or apparently random scatter. In fact, from all of the results thus far, it appears that the scatter increases with the flux.

Finally, Figure 8 shows a G8 star whose behavior is somewhat difficult to categorize, although the variations in its flux are obvious enough.

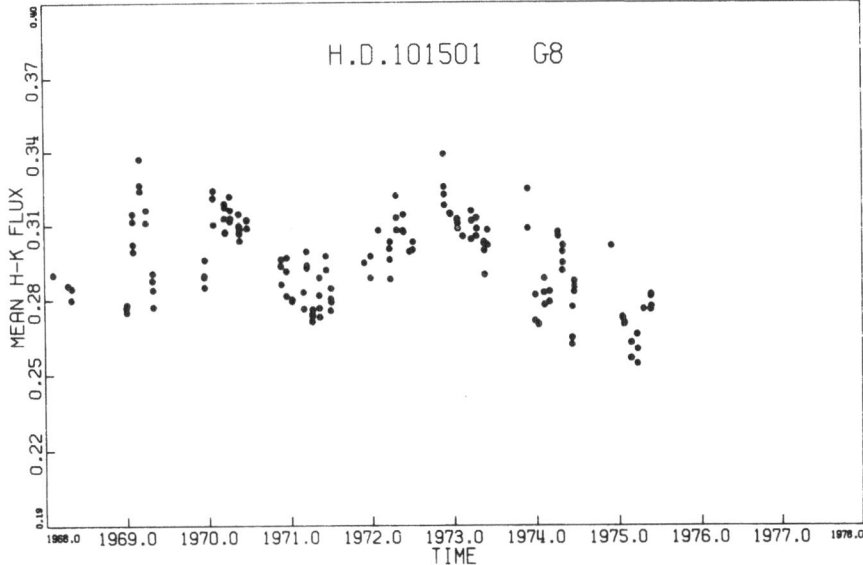

Fig. 8.

The samples given are a fair representation of what has been accomplished thus far. Clearly it is just a beginning, but it is hoped that observation of chromospheric variations in stars may lead to a better understanding of the solar cycle. In conclusion, it should be mentioned that none of the stars in this program are known spectroscopic binaries.

THE INTERNAL STRUCTURE OF THE SUN AND
SOLAR TYPE STARS

IAN W. ROXBURGH

Queen Mary College, University of London
and
High Altitude Observatory, National Centre for Atmospheric Research, Boulder, Colo., U.S.A.

Abstract. Our understanding of the internal structure of the Sun and solar type stars has been undermined by recent observations. In this paper we consider some of the puzzles and possible resolutions; the solar neutrino problem, lithium and beryllium abundance, rotation and calcium emission, variation of the solar constant, solar oscillations and the solar convective zone. The picture that emerges is one of confusion, and so it should be since we have no idea of what is going on inside the Sun and *a fortiori* of what is going on inside other stars.

1. Introduction

A few years ago, the internal structure and evolution of stars was thought to be well understood. The theory could more or less explain the few observations that existed and there was every confidence that other details would also be explained within the same theoretical structure.

This dream world of the stellar theoretician is now ended. The stellar theorist who had only a passing interest in the Sun has been, is being, or should be, forced to re-examine his theoretical speculations in the light of new data from the Sun. Solar physics has forced the stellar physicist to think again, and the results of that re-thinking could have repercussions in many branches of astrophysics and perhaps other branches of physics as well.

Perhaps the most spectacular experiments are the abortive search for solar neutrinos by Davis and co-workers (Davis 1972), measurements of solar oblateness by Dicke and Goldenberg (1967) and the improved measurements by Henry Hill and his co-workers, with their measurements of oscillations of the Sun (Hill *et al.*, 1975). But other observations have added to the confusion; abundance measurements of lithium and beryllium in the Sun and other stars, (Wallerstein and Conti, 1969) observations of Ca^+ emission in the Sun and stars (Wilson, 1963), measurements of the solar wind, geological evidence from Earth, and theoretical work on climate modelling (Hays, 1975). The observations of multi-scale convection on the Sun has raised doubts about the mixing length theory and when all placed together, it is difficult to escape the conclusion that something is wrong with standard ideas on stellar structure. While we have not yet evolved a new theory, the need to accommodate these new results has made the study of the internal structure of the Sun and solar type stars one of the most exciting and challenging areas of research.

In this paper I want to describe some of these dilemmas and proposed resolutions. I cannot give you a new theory for the solar interior for one does not exist – all is confusion. So in contrast to most papers, the purposes of this paper is to confuse you, not to enlighten you. This should be an easy task for one whose papers are normally confusing, even when they are not supposed to be.

Bumba and Kleczek (eds.), Basic Mechanisms of Solar Activity, 453–466. All Rights Reserved.

2. A Simple Man's Guide to the Internal Structure of the Sun

The standard models of stellar structure can be found in many textbooks (e.g. Schwarzschild, 1958). In arriving at these models many assumptions are made which we now have to question; let me list some of them:

(1) The stars are in hydrostatic equilibrium with pressure balancing gravity – there are no significant accelerations.

(2) The stars are in thermal equilibrium – the energy radiated is equal to the energy produced so there are no secular changes, except in the early pre-main sequence contraction phase of solar evolution.

(3) Energy is carried by radiation only in those regions of a star which are stable against convection according to the Schwarzschild criterion – there is no significant transport of energy by waves.

(4) The energy produced is due to nuclear burning of hydrogen to helium.

(5) The only regions where matter is chemically mixed are those regions where convection takes place.

(6) Convective regions can be adequately modelled by the mixing length theory of convection, which is a purely local theory. Thus convective overshooting is considered to be unimportant.

(7) Mass loss, rotation and magnetic fields are unimportant and can be treated as a small perturbation to the standard model.

(8) The Sun was initially homogeneous in composition and its age is the same as the Earth, approximately 4.7×10^9 yr.

(9) The laws of physics are time independent.

With these assumptions, models of solar type (and other) stars can be constructed. They have had moderate success; it is possible to construct a model of the present Sun that agrees with its two main characteristics, the present luminosity and radius of the Sun.

But such agreement with observation is not very significant: the models of the Sun have two degrees of freedom, the helium abundance and the mixing length in the theory of convection. The initial helium abundance is adjusted to obtain the correct luminosity and the mixing length to get the correct radius; the agreement between prediction and observation is, therefore, no check at all! As we shall see, the classical models constructed in this way make other predictions, which at least for the Sun, can be tested at various levels of accuracy – the models do not fare so well!

The classical picture that emerges is that the Sun was formed from the collapse of a gas cloud. The collapse stopped when the matter became ionised and the proto sun had a radius of about $50 R_\odot$, a central temperature of maybe 50 000 K and an effective temperature of about 3500 K. During the Hayashi phase (Hayashi, 1961), the proto sun was fully convective from centre to surface and very luminous, the energy radiated into space causing the star to contract releasing gravitational energy and increasing its internal temperature. As the contraction continues, the convection dies away in a growing region around the centre and the temperature in the mixed convective region never exceeds about 1.5×10^6 K. Eventually, the central regions become hot enough for nuclear reactions to be ignited, this energy release compensates for the energy loss by radiation and the gravitational contraction ceases.

Future evolution is now due to the hydrogen burning in the centre. This increases the helium abundance in the central regions (which are unmixed) and the non-uniform chemical composition produces slow changes in the Sun's internal structure until after 4.7×10^9 yr, it has the present radius and luminosity.

These models predict a number of consequences that can now be subjected to some tests.

(1) Present central temperature of the Sun $\simeq 1.5 \times 10^7$ K – neutrino flux $\simeq$ 6 SNU*.

(2) Luminosity of the Sun has monotonically increased from 0.65 to 1 $L_\odot$ over the last 4×10^9 yr.

(3) The matter in the surface layers has never been at temperatures greater than 1.5×10^6 K.

(4) The Sun is stable to oscillations.

In the next sections, I want to discuss the contradiction between these predictions and the observations and what we can conclude from them. Put briefly, there is evidence that all four predictions are wrong, since

(1) The upper limit on the neutrino flux is 1 SNU.

(2) The paleoclimatological evidence, and theoretical climate modelling indicates $L \simeq (1 \pm 0.05) L_\odot$ over 4×10^9 yr.

(3) The lithium abundance in the surface layers is very low indicating that the matter in the surface layers has been mixed to about 3×10^6 K.

(4) Recent observations by Hill indicate that the Sun is oscillating in its normal modes.

One hope for the orthodox point of view is that the inclusion of neglected effects like rotation and magnetic fields can save the models; the heterodox view which I wish to urge on you is that the models cannot be saved – we need to reconsider our basic physical assumptions. We can make a preliminary guess at the internal structure of the Sun. It could have:

(1) Present central temperature approx 1.2×10^7 K.

(2) A large central region that is chemically mixed on a time scale less than 10^9 yr.

(3) Some mixing process that carries surface layer material down to a temperature of 3×10^6 K but not to 4×10^6 K since beryllium has not been burnt – this could be substantial convective overshooting.

(4) Some mechanism for exciting oscillations – a convective core or some other central instability.

3. The Solar Neutrino Problem

The nuclear reactions in the proton–proton chain that are thought to be responsible for energy production have many branches. The neutrinos that should be detected are produced in the boron–beryllium reaction $^8B \rightarrow {}^8Be + e^+ + \nu$; although only a small fraction of the reactions are along this chain (about 0.02%) this is the important branch for the production of neutrinos. The fraction of reactions on each branch of

---

* 1 SNU $= 10^{-36}$ captures per ^{37}Cl atom per second.

the p-p chain is sensitive to the temperature. At high temperature a large fraction is on the boron–beryllium branch and at low temperatures a smaller fraction: thus the neutrino flux is a sensitive thermometer giving an estimate of the central temperature of the Sun. For a flux of 1 SNU, the temperature has to be about $12–13 \times 10^7$ K.

The cross-section for the absorption of neutrinos is very small and thus almost all the neutrinos escape from the Sun without being absorbed. In neutrinos we are therefore looking at the Sun now (or rather 8 min ago). This is to be contrasted with photon measurements: photons are readily absorbed and the photon diffusion time through the Sun is of the order of 3×10^7 yr. Thus one possibility that has been advocated (Fowler, 1972; Dilke and Gough, 1973) is that the centre of the Sun is indeed at the temperature suggested by the neutrino experiment and is, therefore, too cool to produce much energy – there is a thermal inbalance which will manifest itself over the thermal diffusion time scale of 3×10^7 yr. Before examining this proposal, let us consider a number of other proposed resolutions.

3.1. HOMOGENEOUS SUN

If the Sun has been mixed throughout its lifetime, the chemical composition is still uniform and the hydrogen abundance in the centre is larger than in the standard model – there is more fuel available and therefore it burns at a slower rate – that is, at a lower temperature. This can lower the neutrino flux to 1.5–2 SNU, still above the experimental limit. However, if the Sun were mixed, all the beryllium in the surface layers would be burnt – which it is not. The smaller the fraction of matter that is mixed, the higher the neutrino flux. However, a mixed model gives only a very small change in the Sun's luminosity over its lifetime.

3.2. RAPIDLY SPINNING CORE OR LARGE CENTRAL MAGNETIC FIELD

If there is some other contribution to the support of the Sun than pressure, then it is possible to affect the neutrino flux. In order to lower this flux, but still produce the luminosity observed, then the energy must be produced at a lower temperature over a larger region. This will happen if the ratio of the extra supporting force (centrifugal or magnetic) to gravity decreases away from the centre (Roxburgh, 1974, 1975a). However, such models give a substantial quadrupole moment at variance with the most recent determinations by Hill and co-workers (Hill and Stebbins, 1975). There are other objections: would such a distribution of angular momentum or magnetic flux be stable? The answer is almost certainly no, the magnetic flux would float and the rotation would be unstable to thermal instabilities (e.g. Goldreich and Schubert, 1967; Fricke, 1968). Perhaps a chemical composition gradient could stabilise the rotation, but it is difficult to build a cosmogonical picture of why the centre should spin as rapidly as required, yet have a small quadrupole moment (Roxburgh, 1974). Moreover, if the central region contains any magnetic field, this would link regions of different angular velocity and wipe out the required large inward increase in spin. A large magnetic field, even if stable, would decay on a time scale of order 10^9 yr leaving only the fewest slowly decaying modes which are nowhere near sufficiently centrally condensed to lower the neutrino flux.

3.3. VARYING CONSTANT OF GRAVITY

Various attempts to incorporate Mach's Principle into a theory of gravitation and dynamics predict that the 'constant' of gravity G changes as the Universe expands. Some models (Milne, 1935) give $G \propto t$, others (Dirac, 1936) give $G \propto 1/t$. The effect of such a change is to lower the neutrino flux to 2 SNU in the Milne case (Roxburgh, 1973) and to increase it to 20 SNU in the Dirac case (Shaviv and Bachall, 1969). In both cases the luminosity of the Sun varies significantly since $L \propto G^7$ for the Sun, and even though the diameter of the Earth's orbit changes as G changes, $d \propto 1/G$, the flux of radiation falling on the Earth changes by factors of 4 or more. This is difficult to reconcile with the paleoclimatological evidence that the flux has been remarkably constant over the age of the Earth.

3.4. CHEMICAL ANOMALIES

If the central composition of the Sun is significantly different from the surface layers, in particular, if it has a very high heavy element abundance, Z, then the structure is significantly changed. A recent model by Hoyle (1975) has a core of 0.5 $M_\odot$ with $Z \sim 0.5$, which because of the high opacity is convective, and an envelope of 0.5 $M_\odot$ and normal composition $Z \sim 0.02$. This model gives a low neutrino flux of less than 1 SNU, it also predicts an almost constant solar luminosity. However, it is difficult to understand the cosmogonical history of this model. Hoyle suggests that the inner half of the Sun was formed first from heavy element rich matter produced in a fluctuation in the early stages of the Universe; the rest of the matter was added later some 4.7×10^9 yr ago – but how? Until some plausible scheme for the formation of such a two zone Sun is put forward, this model is highly suspect.

3.5. WAVE TRANSPORT OF ENERGY

Recently both Hoyle (1975) and Hill (1975) have suggested that a significant fraction of the Sun's energy is transported by waves rather than radiation: this would change the structure of the Sun and could possibly lower the central temperature and hence the neutrino flux. No detailed models have been constructed, so it is difficult to say whether this would be significant, indeed, since the Sun has to produce 1 $L_\odot$, wave transport is not necessarily helpful. It is necessary to change the temperature profile in the centre of the Sun in a manner similar to the change produced by a very rapidly spinning core, and it has not been shown that this is possible.

3.6. PERIODIC INSTABILITIES AND THERMAL INBALANCE

It was pointed out by Gough and Dilke that the Sun is possibly unstable to non radial oscillations during its evolution. At the high temperature in the central regions of the Sun, the proton-proton fusion chain goes to completion to ^{4}He, but at lower temperatures away from the centre the reactions only proceed to ^{3}He. After some time, a ^{3}He concentration develops away from the centre and this is potentially an unstable situation. If such ^{3}He rich matter is displaced to regions of higher temperature, it rapidly burns to ^{4}He the energy released enhancing the displacement. Detailed calculations by Christian Dalsgaard *et al.* (1974) show that such an instability sets in very early in the Sun's evolution, after some 2×10^8 yr from the beginning of nuclear burning. The growth rate of the instability is long, and what eventually happens is speculation, but the suggestion is that the central regions are mixed, increasing the ^{3}He burning. The Sun is then out of thermal balance and readjusts on a thermal time scale, the ^{3}He then relaxes to equilibrium and the central temperature is low. The neutrino flux during this stage is also low until the Sun has adjusted again to thermal equilibrium and in a further 2×10^8 yr the process repeats. In this model we just happen to be observing in a period of anomalously low neutrino flux.

At present there are no adequate models of the evolution of the Sun with such a repetitive instability, nor are the consequences of such in instability known. Preliminary, though inappropriate calculations by Ulrich, suggest that a large variation in the Sun's luminosity during the mixing process ($\approx 40\%$) but again it is premature to judge this model. It is certainly very promising; if the inner 40–50% of the Sun was mixed every 2–3×10^8 yr, the average luminosity of the Sun would change very little. However, paleoclimatological research suggests that short time variations of greater than 5% in the Sun's luminosity have not occurred; could it be that the ^{3}He instability drives steady currents slowly mixing the interior? It is worth remarking that if the inner half of stars like the Sun are effectively mixed on a time scale shorter than the evolution time scale, it will significantly affect the time scale and characteristics of stellar evolution. This in turn will affect the estimates of the age of globular clusters, the evolution of galaxies and therefore the correction for galactic evolution in cosmological investigations.

4. Variation of the Solar Constant

While paleoclimatology is still in a very early stage of development, it is worth reporting the tentative conclusions that have been reached so far. The pertinent observations are (a) that the Earth is *not* covered with ice at the present time! (b) the geological record indicates that the Earth has never been completely covered with ice and (c) that life in some form has existed for as long as 3×10^9 yr. The climate models on the other hand indicate that if the Sun's luminosity was less than 95% of its present value, the Earth would be iced over and indeed that it would then remain ice covered if the luminosity increased to its present value, since an ice covered Earth would have a much higher albedo and therefore reflect a large part of the incoming radiation.

Moreover, while it is tempting to identify the ice ages with variations in the solar luminosity, the current climate models indicate that these correspond to at most a 3% variation in the solar constant, but can anyway be explained by plate motion and continental wandering on the Earth's surface without any variation in the Sun's luminosity. While it is perhaps premature to put too much weight on these results, the present conclusions are (Hays, 1975):

$$L = (1 \pm 0.05) L_\odot \qquad t \approx 3 \times 10^9 \text{ yr}$$

$$\Delta L / L \lesssim 0.03 \text{ on time scales } t \approx 3.10^8 \text{ yr.}$$

These results are difficult to reconcile with current models of the Sun and solar type stars. In the standard evolutionary models the Sun's luminosity has increased monotonically from about $0.6 \, L_\odot$ to $1 \, L_\odot$ over its lifetime of 4.7×10^9 yr, and most of this variation has been over the last 3×10^9 yr. If the Sun's luminosity is to vary by less than 5%, then a substantial region in the centre of the Sun has to be mixed on a time scale of less than 10^9 yr; the mixing does not have to be fast (as in a convective core) and periodic mixing on a time scale of 3×10^8 yr is compatible with the climate models provided that during this mixing the Sun's luminosity does not vary by more than 3%. We could, therefore, imagine an evolutionary pattern similar to that suggested by Dilke and Gough (1973); an inhomogeneous evolution until the onset of an instability which mixes the central regions of the Sun, and the pattern repeats. The problem with this mode of evolution is that it seems to indicate a large change in the solar constant during the readjustment period (Ulrich, 1975), so again we are led to suggest a slow steady mixing in the solar interior perhaps a sort of semi-convection keeping the Sun on the verge of stability, lowering the neutrino flux and keeping the solar constant constant.

5. Lithium and Beryllium Abundance, Rotation and Calcium Emission

Turning now to the surface layers of solar type stars, there are other enigmas. The Sun is very low in lithium but normal in beryllium abundance; since lithium burns at about 3×10^6 K and beryllium at 4×10^6 K, this indicates that the matter in the outer regions of the Sun has at some time been at temperatures between these limits. The standard solar evolutionary models predict that the matter in the outer layers has never been hotter than 2×10^6 K.

One possible explanation could be that the Sun has lost a lot of mass due to enhanced activity at an early stage in its lifetime; if, say, the Sun has lost 20% of its mass since it was near the main sequence, then the matter that is now in the convective zone was $\frac{4}{5}$ of the way (by mass) from the centre of the original Sun of mass $1.25 \, M_\odot$ and therefore at a temperature of about 4×10^6 K. This matter could, therefore, be depleted in lithium but not beryllium.

Another possibility is that the convective motions in the outer envelope of the Sun penetrates well inside the stable regions so carrying down to regions where lithium could be burnt, or the convective zone itself could be very much deeper than current models predict (see Section 7). Indeed. recent calculations I have made suggest that the simple mixing length theory is basically incorrect and has to be replaced by a non local theory of turbulent convection which requires substantial overshooting to reduce the convective velocities to zero (Roxburgh, 1975).

But one of the possible clues to the lithium-beryllium problem may be in the correlation between lithium abundance, rotation and calcium emission pointed out by Skumanich (1972); all seem to decrease with age on a time scale of the order of 10^9 yr, although the present solar lithium abundance is anomalously low. The picture suggested by Skumanich (1972) and Durney (1972) is that the Ca^+ emission is an indicator of magnetic field strength B, and that due to angular momentum loss in the solar wind, the Sun's rotation is gradually decreased. This decrease in the rotation results in a lower magnetic field strength due to a reduced dynamo action since the dynamo field B is conjectured to be proportional to the angular velocity Ω. The rate of change of angular momentum H is then:

$$\frac{dH}{dt} \propto \frac{dM}{dt} \Omega r_A^2$$

the mass loss

$$\frac{dM}{dt} = 4\pi \rho_A u_A r_A^2$$

where r_A, the radial distance at which the flow goes Alfénic is given by

$$4\pi \rho u_A^2 = B_A^2$$

hence

$$\frac{dH}{dt} \propto \frac{\Omega B_A^2 r_A^4}{u_A} \propto \frac{\Omega B_\odot^2 R^4}{u_A}$$

since Br^2 is constant for a radial field. If the solar wind has remained unchanged (a dubious assumption since the coronal heating and resulting solar wind is likely to be changed by the changing solar magnetic field) and is therefore constant so

$$\frac{dH}{dt} \propto \Omega B_\odot^2 \propto \Omega^3$$

if the dynamo gives $B \propto \Omega$. The solution of this equation gives approximately

$$\left(\frac{\Omega}{\Omega_\odot}\right)^2 = \frac{5 \times 10^9}{t}$$

which gives the present angular velocity for the Sun, consistent with the Sun rotating approximately uniformly throughout.

This slowing down of the Sun by the magnetic solar wind has been used by Dicke (1972) to explain the lithium-beryllium problem. The wind slows down the surface layers first which being strongly coupled to the convective zone slows down the outer convective zone. This produces a layer of differential rotation beneath the solar convective zone in which slow thermally driven turbulence couples the outer zone down to a depth where lithium is depleted.

The thermally driven turbulence is due to the instability of differential rotation first pointed out by Goldreich and Schubert (1967) and Fricke (1968). A rotating liquid is dynamically unstable if the angular momentum per unit mass decreases outwards (Rayleigh Instability)

$$\frac{\partial}{\partial \varpi}(\Omega \varpi^2) < 0.$$

That this instability is also operative in stably stratified layers was noticed by myself (Roxburgh, 1964) since there will always exist a displacement for which the stabilising effect is negligible, but Goldreich, Schubert and Fricke showed that the

Taylor-Proudman theorem had also to be satisfied, namely

$$\frac{\partial\Omega}{\partial z}=0$$

(ϖ, φ, z) being cylindrical polar coordinates, $\varpi = 0$ the axis of rotation. Moreover, their detailed stability analysis showed that these results are only strictly true for zero Prandtl number $\sigma = \nu/\kappa$. For the Sun $\sigma \approx 10^{-6}$ so there can exist a stable angular velocity gradient beneath the solar convective zone, with

$$\frac{d\Omega}{dr}\approx\frac{5\Omega_{\odot}}{R_{\odot}}.$$

Dicke's picture is that the angular velocity gradient is just slightly in excess of this marginally stable value, so that the turbulence it produces diffuses angular momentum outwards to balance the loss by the solar wind. If the typical turbulent velocity is v and the type scale l then

$$4\pi\rho r^{4}\langle vl\rangle\frac{d\Omega}{dr}=\frac{dH}{dt}$$

giving

$$\langle vl\rangle\approx 10^{4}\,\text{cm}^{2}\,\text{s}^{-1}.$$

The diffusion coefficient $\langle vl\rangle$ determined by the angular momentum then determines the rate of depletion of lithium and if this thermally turbulent layer extends down to 0.56 of the solar radius, then the rate of depletion of lithium agrees tolerably well with the observations of the Hyades, Pleiades and the Sun.

We are therefore led to a picture in which the angular momentum of the core is either carried to the region of 0.56 $R_{\odot}$ by magnetic stresses or is uncoupled from the rest of the star and is still rapidly spinning. This later is Dicke's own picture, but is incompatible with the new limits on the solar oblateness derived by Hill and Stebbins (1975). Their observations give a gravitational quadrupole moment

$$J_2\lesssim\tfrac{1}{10}J_2\text{ (Dicke)}$$

where J_2 (Dicke) $= 2.5\times 10^{-5}$ is the value predicted by Dicke's rapidly rotating core model. However, if the angular momentum of the core is carried outwards by magnetic stress so that $\Omega_{\text{core}}\approx 2\Omega_{\odot}$ then

$$J_2=\tfrac{1}{100}J_2\text{ (Dicke)}$$

and is compatible with the Hill-Stebbins result.

But a number of questions immediately arise; why is the core of 0.56 $R_{\odot}$ isolated from the rest of the Sun? Why does the magnetic field in the core, and indeed that in the envelope, not penetrate into the region $0.56<r/R_{\odot}<0.8$? At present no convincing model exists, but the success of Dicke's calculation in predicting the lithium depletion from the angular momentum loss suggests that the model is worth exploring.

If this region of differential rotation exists, it is perhaps worth asking if the solar dynamo could operate there, since this is at least a region where the magnetic buoyancy time is of the order of years.

6. Oscillations of the Sun

Our understanding of the solar interior is even more confused by the recent observations of Hill et al. (1975) of solar oscillations. As part of their programme to determine the solar oblateness, they have discovered that the Sun is oscillating with frequencies in the range 0.5 to 4 MHz. The frequencies they observe fit tolerably well to the normal modes of oscillation of a detailed solar model calculated by Gough (1975). At present, it is not known whether these are radial or non-radial modes and the spectrum can be fitted with either, but as observations proceed, the nature of the oscillations will become clearer. Already they pose a problem for the theorist; what is the exciting mechanism? Could they be driven by the solar convective zone, or by a slowly mixing core, or are they indicators of the periodic instabilities proposed by Christensen-Dalsgaard et al. (1974)? In any case, do these waves contribute significantly to energy transport in the Sun, so that there is mechanical as well as radiative transport in the interior? Although the surface amplitudes are small $\delta R/R_{\odot} \approx 10^{-6}$ this does not necessarily mean that the internal amplitudes are as small, and if a sufficiently large number of modes are excited even with small amplitude, they could carry a substantial fraction of the solar luminosity. Such mechanical transport has been suggested by Hoyle (1975) and Hill (1975) as being responsible for changing the internal structure of the Sun, and lowering the neutrino flux. While again no detailed models exist, this is an exciting possibility, and a rich field for future research.

7. The Surface Convective Zone

The convective theory used in modelling the Sun and solar-type stars is still the mixing length theory. In this 'theory' the convective heat flux F_c is estimated by the expressions

$$F_c = \langle c_p \rho v T \rangle = c_p \rho v \delta T$$

$$v^2 = \frac{g}{T}\delta Tl, \qquad \delta T = \Delta \nabla l$$

where

$$\Delta \nabla = \frac{dT}{dr} - \left(1 - \frac{1}{\gamma}\right)\frac{T}{P}\frac{dP}{dr}$$

and l the mixing length is taken to be some multiple, usually one, of the pressure scale height $P/(dP/dr)$. As pointed out in Section 2, the Sun provides no test of this theory since the observed solar radius is used to determine the mixing length!

Even for the Sun, the theory is clearly inadequate since it considers only motion on one scale l, whereas we observe the granulation, supergranulation and possibly giant cells; how can these be explained? One possibility is that the larger scales of motion represent a convective instability of the small scale equilibrium. To see how this could work, we note that the turbulent average of the heat equation gives a turbulent heat flux

$$F_c = \kappa \nabla S \qquad \kappa = \langle vl \rangle$$

and a turbulent viscous transport

$$\nu \nabla^2 \mathbf{v} \qquad \nu = \langle vl \rangle .$$

The turbulent transport coefficients κ and ν are then analogous to the kinetic transport coefficients κ, ν, and so the mean flow equations can be unstable to convection just as the mean molecular equations can be unstable to convective motions. The transport coefficients are functions of depth z, so the mean flow Rayleigh number for the convective zone is

$$R = \frac{g\alpha\beta d^4}{\kappa\nu} = R(z)$$

If $R(z)$ is sufficiently large in the surface layers, these layers can be unstable and drive motions in the whole of the layer, whose characteristic scale is the depth of the unstable driving region. Calculations by myself and G. Vickers for convection with varying transport coefficients show that this is a possibility in a Boussinesq fluid, and it is possible to show that the compressible case can be analysed in a similar way by a quasi-adiabatic approximation giving similar results (Roxburgh, 1975b). Indeed, it is possible that the giant cells represent a third scale in which the whole convective zone is unstable to giant cells influenced by rotation as in the Boussinesq calculations of Busse (1970), Durney (1975) and Gilman (1972, 1975), the mean flow equations and the quasi-adiabatic approximation yielding equations similar to those of their Boussinesq treatments (Roxburgh, unpublished).

But is the mixing length theory adequate to represent small scales? I believe the answer is no. The mixing length theory is a *local* order of magnitude theory and it could well be misleading. To see this we write down the full time independent energy equation for a gas in the form

$$\text{div} \left[\frac{\gamma}{\gamma-1} P\mathbf{v} + \frac{1}{2}\rho v v^2 + F_R \right] = 0 .$$

The mixing length theory is obtained by using a turbulent decomposition

$$\rho = \rho_1 + \rho_1, \quad T = T_0 + T_1, \quad v = v_1$$

such that

$$\langle \rho v \rangle = 0 \quad \langle v_1 \rangle = -\langle \rho_1 v_1 \rangle / \rho_0$$

The pressure fluctuations are then taken to be negligible so

$$\left\langle \frac{P_1^2}{P_0^2} \right\rangle \ll \left\langle \frac{\rho_1^2}{\rho_0^2} \right\rangle, \quad \left\langle \frac{T_1^2}{T_0^2} \right\rangle, \qquad \frac{\rho_1}{\rho_0} = -\frac{T_1}{T_0}$$

and then neglecting $\langle \rho v^2 v \rangle$ so that

$$F_c = \frac{\gamma}{\gamma-1}\langle pv \rangle + \tfrac{1}{2}\langle \rho vv^2 \rangle \simeq \frac{\gamma}{\gamma-1}P_0 \langle v_1 \rangle = -\frac{\gamma}{\gamma-1}P_0\frac{\langle \rho_1 v_1 \rangle}{\rho_0}$$

$$= \frac{\gamma}{\gamma-1}\frac{P_0}{T_0}\langle T_1 v_1 \rangle = c_p \rho_0 v_1 T_1 .$$

Then on using

$$T_1 = \Delta \nabla l, \quad \text{and} \quad v_1^2 = \frac{g}{T}T_1 l$$

the standard expressions follow.

However, this is clearly inconsistent since if $\langle P_1^2/P_0^2 \rangle$ can be neglected, then on taking the scalar product of the momentum equation with $\mathbf{v}$ and averaging we find

$$\langle v \rangle \frac{dP_0}{dz} = -\frac{d}{dz}\left\langle \frac{\rho v v^2}{z} \right\rangle$$

so that

$$F_E = \frac{\gamma}{\gamma-1}\langle Pv \rangle = \frac{c_p T}{g}\frac{d}{dz}(\tfrac{1}{2}\rho vv^2) \equiv \frac{c_p T}{g}\frac{dF_k}{dz}$$

which is of the same order as the kinetic energy flux $\langle \tfrac{1}{2}\rho vv^2 \rangle$. Indeed in terms of the total flux F the energy equation has the integral

$$F_\kappa = p^{1-1/\gamma}\int \frac{g}{c_p T}p^{1/\gamma-1}(F-F_R)\,dz$$

so that F_κ cannot be zero at the boundaries where $F = F_R$, nor is F_E, it is only the sum that vanishes. In this case there has to be substantial overshooting into the stable region where $F-F_R$ is negative and we need a non-local theory of convection. Preliminary calculations by the author suggest that this region can be as large as a scale height at the bottom of the convective layer, and about 0.3 of a scale height at the top (Roxburgh, 1975c).

However, it is clear that such a model is not correct either since the pressure fluctuations cannot be neglected; they are necessary both to stop the vertical motions and to drive the horizontal motions and must, therefore, be comparable to the density and temperature fluctuations. In a correct theory, we need to include the correlations $\langle P_1 v_1 \rangle$ and have a third term in the expression for the energy flux so that

$$F_c = \frac{\gamma}{\gamma-1}\langle P_0 v_1 \rangle + \frac{\gamma}{\gamma-1}\langle P_1 v_1 \rangle + \tfrac{1}{2}\langle \rho vv^2 \rangle = F = F_R$$

but the essential part of the previous result remains that $F-F_R$ can be zero without each term in the expression for F_c vanishing. Convective overshooting will still be important.

Clearly much remains to be done before any adequate theory of turbulent convection is available.

8. The Solar Interior

What then is the Sun like? We may conjecture a model consisting of:

(a) A core of ~0.56 $R_\odot$ that is slowly but steadily mixed, the mixing currents driven by the ^{3}He instability.

(b) Energy transport by waves excited by the mixing core.

(c) A trapped magnetic field in the central core.

(d) An intermediate layer of weak thermally driven turbulence between $0.56 R_\odot$ and 0.7 $R_\odot$ diffusing both angular momentum and lithium depleted matter.

(e) A convective overshoot region between 0.7 $R_\odot$ and 0.8 $R_\odot$.

(f) The convective zone proper between 0.8 $R_\odot$ and 1 $R_\odot$.

(g) A convective overshoot region into the photosphere of order 100 km.

(h) The central core having an angular velocity $\sim 2\Omega_0$.

Such a model is highly speculative!

9. Conclusion

In conclusion let me emphasise that as yet we have no satisfactory model of the internal structure of the Sun and solar type stars. If the central regions are mixed, this will have a sigfnificant effect on the evolution of stars, on estimates of the age of globular clusters, the evolution of galaxies, the determination of the deceleration parameter in cosmology and so on the determination of the structure of the Universe in which we live. We no longer believe that the Sun is the centre of the Universe, but the solution of the puzzles posed by the Sun is central to our understanding of stellar evolution and may well influence the picture we have of the Universe we live in.

References

Busse, F. H.: 1970 *Astrophys. J.* **159**, 629.

Christensen-Dalsgaard, J., Dilke, F. W. W., and Gough, D. O.: 1974, *Monthly Notices Roy. Astron. Soc.* **169**, 429.

Davis, R.: *Bull. Amer. Phys. Soc.* **17**, 527.

Dicke, R. H.: 1972, *Astrophys. J.* **171**, 331.

Dicke, R. H. and Goldenberg, H. M.: 1967 *Phys. Rev. Letters* **18**, 313.

Dicke, R. H. and Goldenberg, H. M.: 1974 *Astrophys. J. Suppl.* **27**, 131.

Dilke, F. W. W. and Gough, D. O.: 1972, *Nature* **240**, 262.

Dirac, P. A. M.: 1938, *Proc. Royal Soc. A*, **165**, 199.

Durney, B. R.: 1971, *Astrophys. J.* **163**, 353.

Durney, B. R.: 1972, *Proceedings of the 1971 Asilomar Conference on the Solar Wind*, p. 282 NASA publ.

Durney, B. R.: 1975, this volume, p. 243.

Fowler, W. A.: 1972, *Nature* **238**, 24.

Fricke, K.: 1968 *Z. Astrophysik* **68**, 317.

Goldreich, P. and Schubert, G.: 1967, *Astrophys. J.* **150**, 571.

Gilman, P. A.: 1972, *Solar Phys.* **27**, 3.

Gilman, P. A.: 1975, this volume, p. 207.

Hayashi, C.: 1961, *Proc. Astron. Soc. Japan* **13**, 450.

Hayes, J.: 1975, Proceedings of the Solar Constant Workshop (in press).

Hill, H. A. and Stebbins, R. T.: 1975, *Ann. New York Acad. Sci.* (in press).

Hill, H. A., Stebbins, R. T., and Brown, T. M.: SCLERA Report 1975.

Hill, H. A., and McCullen, J. D., Brown, T. M., and Stebbins, R. T.: SCLERA Report 1975.

Hoyle, F.: 1975 *Astrophys. J.* **197**, L127.
Milne, E. A.: 1935, *Gravitation and World Structure*, Oxford University Press.
Roxburgh, I. W.: 1964, *Monthly Notices Roy. Astron. Soc.* **128**, 157.
Roxburgh, I. W.: 1973 (unpublished note).
Roxburgh, I. W.: 1974. *Nature* **248**, 209.
Roxburgh, I. W.: 1975a, *Monthly Notices Roy. Astron. Soc.* **170**, 35p.
Roxburgh, I. W.: 1975b (to be published).
Roxburgh, I. W.: 1975c (to be published).
Schwarzschild, M.: 1968 *Structure and Evolution of the Stars*, Princeton Univ. Press.
Shaviv, G. and Buchall, J. N.: 1969 *Astrophys. J.* **155**, 135.
Skumanich, A.: 1972, *Astrophys. J.* **171**, 565.
Ulrich, R. K.: 1975 (to be published).
Wallerstein, G. and Conti, P.: 1969, *Ann. Rev. Astron. Astrophys.* 99.
Wilson, O. C.: 1963, *Astrophys. J.* **140**, 1401.

DISCUSSION

Parker: There are two comments that come to mind in your review of possible solutions to the dilemma presented by the Sun. One is that any non-uniform rotation participating in the solar dynamo must be driven by thermal convective effects. There is not enough energy in solar rotation to drive the present dynamo for the life of the Sun. Second, the fields that might be trapped in the central core of the Sun cannot be very strong, because very strong fields are lost by magnetic buoyancy. For instance, the 10^7 G that some have postulated to resolve the neutrino dilemma would be lost in a period of the order of only 10^1 yr.

Mestel: I am certainly not against primeval magnetic fields, and in fact think it quite likely that beneath the solar convection zone there is zone trapped flux (quite distinct from the solar dynamo field). However, I doubt if it would stay confined to your central region: I would expect it to diffuse so as to couple the base of the convection zone with the core, so probably wiping out your Ω gradient. There are other dynamical consequences of a primeval magnetic field which may be relevant. Unless such a field has a high degree of symmetry with respect to the rotation axis it tends to set up oscillatory rotations, with the period long compared with the rotation period but short compared with the thermal time scale. The amplitude of these rotations is small, but they may still lead by virtue of finite resistivity to a steady diffusion of material that would be significant within the leisurely time-scale of stellar evolution. Conceivably this could be relevant to both the neutrino and lithium problems. To Dr Parker's remarks on magnetic buoyancy: in a radiative zone the rate at which flux tubes rise is determined by heat-flow essentially an Eddington-Sweet effect; and the motion may very well be suppressed altogether by a slight gradient of mean molecular weight. Of course, all these remarks presume the hypothetical primeval field has a dynamically stable structure – e.g. that it has toroidal flux tubes linking the poloidal loops.

Gilman: The suggestion that there could be dynamo action in the overshoot region below the convection zone is interesting, but if the rotation rate increases as fast with depth there, as you suggest, we should expect to see large scale magnetic features at surface rotating much faster than they do.

Weiss: You quoted the generally accepted result that the Sun was fully convective before approaching the main sequence. This was found by Hayashi but Larson, starting from different initial conditions, finds that only outer 50% by radius was convective. So we have yet another uncertainty. Moreover the central magnetic field may remain distinct from that in the outer convective zone.

THE SUN AS A MAGNETIC STAR

JOHN M. WILCOX

and

LEIF SVALGAARD

Institute for Plasma Research, Stanford University, Stanford, Calif. 94305, U.S.A.

Abstract. The Sun as a magnetic star is described on the basis of recent work on solar magnetism. Observations at an arbitrary angle to the rotation axis would show a 22-year polar field variation and a 25-day equatorial sector variation. The sector variation would be similar to an oblique rotator with an angle of 90° between the magnetic and rotational axes.

To be published in Proceedings of IAU Colloquium No. 32.

Bumba and Kleczek (eds.), *Basic Mechanisms of Solar Activity*, 467. *All Rights Reserved.*

CHROMOSPHERIC ACTIVITY IN F- AND G-STARS

DAINIS DRAVINS

*Lund Observatory, S-222 24 Lund, Sweden**

Summary. Chromospheric activity, as seen by the presence of a central K_2 emission in the Ca II K λ 393.4 nm line, is studied in F- and G-stars of all luminosity classes. K_2 equivalent widths have been measured and corresponding absolute K_2 fluxes (power per unit area at the stellar surface) calculated. Large K_2 equivalent widths and thus easily detectable chromospheres occur only for stars later than about F8 but the absolute level of K_2 flux does not significantly change between F0 and K0; *there is no break in chromospheric activity between F- and G-stars.* The apparent break, cited in the past (e.g. Wilson, 1973) is illusory and due to the observational threshold being different for different spectral classes.

90 stars have been observed, covering spectral types F0 to K0. Coudé spectra at dispersions 0.3 or 1.2 nm mm^{-1} were obtained with the plates heavily exposed to show details in the K-line core. To reduce photographic grain noise, the spectra are well widened, typically to 1 mm. Through digital microphotometry, followed by spatial frequency filtering, the plate densities have been converted to intensity profiles using the detailed shape of the characteristic curve of the emulsion. The resulting grain noise even at the lower dispersion typically corresponds to a K_2 emission equivalent width $\simeq 0.05$ pm (1 pm = 10 mÅ) which is less than the solar K_2. On highly widened high-dispersion spectra, the grain noise is again only a fraction of this. The main source of uncertainty in assessing the K_2 emission is now the lack of knowledge of the exact shape of the undisturbed absorption line.

The K_2 equivalent widths were obtained by extrapolating the absorption line profile to the line center and then measuring the emission component as the difference to the observed profile. Conservative estimates were aimed at and only rarely was the absorption line extended below the lowest observed intensity point and then only in cases with an obvious central emission peak. Thus the values below are probably on the low side and will not take into account such possible emission that contributes to a symmetrical and smooth infilling of the K-line. The apparent local 'continuum' was defined as the point of highest intensity between the Ca II H and K lines.

In Figure 1 the K_2 equivalent width is given as a function of spectral class. Late G-giants have the largest equivalent widths and easily the most prominent emission profiles while all stars earlier than F8 have emission <6 pm. A given equivalent width is in addition more difficult to see in earlier-type stars since the background K-line is narrower in wavelength and the emission components are frequently seen only as undulations in the line wings rather than isolated emission peaks in the line bottom. Figure 1 thus shows the apparent cessation of K-line emission as seen on moderate resolution spectrograms for spectral types earlier than about F8. However this gives a misleading impression of the level of chromospheric activity since the equivalent

* Work based on observations made at European Southern Observatory, La Silla, Chile.

Bumba and Kleczek (eds.), Basic Mechanisms of Solar Activity, 469–471. All Rights Reserved.
Copyright © 1976 by the IAU.

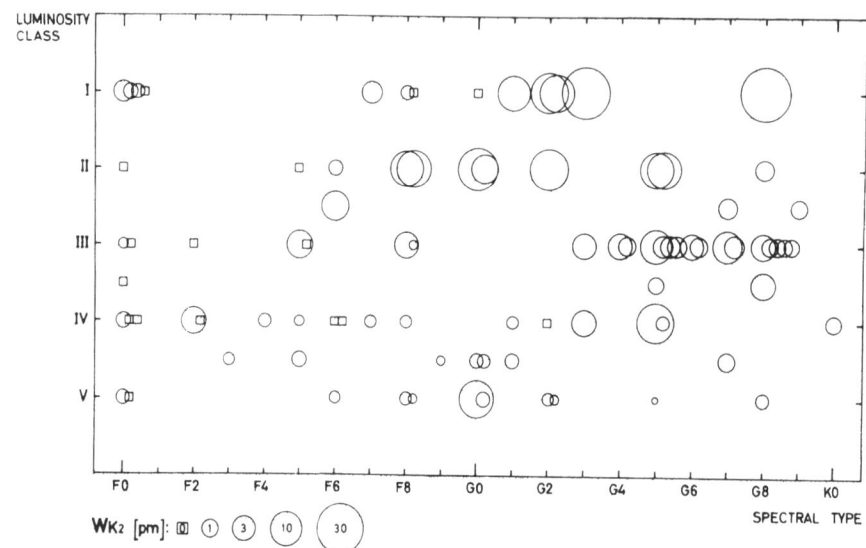

Fig. 1. Ca II K₂ emission equivalent width as function of spectral class. The diameter of a circle is proportional to the cube root of the equivalent width. 1 pm = 10 mÅ.

widths are measured relative to a local continuum whose absolute level is greatly different in different spectral types.

Absolute fluxes were therefore calculated, using standard fluxes given by Lamla (1959) for the smeared-out spectra of different types of stars. The intensity averaged over a 2 nm wavelength interval between the Ca II H and K lines was taken as the value for the local smeared-out spectrum and the absolute K_2 flux then obtained from its equivalent width and knowing the absolute flux per wavelength interval for that spectral type; Figure 2.

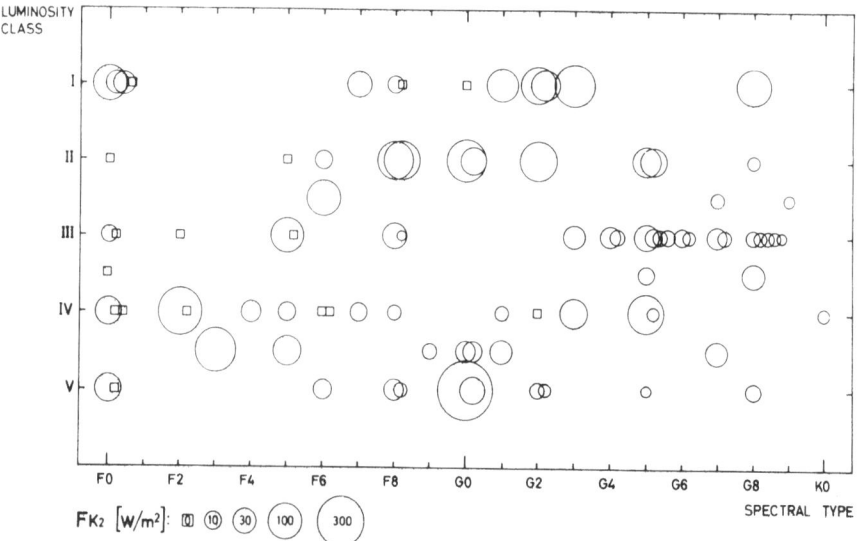

Fig. 2. Ca II K₂ emission flux in absolute units (W m⁻²) at the stellar surface as function of spectral class.

Now there are no special differences in the level of K_2 flux to the left and to the right of a vertical line at F8. The previously prominent emission from late G giants dwindles because the absolute surface fluxes of these cool stars are low and the previously modest emission in earlier F-stars becomes more prominent. There is no clear tendency for the K_2 flux to decrease with earlier spectral type and chromospheres should continue beyond F0. It seems that the few F- and A-stars previously reported with K_2 emission mark the rule, not the exception. If one assumes that the amount of K_2 emission is directly connected with the mechanical flux from the convection zone, this is fully consistent with the calculations by de Loore (1970).

References

de Loore, C.: 1970, *Astrophys. Space Sci.* **6**, 60.
Lamla, E.: 1959, *Astron. Nachr.* **285**, 12.
Wilson, O. C.: 1973, in S. D. Jordan and E. H. Avrett (eds.), *Stellar Chromospheres*, IAU Colloquium No. 19, NASA SP-317, p. 305.

ABSOLUTE FLUXES OF K CHROMOSPHERIC EMISSION
ON THE H–R DIAGRAM

C. BLANCO, S. CATALANO* and E. MARILLI

Catania Astrophysical Observatory and Astronomical Institute, University of Catania, Italy

Abstract. Continuing our previous analysis of the chromospheric emission (Blanco *et al.*, 1974), absolute fluxes of the K emission line have been evaluated from 10 Å mm^{-1} spectrograms of the O. C. Wilson collection for 31 F5–K7 main sequence stars and 172 G2-M5 giants.

Our previous main results concerning the fluxes of main sequence stars are confirmed. Values for stars common to our previous and present set of data compare with one another within $\pm 10\%$, apart from the values of 61 Cyg A, for which a scatter as large as $\pm 20\%$, was found. These discordancies are discussed in terms of the accuracy of the data and of real variations.

The K line flux of giant stars is found to be about $10^5 \text{ erg cm}^{-2} \text{ s}^{-1}$ for early G type and decreases by two orders of magnitude from the spectral type G2 to M5. Differences in the flux are evident for giants earlier than K0, but no clear age dependence could be found. The average flux of giants is found to be lower by one order of magnitude than the flux of main sequence stars of the same spectral type.

The lines of iso-chromospheric emission on the H–R diagram are roughly vertical lines with a slight inclination to the left for earlier spectral type. In any case they display a rather different trend from the lines of iso-acoustic flux.

The ratio of K line flux to the bolometric flux $B(T)$ assumed as an index of the chromospheric activity shows a maximum at $T_{eff} \approx 4500 \text{ K}$ for giant stars, and at $T_{eff} \approx 5000 \text{ K}$ for main sequence stars.

Finally the observed decline in the equivalent width of K emission on giant stars later than M0 is discussed and interpreted in terms of circumstellar absorption.

Reference

Blanco, C., Catalano, S., Marilli, E., and Rodonò, M.: 1974, *Astron. Astrophys.* **33**, 257.

---

* Guest investigator at the Hale Observatories, Astronomical Institute of Catania University.

Bumba and Kleczek (eds.), Basic Mechanisms of Solar Activity, 473. All Rights Reserved.
Copyright © 1976 by the IAU.

DOUBLE-PEAKED FLARE EVENTS ON RED DWARF STARS AND SOLAR 'SYMPATHETIC' FLARES

M. RODONÒ

Catania Astrophysical Observatory and Institute of Mathematics,
University of Messina, Italy

Abstract. About 50% of the flare events observed on red dwarfs are at least double-peaked. As the majority of flare stars are members of double or multiple systems, the possibility that time-overlapping flares originate quasi-simultaneously on the individual components is discussed.

Assuming a poissonian occurrence of flares in both components, the expected probability of observing double-peaked flares is lower than 1% for the most active binary systems.

However, from photometric observations of the double flare star EQ Peg (BD +19°5116 AB) carried out by the author with an area scanner (the components' angular separation is 3.7″) about 20% of the observed flares have been found to be double-peaked flares resulting from separate flares, one in each component. A direct flare triggering of the following flare by the preceding one can be ruled out since the light travel-time between the two components is 3.5 h, while the observed time delay between the flare peaks is about 10 min. Moreover, the proximity effect does not seem to play an important triggering role.

It is concluded that, although the analogy with solar 'sympathetic' flares is not always applicable, it is the most promising framework within which the majority of double-peaked flare events on red dwarfs must be interpreted.

The results here presented are included in the paper 'Flare Activity on Binary Systems' submitted to *Astronomy and Astrophysics*.

CONCLUSIONS

SUMMARY OF THE FINAL DISCUSSION ON AUGUST 29

(Chairman: G. Newkirk)

B. R. DURNEY, P. A. GILMAN, and M. STIX

National Center for Atmospheric Research, Boulder, Colo. 80303, U.S.A.

The aim of this summary is to convey to the reader only the highlights and the 'spirit' of the discussion. We have decided, therefore, to give an assimilated version of the discussion and realize that in doing so we have perhaps introduced our personal opinions; we have tried, however, to give due prominence to ideas different from our own.

As was pertinent to a symposium dedicated to the Basic Mechanisms of Solar Activity, much of the discussion was centered about problems concerning the solar dynamo and differential rotation. In relation to the Sun's differential rotation those problems relevant to theories of the solar activity cycle were especially emphasized.

Whereas the most favored explanation of the Sun's activity cycle is based on a dynamo theory of this phenomenon, this explanation is not accepted by everybody. The main objections are two-fold:

(a) There is not conclusive observational proof that turbulent diffusivity of the magnitude required by dynamo theories is present in the Sun;

(b) Dynamo theories describe magnetic fields that vary slowly in space and time whereas the observations show that most of the solar magnetic field is concentrated in small bundles of high field strength; how then can one talk about a mean magnetic field so essential for dynamo theories? These represent the most outstanding difficulties which have led some scientists to doubt the validity of dynamo theories. Others can also be found; for example:

(c) The variations of the angular velocity with depth and latitude are not known; and the dynamo responds in a complicated way to this dependence. In particular, most dynamo models require that the angular velocity *increase* with depth, while most theories of differential rotation predict that the angular velocity should *decrease* with depth because rotational constraints tend to make the angular velocity constant on cylinders;

(d) It appears that the toroidal magnetic field must be amplified *deep* inside the convection zone, because elsewhere (in the convection zone) magnetic buoyancy is so strong that it does not allow fields to remain long enough to be amplified to the strength observed at the surface. Such a location of the amplification zone could make the dynamo sensitive to the structure of the deep solar interior which the neutrino deficit emphasizes as a poorly understood region.

(e) Statistical averages play an essential role in the development of mean field electrodynamics, and the relation to the observational averages is not straight-forward;

(f) Second-order terms in the equation for the fluctuating magnetic fields are neglected with only marginal justification;

(g) The variation within the Sun of the parameters α or R, which describe the regeneration of the poloidal magnetic field (from the toroidal magnetic field)

Bumba and Kleczek (eds.), Basic Mechanisms of Solar Activity, 479–481. All Rights Reserved.
Copyright © 1976 by the IAU.

depends on the form of the convective motions, rotation, shear and density stratification; to calculate the values of these parameters one must make simplifying assumptions which are not always justified.

Some of the difficulties listed above are readily acknowledged and even put forward by proponents of the dynamo theory; the existence of such dilemmas merely indicates that our knowledge of this complex phenomenon is far from complete. Models of the solar cycle should therefore be developed more as a way to understand the physical processes leading to the solar activity cycle than to exactly reproduce observed details. The main difficulties (a) and (b) are answered by supporters of the dynamo theories as follows:

(a) The existence of turbulent diffusion can be derived from Maxwell's equations with the help of reasonable approximations; therefore, to doubt the existence of turbulent diffusion is to doubt the validity of Maxwell's equations. Ohmic dissipation, or field line reconnection, *will* take place at small scales, probably assisted by dynamic effects such as Petschek's mechanism;

(b) No matter how concentrated the magnetic filaments are, the mean field can always be defined. It is the average of the whole field, filamentary *and* (if there is any) interfilamentary.

Whereas there can be no doubt about the existence of the magnetic filaments, it is not understood how they are formed (a convergent velocity field is not sufficient). Do they exist throughout the convection zone as some observers argued? Or, is the large flux density limited to the surface as is, perhaps, more plausible? The persistence of these filaments is also difficult to understand; to achieve equilibrium an efficient cooling mechanism is needed; and one would expect the equilibrium to be rapidly destroyed by exchange instabilities. The idea that *twisting* magnetic flux tubes concentrates them is heavily disputed.

We have given above the essence of the arguments presented in favor, as well as the criticisms, of dynamo theories. Some of those who do not believe in dynamo theories advocate an explanation of the solar activity cycle based on a steady primordial field. A necessary condition for the validity of this idea is that turbulent diffusion and the resulting reconnection of field lines (at small scales) in the convection zone must be so slow that the primordial field has not been destroyed in the lifetime of the Sun. Also, the observed reversals of the polar fields near the maxima of the last two cycles, if real, are in contradiction to the existence of the primordial field.

It is commonly accepted that the amplification of the toroidal magnetic field is achieved by the differential rotation of the Sun's convection zone. Theories of differential rotation and dynamo theories are therefore closely linked. This bond puts severe constraints on both theories. It is, therefore, of great importance to establish firmly that the dynamo operates, indeed, in the solar convection zone and thus demonstrate that differential rotation is not a surface phenomenon but is significant over the entire convection zone. If this is the case, theories of differential rotation find it, at present, difficult to account for the observed latitudinal variations in the angular velocity without significant variations in the energy flux.

Can the dynamo be located *below* the convection zone? In this case the shear necessary to amplify the toroidal magnetic field would exist as a primordial feature;

the torque exerted by the solar wind would uniformly slow down the convective shell (because of its large viscosity) and leave a rapidly spinning core. This idea encounters serious difficulties: for example, an alternating field cannot penetrate into a region of high conductivity such as the radiative core (the penetration depth is only a few hundred kilometers if the diffusivity is determined by turbulence generated by the Goldreich-Schubert-Fricke instability). Furthermore, the dynamo would tend to choke itself off by the action of Lorentz forces. Finally, questions arise concerning the energy available since the energy for the dynamo must come from the rapidly rotating core and it is not clear that this energy is sufficient to have run the dynamo since the formation of the Sun. An important overshooting of convective motions therefore seems necessary, if the dynamo is to be located in the radiative region. However, everything considered, the most plausible region for the operation of the solar dynamo appears to be the convection zone.

As mentioned in (c), differential rotation theories favor angular velocity decreasing with depth while most dynamo models require the opposite. It is not clear that a real contradiction exists because the approximations used in both the differential rotation and the dynamo models are not well justified for the Sun. In this context, the comparison of the observed and theoretically predicted phase relations between the radial and azimuthal components of the mean field could put an important constraint on how the angular velocity must vary with depth.

Increased efforts to observe velocities on global scales on the Sun should provide useful guidance in the development of more advanced circulation models. Among the important effects which should be studied are (a) variations in the rotation rate with time, (b) structure of additional global-scale velocity anomalies to see if they represent convection, (c) comparisons of velocities measured simultaneously by doppler and tracer techniques, (d) differences in velocity between low and high latitudes, (e) persistence of velocity patterns from one rotation to the next, (f) correlations between velocity fluctuations in northern and southern hemispheres, (g) comparisons of velocity anomalies and time variations with brightness changes, (h) separation of east–west from north–south velocities, so that transport properties can be calculated, (i) solar cycle length variations in velocities, to see whether the magnetic fields are strong enough to react back upon the flow which induced them.